Test Item File

Gary Hollis
Roanoke College

Salim Diab

Organic Chemistry

SECOND EDITION

Paula Yurkanis Bruice
University of California, Santa Barbara

PRENTICE HALL, Upper Saddle River, NJ 07458

Associate Editor: Mary Hornby
Special Projects Manager: Barbara A. Murray
Production Editor: Carole Suraci
Supplement Cover Manager: Paul Gourhan
Supplement Cover Designer: Liz Nemeth
Manufacturing Buyer: Ben Smith

Upper Saddle River, NJ 07458

Printed in the United States of America

10 9 8 7 6 5 4 3

ISBN 0-13-889320-9

Prentice-Hall International (UK) Limited,London
Prentice-Hall of Australia Pty. Limited, Sydney
Prentice-Hall Canada Inc., Toronto
Prentice-Hall Hispanoamericana, S.A., Mexico
Prentice-Hall of India Private Limited, New Delhi
Prentice-Hall of Japan, Inc., Tokyo
Pearson Education Asia Pte. Ltd., Singapore
Editora Prentice-Hall do Brasil, Ltda., Rio de Janeiro

CONTENTS

Prentice Hall: Instructor Support for Test Item Files

This hard copy Test Item File is just one part of Prentice Hall's comprehensive testing support service, which also includes:

1. **Prentice Hall Custom Test:** This powerful computerized testing package is designed to operate on the DOS, WINDOWS and MACINTOSH platforms. It offers full mouse support, complete question editing capabilities, random test generation, graphics, and printing capabilities.

Prentice Hall Custom Test has a unique two-track design -- *Easy test f*or the novice computer user, and *Fulltest* for those who wish to write their own questions and create their own graphics.

In addition to traditional printing capabilities, Prentice Hall Custom Test also offers the On-Line Testing System -- the most efficient, time-saving examination aid on the market. With just a few keystrokes, the instructor can administer, correct, record and return computerized exams over a variety of networks.

Prentice Hall Custom Test is designed to assist educators in the recording and processing of results from student exams and assignments. Much more than a computerized gradebook, it combines a powerful database with analytical capabilities so the instructor can generate a full set of statistics. There is no grading system more complete or easier to use.

The Prentice Hall Custom Test is free. To order a specific Prentice Hall Custom Test title you may contact your local rep or call our Faculty Support Services Department at 1-800-526-0485. Please identify the main text author and title.

Toll-free **technical support** is offered to all users at **1-800-550-1701**.

2. For those instructors without access to a computer, we offer the popular **Prentice Hall Telephone Testing Service:** It's simple, fast and efficient. Simply pick the questions you'd like on your test from this bank, and call the Simon & Schuster Testing Service at 1-800-550-1701; outside the US and Canada, call 612-550-1705.

Identify the main text and test questions you'd like, as well as any special instructions. We will create the test (or multiple versions, if you wish) and send you a master copy for duplication within 48 hours. Free adopters for life of text use.

Chapter 1: Electronic Structure and Bonding, Acids and Bases

MULTIPLE CHOICE

1. Which of the following elements does this electronic configuration represent?

$$1s^2\ 2s^2\ 2p^5$$

a. F
b. C
c. N
d. Al
e. O

Answer: a Section: 2

2. Which carbon(s) in the following molecule is (are) sp hybridized?

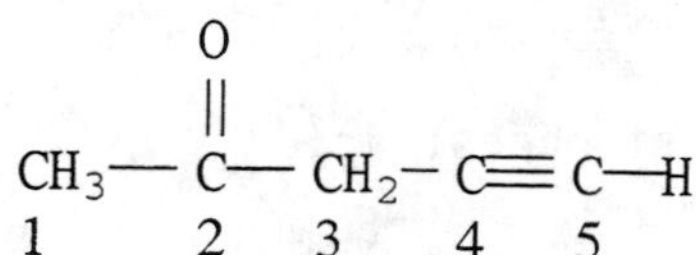

a. carbon 1
b. carbon 2
c. carbon 1, 3
d. carbon 4
e. carbon 4, 5

Answer: e Section: 9

3. Which of the following molecules does not exhibit a net dipole moment of zero?

a. CO_2
b. CH_4
c. CCl_4
d. H_2O
e. SO_3

Answer: d Section: 15

4. The conjugate acid of H_2O is:

   a. $H_2\ddot{O}:$

   b. $H_3O:$

   c. $\overset{\oplus}{H_3O:}$

   d. $^{-}:\ddot{\underset{..}{O}}H$

   e. $\overset{\oplus}{H_2\ddot{O}:}$

   Answer: c Section: 16

5. Which of the following is the strongest acid?

   a. 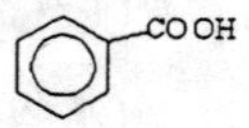 COOH

   b. 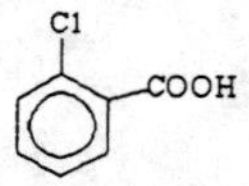 Cl, COOH

   c. 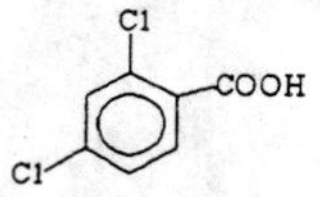 Cl, COOH, Cl

   d. 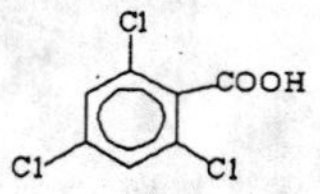 Cl, COOH, Cl, Cl

   e. 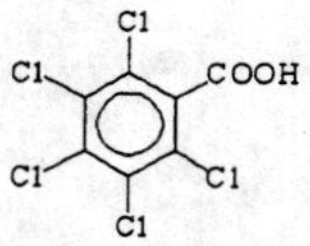 Cl, Cl, COOH, Cl, Cl, Cl

   Answer: e Section: 18

6. Which of the following is an sp^2 hybridized carbon?

   a. $^{\oplus}CH_3$
   b. $\cdot CH_3$
   c. $^{\ominus}:CH_3$
   d. a and b
   e. a, b and c

   Answer: d Section: 10

7. Which bond in the following molecule is the shortest?

$$CH_3 \overset{1}{—} CH_2 \overset{2}{—} CH \overset{3}{=} CH — C \overset{4}{\equiv} C — \overset{O}{\overset{||}{C}} \overset{5}{—} H$$

a. bond 1
b. bond 2
c. bond 3
d. bond 4
e. bond 5

Answer: d Section: 14

8. Which of the following molecules has a net dipole moment of zero?

a. $(Cl)(H)C=C(Cl)(H)$ — Cl and Cl on top, H and H on bottom

b. $(H)(Cl)C=C(Cl)(H)$ — H and Cl on top, Cl and H on bottom

c. $(H)(H)C=C(Cl)(Cl)$ — H and Cl on top, H and Cl on bottom

d. $(Cl)(H)C=C(Cl)(Cl)$ — Cl and Cl on top, H and Cl on bottom

e. $(H)(Cl)C=C(H)(H)$ — H and H on top, Cl and H on bottom

Answer: b Section: 15

9. Which of the following molecules has the <u>smallest</u> dipole moment?

a. Br_2
b. NH_3
c. HCl
d. HBr
e. HI

Answer: a Section: 15

10. Which of the following <u>is not</u> a conjugate acid-base pair?

a. $H_2\ddot{O}:$, $^{-}:\ddot{O}H$

b. $H_2\ddot{O}:$, $H_3\ddot{O}^+$

c. $HSO^-{}_4$, H_2SO_4
d. ^-OH, O^{2-}
e. $NO_3{}^-$, $NO_2{}^-$

Answer: e Section: 16

11. What is the product formed from the following acid-base reaction?

$$CH_3\ddot{O}H + :NH_3 \rightleftharpoons$$

a. $CH_3O^- + {}^+NH_4$
b. $CH_2OH + {}^+NH_3$
c. $CH_3OH_2{}^+ + {}^-NH_2$
d. $CH_3NH_2 + H_2O$
e. $CH_4 + NH_2OH$

Answer: a Section: 16

12. What is the pH of a 0.1 M solution of HCl? (note: pK_a for HCl is -6)

a. 6
b. -6
c. 1
d. -8
e. -1

Answer: c Section: 17

13. How many unpaired electrons are present in the isolated carbon atom (atomic number = 6)?

    a. none
    b. one
    c. two
    d. three
    e. four

    Answer: c Section: 2

14. Which of the following is the most likely electronic structure for C_2H_2?

    a. H — C = C:

           |
           H

    b. H — C̈ — C̤̈:

           |
           H

    c. H — C̈ — C̈ — H
    d. H — C ≡ C — H
    e. H — C⊕ = C̈ — H

    Answer: d Section: 4

15. Which of the following is closest to the C-O-C bond angle in CH_3-O-CH_3?

    a. 180°
    b. 120°
    c. 109.5°
    d. 90°
    e. 160°

    Answer: c Section: 11

16. What is the predicted shape, bond angle, and hybridization for $^+CH_3$?

    a. Trigonal planar, 120°, sp^2
    b. Trigonal planar, 120°, sp^3
    c. Trigonal planar, 109.5°, sp^2
    d. Trigonal pyramidal, 120°, sp^2
    e. Trigonal pyramidal, 109.5°, sp^2

    Answer: a Section: 10

17. Which of the following is the strongest acid?

a. HF
b. H_2O
c. $:NH_3$
d. CH_4
e. CH_3OH

Answer: a Section: 18

18. Which of the following is the strongest acid?

a. CH_3CH_2OH
b. CH_3OCH_3
c. CH_3-NH-CH_3
d. $CH_3-C{\equiv}CH$
e. $CH_3-CH{=}CH_2$

Answer: a Section: 18

19. Which of the following ions is the strongest acid?

a. $:H^{\ominus}$
b. $H\ddot{\underset{..}{O}}:^{\ominus}$
c. $HSO_4^{\ominus}$
d. H_2O
e. $H_3O^{\oplus}$

Answer: e Section: 16

20. Which species act as bases in the following reaction?

$$\underset{1}{H_2SO_4} + \underset{2}{HNO_3} \rightleftharpoons \underset{3}{H_2NO_3^{\oplus}} + \underset{4}{HSO_4^{\ominus}}$$

a. 1 and 2
b. 3 and 4
c. 2 and 4
d. 1 and 3
e. 2 and 3

Answer: c Section: 16

21. What is the conjugate acid of NH_3?

a. $^{+}NH_3$
b. ^{-}NH
c. $^{+}NH_4$
d. $^{-}NH_2$
e. $^{+}NH_2$

Answer: c Section: 16

22. Which of the following species have tetrahedral bond angles?

NH_3	BF_3	$^{\oplus}CH_3$	$^{\ominus}CH_3$	H_2O	$H_3O^{\oplus}$
A	B	C	D	E	F

a. A, D and E
b. A, D, E and F
c. A and E
d. D only
e. A, B and E

Answer: b Section: 14

23. Which of the following is the electronic configuration of the element Fe?

a. $1s^2\ 2s^2\ 2p^6\ 3s^2\ 3p^6\ 4s^2\ 3d^6$
b. $1s^2\ 2s^2\ 2p^6\ 3s^2\ 3p^8\ 3d^6$
c. $1s^2\ 2s^2\ 2p^8\ 3s^2\ 3p^6\ 4s^2\ 3d^6$
d. $1s^2\ 2s^2\ 2p^6\ 3s^2\ 3p^6\ 4s^2\ 4d^6$
e. $1s^2\ 2s^2\ 2p^6\ 3s^2\ 3p^6\ 4s^2\ 4p^6$

Answer: a Section: 2

24. Which of the following structures, including formal charges, is correct for diazomethane, CH_2N_2?

a. $:CH_2—N{\equiv}N:$ (⊖ on N, ⊕ on N)

b. $CH_2{=}N{=}\ddot{N}:$ (⊕ on N, ⊖ on N)

c. $CH_2{=}N{\equiv}N:$ (⊕ on N, ⊖ on N)

d. $CH_2{=}N{=}\ddot{N}:$

e. $CH_2—\ddot{N}—\ddot{N}:$ (+3 on N, −3 on N)

Answer: d Section: 4

25. What are the formal charges on nitrogen and the starred oxygen atom in the following molecule?

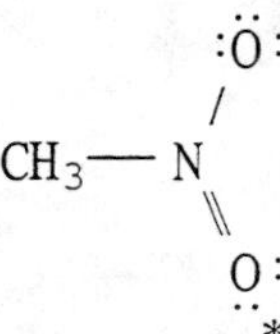

a. N = -1, O = 0
b. N = +1, O = -1
c. N = +1, O = +1
d. N = -1, O = -1
e. N= +1, O = 0

Answer: e Section: 4

SHORT ANSWER

26. Ar, K^+, Cl^- are isoelectronic elements (elements with the same number of electrons). What orbital does the last electron occupy?

Answer: 3p orbital Section: 2

27. Using the symbol δ^+ and $\delta-$, indicate the direction of the polarity in the indicated bond.

$CH_3\ddot{O}$—H

Answer:

$\delta-$ $\delta+$

$CH_3\ddot{O}$—H

Section: 3

28. Give the conjugate acid and the conjugate base for HSO_4^-.

Answer:
conjugate acid: H_2SO_4

conjugate base: SO_4^{-2}

Section: 16

29. If H_2O has a pK_a value of 15.7 and HF has a pK_a value of 3.2, which is a stronger base, HO^- or F^-? Explain.

Answer:
HO^- is a stronger base than F^- because HF is a stronger acid than H_2O, and the stronger the acid the weaker its conjugate base.

Section: 17

30. Draw the Kekulé structure for each of the following:

a. CH_3CH_2OH b. CH_3CHO c. $(CH_3)_3C^+$

Answer:

```
        H     H
        |     |
a.  H — C  —  C — OH
        |     |
        H     H

        H     Ö:
        |     ||
b.  H — C  —  C — H
        |
        H

          CH3
           |
c.  CH3 — C⊕
           |
          CH3
```

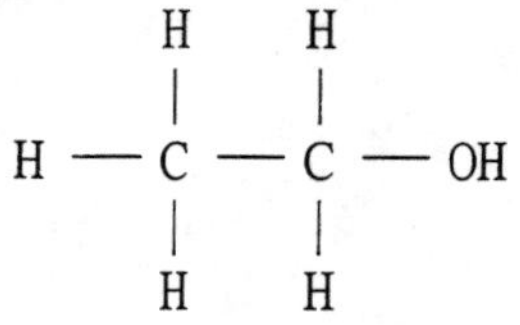

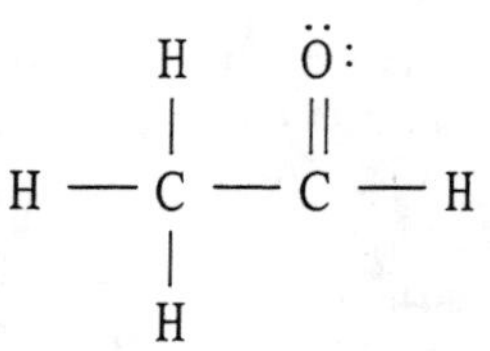

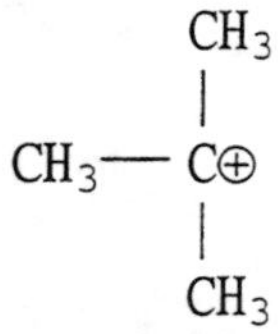

Section: 4

31. Draw the Kekulé structure and show the direction of the dipole moment for CH_2Cl_2.

Answer:

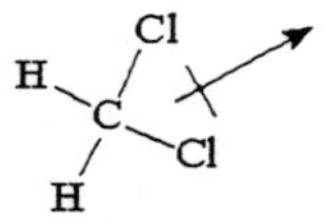

Cl
H
C
Cl
H

Section: 15

32. BF_3 has a diplole moment of zero. Propose a structure for BF_3 that is consistent with this information.

Answer:
BF_3 is trigonal planar.

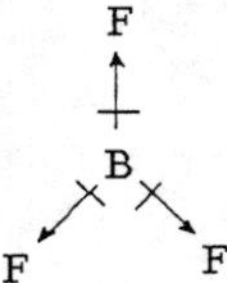

Section: 15

33. Both sigma (σ) and pi (π) bonds can be formed by overlapping p orbitals. Describe the difference.

Answer:
Sigma bonds are formed from the overlap of atomic orbitals along a circular axis of symmetrical nature, i.e., head-on overlap. All single bonds are sigma bonds.

Pi bonds are formed from the overlap of atomic orbitals along a non-symmetrical (parallel) axis, i.e., side-to-side overlap. Double and triple bonds are pi bonds.

Section: 6

34. Explain why $:NF_3$ is a weaker base than $:NH_3$.

Answer:
Fluorine has an electron withdrawing effect that reduces the availability of the pair of electrons on nitrogen. Thus the basicity of $:NF_3$ is less than that of $:NH_3$.

Section: 18

35. Explain why $AlCl_3$ is a Lewis acid.

Answer:
A Lewis acid is an electron pair acceptor. Aluminum in $AlCl_3$ has an empty p orbital that can accommodate the pair of electrons provided by a Lewis base.

Section: 20

MULTIPLE CHOICE

36. The atomic number of boron is 5. The correct electronic configuration of boron is:

    a. $1s^2 2s^3$
    b. $1s^2 2p^3$
    c. $1s^2 2s^2 2p^1$
    d. $2s^2 2p^3$
    e. $1s^2 2s^2 3s^1$

    Answer: c Section: 2

SHORT ANSWER

37. Atoms with the same number of protons but different numbers of neutrons are called ____________.

    Answer: isotopes Section: 1

38. Draw the shape of a 2p orbital.

    Answer:

    Section: 5

MULTIPLE CHOICE

39. How many distinct p orbitals exist in the second electron shell, where n = 2?

    a. 0
    b. 1
    c. 2
    d. 3
    e. 4

    Answer: d Section: 5

SHORT ANSWER

40. Covalent bonds may be polar or nonpolar. What property of the atoms forming a given bond determines this?

    Answer: Electronegativity Section: 3

MULTIPLE CHOICE

41. The compound methylamine, CH_3NH_2, contains a C-N bond. In this bond, which of the following best describes the charge on the nitrogen atom?

    a. +1
    b. slightly positive
    c. uncharged
    d. slightly negative
    e. -1

    Answer: d Section: 3

42. The formal charge on nitrogen in the compound below is _____.

    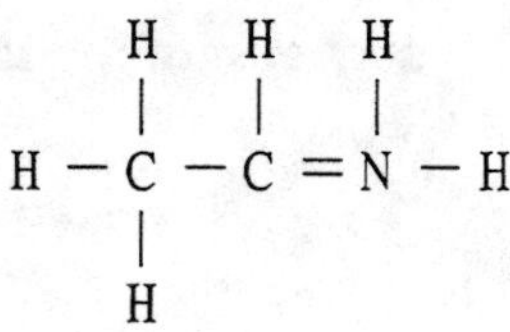

    A. +2
    B. +1
    C. 0
    D. -1
    E. -2

    Answer: b Section: 4

SHORT ANSWER

43. The Kekulé structure of pentane is shown below. Draw the condensed structural formula which corresponds to this Lewis structure.

```
     H   H   H   H   H
     |   |   |   |   |
 H — C — C — C — C — C — H
     |   |   |   |   |
     H   H   H   H   H
```

    Answer: $CH_3(CH_2)_3CH_3$ Section: 4

44. Write a completed equation for the acid-base pair shown below.
    $HCO_2H + {}^{-}NH_2 \longrightarrow$

    Answer: $HCO_2H + {}^{-}NH_2 \longrightarrow HCO_2^{-} + NH_3$ Section: 16

45. Consider the set of compounds, NH_3, HF, and H_2O. Rank these compounds in order of increasing acidity and discuss your rationale.

Answer:
$NH_3 < H_2O < HF$

When determining relative acidity, it is often useful to look at the relative basicity of the conjugate bases. The stronger the acid, the weaker (more stable, less reactive) the conjugate base. In this case, one would look at the relative basicity of F^-, OH^-, and NH_2^-. The relative strengths of these species can be gauged based on the electronegativity of the charged atom in each. Since fluorine is the most electronegative, F^- is the most stable, least reactive base in the group. This means that its conjugate acid, HF, is the strongest.

Section: 18

46. Would you predict trifluoromethanesulfonic acid, CF_3SO_3H, to be a stronger or weaker acid than methanesulfonic acid, CH_3SO_3H? Explain your reasoning.

Answer:
Trifluoromethanesulfonic acid is a stronger acid. Compare the strengths of the conjugate bases and remember that the weaker the base, the stronger the conjugate acid. In the case of the trifluoro derivative, the presence of the highly electronegative fluorine atoms serves to delocalize the negative charge to a greater extent. This additional delocalization makes trifluoromethanesulfonate a weaker base.

Section: 18

MULTIPLE CHOICE

47. Which of the compounds below bond predominantly via ionic bonding?

A. KCl
B. CF_4
C. NH_3
D. both a and b
E. both b and c
F. all of the above

Answer: a Section: 3

SHORT ANSWER

48. Draw condensed structures for the four compounds with formula C_3H_9N.

Answer:
$CH_3CH_2CH_2NH_2$
$CH_3CH_2NHCH_3$
$(CH_3)_2CHNH_2$
$(CH_3)_3N$

Section: 4

MULTIPLE CHOICE

49. Consider the interaction of two hydrogen 1s atomic orbitals of the same phase. Which of the statements below is an incorrect description of this interaction?

a. A sigma bonding molecular orbital is formed.
b. The molecular orbital formed is lower in energy than a hydrogen 1s atomic orbital.
c. The molecular orbital formed has a node between the atoms.
d. The molecular orbital formed is cylindrically symmetric.
e. A maximum of two electrons may occupy the molecular orbital formed.

Answer: c Section: 6

SHORT ANSWER

50. What kind of molecular orbital (σ, σ^*, π, or π^*) results when the two atomic orbitals shown below interact in the manner indicated?

Answer: σ^* Section: 6

51. What kind of molecular orbital (σ, σ^*, π, or π^*) results when the two atomic orbitals shown below interact in the manner indicated?

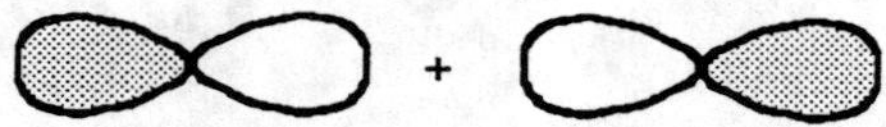

Answer: σ Section: 6

52. What kind of molecular orbital (σ, σ^*, π, or π^*) results when the two atomic orbitals shown below interact in the manner indicated?

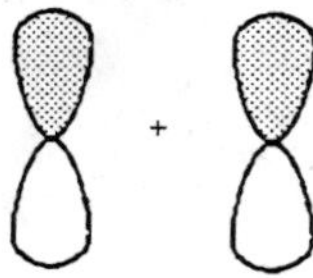

Answer: π Section: 6

53. What kind of molecular orbital (σ, σ^*, π, or π^*) results when the two atomic orbitals shown below interact in the manner indicated?

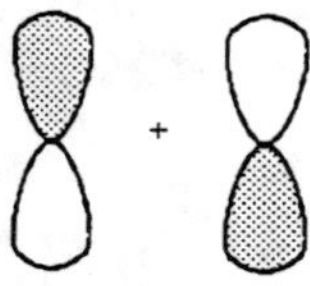

Answer: π^* Section: 6

54. What kind of molecular orbital (σ, σ^*, π, or π^*) results when the two atomic orbitals shown below interact in the manner indicated?

Answer: σ^* Section: 6

MULTIPLE CHOICE

55. Triethylamine $[(CH_3CH_2)_3N]$ is a molecule in which the nitrogen atom is ______ hybridized and the CNC bond angle is ______.

A. sp^2, >109.5°
B. sp^2, <109.5°
C. sp^3, >109.5°
D. sp^3, <109.5°
E. sp, 109.5°

Answer: d Section: 12

56. Choose the correct hybridization for the atom indicated in the molecule below.

$CH_3CH_2CH_2CH_3$
↑

a. sp
b. sp^2
c. sp^3
d. none of the above

Answer: c Section: 7

57. What type of bonding is <u>most</u> inportant in $CH_3CH_2CH_2CH_2CH_2CH_3$?

a. ionic
b. hydrogen
c. covalent
d. polar

Answer: c Section: 3

58. What is the conjugate acid of CH_3NH_2?

a. $CH_3NH_3^+$
b. CH_3NH^-
c. NH_4^+
d. NH_2^-

Answer: a Section: 12

59. What is the conjugate base of CH_3NH_2?

a. $CH_3NH_3^+$
b. CH_3NH^-
c. NH_4^+
d. NH_2^-

Answer: b Section: 12

SHORT ANSWER

60. Write Lewis structures for the molecule given and show all formal charges.

CH_2CO

Answer:
H
C::Ö:
H

Section: 4

Chapter 2: An Introduction to Organic Compounds

MULTIPLE CHOICE

1. What is the common name for the following structure?

$$CH_3\text{—}\underset{\substack{|\\ CH_3}}{CH}\text{—}CH_3$$

a. Isobutane
b. Isopropylmethane
c. t-Butane
d. n-Butane
e. sec-Butane

Answer: a Section: 2

2. What is the common name for the following structure?

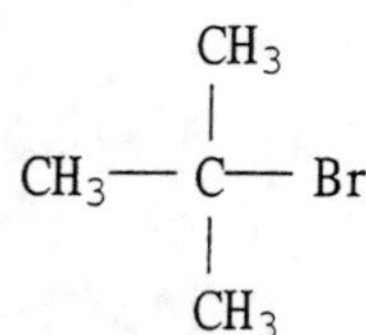

a. Isobutyl bromide
b. t-Butyl bromide
c. Neobutyl bromide
d. sec-Butyl bromide
e. Isopropyl methyl bromide

Answer: b Section: 4

3. How many constitutional isomers are possible for C_6H_{12}?

a. 4
b. 5
c. 6
d. 7
e. 8

Answer: b

4. Give the IUPAC name for the following structure:

   a. 2-methyl-3-ethylheptane
   b. 3-ethyl-2-methylheptane
   c. 5-Isopropyloctane
   d. 4-Isopropyloctane
   e. 2-methyl-3-propylheptane

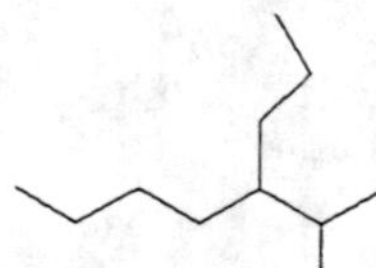

Answer: d Section: 2

5. Give the IUPAC name for the following compound:

   a. 1-chloro-2-methylcyclohexane
   b. 1-methyl-2-chlorocyclohexane
   c. 1-chloro-5-methylcyclohexane
   d. 1-methyl-5-chlorocyclohexane
   e. 1,2-chloromethylcyclohexane

CH_3

Cl

Answer: a Section: 4

6. Which of the following is diisopropyl ether?

   a. $CH_3CH_2CH_2-O-CH_2CH_2CH_3$
   b. $CH_3CH(CH_3)-O-CH_2CH_2CH_3$
   c. $CH_3CH_2CH_2-O-CHCH_3(CH_3)$
   d. $CH_3-C(CH_3)_2-O-C(CH_3)_2-CH_3$
   e. $(CH_3)_2CH-O-CH(CH_3)-CH_3$

Answer: e Section: 5

7. Give the IUPAC name for the following structure:

   a. 3-chloro-2-methylcyclohexanol
   b. 2-methyl-5-chlorocyclohexanol
   c. 1-chloro-4-methylcyclohexanol
   d. 5-chloro-2-methylcyclohexanol
   e. 2-methyl-3-chlorocyclohexanol

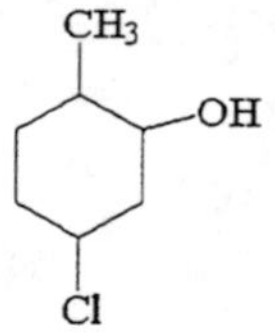

CH_3
OH
Cl

Answer: d Section: 6

8. Which of the following is a tertiary amine?

   a. $CH_3 - CH(CH_3) - CH_2 - \ddot{N}H_2$
   b. $CH_3 - CH(CH_3) - CH(\ddot{N}H_2) - CH_3$
   c. $CH_3 - C(CH_3)(\ddot{N}H_2) - CH_3$
   d. $CH_3 - C(CH_3)(\ddot{N}H - CH_3) - CH_3$
   e. $CH_3 - \ddot{N}(CH_3) - CH_3$

Answer: e Section: 1

9. Which of the following will have the <u>lowest</u> boiling point?

   a. CH_3Cl
   b. CH_4
   c. CH_2Cl_2
   d. $CHCl_3$
   e. CCl_4

Answer: b Section: 9

10. Which of the following has the <u>greatest</u> Van der Waal's interaction between molecules of the same kind?

a. $CH_3-CH-CH_2CH_3$ (with CH_3 on CH)
b. $CH_3CH_2CH_2CH_3$
c. CH_3-C-CH_3 (with CH_3 above and CH_3 below C)
d. $CH_3CH_2CH_2CH_2CH_3$
e. $CH_3-CH-CH-CH_3$ (with CH_3 on each CH)

Answer: d Section: 9

11. Which of the following has the <u>lowest</u> boiling point?

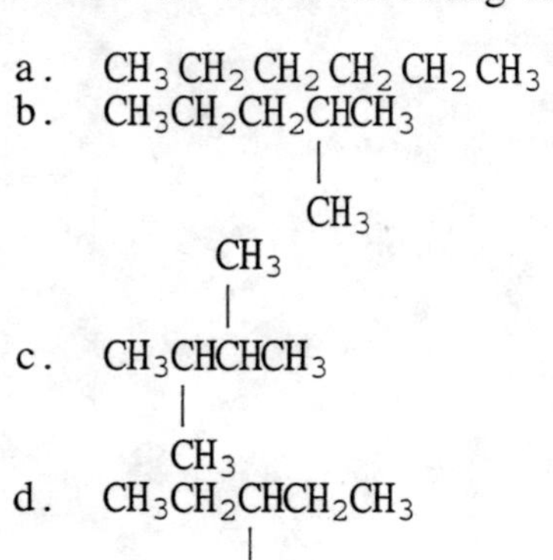

a. $CH_3CH_2CH_2CH_2CH_2CH_3$
b. $CH_3CH_2CH_2CHCH_3$ (with CH_3 on CH)
c. $CH_3CHCHCH_3$ (with CH_3 above the first CH and CH_3 below the second CH)
d. $CH_3CH_2CHCH_2CH_3$ (with CH_3 above and CH_3 below CH)
e. $CH_3CH_2CH_2CH_2CH_2CH_2CH_3$

Answer: c Section: 9

12. Which of the following has the <u>greatest</u> solubility in $CH_3CH_2CH_2CH_3$?

a. CH_3OH
b. $CH_3O^-Na^+$
c. CH_3NH_2
d. CH_3OCH_3
e. $(CH_3)_3CH$

Answer: e Section: 9

13. Which of the following is the most soluble in H_2O?

a. CH_3OCH_3
b. CH_3CH_2OH
c. CH_3CH_2Cl
d. $CH_3CH_2CH_3$
e. CH_3CHO

Answer: b Section: 9

14. Which of the following would have the highest boiling point?

a. $CH_3CH_2—O—CH_2CH_2—O—CH_3$
b. $CH_3—O—CH_2CH_2CH_2—O—CH_3$
c. $HO—CH_2CH_2CH_2CH_2—OH$
d. $CH_3CH_2—O—CH_2—O—CH_2CH_3$
e. $CH_3—O—CH_2CH(CH_3)—O—CH_3$

Answer: c Section: 9

15. The eclipsed and staggered forms of ethane are said to differ in:

a. molecular formula.
b. configuration.
c. conformation.
d. constitution.
e. structure.

Answer: c Section: 11

16. Which of the following is the staggered conformation for rotation about the C_1-C_2 bond in the following structure?

CH_3
$CH_3CHCH_2CH_3$
1 2 3 4

a. 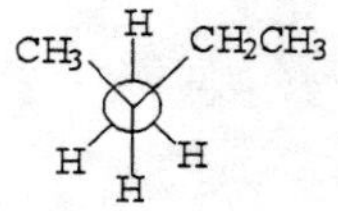 CH_3, H, CH_2CH_3, H, H, H

b. 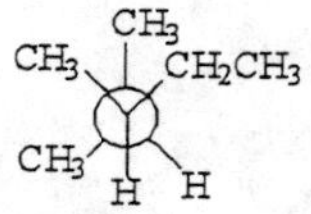 CH_3, CH_3, CH_2CH_3, CH_3, H, H

c. 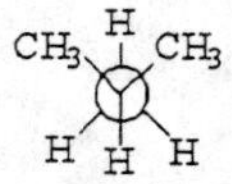 CH_3, H, CH_3, H, H, H

d. 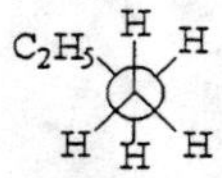 C_2H_5, H, H, H, H, H

e. 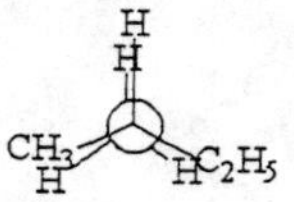 H, H, CH_3, H, H, C_2H_5

Answer: a Section: 11

17. Which of the following is the <u>most</u> stable conformation of bromocyclohexane?

a. 

b. 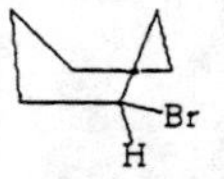 Br, H

c. 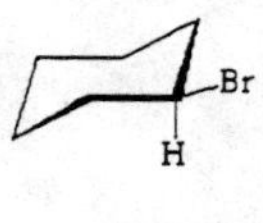 Br, H

d. 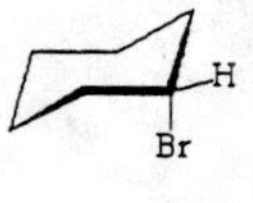 H, Br

e.  Br, H

Answer: c Section: 13

SHORT ANSWER

18. There is something wrong with the following name. Write the structure and correct the name: 2-ethylpropane.

Answer:

```
CH3— CH — CH3
      |
     CH2
     |
    CH3
```

The correct name is 2-methyblutane.

Section: 2

19. Give the structure of isopentyl alcohol.

Answer:

```
CH3
   \
    CH—CH2 CH2 OH
   /
CH3
```

Section: 6

20. Give the structure of tetramethylammonium chloride.

Answer:

```
         CH3
         |
CH3  — N(+)— CH3Cl(-)
         |
         CH3
```

Section: 7

21. Fluorine is more electronegative than chlorine yet the carbon-fluorine bond in CH_3-F is shorter than CH_3-Cl. Explain.

Answer:
Chlorine is a larger atom than fluorine and occupies a 3p rather than 2p orbital. The overlap of a $2sp^3$ orbital with a 3p orbital is not as good as the overlap of a $2sp^3$ orbital with a 2p orbital, causing the bond to be longer and weaker.

Section: 8

22. Explain why CH_3CHCH_3 with a CH_3 branch (isobutane) has a lower boiling point than $CH_3CH_2CH_2CH_3$.

Answer:
$CH_3CH_2CH_2CH_3$ has greater van der Waals forces because it has a greater contact area than isobutane. Therefore, the boiling point of $CH_3CH_2CH_2CH_3$ is higher.

Section: 9

23. Primary and secondary amines exhibit hydrogen bonding; tertiary amines do not. Explain.

Answer:
The nitrogen in a tertiary amine is not attached to a hydrogen. Recall that for a molecule to exhibit hydrogen bonding it must have a hydrogen attached to a highly electronegative atom such as F, N, or O.

Section: 9

24. Draw the Newman structure for the most stable conformation of 1-bromopropane considering rotation about the C_1-C_2 bond.

Answer:

Br
H H
H H
CH_3

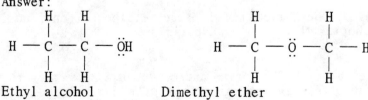

Section: 11

25. Draw all possible constitutional isomers for C_2H_6O and give common names for each structure.

Answer:

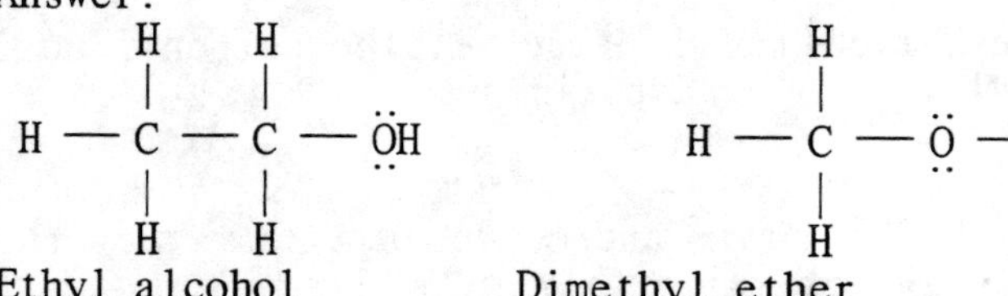

Ethyl alcohol Dimethyl ether

Section: 5

26. Explain why trimethylamine, $(CH_3)_3N$:, has a considerably lower boiling point than propylamine $CH_3CH_2CH_2NH_2$, even though both compounds have the same molecular formula.

Answer:
Since hydrogen bonding is possible for propylamine and not for trimethylamine, the boiling point is higher for propylamine.

Section: 9

27. Which of the molecules below has the higher boiling point? Briefly explain your choice.

$CH_3CH_2CH_2OH$ or $CH_3CH_2OCH_3$

Answer:
$CH_3CH_2CH_2OH$ has the higher boiling point since it is capable of intermolecular hydrogen bonding.

Section: 9

28. Would you expect sodium chloride (NaCl) to be highly soluble in the organic solvent hexane ($CH_3CH_2CH_2CH_2Ch_2CH_3$)? Briefly explain your answer.

Answer:
One would not expect NaCl to be highly soluble in hexane. NaCl is an ionic solid (ie, a vary polar material) while hexane is nonpolar. Nonpolar solvent molecules do not solvate ions well. The attractions of oppositely charged ions to each other are vastly greater than the weak attractions of the ions for the solvent.

Section: 9

29. Which compound is more soluble in water? Briefly explain your choice.

CH_3OCH_3 or CH_3CH_2OH

Answer:
CH_3CH_2OH is more soluble in water since it can donate a hydrogen bond to water and accept a hydrogen bond from water. CH_3OCH_3 can only accept a hydrogen bond from water; it has no hydrogen which can hydrogen bond to water.

Section: 9

30. Which compound is more soluble in water? Briefly explain your choice.

$(CH_3)_2NH$ or $CH_3CH_2CH_3$

Answer:
$(CH_3)_2NH$ is more soluble in water since it can hydrogen bond with water. Alkanes are not capable of hydrogen bonding with water.
Section: 9

MULTIPLE CHOICE

31. If an acyclic alkane hydrocarbon contains n carbon atoms, how many hydrogen atoms must it also contain?

a. n
b. n + 2
c. n - 2
d. 2n
e. 2n + 2
F. 2n - 2

Answer: e

SHORT ANSWER

32. Give structures for the three isomers with molecular formula C_5H_{12} and provide the common name of each.

Answer:

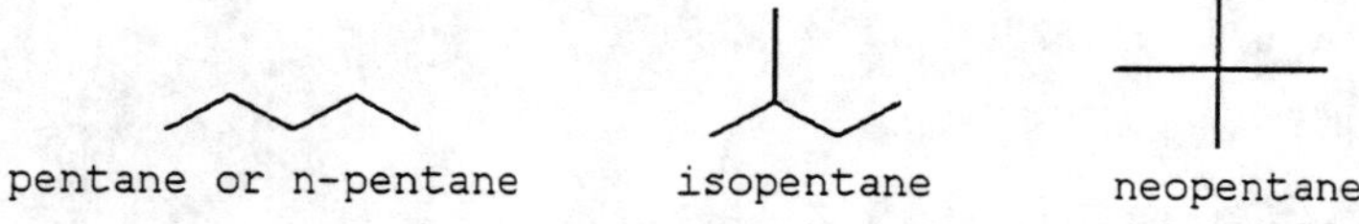

pentane or n-pentane isopentane neopentane

Section: 2

33. Provide an acceptable name for the alkane shown below.

$CH_3CH_2CH_2CH_2CH_2CH_3$

Answer: hexane <u>or</u> n-hexane Section: 2

34. Provide an acceptable name for the alkane shown below.

$CH_3-CH(CH_2CH_3)-CH_2CH_2CH(CH_3)_2$

Answer: 2, 5-dimethylheptane Section: 2

35. Provide an acceptable name for the alkane shown below.

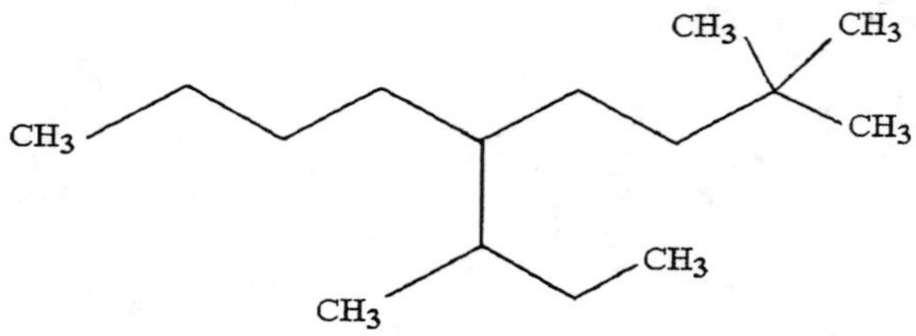

Answer:
5-sec-butyl-2, 2-dimethylnonane <u>or</u>
2, 2-dimethyl-5-(1-methylpropyl) nonane
Section: 2

36. Provide an acceptable name for the alkane shown below.

$$\begin{array}{c} \qquad\qquad\qquad\qquad\quad CH(CH_3)_2 \\ \qquad\qquad\qquad\qquad\quad | \\ CH_3CH_2CH_2CH_2CH_2CH_2CHCH_2CH_2CH_3 \end{array}$$

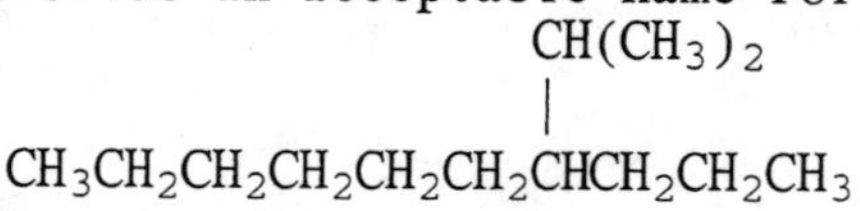

Answer: 4-isopropyldecane <u>or</u> 4-(1-methlyethyl) decane Section: 2

37. Provide an acceptable name for the alkane shown below.

$$CH_3CH_2CH_2CH(CH_3)CH(CH_2CH_2CH_3)CH_2CH(CH_2CH_3)CH_2CH_3$$

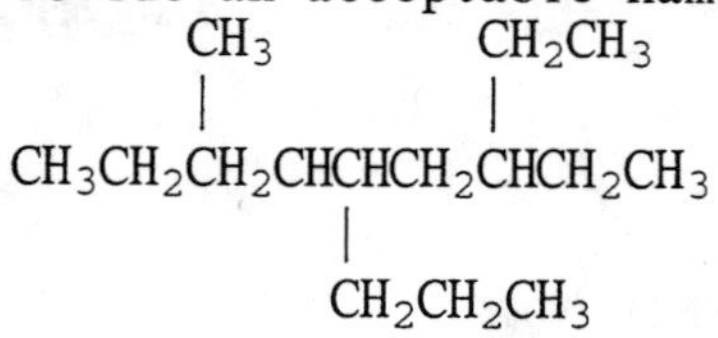

Answer: 3-ethyl-6-methyl-5-propylnonane Section: 2

38. Provide an acceptable name for the alkane shown below.

$$CH_3CH_2CH_2—C(CH_3)(CH_3)—C(CH_2CH_3)(CH_2CH_3)—H$$

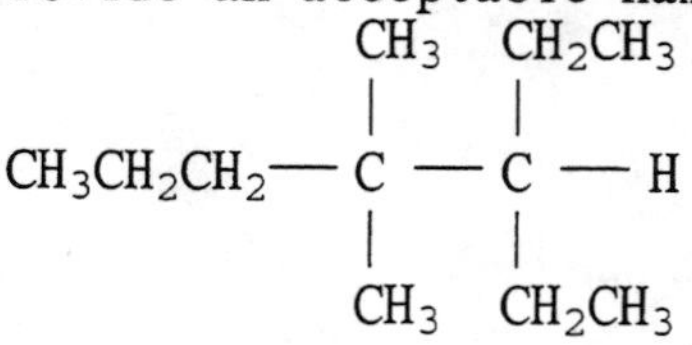

Answer: 3 Ethyl-4, 4-dimethylheptane Section: 2

39. Draw an acceptable structure for 4-t-butyloctane.

Answer:

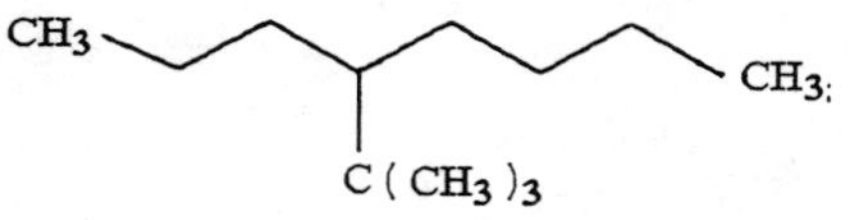

Section: 2

40. Draw an acceptable structure for 3-ethyl-3-methylhexane.

Answer:

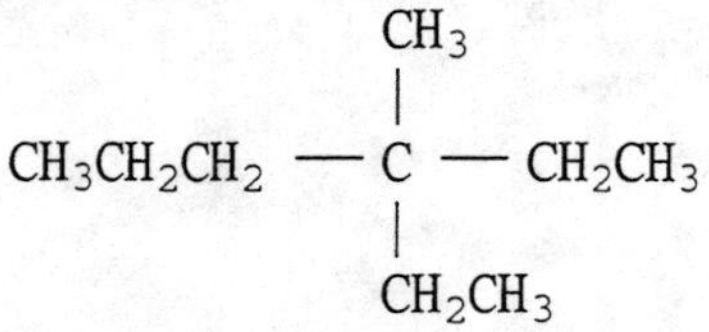

$$CH_3CH_2CH_2-\overset{\displaystyle CH_3}{\underset{\displaystyle CH_2CH_3}{C}}-CH_2CH_3$$

Section: 2

41. Draw an acceptable structure for 4-isopropyl-2-methlheptane.

Answer:

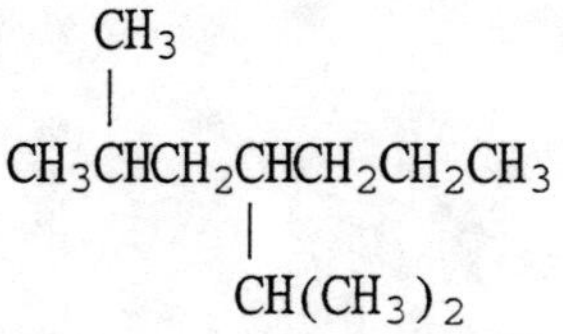

$$CH_3\overset{\displaystyle CH_3}{CH}CH_2\underset{\displaystyle CH(CH_3)_2}{CH}CH_2CH_2CH_3$$

Section: 2

42. Draw an acceptable structure for 6-ethyl-2,6,7-trimethyl-5-propylnonane.

Answer:

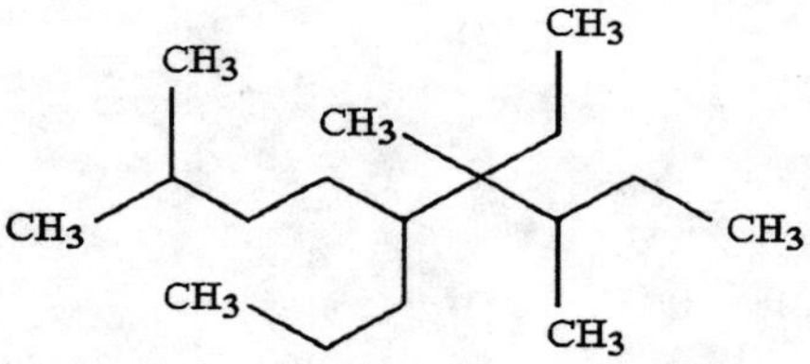

CH3 CH3 CH3 CH3 CH3 CH3 CH3

Section: 2

43. Provide an acceptable name for the alkane shown below.

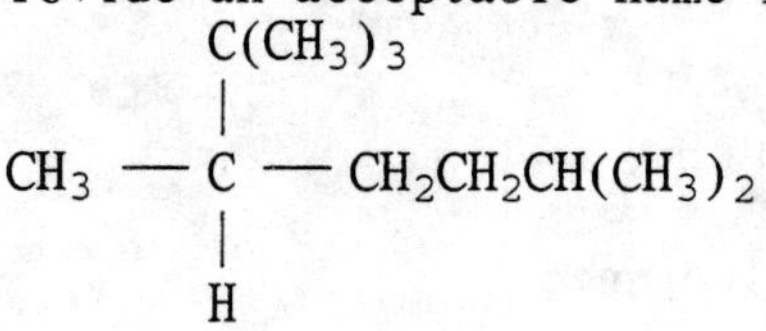

$$CH_3-\overset{\displaystyle C(CH_3)_3}{\underset{\displaystyle H}{C}}-CH_2CH_2CH(CH_3)_2$$

Answer: 2,2,3,6-tetramethylheptane Section: 2

44. Provide an acceptable name for the alkane shown below.

$$\begin{array}{ccccc} & H & & CH_2CH_2CH(CH_3)_2 & \\ & | & & | & \\ CH_3CH_2CH_2CH_2— & C & — & C & — CH_2CH_2CH_3 \\ & | & & | & \\ & CH_3CH_2 & & H & \end{array}$$

Answer: 6-ethyl-2-methyl-5-propyldecane Section: 2

45. Which intermolecular force is primarily responsible for the interactions among alkane molecules?

Answer: van der Waals <u>or</u> London forces Section: 9

MULTIPLE CHOICE

46. Consider the three isomeric alkanes n-hexane, 2,3-dimethylbutane, and 2-methylpentane. Which of the following correctly lists these compounds in order of increasing boiling point?

a. 2,3-dimethylbutane < 2-methylpentane < n-hexane
b. 2-methylpentane < n-hexane < 2,3-dimethylbutane
c. 2-methylpentane < 2,3-dimethylbutane < n-hexane
d. n-hexane < 2-methylpentane < 2, 3-dimethylbutane
e. n-hexane < 2,3-dimethylbutane < 2-methylpentane

Answer: a Section: 9

SHORT ANSWER

47. Draw a Newman projection of the most stable conformation of 2-methylpropane.

Answer:

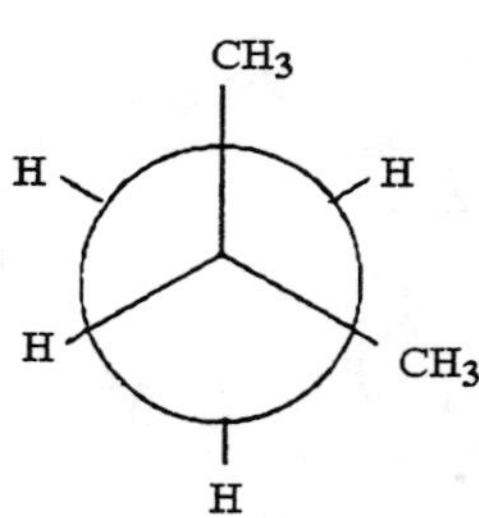

Section: 11

48. Define the term <u>conformation</u>.

Answer:
Conformations are different arrangements of the same molecule formed by rotations about single bonds.

Section: 11

49. Use a sawhorse structure to depicit the eclipsed conformer of ethane.

Answer:

Section: 11

50. View a butane molecule along the C_2-C_3 bond and provide a Newman projection of the lowest energy conformer.

Answer:

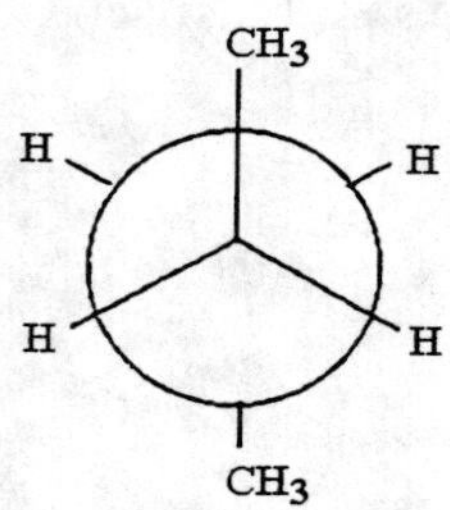

Section: 11

51. Provide a representation of the gauche conformer of butane.

Answer:

Section: 11

MULTIPLE CHOICE

52. Among the butane conformers, which occur at energy minima on a graph of potential energy versus dihedral angle?

    a. gauche only
    b. eclipsed and totally eclipsed
    c. gauche and anti
    d. eclipsed only
    e. anti only

    Answer: c Section: 11

53. Which of the following cycloalkanes exhibits the greatest molar heat of combustion?

    a. cyclopropane
    b. cyclobutane
    c. cyclopentane
    d. cyclohexane
    e. cycloheptane

    Answer: e Section: 10

54. Which of the following correctly ranks the cycloalkanes in order of increasing ring strain per methlene?

    a. cyclopropane < cyclobutane < cyclohexane < cycloheptane
    b. cyclohexane < cyclopentane < cyclobutane < cyclopropane
    c. cyclopentane < cyclobutane < cyclopentane < cyclopropane
    d. cyclopentane < cyclopropane < cyclobutane < cyclohexane
    e. cyclopropane < cyclopentane < cyclobutane < cyclohexane

    Answer: b Section: 12

SHORT ANSWER

55. Describe the source of angle strain and torsional strain present in cyclopropane.

    Answer:
    The angle strain arises from the compression of the ideal tetrahedral bond angle of 109.5° to 60°. The large torsional strain occurs since all C-H bonds on adjacent carbons are eclipsed.

    Section: 2

56. Draw the chair conformer of cyclohexane. Label the axial hydrogens (H_a) and the equatorial hydrogens (H_e).

Answer:

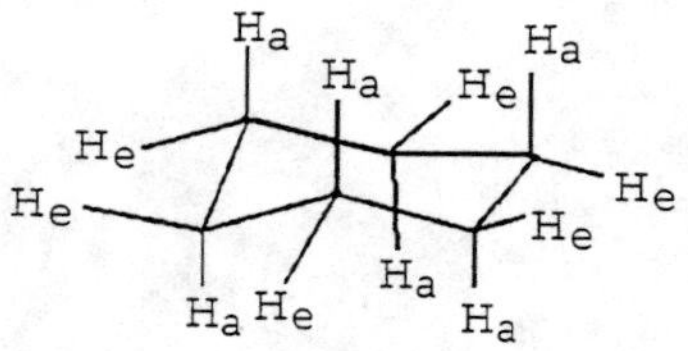

Section: 13

MULTIPLE CHOICE

57. Which of the following correctly lists the conformations of cyclohexane in order of increasing energy?

a. chair < boat < twist-boat < half-chair
b. half-chair < boat < twist-boat < chair
c. chair < twist-boat < half-chair < boat
d. chair < twist-boat < boat < half-chair
e. half-chair < twist-boat < boat < chair

Answer: d Section: 13

58. In the boat conformation of cyclohexane, the "flagpole" hydrogens are located:

a. on the same carbon.
b. on adjacent carbons.
c. on C-1 and C-3.
d. on C-1 and C-4.
e. none of the above

Answer: d Section: 13

59. Which of the following best explains the relative stabilities of the eclipsed and staggered forms of ethane? The _____ form has the most _____ strain.

a. eclipsed; steric
b. eclipsed; torsional
c. staggered; steric
d. staggered; torsional

Answer: b Section: 11

60. Which of the following best explains the reason for the relative stabilities of the conformers shown?

    a. I has more torsional strain.
    b. I has more steric strain.
    c. II has more torsional strain.
    d. II has more steric strain.

    I: CH_3, H, H, H, H, CH_3

    II: CH_3, H, CH_3, H, H, H

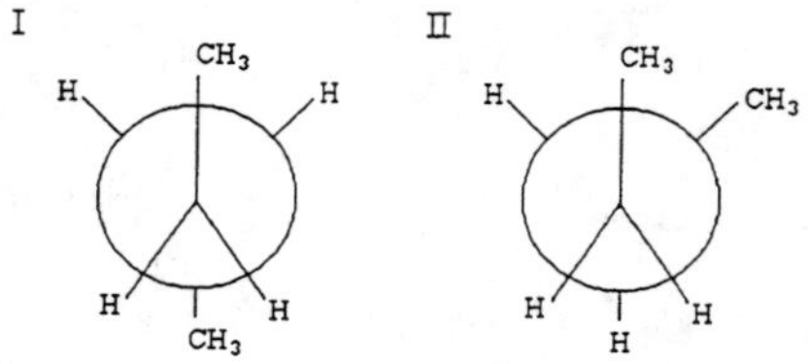

    Answer: d Section: 11

Chapter 3: Reactions of Alkenes, Thermodynamics and Kinetics

MULTIPLE CHOICE

1. What is the common name for the following compound?

$$CH_3-\overset{\displaystyle CH_3 \atop |}{C}=CH_2$$

a. t-Butylene
b. sec-Butylene
c. Isobutylene
d. Butylene
e. Methylpropylene

Answer: c Section: 2

2. What is the IUPAC name for the following compound?

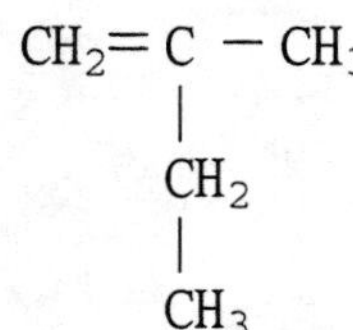

a. 2-methyl-1-butene
b. Isopentene
c. 2-methybutene
d. 2-Ethylpropene
e. 3-methyl-3-butene

Answer: a Section: 2

3. What is the IUPAC name for the following compound?

a. 5-methylcyclohexene
b. 4-methylcyclohexene
c. 1-methyl-3-cyclohexene
d. 1-methyl-4-cyclohexene
e. methylcyclohexene

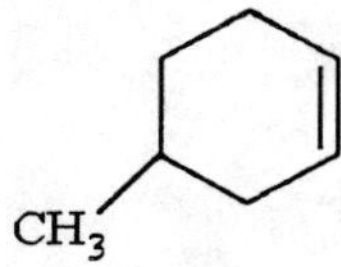

Answer: b Section: 2

4. Which of the following is vinyl chloride?

a. CH_3CH_2Cl

b. $CH_2{=}CHCH_2Cl$

c. $CH_2{=}CHCl$

d. C_6H_5—CH=CHCl

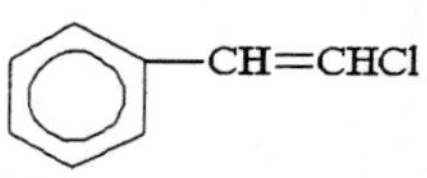

e. $\underset{\displaystyle Cl}{CH}{=}CHCl$

Answer: c Section: 2

5. Which of the following is capable of exhibiting cis-trans isomerism?

a. 1-butene
b. 1-pentene
c. cyclohexene
d. ethene
e. 2-butene

Answer: e Section: 2

6. Which of the following is <u>not</u> an electrophile?

a. H^+
b. BF_3
c. $^+NO_2$

d. Fe^{+3}
e. $CH_2{=}CH_2$

Answer: e Section: 6

7. Which of the following is <u>not</u> a nucleophile?

a. $FeBr_3$
b. Br^-

c. NH_3

d. benzene ring (structure)

e. CH_3OCH_3

Answer: a Section: 6

8. Which of the following statements about ethene, C_2H_4, is incorrect?

a. The H-C-H bond angles are approximately 109.5° .
b. All of the hydrogen atoms are in the same plane.
c. There is a total of five sigma bonds.
d. The carbon atoms are sp^2 hybridized.
e. The H-C-H bond angles are approximately 120° .

Answer: a Section: 3

9. Which of the following statements about propene, $CH_3CH{=}CH_2$, is correct?

a. All nine atoms lie in the same plane.
b. The compound has a cis and trans isomer.
c. It generally acts as a Lewis acid.
d. There is a total of eight sigma bonds.
e. All the carbon atoms are sp^2 hybridized.

Answer: d Section: 1

10. What is the value of ΔH in kcal/mole for the reaction shown?

$$(CH_3)_3{-}C{-}H + Cl{-}Cl \longrightarrow (CH_3)_3{-}C{-}Cl + H{-}Cl$$

Bond energies are: $(CH_3){-}C{-}H$ = 91 kcal/mole
$(CH_3)_3{-}C{-}Cl$ = 78.5 kcal/mole
$Cl{-}Cl$ = 58 kcal/mole
$H{-}Cl$ = 103 kcal/mole

a. +32.5
b. -57.5
c. -32.5
d. +57.5
e. -8.5

Answer: c Section: 7

11. How many transition states are present in the following reaction diagram?

    a. 3
    b. 4
    c. 5
    d. 2
    e. 1

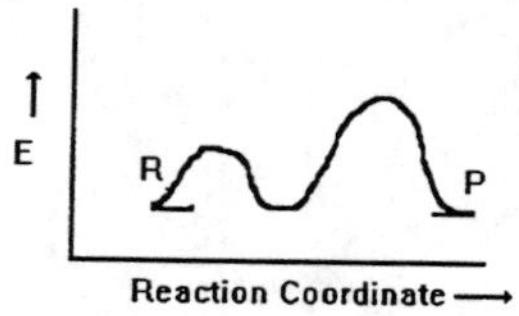

E ↑ R P

Reaction Coordinate ⟶

Answer: d Section: 7

12. An <u>increase</u> in which of the following results in a <u>decrease</u> in the rate of the chemical reaction?

    a. Temperature
    b. Concentration
    c. Collision frequency
    d. Energy of activation
    e. Fraction of collisions with proper orientation

Answer: d Section: 7

13. An increase in which of the following will occur if the reaction temperature is <u>increased</u>?

    I. Energy of activation
    II. Collision frequency
    III. Fraction of collisions with sufficient energy

    a. I and II
    b. I and III
    c. II and III
    d. I, II, and III
    e. I

Answer: c Section: 7

14. What is the activation energy for the reaction B → A in the following diagram?

a. A
b. B
c. C
d. D
e. E

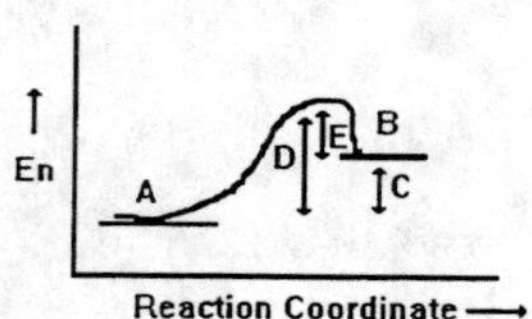

Answer: e Section: 7

15. Which of the following reagents gives the reaction shown below?

$$CH_3CH{=}CH_2 + ? \longrightarrow CH_3CH_2CH_3$$

a. H_2/HCl
b. H_2/H_2SO_4
c. H_2/Ni
d. H_2O/Ni
e. H_2O/H_2SO_4

Answer: c Section: 19

16. Which of the following is the most stable alkene?

a. $CH_3-CH(CH_3)CH=CH_2$ (CH_3 on the second carbon)

b. $CH_3CH_2C(CH_3)=CH_2$

c. $CH_3C(CH_3)=CH_2$

d. $(H)(CH_3)C=C(CH_3)(CH_3)$

e. $(CH_3CH_2)(CH_3)C=C(CH_3)(CH_3)$

Answer: e Section: 19

17. Which of the following alkenes reacts with HCl at the slowest rate?

a. $CH_3CHCH=CH_2$ (with CH_3 on the second carbon)

b. $CH_3CH_2C=CH_2$ (with CH_3 on the third carbon)

c. $CH_3C=CH_2$ (with CH_3 on the second carbon)

d. $(H)(CH_3)C=C(CH_3)(CH_3)$

e. $(CH_3CH_2)(CH_3)C=C(CH_3)(CH_3)$

Answer: a Section: 10

18. Which of the following is the most stable carbocation?

a. CH_3-substituted cyclohexane, with ⊕ on C-4

b. CH_3-substituted cyclohexane, with ⊕ on C-3

c. CH_3-substituted cyclohexane, with ⊕ on C-2

d. CH_3-substituted cyclohexane, with ⊕ on C-1

e. cyclohexyl with ⊕CH_2

Answer: d Section: 10

19. Which of the following is a step in the mechanism of the reaction shown?

$$CH_2{=}CHCH_3 + HBr \xrightarrow{\text{Peroxide}} \underset{\displaystyle Br}{\underset{|}{C}}H_2CH_2CH_3$$

a. $\underset{\displaystyle Br}{\underset{|}{C}}H_2\dot{C}HCH_2 + HBr$

b. $\dot{C}H_2CH_2CH_3 + HBr$

c. $\underset{\displaystyle Br}{\underset{|}{C}}H_2\overset{\oplus}{C}HCH_3 + HBr$

d. $\underset{\oplus}{C}H_2CH_2CH_3 + HBr$

e. $\underset{\displaystyle Br}{\underset{|}{C}}H_2\dot{C}HCH_3 + H\cdot$

Answer: a Section: 18

20. Which of the following carbocations is likely to rearrange?

a. 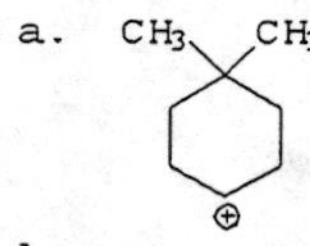

b. CH3 CH3 ⊕

c. 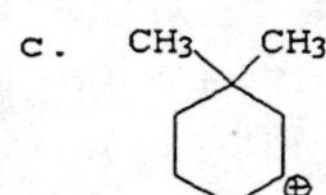

d. CH3 ⊕CH2

e. b and d

Answer: e Section: 14

21. What is the major product of the following reaction?

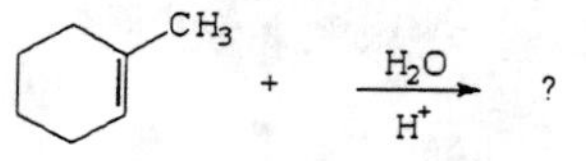

CH_3 + $\xrightarrow[H^+]{H_2O}$?

a. 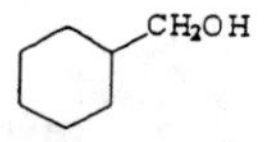 CH_2OH

b. 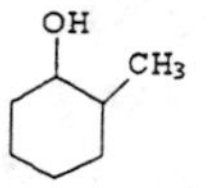 OH CH_3

c. 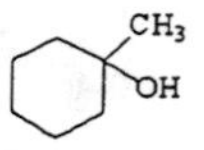 CH_3 OH

d. 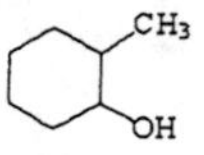 CH_3 OH

e. 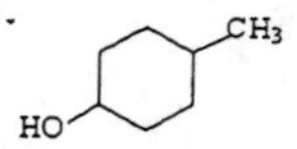 CH_3 HO

Answer: c Section: 13

22. What is the major product of the following reaction?

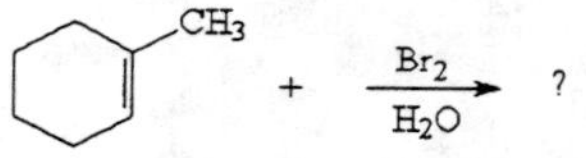

CH_3 + $\xrightarrow[H_2O]{Br_2}$?

a. 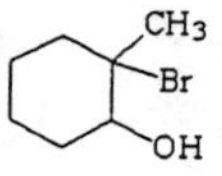 CH_3 Br OH

b. 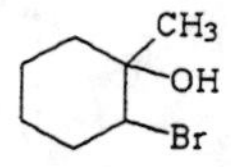 CH_3 OH Br

c. 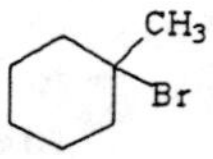 CH_3 Br

d. 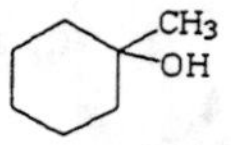 CH_3 OH

e. 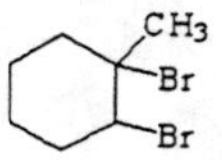 CH_3 Br Br

Answer: b Section: 15

23. What reagents are needed to accomplish the following transformation?

a. H_2O/H^+
b. H_2O/Peroxide
c. OH-
d. BH_3
e. 1. BH_3 / 2. HO^-, H_2O_2, H_2O

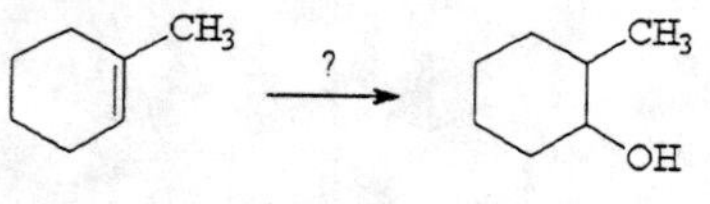

CH3 ? CH3 OH

Answer: e Section: 17

24. Which reaction intermediate is formed when Br_2/CCl_4 reacts with cyclohexene?

a. Br ⊕

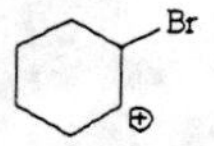

b. ⊕ Br

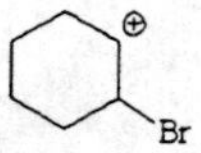

c. Br •

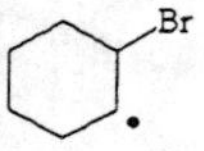

d. Br⊕

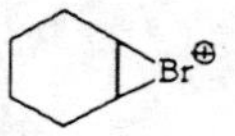

e. Br Br

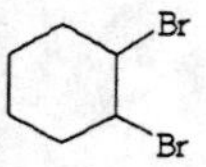

Answer: d Section: 15

SHORT ANSWER

25. Under what conditions is ΔG° equal to ΔH° for a chemical reaction?

Answer: Since ΔG° = ΔH° - TΔS°, then ΔG° is equal to ΔH° when TΔS° is zero.

Section: 7

26. Draw all the possible constitutional isomers of C_4H_8.

Answer:

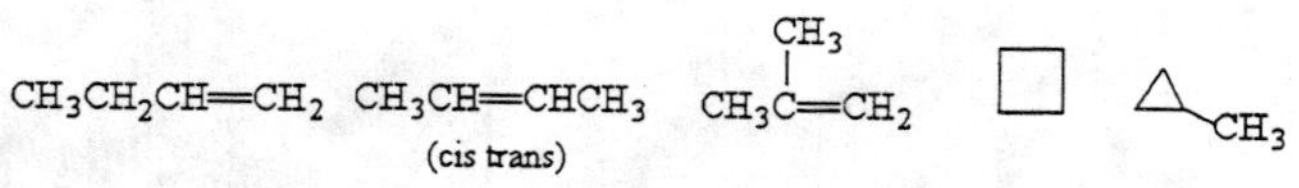

$CH_3CH_2CH{=}CH_2$ $CH_3CH{=}CHCH_3$ (cis trans) $CH_3C(CH_3){=}CH_2$ □ △CH_3

Section: 1

27. Complete the following tree of reactions by giving the major products:

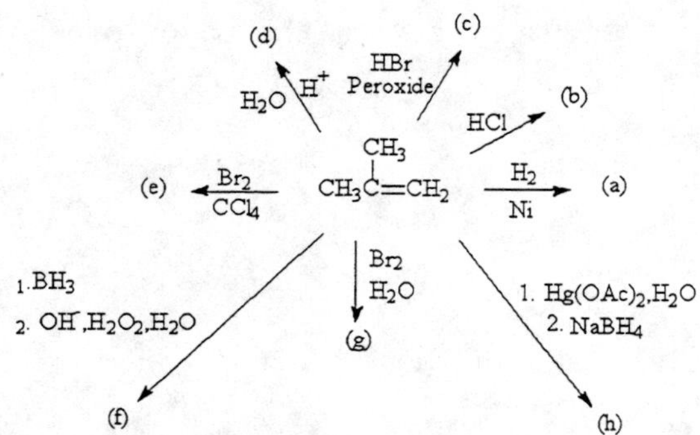

Answer: a. CH_3CHCH_3
|
CH_3

b. CH_3
|
CH_3CCH_3
|
Cl

c. Br
|
CH_3CHCH_2
|
CH_3

d. CH_3
|
CH_3CCH_3
|
OH

e. CH_3
|
CH_3-C-CH_2
| |
Br Br

f. CH_3
|
CH_3CHCH_2
|
OH

Answer:

g. $CH_3-C(CH_3)(OH)-CH_2Br$

```
        CH3
         |
g. CH3 — C — CH2
         |    |
        OH   Br
```

h. $CH_3C(CH_3)(OH)CH_3$

```
      CH3
       |
h. CH3CCH3
       |
      OH
```

Section: 9

28. Which of the following alkenes would have the smallest heat of hydrogenation? Explain.

$$CH_2{=}CH_2 \text{ or } CH_3CH{=}CH_2$$

Answer:
$CH_3CH{=}CH_2$ would have the smallest heat of hydrogenation because the double bond is more stable than that of $CH_2{=}CH_2$. The greater stability is due to the fact that the double bond in propane is more alkyl substituted than that of ethene.

Section: 19

29. Based on the following energy diagram, which compound, A or C, is formed faster from B? Which is more stable, A or C? Explain.

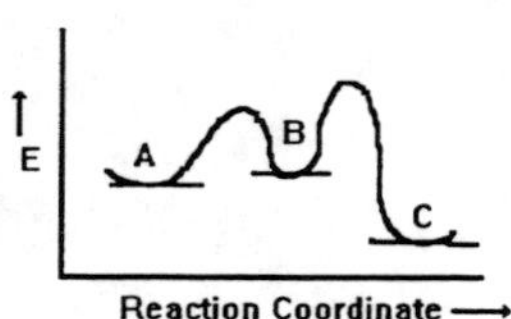

Answer:
A is formed faster since the pathway for its formation has the smaller activation energy. C is more stable than A because it has a lower energy.

Section: 7

30. Assign the E or Z configuration to the following molecule:

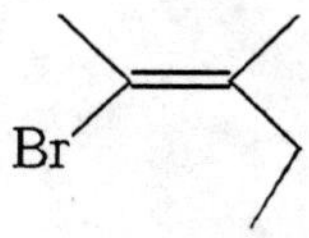

Br

Answer:
Therefore, the molecule has a Z configuration based on the priority rule established by Cahn-Ingold-Prelog.

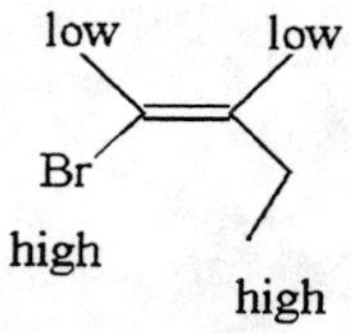

low low

Br

high

high

Section: 5

31. Propose a mechanism for the following reaction:

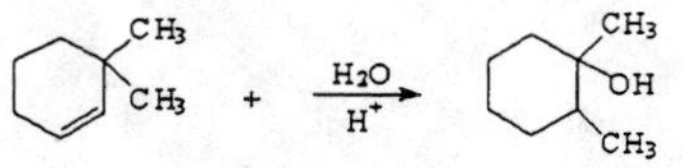

CH_3 CH_3 + $\xrightarrow[H^+]{H_2O}$ CH_3 OH CH_3

Answer:

(i) + H–Ö–H (⊕) H → ⊕ H + $H_2\ddot{O}$:

(ii) ⊕ H 1,2 -methyl shift → ⊕ H

(iii) ⊕ H + $H_2\ddot{O}$ → H ⊕ O H H

(iv) H ⊕ O H H + $H_2\ddot{O}$: → OH + H_3O^+

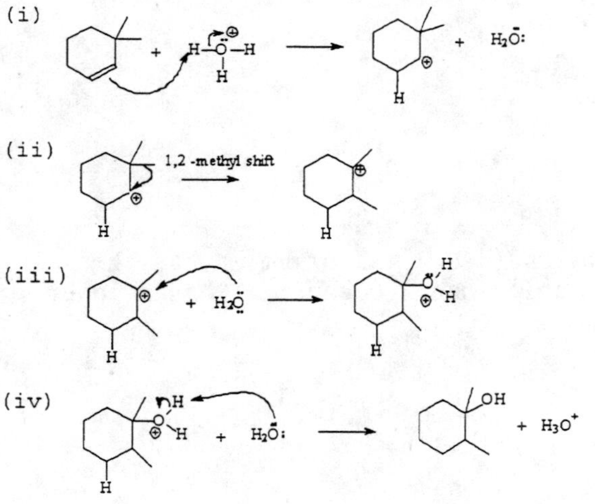

Section: 13

32. Draw the major organic product generated in the reaction below.

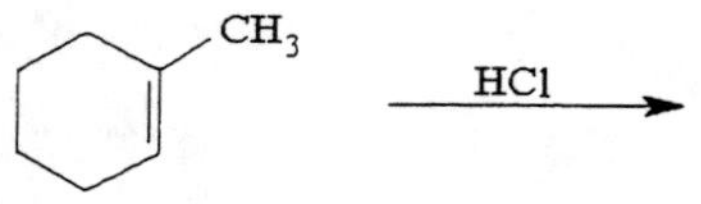

CH_3 HCl

Answer:

CH_3 Cl

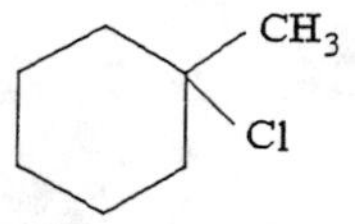

Section: 12

33. Draw the major organic product generated in the reaction below.

HCl

Answer:

Cl

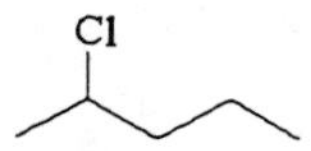

Section: 12

34. Draw the major organic product generated in the reaction below.

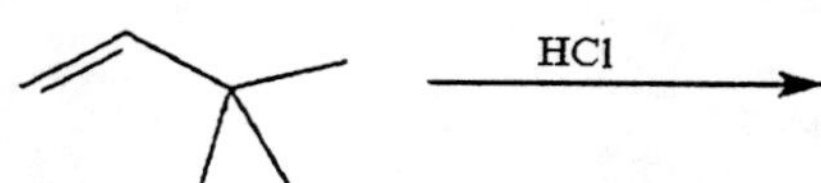

Answer:

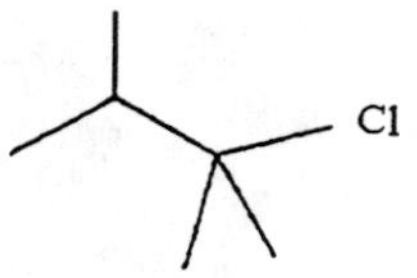

Cl

Section: 12

35. Draw the major organic product generated in the reaction below.

HCl

Answer:

Cl

Section: 12

36. Draw the major organic product generated in the reaction below.

HBr

Answer:

Br

Section: 12

37. Draw the major organic product generated in the reaction below.

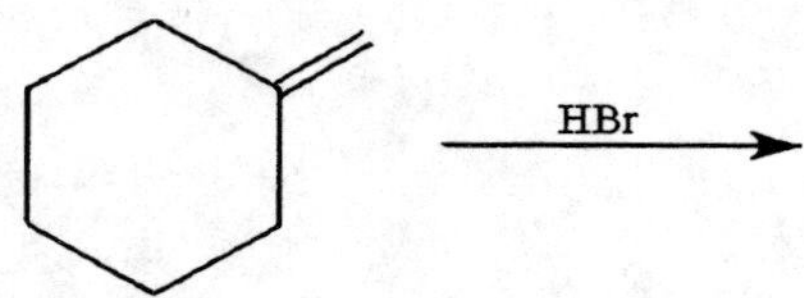

Answer:

Br

Section: 12

38. Draw the major organic product generated in the reaction below.

$\xrightarrow[\text{dilute aqueous acid}]{H^+, H_2O}$

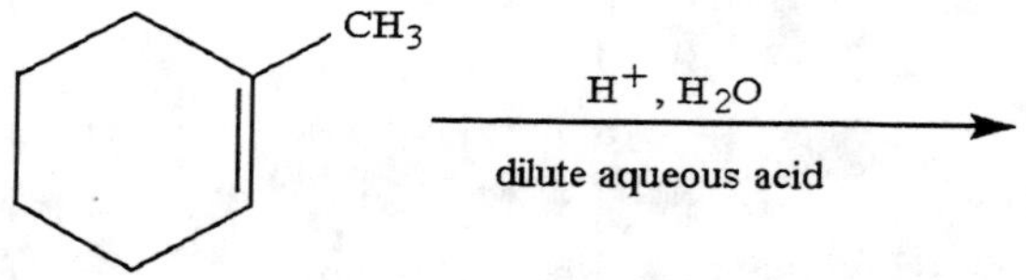

Answer:

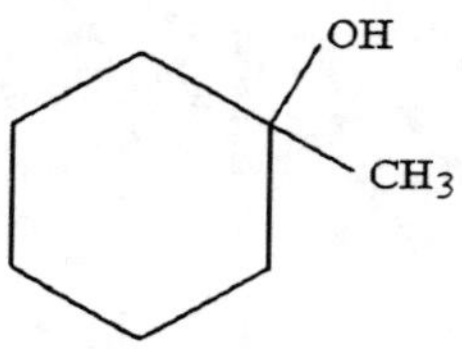

Section: 13

39. Draw the major organic product generated in the reaction below.

$\xrightarrow[\text{dilute aqueous acid}]{H^+, H_2O}$

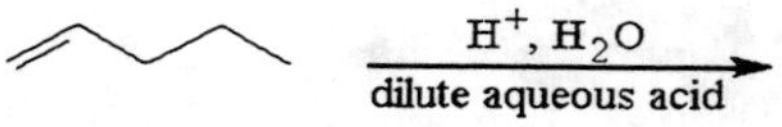

Answer:

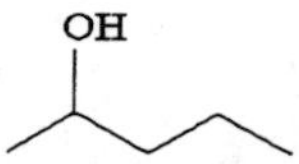

Section: 13

40. Draw the major organic product generated in the reaction below.

$\xrightarrow[\text{dilute aqueous acid}]{H^+, H_2O}$

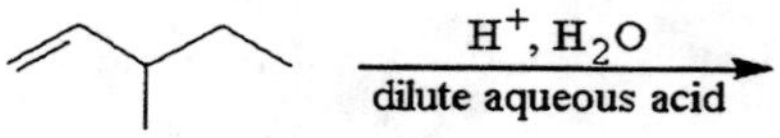

Answer:

Section: 13

41. Draw the major organic product generated in the reaction below.

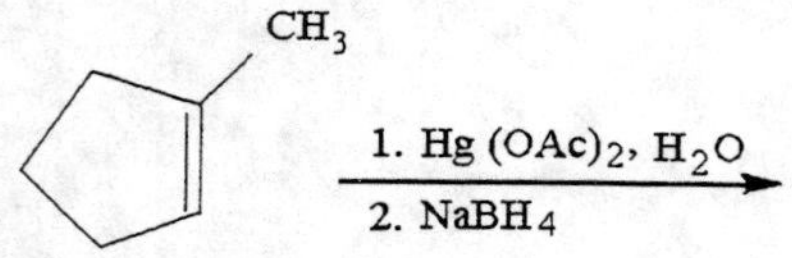

CH_3

1. Hg $(OAc)_2$, H_2O
2. $NaBH_4$

Answer:

OH

CH_3

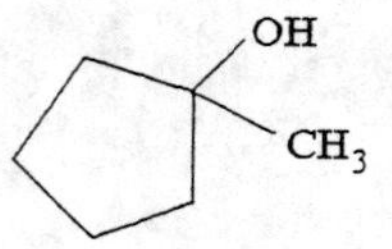

Section: 16

42. Draw the major organic product generated in the reaction below.

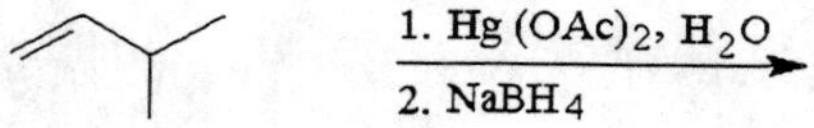

1. Hg $(OAc)_2$, H_2O
2. $NaBH_4$

Answer:

OH

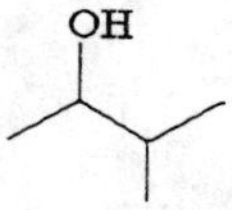

Section: 16

43. Draw the major organic product generated in the reaction below.

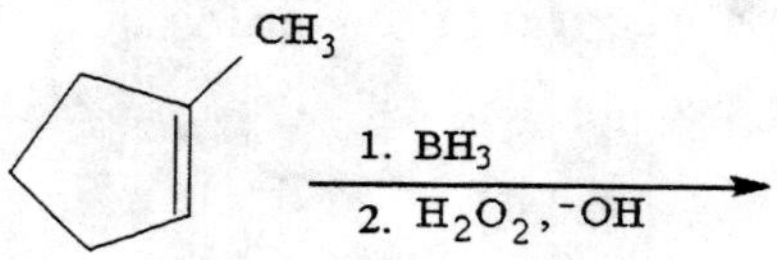

CH_3

1. BH_3
2. H_2O_2, ^-OH

Answer:

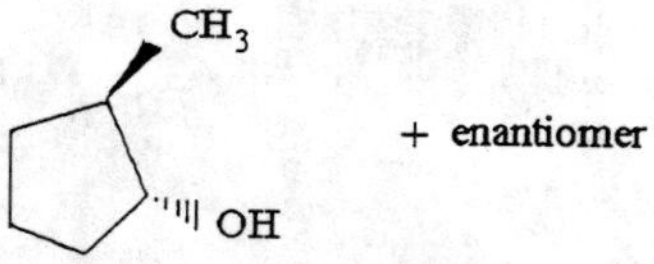

CH_3

OH

+ enantiomer

Section: 17

44. Draw the major organic product generated in the reaction below.

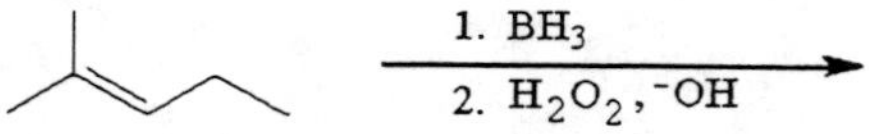

1. BH_3
2. H_2O_2, ^{-}OH

Answer:

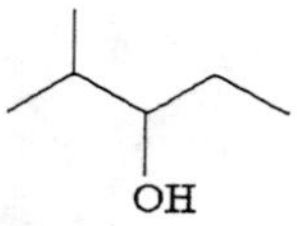

OH

Section: 17

45. Draw the major organic product generated in the reaction below.

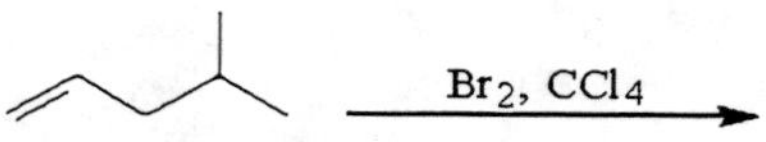

Br_2, CCl_4

Answer:

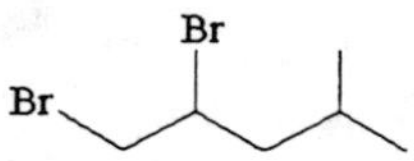

Br
Br

Section: 15

46. Draw the major organic product generated in the reaction below.

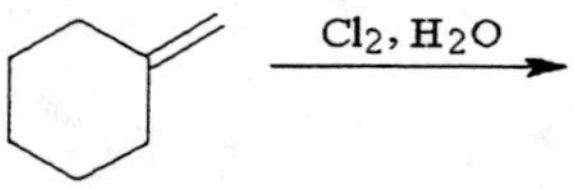

Cl_2, H_2O

Answer:

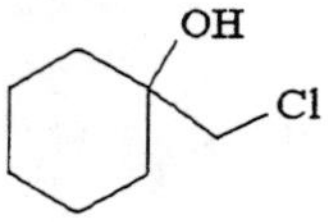

OH
Cl

Section: 15

47. Complete the following reaction and provide a detailed, step-by-step mechanism for the process.

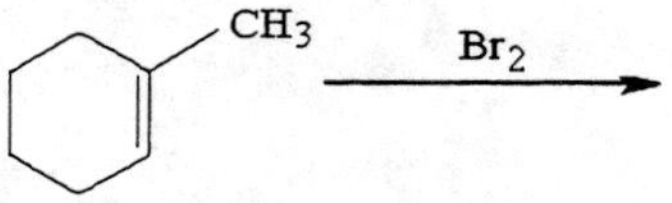

CH_3 Br_2

Answer:

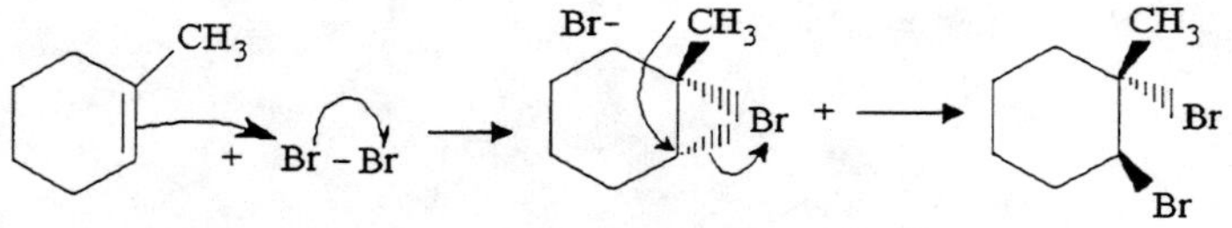

CH_3 + Br - Br → Br- CH_3 Br + → CH_3 Br Br

Section: 15

48. Complete the following reaction and provide a detailed, step-by-step mechanism for the process.

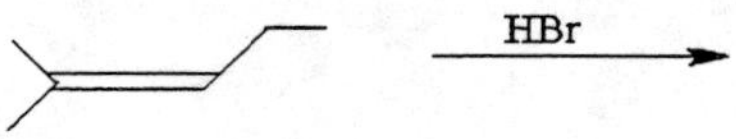

HBr

Answer:

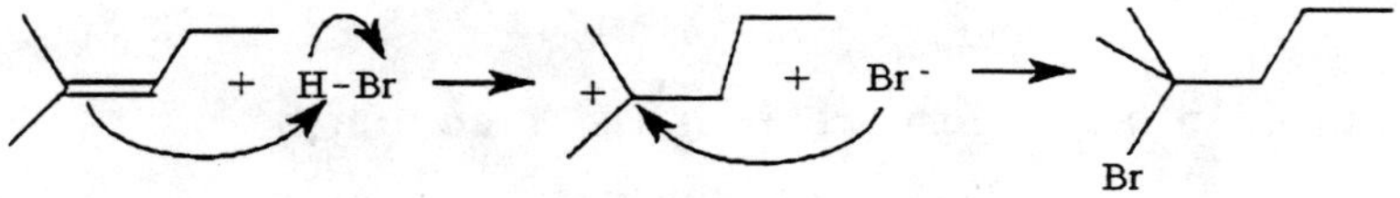

+ H-Br → + + Br^- → Br

Section: 9

49. Complete the following reaction and provide a detailed, step-by-step mechanism for the process.

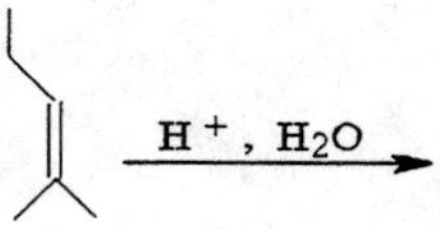

H^+, H_2O

Answer:

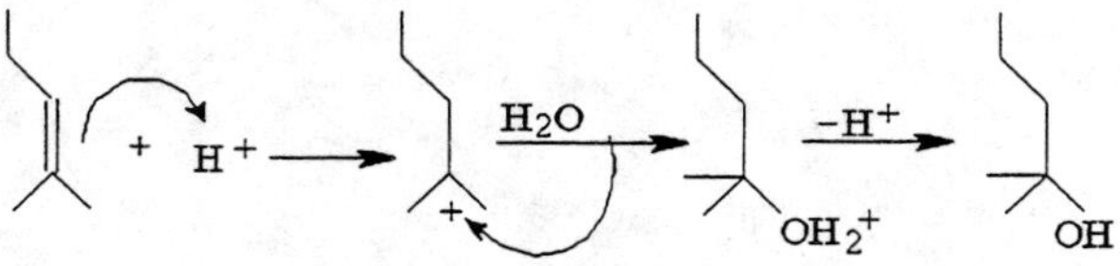

+ H^+ → + H_2O → OH_2^+ $-H^+$ → OH

Section: 13

50. Provide a detailed, step-by-step mechanism for the reaction shown below.

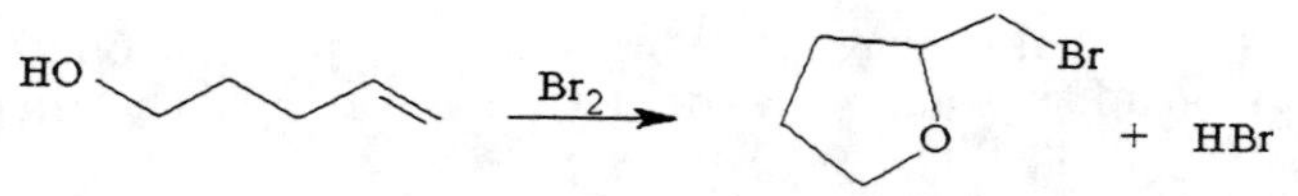

HO — Br_2 → Br + HBr

Answer:

HO Br–Br HO Br + → Br + Br + Br^- O+ H

HBr + Br

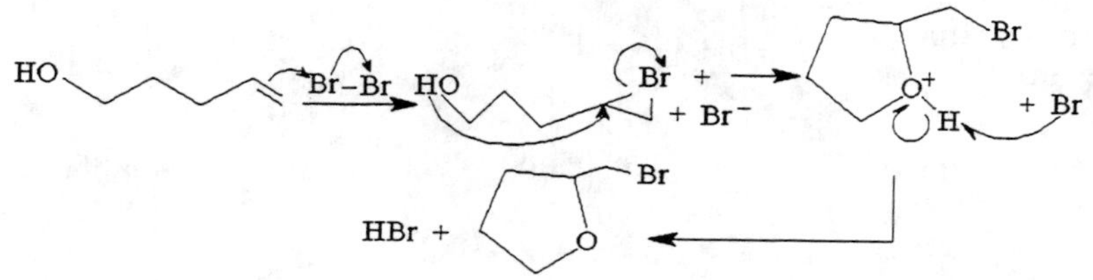

Section: 15

51. Based on the relative stabilities of the intermediates involved, explain the basis for Markovnikov's rule in the addition of hydrogen halides to alkenes.

Answer:
The rate-determining step in this reaction is the production of a carbocation intermediate. Since this step is endothermic, Hammond's postulate allows one to gauge the relative stabilities of the transition states by comparing the relative stabilities of the carbocation intermediates. The reaction pathway which produces the more substituted carbocation will thus occur more rapidly.

Section: 11

52. Explain the regioselectivity observed in the radical addition of HBr to 2-methylpropene.

Answer:
The reaction proceeds via the addition of Br• to the alkene. Two competing pathways are possible, but the transition state leading to the more substituted alkyl radical is lower in energy. This process ultimately makes the addition anti-Markovnikov in nature.

Section: 18

MULTIPLE CHOICE

53. Which of the following is the best reaction sequence to use if one wants to accomplish a Markovnikov addition of water to an alkene with minimal skeletal rearrangement?

    a. water + dilute acid
    b. water + concentrated acid
    c. oxymercuration-demercuration
    d. hydroboration-oxidation
    e. none of the above

    Answer: c Section: 16

SHORT ANSWER

54. Given an activation energy of 15 kcal/mol, use the Arrhenius equation to estimate how much faster the reaction will occur if the temperature is increased from 100° C to 120° C. R = 1.987 cal/mol•K.

    Answer: The reaction will occur about 2.8 times faster. Section: 7

55. Consider the conversion of C to D via a one-step mechanism. The activation energy of this conversion is 3 kcal/mol. The energy difference between D and the transition state of the reaction is 7 kcal/mol. Estimate $\Delta H°$ for the reaction C ----> D.

    Answer: 4 kcal/mol Section: 7

56. Consider the one-step conversion of F to G. Given that the reaction is endothermic by 5 kcal/mol and that the energy difference between G and the transition state for the process is 15 kcal/mol, sketch a reaction-energy profile for this reaction. Make sure to show how the given energy differences are consistent with your sketch.

    Answer:

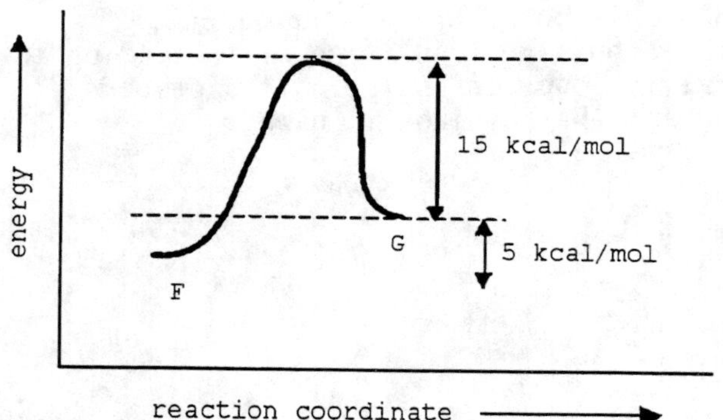

Section: 7

57. Provide the proper IUPAC name for the alkene shown below.

```
CH3        CH2 CH2 CH3
   \      /
     C=C
   /      \
CH3 CH2     CH3
```

Answer: (E)-3,4-dimethyl-3-heptene Section: 2

58. Provide the proper IUPAC name for the alkene shown below.

Br

CH_3

Answer: 6-bromo-1-methylcyclohexene Section: 2

59. Draw and name the six alkenes which have the molecular formula C_5H_{10}.

Answer:

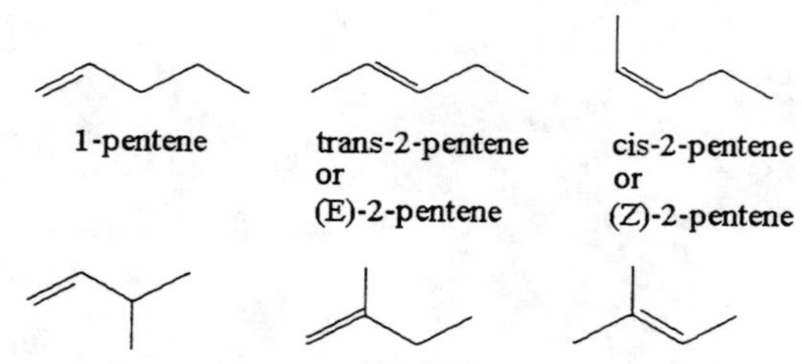

Section: 2

Chapter 3: Reactions of Alkenes, Thermodynamics and Kinetics

MULTIPLE CHOICE

60. Which of the following best describes the geometry about the carbon-carbon double bond in the alkene below?

```
CH3     Cl
   \   /
    C=C
   /   \
CH3     Cl
```

a. E
b. Z
c. neither E nor Z

Answer: c Section: 5

Chapter 4: Stereochemistry

MULTIPLE CHOICE

1. What type of isomers are the staggered and eclipsed forms of ethane?

   a. Constitutional
   b. Structural
   c. Configurational
   d. Conformational
   e. Stereochemical

   Answer: d

2. What is the relationship between the following compounds?

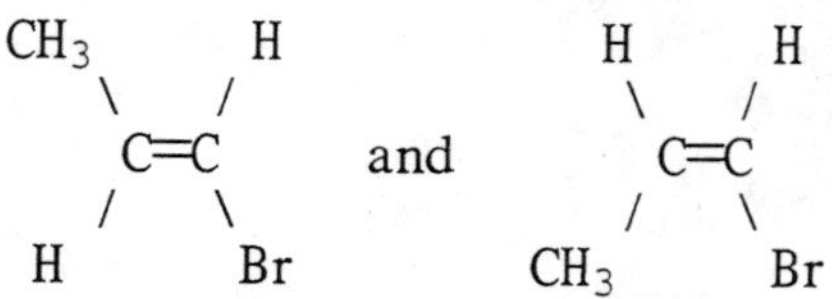

   a. Configurational isomers
   b. Conformational isomers
   c. Constitutional isomers
   d. Structural isomers
   e. Positional isomers

   Answer: a Section: 2

3. Which of the following compounds has a chirality center?

a. 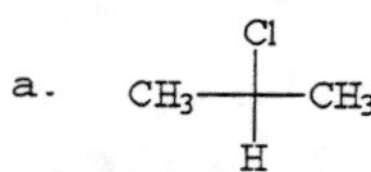

b. 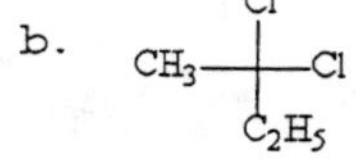

c. 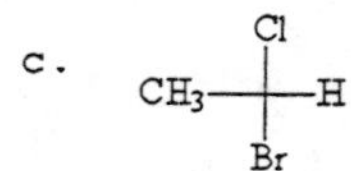

d. Cl
Cl—|—H
CH3

e.

   Answer: c Section: 3

4. Which of the following compounds is never chiral?

   a. 2,3-dibromobutane
   b. 1,3-dibromobutane
   c. 1,2-dichlorobutane
   d. 1,4-dibromobutane
   e. 1-bromo-2-chlorobutane

   Answer: d Section: 8

5. A and B are stereoisomers. They are nonsuperimposable and are mirror images of one another. Which of the following best describes the relationship between A and B?

   a. Structural isomers
   b. Enantiomers
   c. Confrontational isomers
   d. Diastereomers
   e. Constitutional isomers

   Answer: b Section: 3

6. What is the relationship between the following compounds?

   a. Superimposable without bond rotation
   b. Constitutional isomers
   c. Conformational isomers
   d. Diastereomers
   e. Enantiomers

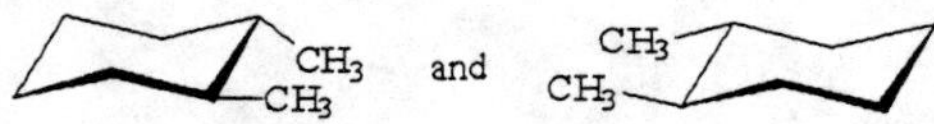

   Answer: e Section: 8

7. What is the relationship between the following compounds?

   a. Constitutional isomers
   b. Enantiomers
   c. Diastereomers
   d. Conformational isomers
   e. Superimposable without bond rotation

CH_3
H–C–Cl
CH_3

and

CH_3
C–H
H
CH_2Cl

   Answer: a

8. Which of the following compounds has an S configuration?

a. Fischer projection: top Cl; left H; right CH_3; bottom CH_2Cl

b. Fischer projection: top CH_2Cl; left H; right CH_3; bottom Cl

c. Fischer projection: top CH_2Cl; left CH_3; right H; bottom Cl

d. Fischer projection: top H; left CH_3; right Cl; bottom CH_2Cl

e. Fischer projection: top CH_2Cl; left Cl; right CH_3; bottom H

Answer: b Section: 5

9. Which of the following compounds has an R configuration?

a. Fischer projection: top H; left CH_3CH_2; right Br; bottom $CH=CH_2$

b. Fischer projection: top $CH=CH_2$; left Br; right H; bottom CH_2CH_3

c. Fischer projection: top CH_2CH_3; left Br; right $CH=CH_2$; bottom H

d. Fischer projection: top CH_2CH_3; left $CH_2=CH$; right H; bottom Br

e. Fischer projection: top $CH=CH_2$; left CH_3CH_2; right Br; bottom H

Answer: a Section: 5

10. Which of the following groups has the highest priority using the Cahn, Ingold, Prelog rules?

a. $-\overset{\overset{\displaystyle O}{\|}}{C}-OH$

b. $-\overset{\overset{\displaystyle O}{\|}}{C}-H$

c. $-OH$

d. $-O-CH_3$

e. $-\overset{\overset{\displaystyle O}{\|}}{C}-O-CH_3$

Answer: d Section: 5

11. What is the relationship between the structures shown below?

a. Structural isomers
b. Constitutional isomers
c. Conformational isomers
d. Configurational isomers
e. Enantiomers

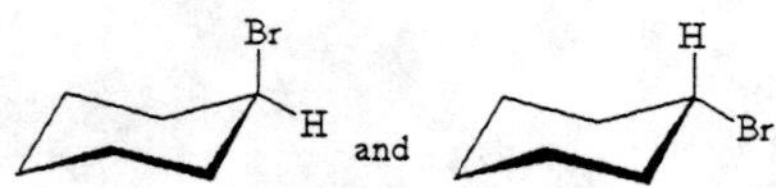

Answer: c

12. What is the relationship between the structures shown below?

a. Diastereomers
b. Constitutional isomers
c. Conformational isomers
d. Configurational isomers
e. Enantiomers

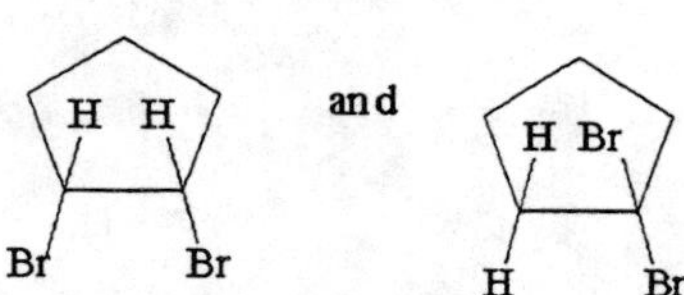

Answer: b

13. What is the relationship between the structures shown below?

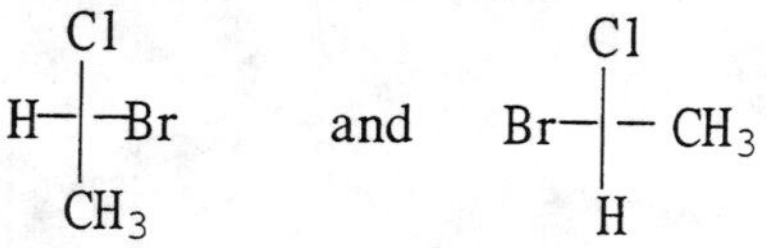

Cl
H—|—Br
CH_3

and

Cl
Br—|—CH_3
H

a. Enantiomers
b. Diastereomers
c. Configurational isomers
d. Identical compounds
e. Constitutional isomers

Answer: d Section: 3

14. Which of the following compounds is an enantiomer of the structure above?

a. 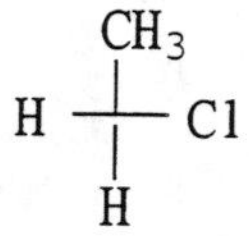
CH_3
H —|— Cl
H

b. 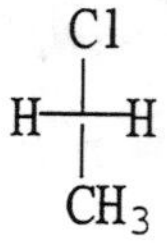
Cl
H—|—H
CH_3

c. 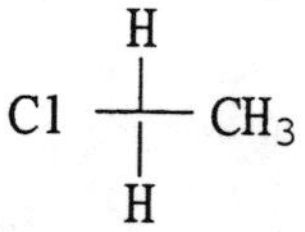
H
Cl —|— CH_3
H

d. a and b
e. It does not have an enantiomer.

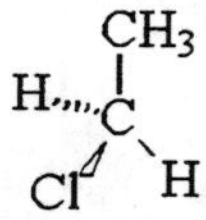

CH_3
H.....C
Cl H

Answer: e Section: 3

15. Which of the following is not true of enantiomers?

a. They have the same melting point.
b. They have the same boiling point.
c. They have the same chemical reactivity with non-chiral reagents.
d. They have the same density.
e. They have the same specific rotation.

Answer: e Section: 3

16. The specific rotation of a pure substance is 1.68° . What is the specific rotation of a mixture containing 75% of this isomer and 25% of the (-) isomer?

a. +1.68°
b. 0°
c. +1.26°
d. +0.84°
e. +.042°

Answer: d Section: 7

17. The specific rotation of a pure substance is -5.90° . What is the percentage of this isomer in a mixture with an observed specific rotation of -2.95° ?

a. 25%
b. 50%
c. 75%
d. 80%
e. 0%

Answer: c Section: 7

18. What is the relationship between the structures shown below?

a. Enantiomers
b. Diastereomers
c. Configurational isomers
d. Identical compounds
e. Constitutional isomers

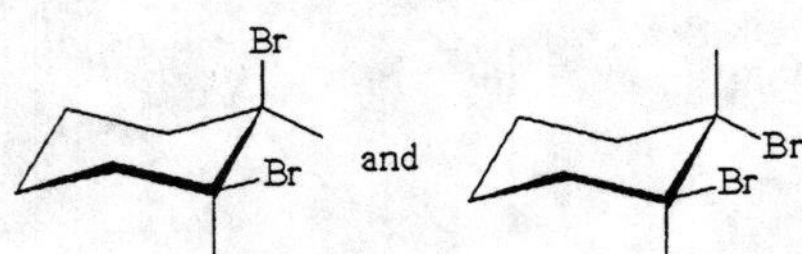

Answer: b Section: 8

19. Which of the following compounds is chiral?

a.

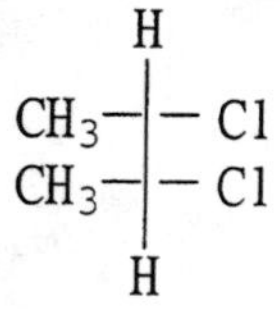

H

CH_3 — | — Cl

CH_3 — | — Cl

H

b.

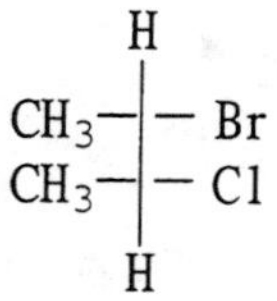

H

CH_3 — | — Br

CH_3 — | — Cl

H

c.

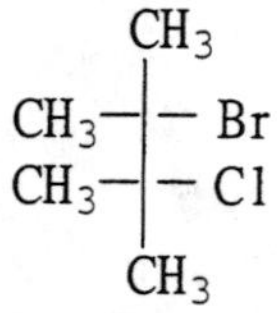

CH_3

CH_3 — | — Br

CH_3 — | — Cl

CH_3

d.

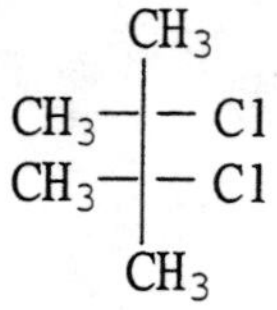

CH_3

CH_3 — | — Cl

CH_3 — | — Cl

CH_3

e.

CH_3

Cl — | — Cl

Br — | — Br

CH_3

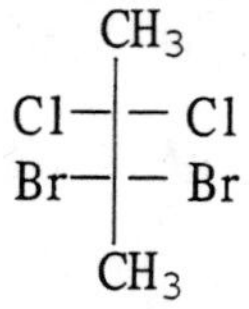

Answer: b Section: 3

20. Which of the following is a meso compound?

a.
CH_3
H— —Cl
H— —Cl
CH_3

b.
CH_3
H— —Cl
Cl— —H
CH_3

c.
CH_3
H— —Cl
H— —Cl
CH_2CH_3

d.
CH_3
H— —Cl
Cl— —H
CH_2CH_3

e.
CH_3
Br— —Cl
Br— —Cl
CH_2Cl

Answer: a Section: 9

21. What is the configuration of the following compound?

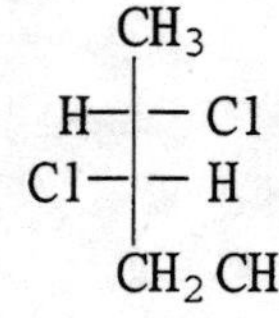

a. 2S, 3R
b. 3R, 3S
c. 2S, 3S
d. 2R, 3R
e. Can't do R and S; the compound is achiral.

Answer: c Section: 10

22. What is the relationship between the following compounds?

```
     CH3                      CH3
 Cl—|— H                  H—|— Cl
 Cl—|— H        and       H—|— Cl
     CH3                      CH3
```

a. Enantiomers
b. Diastereomers
c. Constitutional isomers
d. Conformational isomers
e. Identical compounds

Answer: e Section: 9

23. What is the relationship between the following compounds?

```
     CH3                      CH3
 Cl—|— H                  H—|— Cl
  H—|— Cl       and      Cl—|— H
     CH3                      CH3
```

a. Enantiomers
b. Diastereomers
c. Constitutional isomers
d. Conformational isomers
e. Identical compounds

Answer: a Section: 8

24. What is the relationship between the following compounds?

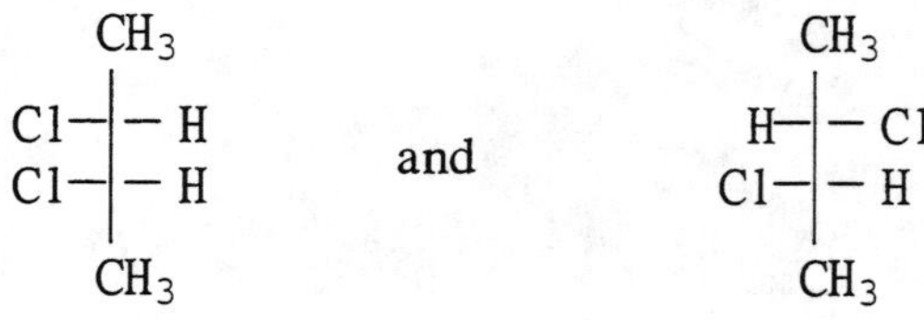

a. Enantiomers
b. Diastereomers
c. Constitutional isomers
d. Conformational isomers
e. Identical compounds

Answer: b Section: 8

25. A sample of (R)-2-chlorobutane,

```
        H
        |
CH3 ----+---- Cl
        |
      CH2CH3
```

reacts with Br_2 in the presence of light, and all the products having the formula C_4H_8BrCl were isolated. Two possible isomers are shown below:

```
      CH3                          CH3
       |                            |
Br ----+---- Cl      and     Cl ----+---- Br
       |                            |
     CH2 CH3                      CH2 CH3

       I                            II
```

Of these:

a. only I was formed
b. both I and II were formed in equal amounts
c. both I and II were formed in unequal amounts
d. only II was formed
e. neither I nor II was formed

Answer: b Section: 12

26. Which of the following is/are optically inactive?

a. A 50-50 mixture of R and S enantiomers
b. A meso compound
c. Every achiral compound
d. A racemic mixture
e. All the above

Answer: e Section: 6

27. The configuration of R-(+)-glyceraldehyde is as follows:

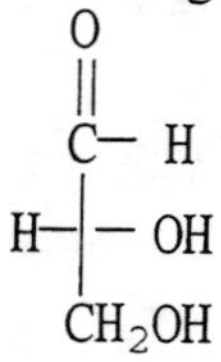

O
||
C— H
H—|— OH
CH_2OH

What is the absolute configuration of (-)-lactic acid?

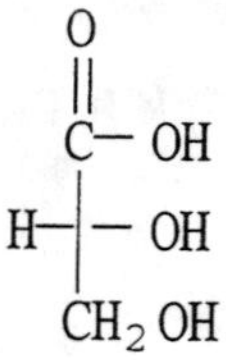

O
||
C— OH
H—|— OH
CH_2OH

a. R configuration
b. L configuration
c. S configuration
d. R and S configuration
e. D and L configuration

Answer: a Section: 5

28. Which of the following compounds contain hydrogens labeled H_a and H_b that are enantiotopic?

a. 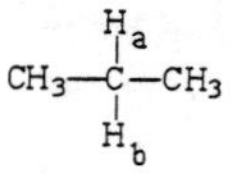 CH_3—C(H_a)(H_b)—CH_3

b. [cyclohexane with H_a, H_b]

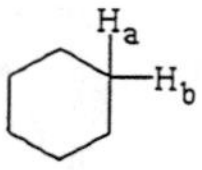

c. 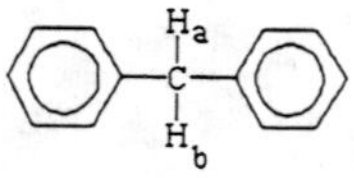 [Ph—C(H_a)(H_b)—Ph]

d. 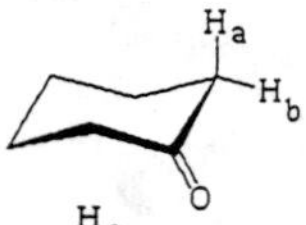 [cyclohexanone with H_a, H_b]

e. 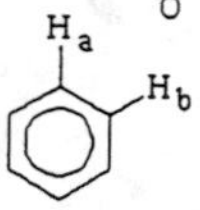 [benzene ring with H_a, H_b]

Answer: d Section: 14

29. Which of the following is a meso compound?

    a. trans-1,4-dimethylcyclohexane
    b. cis-1,3-dimethylcyclohexane
    c. trans-1,3-dimethylcyclohexane
    d. cis-1,4-dimethylcyclohexane
    e. trans-1,2-dimethylcyclohexane

    Answer: b Section: 9

30. Which of the following describes the most stable conformation of trans-1-tert-butyl-3-methylcyclohexane?

    a. Both groups are equatorial.
    b. Both groups are axial.
    c. The tert-butyl group is equatorial.
    d. The tert-butyl group is axial and the methyl group is equatorial.
    e. None of the above

    Answer: c Section: 16

SHORT ANSWER

31. Indicate whether each of the following structures has the R or S configuration. What is the relationship between the two structures?

Br
HO—|— CH_3
CH_2CH_3

and

OH
CH_3CH_2—|—Br
CH_3

assign priorities to each group.

Answer:

Br^1
HO^2—|— CH_3 4
CH_2CH_3
3

Configuration is R

OH
CH_3CH_2—|—Br
CH_3

Configuration is S

Therefore, the two compounds are enantiomers.
Section: 3

32. A solution containing 0.96 g of 2-bromooctane in 10ml ether solution gave an observed rotation of -1.8° in a 10cm cell at 20° C. Calculate the specific rotation of this solution.

Answer:

$$[\alpha]_D^T = \frac{\alpha}{lc}$$

$$[\alpha]_D^{20°C} = \frac{1.8°}{1dm \times \frac{0.96g}{10ml}} = -18.75° \text{ in ether}$$

Section: 6

33. Consider the molecule $C_2H_2Br_2Cl_2$.
   (a) Draw a structure that is optically inactive because it does not have a chirality center.
   (b) Draw a structure that is optically inactive because it has a meso compound.
   (c) Draw a structure that is optically active because it is chiral.

Answer:

a.
```
     Cl
 Cl—|—H
 Br—|—H    no stereogenic center
     Br
```

b.
```
     Cl
  H—|—Br
  H—|—Br   meso compound
     Cl
```

c.
```
     Br
 Cl—|—H
  H—|—Cl   enantiomer
     Br
```

Section: 6

34. Is the molecule shown below chiral or achiral?

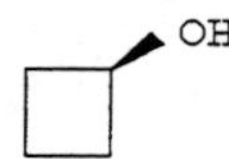

Answer: achiral Section: 3

35. Is the molecule shown below chiral or achiral?

Answer: chiral Section: 8

36. Is the molecule shown below chiral or achiral?

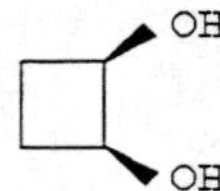

Answer: achiral Section: 8

37. is the molecule shown below chiral or achiral?

Answer: achiral Section: 3

38. Which of the following terms best describes the pair of compounds shown: enantiomers, diastereomers, or the same compound?

Answer: the same compound Section: 3

39. Which of the following terms best describes the pair of compounds shown: enantiomers, diastereomers, or the same compound?

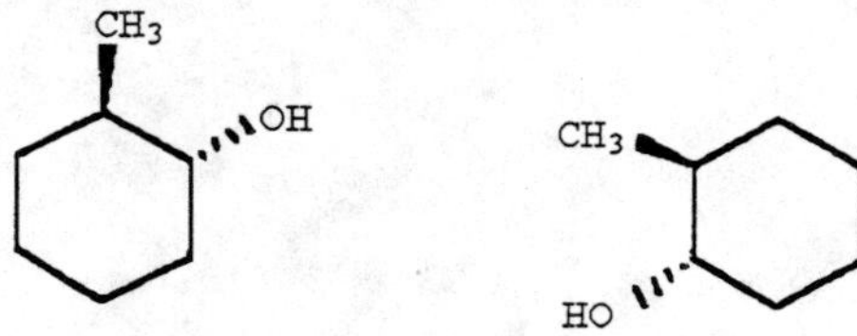

Answer: enantiomers Section: 8

40. Which of the following terms best describes the pair of compounds shown: enantiomers, diastereomers, or the same compound?

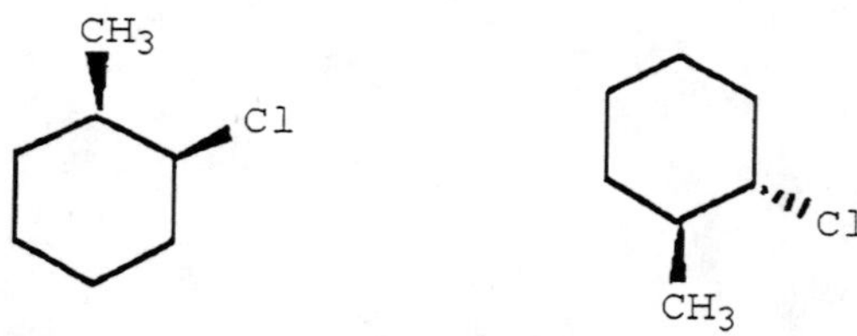

Answer: disatereomers Section: 8

41. Which of the following terms best describes the pair of compounds shown: enantiomers, diastereomers, or the same compound?

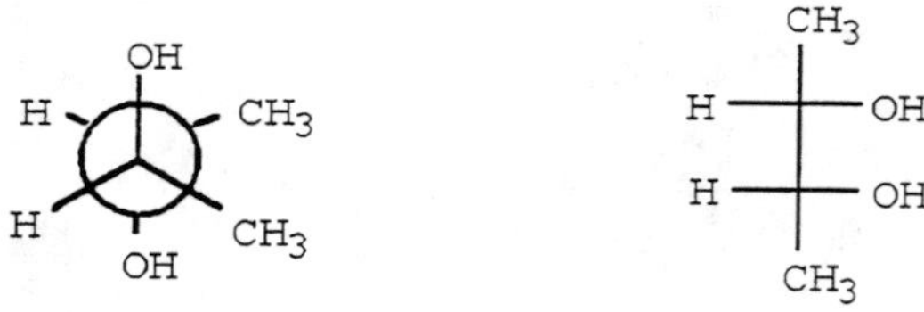

Answer: diastereomers Section: 8

42. Label each chiral carbon in the compound below as R or S.

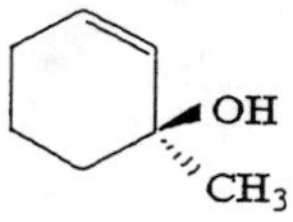

Answer:

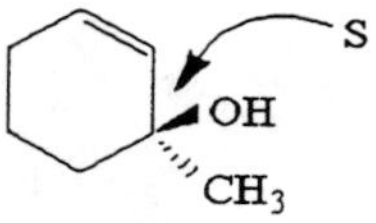

Section: 5

43. Draw the structure of (2R, 3S)-dichloropentane. Take particular care to indicate three-dimensional sterochemical detail properly.

Answer:

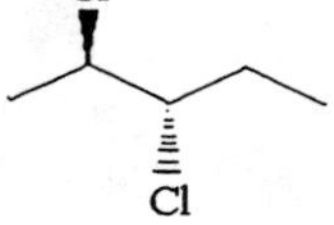

Section: 8

44. Draw the structure of (2R, 3S)-dichloropentane. Take particular care to indicate three-dimensional sterochemical detail properly.

Answer:

Cl

Cl

Section: 8

45. Draw the structure of any diastereomer of (2R, 3S)-dichloropentane. Take particular care to indicate three-dimensional sterochemical detail properly.

Answer:

Cl Cl

or

Cl Cl

Section: 8

46. Draw the structure of a meso form of 1,3-dichlorocyclopentane. Take particular care to indicate three-dimensional sterochemical detail properly.

Answer:

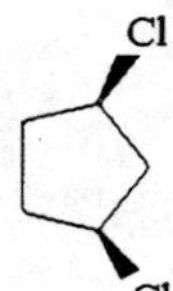

Section: 9

MULTIPLE CHOICE

47. Which of the statements below correctly describes an achiral molecule?

    a. The molecule has a nonsuperimposable mirror image.
    b. The molecule exhibits optical activity when it interacts with plane-polarized light.
    c. The molecule has an enantiomer.
    d. The molecule might be a meso form.
    e. None of the above.

    Answer: d Section: 3

48. Which of the following statements correctly pertains to a pair of enantiomers?

    a. They rotate the plane of polarized light by exactly the same amount and in opposite directions.
    b. They rotate the plane of polarized light by differing amounts and in opposite directions.
    c. They rotate the plane of polarized light by differing amounts and in the same direction.
    d. They have different melting points.
    e. They have the same melting point, but they have different boiling points.

    Answer: a Section: 6

SHORT ANSWER

49. Can one predict whether a compound with a single chiral carbon is dextro- or levorotatory based on the R/S assignment at this chiral carbon? Explain briefly.

    Answer:
    No. R/S assignment is purely a convention of nomenclature and is completely independent of the direction in which plane-polairzed light is rotated by the compound.
    Section: 13

MULTIPLE CHOICE

50. If (S) -glyceraldehyde has a specific rotation of -8.7° , what is the specific rotation of (R) -glyceraldehyde?

    a. -8.7°
    b. +8.7°
    c. 0.0°
    d. cannot be determined from the information given

    Answer: b Section: 6

Chapter 4: Stereochemistry

SHORT ANSWER

51. A newly isolated natural product was shown to be optically active. If a solution of 2.0 g in 10 mL of ethanol in a 50 cm tube gives a rotation of +2.57°, what is the specific rotation of this natural product?

Answer: +2.57° Section: 6

52. Steroisomers which are not mirror image isomers are ________.

Answer: diastereomers Section: 11

MULTIPLE CHOICE

53. A mixture of equal amounts of two enantiomers ________.

a. is called a racemic mixture
b. is optically inactive
c. implies that the enantiomers are meso forms
d. both a and b
e. none of the above

Answer: d Section: 6

SHORT ANSWER

54. Briefly describe how two enantiomers might be separated.

Answer:
The two compounds can be converted to diastereomers, separated based on different physical properties of these diastereomers, and subsequently returned to their original forms after chromatographic separation using a chiral stationary phase.

Section: 11

55. Which of the following terms best describes the pair of compounds shown: enantiomers, diastereomers, or the same compound?

Answer: the same compound Section: 8

56. Which of the following terms best describes the pair of compounds shown: enantiomers, diastereomers, or the same compound?

Answer: the same compound Section: 8

57. Which of the following terms best describes the pair of compounds shown: enantiomers, diastereomers, or the same compound?

Answer: enantiomers Section: 8

58. Label each chiral carbon in the compound below as R or S.

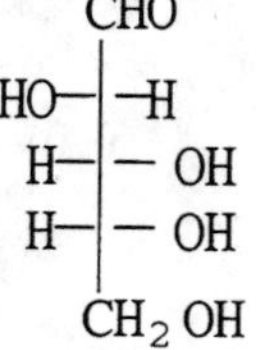

Answer:

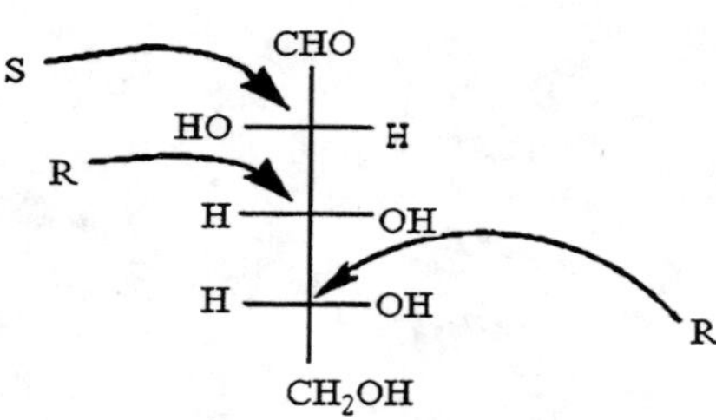

Section: 5

59. Draw the structure of (S) -1-bromo-1-chloropropane. Take particular care to indicate three-dimensional sterochemical detail properly.

Answer:

Section: 5

60. How many chiral carbons are present in the compound below?

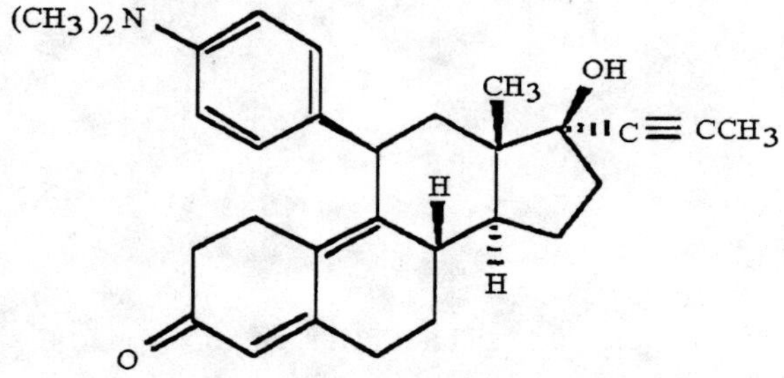

$(CH_3)_2N$ ... CH_3 ... OH ... $C\equiv CCH_3$ ... H ... H ... O

Answer: 5 Section: 3

Chapter 5: Reactions of Alkynes, Introduction to Multistep Synthesis

MULTIPLE CHOICE

1. What is the hybridization of the carbon atoms numbered 1 and 2 respectively in the following structure?

   a. sp^3, sp^2
   b. sp^2, sp^2
   c. sp, sp
   d. sp^2, sp
   e. sp, sp^2

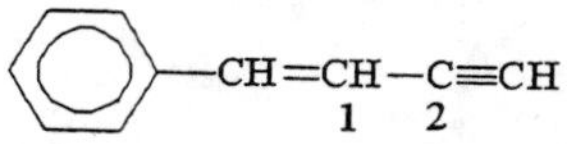

   C_6H_5–CH=CH–C≡CH (carbon 1 = second CH of CH=CH; carbon 2 = C of C≡CH)

   Answer: d Section: 3

2. Which of the following statements is not true about propyne, HC≡C–CH_3?

   a. It contains six sigma bonds.
   b. It contains three pi bonds.
   c. The H-C-H bond angle is about 109.5°.
   d. The C-C-C bond angle is 180°.
   e. The pi bond is weaker than the sigma bond.

   Answer: b Section: 3

3. What is the IUPAC name for the following alkyne?

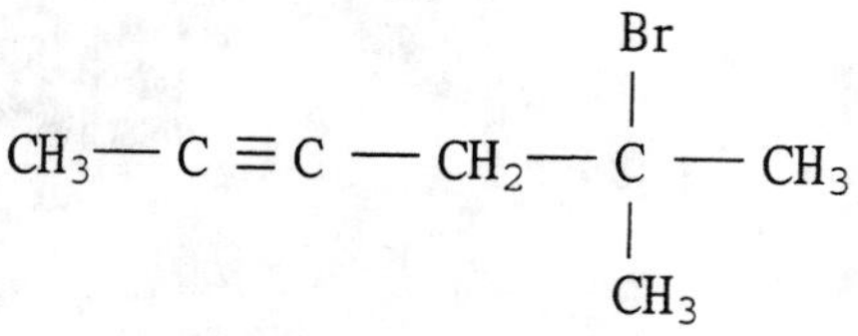

   CH_3–C≡C–CH_2–C(Br)(CH_3)–CH_3

   a. 5-Bromo-2-heptyne
   b. 3-Bromo-5-heptyne
   c. 2-Bromo-2-methyl-4-hexyne
   d. 5-Bromo-5,5-dimethylhexyne
   e. 5-Bromo-5-methyl-2-hexyne

   Answer: e Section: 1

4. Which is the correct order of decreasing acidity in the following compounds?

H_2O	CH_3CH_3	NH_3	$CH_2{=}CH_2$	$HC{\equiv}CH$
A	B	C	D	E

a. A>E>C>D>B
b. A>E>D>B>C
c. E>A>C>B>D
d. A>C>E>D>B
e. E>D>B>A>C

Answer: a Section: 9

5. What is the major product of the following reaction?

$$CH_3C\equiv CH \xrightarrow[\text{excess}]{HCl} ?$$

a. $CH_3CH_2CH(Cl)Cl$ (CH bearing two Cl)
b. $CH_3-C(Cl)(Cl)-CH_3$
c. $CH_3CH=CH-Cl$
d. $CH_3C(Cl)=CH_2$
e. $CH_3C(Cl)=CH(Cl)$

Answer: b Section: 5

6. Which of the following is the best synthesis of 2,2-dibromopropane

$$CH_3-\underset{\underset{Br}{|}}{\overset{\overset{Br}{|}}{C}}-CH_3 \quad ?$$

a. $CH_3CH=CH_2 \xrightarrow[CCl_4]{Br_2}$

b. $CH_3CH=CH_2 \xrightarrow[light]{Br_2}$

c. $CH_3C\equiv CH \xrightarrow{2HBr}$

d. $CH_3C\equiv CH \xrightarrow[peroxide]{2\ HBr}$

e. $CH_3CH=CH_2 \xrightarrow[H_2O]{Br_2}$

Answer: c Section: 5

7. Which of the following is the best synthesis of 1,1-dibromopropane,

$$CH_3CH_2\underset{\underset{Br}{|}}{\overset{\overset{Br}{|}}{CH}} \quad ?$$

a. $CH_3CH=CH_2 \xrightarrow[CCl_4]{Br_2}$

b. $CH_3CH=CH_2 \xrightarrow[light]{Br_2}$

c. $CH_3C\equiv CH \xrightarrow{2HBr}$

d. $CH_3C\equiv CH \xrightarrow[peroxide]{2\ HBr}$

e. $CH_3CH=CH_2 \xrightarrow[H_2O]{Br_2}$

Answer: d Section: 5

8. Which of the following are enol forms of 2-butanone,

$$CH_3\overset{\overset{\displaystyle O}{\|}}{C}CH_2CH_3$$

a. $CH_3\underset{OH}{\underset{|}{C}}=CHCH_3$ and $CH_3\underset{OH}{\underset{|}{C}}HCH=CH_2$

b. $CH_3\underset{OH}{\underset{|}{C}}HCH=CH_2$ and $CH_2=\underset{OH}{\underset{|}{C}}CH_2CH_3$

c. $CH_3\underset{OH}{\underset{|}{C}}=CHCH_3$, $CH_3\underset{OH}{\underset{|}{C}}HCH=CH_2$ and $CH_2=\underset{OH}{\underset{|}{C}}CH_2CH_3$

d. $CH_3\underset{OH}{\underset{|}{C}}HCH_2CH_3$ and $CH_3\underset{OH}{\underset{|}{C}}=CHCH_3$

e. $CH_3\underset{OH}{\underset{|}{C}}=CHCH_3$ and $CH_2=\underset{OH}{\underset{|}{C}}CH_2CH_3$

Answer: e Section: 6

9. Which of the following is the final and major product of this reaction?

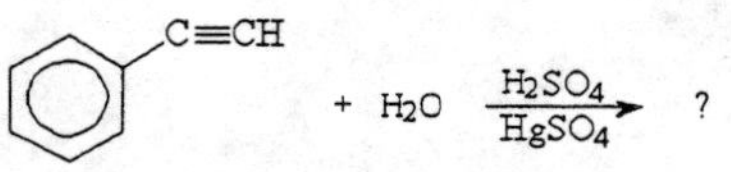

$$C_6H_5C\equiv CH + H_2O \xrightarrow[HgSO_4]{H_2SO_4} ?$$

a. 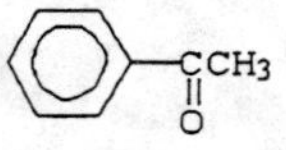 $C_6H_5\overset{}{C}(=O)CH_3$

b. $C_6H_5C(OH)=CH_2$

c. $C_6H_5CH_2CH=O$

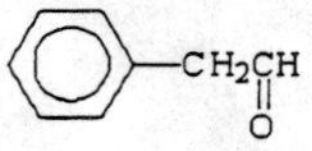

d. 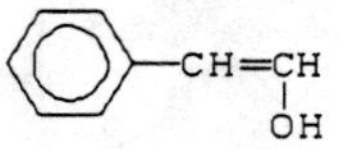 $C_6H_5CH=CH(OH)$

e. 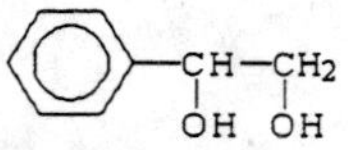 $C_6H_5CH(OH)-CH_2(OH)$

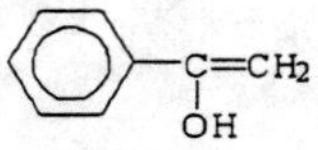

Answer: a Section: 6

10. Which of the following is the final and major product of this reaction?

$C_6H_5C{\equiv}CH \xrightarrow[\text{2. } HO^-, H_2O_2, H_2O]{\text{1. disiamylborane}} ?$

a. $C_6H_5C(=O)CH_3$

b. $C_6H_5C(OH)=CH_2$

c. $C_6H_5CH_2CH(=O)$

d. $C_6H_5CH=CH(OH)$

e. $C_6H_5CH(OH)-CH_2(OH)$

Answer: c Section: 7

11. The reagent needed to convert 2-butyne to cis-2-butene is:

a. H_2 /Pt
b. H_2 /Lindlar's catalyst
c. Li /NH_3
d. Na /NH_3
e. H^+ /Zinc dust

Answer: b Section: 8

12. The reagent needed to convert 2-butyne to trans-2-butene is:

a. H_2 /Pt
b. H_2 /Lindlar's catalyst
c. Li /NH_3
d. Na /NH_3
e. c or d

Answer: e Section: 8

13. What are the products of the following reaction,

$$CH_3OH + CH_3C{\equiv}C^-Na^+ \longrightarrow ?$$

a. $CH_3C{\equiv}CCH_3$ + NaOH
b. $CH_3C{\equiv}COCH_3$ + NaOH
c. $CH_3C{\equiv}CH$ + $CH_3O^-Na^+$
d. $CH_3OC{\equiv}CH$ + $NaCH_3$
e. No reaction

Answer: c Section: 9

14. What is/are the major organic product(s) of the following reaction,

$$HC{\equiv}\bar{C}: + CH_3CH_2Br \longrightarrow ?$$

a. $CH_2{=}CH_2$ + $HC{\equiv}CH$
b. $CH_3CH_2C{\equiv}CH$
c. $HC{\equiv}CBr$
d. $HC{\equiv}CCH_2CH_2Br$
e. $CH_3C{\equiv}CCH_3$

Answer: b Section: 10

SHORT ANSWER

15. Write structures and give IUPAC names for all alkynes with molecular formula C_5H_8.

Answer:

$CH_3CH_2CH_2C\equiv CH$
1-pentyne

$CH_3CH_2C\equiv CCH_3$
2-pentyne

$CH_3CH(CH_3)C\equiv CH$
3-methyl-butyne

Section: 1

16. Complete the following tree of reactions by giving the major products.

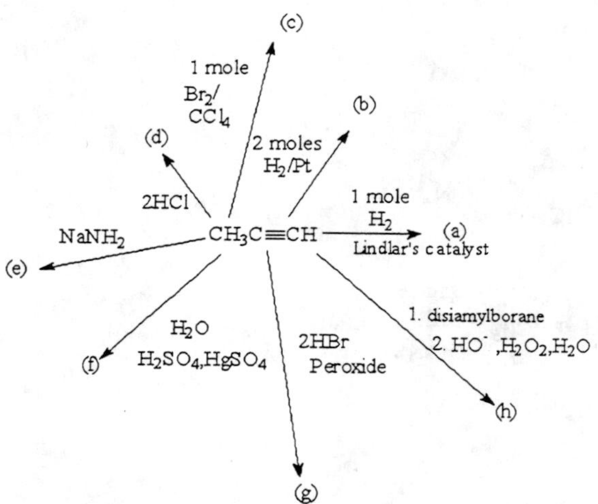

Answer:

a. $CH_3CH{=}CH_2$

b. $CH_3CH_2CH_3$

c. trans $BrC(CH_3){=}CHBr$ (Br and CH_3 on one carbon; H and Br on the other)

$$\begin{array}{ccc} Br & & H \\ & C = C & \\ CH_3 & & Br \end{array}$$

trans

d. $CH_3-CCl_2-CH_3$

$$\begin{array}{c} Cl \\ | \\ CH_3 - C - CH_3 \\ | \\ Cl \end{array}$$

e. $CH_3C \equiv C^-Na^+ + NH_3$

f. CH_3CCH_3 with $=O$ on the central C

$$\begin{array}{c} CH_3CCH_3 \\ \| \\ O \end{array}$$

g. $CH_3CH_2CHBr_2$

$$\begin{array}{c} Br \\ | \\ CH_3CH_2CH \\ | \\ Br \end{array}$$

h. CH_3CH_2CH with $=O$

$$\begin{array}{c} CH_3CH_2CH \\ \| \\ O \end{array}$$

Section: 5

17. Suggest a plausible scheme for the following synthesis

$$\underset{\displaystyle Br}{CH_3CH_2\overset{|}{C}HCH_3} \quad \text{from} \quad HC\equiv CH$$

Answer:

$$HC\equiv CH + NaNH_2 \longrightarrow HC\equiv C:^- \xrightarrow{CH_3I} HC\equiv CCH_3 \xrightarrow{NaNH_2} :^-C\equiv CCH_3 \xrightarrow{CH_3I} CH_3C\equiv CCH_3$$

$$CH_3C\equiv CCH_3 \xrightarrow[\text{Lindlar's catalyst}]{\text{1 mole } H_2} CH_3CH{=}CHCH_3 \xrightarrow{HBr} CH_3CH_2CHBrCH_3 \text{ (product)}$$

Section: 11

18. Explain why $HC\equiv N$ is a stronger acid than $HC\equiv CH$.

Answer: Because nitrogen is more electronegative than carbon. Section: 9

19. An optically active compound (A), C_6H_{10}, reacts with H_2/Ni to produce compound (B), C_6H_{14}. (B) is optically inactive. Deduce the structures of (A) and (B).

Answer:
A.

$$HC\equiv C\overset{\displaystyle CH_3}{\overset{|}{C}}HCH_2CH_3$$

optically active

B.

$$CH_3CH_2\overset{\displaystyle CH_3}{\overset{|}{C}}HCH_2CH_3$$

optically inactive

Section: 8

20. Although you might expect 1-propyne to be more reactive than 1-propene in electrophilic addition reactions, the reverse is true. Explain this using your knowledge of the mechanism of electrophilic additions to both alkynes and alkenes.

Answer:
When HCl is added to an alkene, a secondary carbocation intermediate is formed, $CH_3\overset{\oplus}{C}HCH_3$. When HCl is added to an alkyne, a vinylic cation is formed as an intermediate, $CH_3\overset{\oplus}{C}{=}CH_2$. Alkynes are less stable than alkenes but vinylic carbocations are less stable than secondary carbocations. Thus, alkenes are more reactive than alkynes since the difference in stabilities between the carbocations is greater than the difference in stabilities between the alkene and alkyne.
Section: 4

21. Which hydrogens are more acidic, those of benzene or those of cyclohexane?

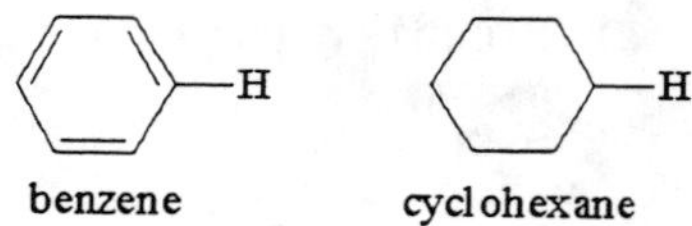

Answer:

Benzene. The carbon hybridization of benzene is sp^2 while that of cyclohexane is sp^3. The more s character in benzene causes its hydrogens to be more acidic than those of cyclohexane.
Section: 9

22. Give the IUPAC name for $(CH_3)_2C(CH_2CH_3)CCCH(CH_3)_2$.

Answer: 2,5,5-trimethyl-3-heptyne Section: 1

23. Give the IUPAC name for $HCCCH_2CH_2CH_3$.

Answer: 1-pentyne Section: 1

24. Give the IUPAC name for $BrCH_2CH_2CCCH_2CH_3$.

Answer: 1-bromo-3-hexyne Section: 1

25. Draw an acceptable structure for acetylene.

Answer: H—C≡C—H Section: 1

26. Draw an acceptable structure for 3-sec-butyl-1-heptyne.

Answer:

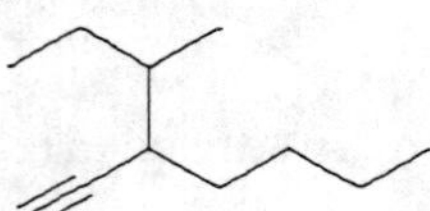

Section: 1

27. Draw an acceptable structure for 2-hexyne.

Answer:

Section: 1

MULTIPLE CHOICE

28. How many distinct terminal alkynes exist with a molecular formula of C_5H_8?

a. 1
b. 2
c. 3
d. 4
e. 5

Answer: b Section: 1

29. The carbon-carbon triple bond of an alkyne is composed of __________.

a. three σ bonds
b. two σ bonds and one π bond
c. one σ bond and two π bonds
d. three π bonds

Answer: c Section: 3

30. Which of the following <u>improperly</u> describes the physical properties of an alkyne?

a. relatively nonpolar
b. nearly insoluble in water
c. less dense than water
d. insoluble in most organic solvents

Answer: d Section: 2

SHORT ANSWER

31. Why are terminal alkynes more acidic than other hydrocarbons?

    Answer:
    The carbanion which results upon deprotonation of a terminal alkyne has the lone pair of electrons in an sp hybrid orbital. The greater % s character of this orbital gives the orbital a significantly lower energy.

    Section: 9

32. Provide the structure of the major organic product(s) in the reaction below.

    $$CH_3CH_2 \equiv CH \xrightarrow[\text{2. } PhCH_2Br]{\text{1. } NaNH_2}$$

    Answer: $CH_3CH_2 \equiv CCH_2Ph$ Section: 10

MULTIPLE CHOICE

33. Which of the species below is less basic than acetylide?

    a. CH_3Li
    b. CH_3ONa
    c. NaOH
    d. both b and c
    e. all of the above

    Answer: d Section: 9

34. A mixture of 1-heptyne, 2-heptyne, and 3-heptyne was hydrogenated in the presence of a platinum catalyst until hydrogen uptake ceased. If one assumes that the hydrogenation went to completion, how many <u>different</u> seven-carbon hydrocarbons were produced?

    a. 1
    b. 2
    c. 3
    d. 6
    e. 8

    Answer: a Section: 8

SHORT ANSWER

35. Provide the structure of the major organic product(s) in the reaction below.

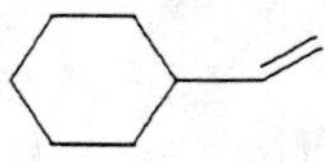

Answer:

Section: 8

MULTIPLE CHOICE

36. In the reduction of alkynes using sodium in liquid ammonia, which of the species below is <u>not</u> believed to be an intermediate in the commonly accepted mechanism?

a. vinyl anion
b. vinyl radical
c. radical anion
d. vinyl cation

Answer: d Section: 8

SHORT ANSWER

37. Provide the structure of the major organic product(s) in the reaction below.

$CH_3CH_2-C\equiv C-CH_3 \xrightarrow{Na,\ NH_3}$

Answer:

CH_3CH_2 H

H CH_3

Section: 8

38. Provide the structure of the major organic product(s) in the reaction below.

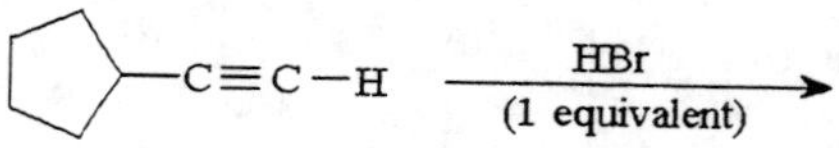

Cyclopentyl–C≡C–H $\xrightarrow[\text{(1 equivalent)}]{\text{HBr}}$

Answer:

(Cyclopentyl)(Br)C=CH$_2$

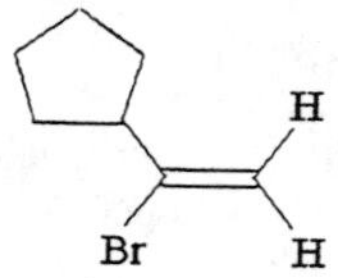

Section: 5

MULTIPLE CHOICE

39. In the addition of hydrogen bromide to alkynes in the absence of peroxides, which of the following species is believed to be an intermediate?

a. vinyl anion
b. vinyl cation
c. vinyl radical
d. carbene
e. none of the above

Answer: b Section: 5

SHORT ANSWER

40. Provide the structure of the major organic product(s) in the reaction below.

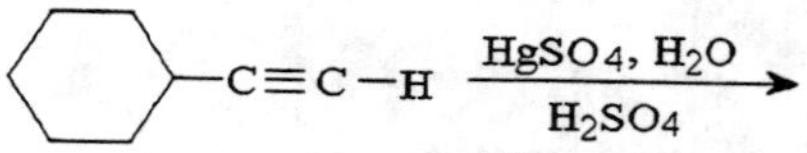

Cyclohexyl–C≡C–H $\xrightarrow[H_2SO_4]{HgSO_4,\ H_2O}$

Answer:

Cyclohexyl–C(=O)–CH_3

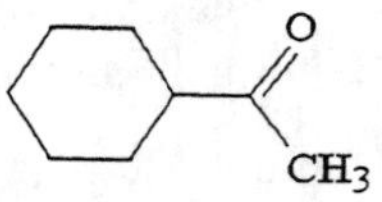

Section: 6

41. Provide the structure of the major organic product(s) in the reaction below.

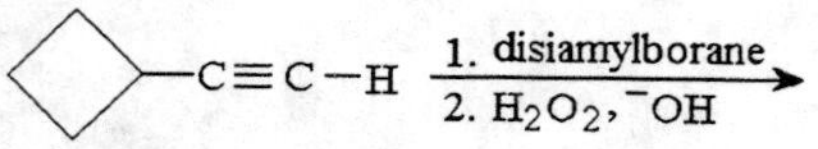

(cyclobutyl)$-C\equiv C-H$ $\xrightarrow[2.\ H_2O_2,\ ^{-}OH]{1.\ \text{disiamylborane}}$

Answer:

(cyclobutyl)$-CH_2-$COH

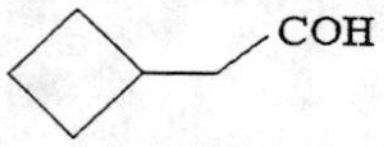

Section: 7

MULTIPLE CHOICE

42. Among the compounds water, 1-butyne, 2-butyne, and ethane, which are stronger acids than ammonia?

a. 1-butyne and ethane
b. water and 1-butyne
c. water and ethane
d. 1-butyne and 2-butyne

Answer: b Section: 9

43. ______________ is produced when 1 equivalent of HBr is added to 1-hexyne in the presence of peroxides.

a. 2-Bromo-1-hexene
b. E-1-Bromo-1-hexene
c. Z-1-Bromo-1-hexene
d. A mixture of E and Z isomers of 1-bromo-1-hexene
e. E-2-Bromo-2-hexene

Answer: d Section: 5

44. Which of the alkyne addition reactions below involves an enol intermediate?

a. hydroboration/oxidation
b. treatment with $HgSO_4$ in dilute H_2SO_4
c. hydrogenation
d. both a and b
e. none of the above

Answer: d Section: 5

SHORT ANSWER

45. Describe a sequence of reactions by which 3-heptyne can be straightforwardly prepared from acetylene.

Answer:
1. $NaNH_2$
2. CH_3CH_2Br
3. $NaNH_2$
4. $CH_3CH_2CH_2Br$

Section: 11

46. Describe a sequence of reactions by which meso-2,3-dibromobutane can be straightforwardly prepared from propyne.

Answer:
1. $NaNH_2$
2. CH_3Br
3. Na, NH_3
4. Br_2, CCl_4

Section: 11

47. Describe a sequence of reactions by which butylbenzene can be straightforwardly prepared from phenylacetylene.

Answer:
1. $NaNH_2$
2. CH_3CH_2Br
3. H_2, Pt

Section: 11

48. Describe a sequence of reactions by which $CH_3CH_2CH_2COCH_2CH_3$ can be straightforwardly prepared from 1-butyne.

Answer:
1. $NaNH_2$
2. CH_3CH_2Br
3. $HgSO_4$, H_2O, H_2SO_4

Section: 11

49. Describe a sequence of reactions by which <u>trans</u>-2-pentene can be straightforwardly prepared from propyne.

Answer:
1. $NaNH_2$
2. CH_3CH_2Br
3. Na, NH_3

Section: 11

MULTIPLE CHOICE

50. What are the hybridizations of the carbon atoms numbered 1 and 2 in the structure below?

$$H_3C-CH=CH-\overset{1}{C}\equiv\overset{2}{C}-H$$

a. sp^3, sp^2
b. sp^2, sp^2
c. sp^2, sp
d. sp, sp

Answer: d Section: 3

SHORT ANSWER

51. For the reaction below indicate whether the equilbrium constant will be greater than 1 or less than 1.

$CH_3Li + CH_3C{\equiv}CH \rightleftharpoons CH_4 + CH_3C{\equiv}CLi$

Answer: >1 Section: 9

52. For the reaction below indicate whether the equilbrium constant will be greater than 1 or less than 1.

$CH_3CH_2CHO \rightleftharpoons CH_3CH{=}CHOH$

Answer: <1 Section: 6

53. For the reaction below indicate whether the equilbrium constant will be greater than 1 or less than 1.

$KOH + CH_3C{\equiv}CH \rightleftharpoons H_2O + CH_3C{\equiv}CK$

Answer: <1 Section: 9

54. Complete the following reaction

? $\xrightarrow[\text{2) } H_2O_2, OH^-]{\text{1) disiamylborane}}$ [cyclobutane]$-CH_2CHO$

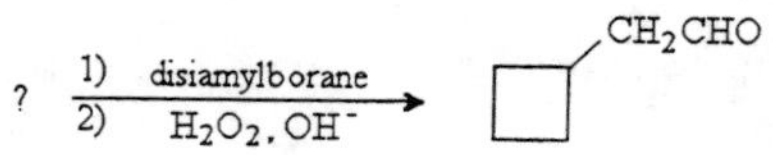

Answer: ethynylcyclobutane Section: 7

55. Complete the following reaction.

3-hexyne $\xrightarrow{Na,\ NH_3}$

Answer: trans-3-heptene Section: 8

MULTIPLE CHOICE

56. How many distinct internal alkynes exist with a molecular formula of C_6H_{10}?

a. 1
b. 2
c. 3
d. 4
e. 5

Answer: c Section: 1

57. What is the common name for the following alkyne?

$$CH_3-\overset{\overset{\large CH_3}{|}}{\underset{\underset{\large CH_3}{|}}{C}}-C\equiv CH$$

a. Neohexyne
b. Trimethylpropyne
c. Trimethylacetylene
d. tert-Butylacetylene
e. Isopropylacetylene

Answer: d Section: 1

Chapter 6: Electron Delocalization, Resonance, and Aromaticity

MULTIPLE CHOICE

1. Which of the following pairs are resonance structures?

a. 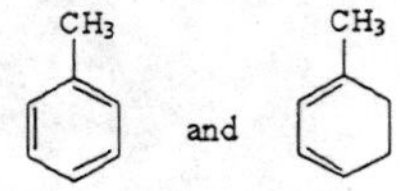 CH_3 and CH_3

b. 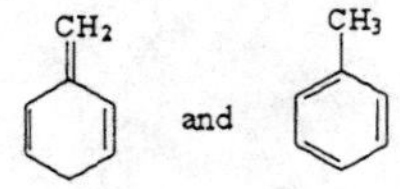 CH_2 and CH_3

c. 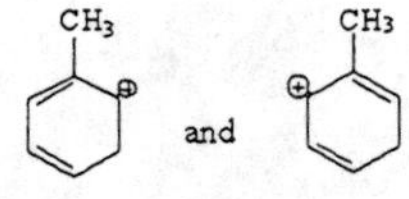 CH_3 ⊕ and ⊕ CH_3

d. 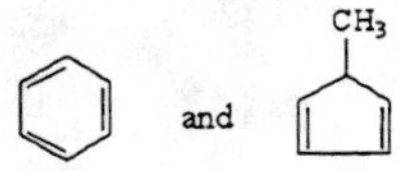 and CH_3

e. 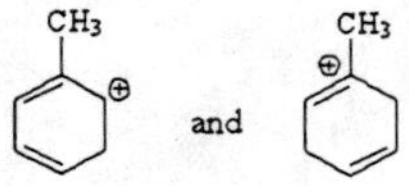 CH_3 ⊕ and ⊕ CH_3

Answer: c Section: 3

2. Which of the following pairs are resonance structures?

a. $CH_2{=}CHCH_3$ and

b. 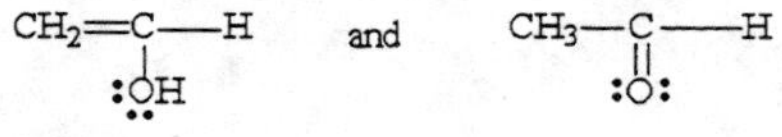 $CH_2{=}C{-}H$ (with $:\ddot{O}H$) and $CH_3{-}C{-}H$ (with $=\ddot{O}:$)

c. $CH_3{-}C{-}H$ (with $=\ddot{O}:$) and $CH_3{-}\overset{\oplus}{C}{-}H$ (with $:\ddot{O}:^{\ominus}$)

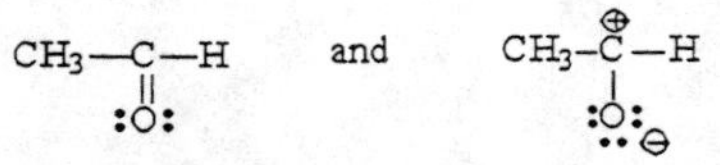

d. 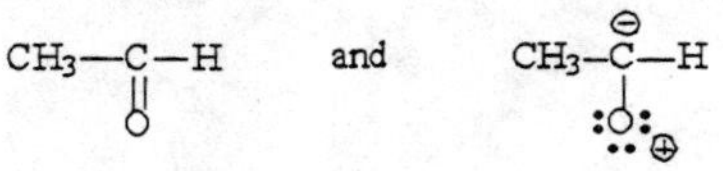 $CH_3{-}C{-}H$ (with $=O$) and $CH_3{-}\overset{\ominus}{C}{-}H$ (with $:\ddot{O}:^{\oplus}$)

e. 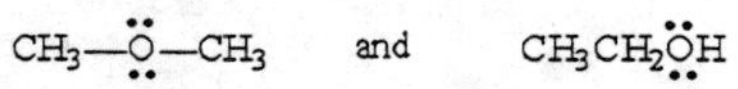 $CH_3{-}\ddot{O}{-}CH_3$ and $CH_3CH_2\ddot{O}H$

Answer: c Section: 3

3. Which of the following statements is incorrect about benzene?

   a. All of the carbon atoms are sp hybridized.
   b. It has delocalized electrons.
   c. The carbon-carbon bond lengths are all the same.
   d. The carbon-hydrogen bond lengths are all the same.
   e. All twelve atoms lie in the same plane.

   Answer: a Section: 2

4. Which of the following is a resonance structure of the species shown?

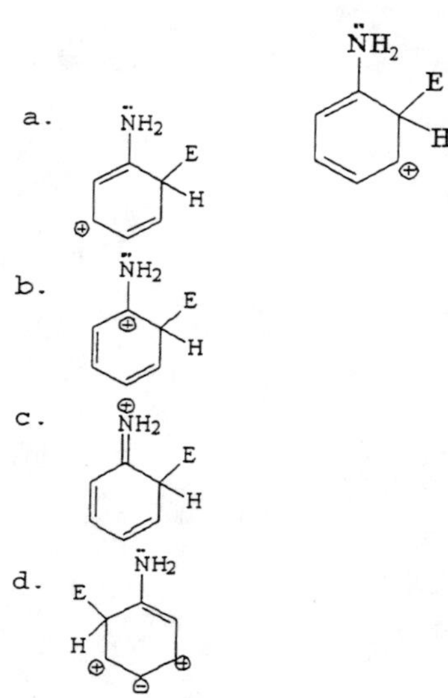

   e. All of the above

   Answer: e Section: 4

5. Which of the following statements about benzene is/are correct?

   a. All of the carbon atoms are sp^3 hybridized.
   b. It has no delocalized electrons.
   c. The carbon-carbon bond length is longer than that of ethane.
   d. It is a planar molecule.
   e. The carbon-hydrogen bonds are not the same length.

   Answer: d Section: 1

6. Which of the following is the most stable resonance contributor to acetic acid?

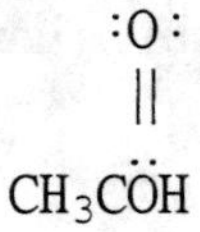

$$CH_3C(=\ddot{O}:)\ddot{O}H$$

a. 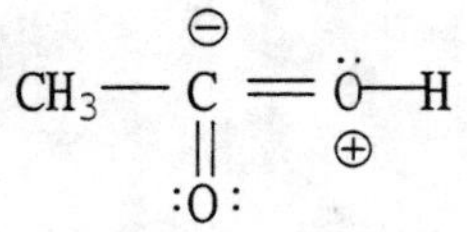 $CH_3-\overset{\ominus}{C}(=\ddot{O}:)=\overset{\oplus}{\ddot{O}}-H$

b. $CH_3-C(=\ddot{O}:)-\overset{\oplus}{\ddot{O}}=\overset{\ominus}{H}$

c. $\overset{\ominus}{CH_3}=C(-\ddot{O}:)-\ddot{O}-H$

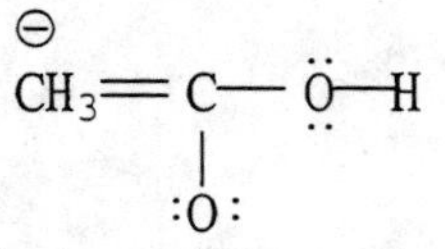

d. 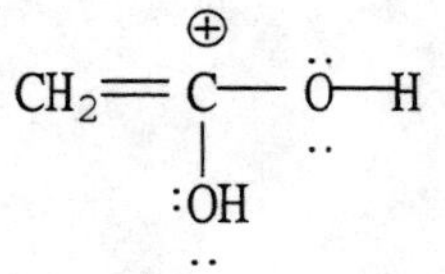 $CH_2=\overset{\oplus}{C}(-\ddot{O}H)-\ddot{O}-H$

e. 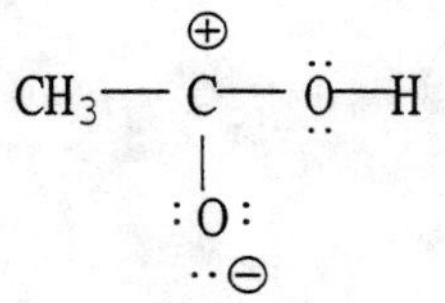 $CH_3-\overset{\oplus}{C}(-\ddot{O}:^{\ominus})-\ddot{O}-H$

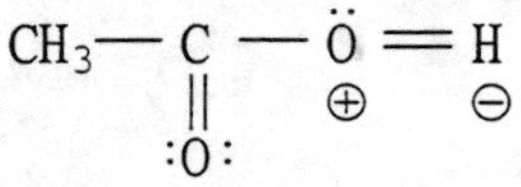

Answer: e Section: 4

7. Which of the following is an allylic cation?

a. $CH_3\overset{\oplus}{C}H=CH_2$

b. $CH_3CH=CH-\overset{\oplus}{C}H_2$

c. $CH_3\overset{\oplus}{C}HCH_3$

d. $CH_3\overset{\oplus}{C}HCH_2CH=CH_2$

e. 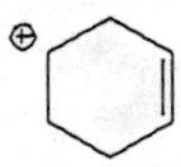 ⊕ (cyclohexene ring with positive charge on a carbon not adjacent to the double bond)

Answer: b Section: 7

8. Which of the following is a benzylic cation?

a.  (cyclohexene)

b. 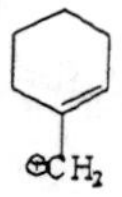 (cyclohexene with $\oplus CH_2$)

c. 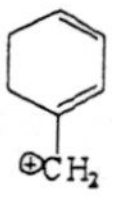 (cyclohexadiene with $\oplus CH_2$)

d.  (benzene with $\oplus CH_2$)

e.  (cyclohexadiene with $\oplus$ and CH_3)

Answer: d Section: 7

9. Which of the following is the <u>most</u> stable cation?

a. 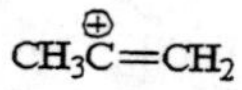 $CH_3\overset{\oplus}{C}{=}CH_2$

b. $CH_3CH{=}\overset{\oplus}{C}{-}C_6H_5$

c. $CH_3{-}\underset{\oplus}{\overset{CH_3}{C}}{-}C_6H_5$

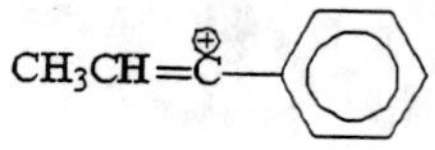

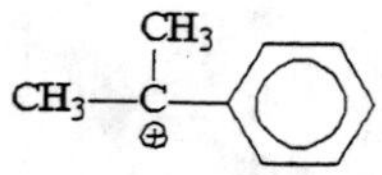

d. 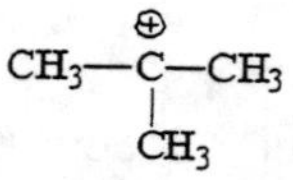 $CH_3{-}\overset{\oplus}{\underset{CH_3}{C}}{-}CH_3$

e. 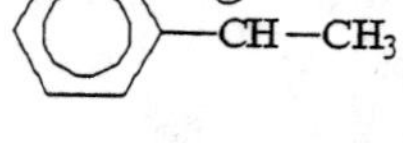 $C_6H_5{-}\overset{\oplus}{C}H{-}CH_3$

Answer: c Section: 7

10. Which of the following is the <u>most</u> stable radical?

a. 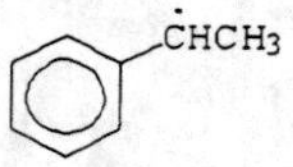 [benzene ring with $\dot{C}HCH_3$ substituent]

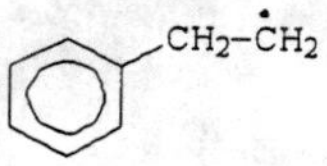

b. [benzene ring with $CH_2-\dot{C}H_2$ substituent]

c. [benzene ring with CH_2CH_3 substituent, radical on ring]

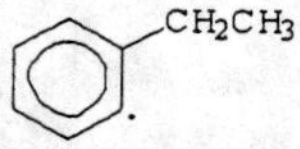

d. 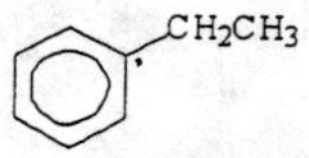 [benzene ring with CH_2CH_3 substituent, radical on ring]

e. 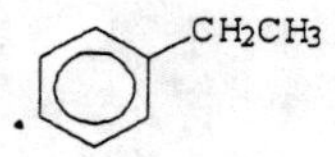 [benzene ring with CH_2CH_3 substituent, radical on ring]

Answer: a Section: 8

11. How many allylic hydrogen atoms are present in the following molecule?

a. 2
b. 3
c. 4
d. 5
e. zero

$CH_2{=}CH{-}CH{=}CH{-}C_6H_5$

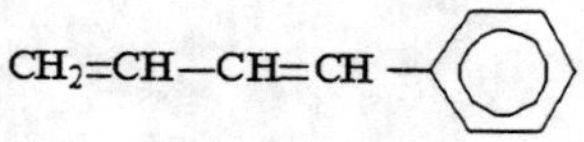

Answer: e Section: 7

12. How many benzylic hydrogens are present in the following molecule?

a. 2
b. 3
c. 4
d. 5
e. zero

$CH_2{=}CH{-}CH{=}CH{-}C_6H_5$

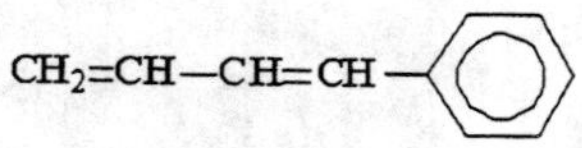

Answer: e Section: 7

13. Which of the following is/are the main product(s) of the following reaction?

$CH_2 = CH - CH = CH_2 \xrightarrow{HBr} ?$

a. $CH_2CH = CHCH_3$ (Br on the first carbon)

b. $CH_2CH_2CH = CH_2$ (Br on the first carbon)

c. $CH_2 = CHCHCH_3$ (Br on the third carbon)

d. a and b
e. a and c

Answer: e Section: 9

14. Which of the following is/are the major product(s) of the following reaction?

a. I.
b. I. and II.
c. I. and III.
d. II.
e. III.

\+ HI → ?

I.

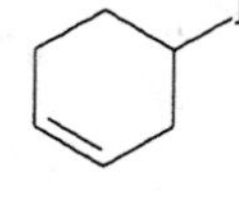

II.

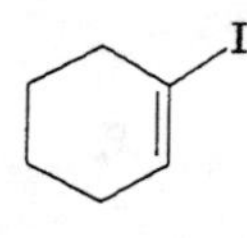

III.

Answer: a Section: 9

15. Which of the following is/are the major product(s) of the following reaction?

a. I
b. I and II
c. II
d. II and III
e. III

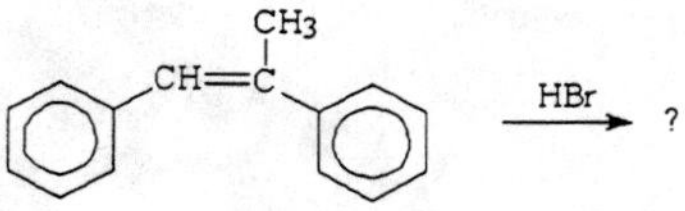

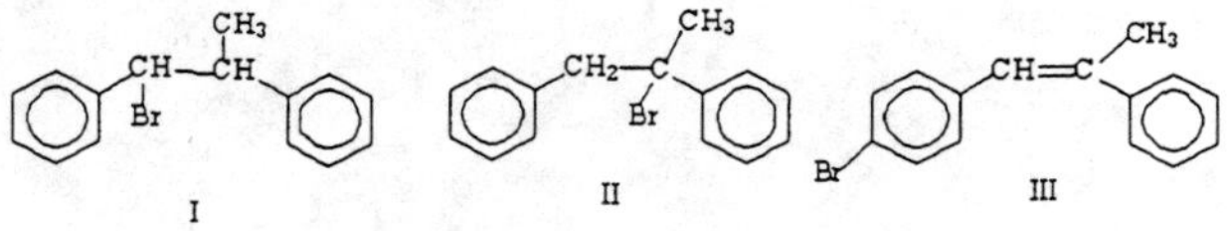

Answer: c Section: 9

16. Which of the following is the strongest acid?

a. CH_3CH_2OH
b. CH_3OCH_3
c. CH_3CH (with =O on the CH)
d. CH_3COCH_3 (with =O on the C)
e. CH_3COH (with =O on the C)

Answer: e Section: 10

17. Which of the following is the <u>strongest</u> acid?

a. 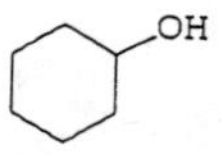

b. 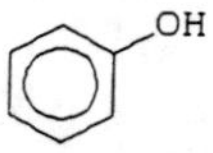

c. 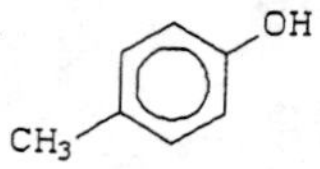

d. 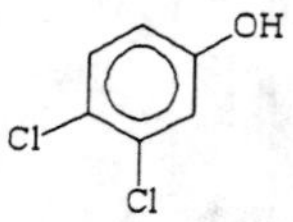

e. 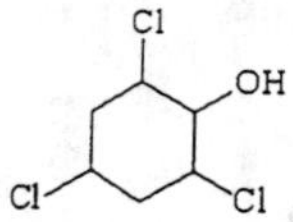

Answer: d Section: 10

18. Which of the following compounds is aromatic?

a. 

b. 

c. 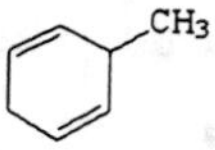

d. 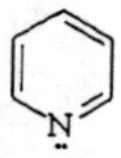

e. 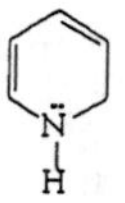

Answer: d Section: 12

19. Which of the following is aromatic?

a.

b.

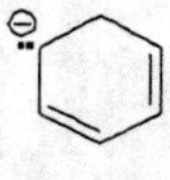

c.

d.

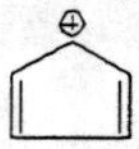

e.

Answer: c Section: 12

20. Which of the following is aromatic?

a.

b.

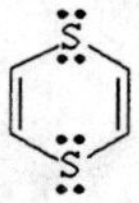

c.

d.

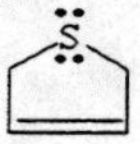

e.

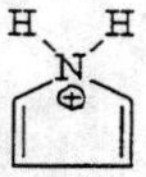

Answer: a Section: 12

21. Which of the following is aromatic?

a. 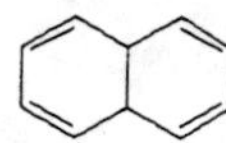

b. 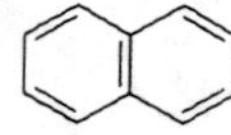

c.

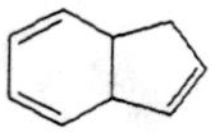

d. 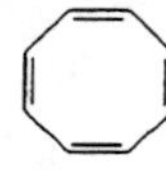

e.

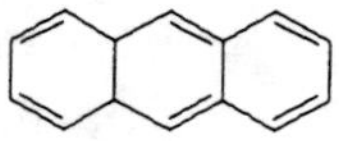

Answer: b Section: 12

22. Which of the following is the most acidic?

a. CH_3CH_2-H

b. CH_2=CH—H

c. CH≡C—H

d.

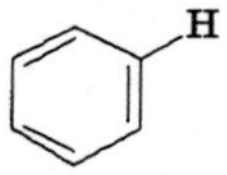

e.

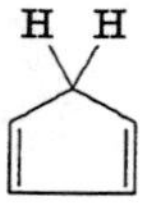

Answer: e Section: 13

23. Which of the following is antiaromatic?

a. 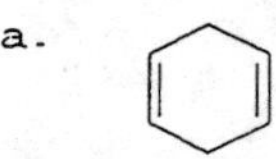

b. 

c. 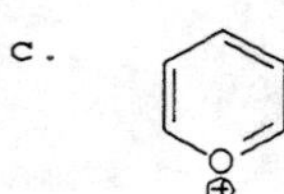

d.

e.

Answer: d Section: 14

24. Which species is represented by the following distribution of π electrons in the molecular orbitals and energy levels diagram?

a. 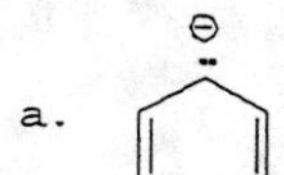

b. $CH_2{=}CH{-}\overset{\oplus}{C}H_2$

c. $CH_2{=}CH{-}\dot{C}H_2$

d. $CH_2{=}CH{-}\overset{\cdot\ominus}{C}H_2$

e.

Answer: c Section: 15

25. Which species is represented by the following distribution of π electrons in the molecular energy diagram?

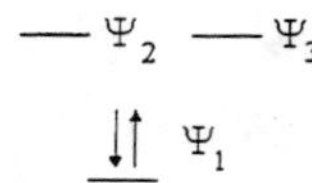

a.

b.

c.

d.

e.

Answer: c Section: 16

SHORT ANSWER

26. Draw two other resonance contributors for the following compound:

Answer:

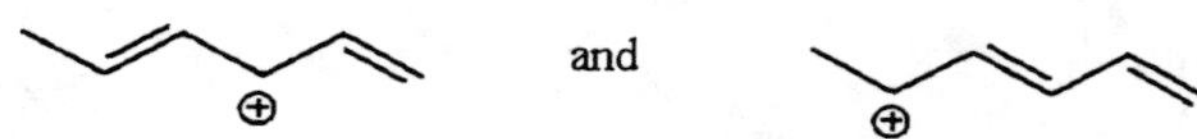

Section: 4

27. Are the following pairs of structures resonance contributors or different compounds?

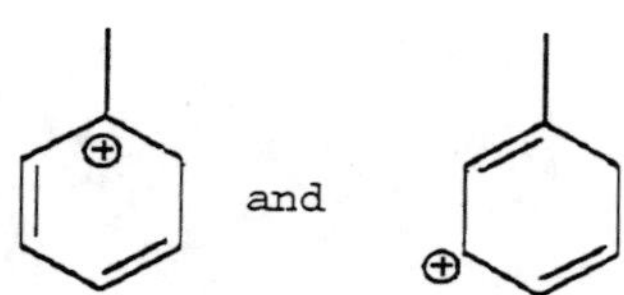

Answer: They are resonance contributors. Section: 4

28. Draw four resonance contributors for the following structure and indicate which is the <u>most</u> important contributor. Explain.

$$\left[CH_2 \text{---} \overset{\overset{\displaystyle O}{\|}}{N} \text{==} O \right]^{\ominus}$$

a. $^{\oplus}CH_2-N(-\ddot{O}:^{\ominus})-\ddot{O}:^{\ominus}$

b. $^{\ominus}CH_2-\overset{\oplus}{N}(=\ddot{O}:)-\ddot{O}:^{\ominus}$

c. $CH_2=\overset{\oplus}{N}(-\ddot{O}:^{\ominus})-\ddot{O}:^{\ominus}$

d. $^{\ominus}\ddot{C}H_2-\overset{\oplus}{N}(-\ddot{O}:^{\ominus})=\ddot{O}:$

Answer:
C is the most important contributor because all the atoms have complete octets and the negative charges are on the most electronegative atoms.

Section: 5

29. Which of the following numbered hydrogens is most easily abstracted by heterolytic cleavage? Explain.

```
            2
            H
   H        |    H
    \       |    |
      C  =  C  - C - H3
    /            |
   H             H
   1
```

Answer:
Hydrogen #3 is most easily abstracted by heterolytic cleavage because the carbocation produced from the cleavage is allylic.

Section: 7

30. The following molecule is not aromatic. Why?

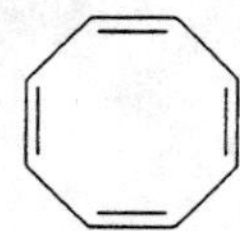

Answer:
It does not have an odd number of pairs of pi electrons (i.e., it does not obey Huckle's rule) and it is not planar.

Section: 11

31. Rate the following molecules in decreasing order of stability:

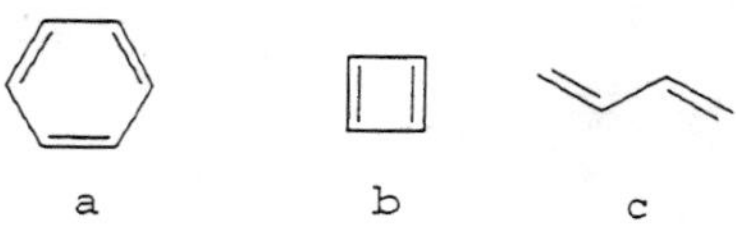

Answer:
a>c>b
aromatic > linear > antiaromatic.
Section: 5

32. One of the following compounds is aromatic and the other is antiaromatic. Which is which?

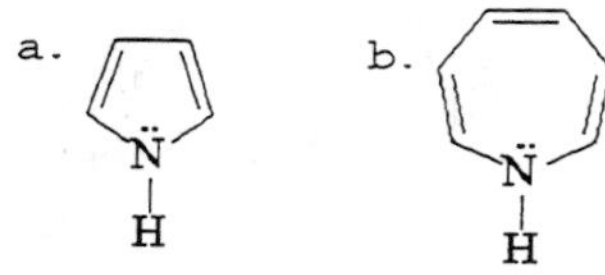

Answer: (a) is aromatic, (b) is antiaromatic. Section: 12

33. Which of the following compounds is more stable? Explain.

a. $\overset{\oplus}{C}H_2$ or b. $\overset{\oplus}{C}HCH_3$

Answer:
(a) and (b) are both benzylic cations. (b) is more stable because it is more alkyl substituted.

Section: 7

34. Which of the following compounds is more stable? Explain.

a. CH_3, C—CH_3 or b. CH_3, C—H

Answer:
(a) is more stable than (b). Although both species are benzylic, (a) is more alkyl substituted than (b), making it more stable.

Section: 8

35. Why is phenol a stronger acid than cyclohexanol?

OH OH

phenol cyclohexanol

Answer:
The anion of phenol is stabilized by resonance, causing the proton to be more easily abstracted than that of cyclohexanol.

Section: 10

36. Is the following molecule aromatic? Explain.

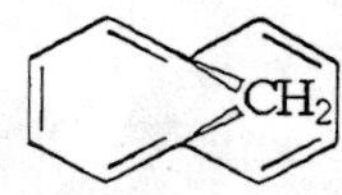

Answer:
Yes, it is aromatic since the molecule is planar, monocyclic with 5 pairs of π electrons.

Section: 11

37. Is the following molecule aromatic? Explain.

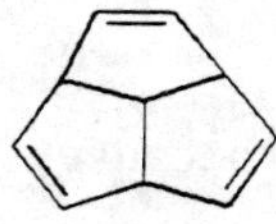

Answer: No, it is not aromatic since the molecule is not a conjugated system.

Section: 11

38. Draw the distribution of π electrons in the molecular orbitals of cyclobutadiene.

Answer:

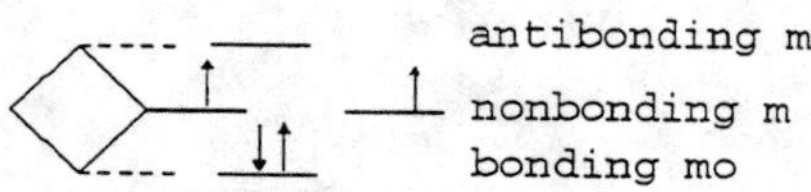

Section: 16

39. Draw the important resonance structures of:

$H_2\bar{C} - \overset{+}{N} \equiv N:$

Answer:

$H_2\bar{C}-\overset{+}{N}\equiv N \longleftrightarrow H_2C=\overset{+}{N}=\ddot{N}:^-$

Section: 4

40. Draw the important resonance structures of:

$CH_3CH = CH\overset{+}{C}HCH_2CH_3$

Answer:

$CH_3CH = CH\overset{+}{C}HCH_2CH_3$ and $CH_3\overset{+}{C}HCH = CHCH_2CH_3$

Section: 4

41. Draw the important resonance structures of:

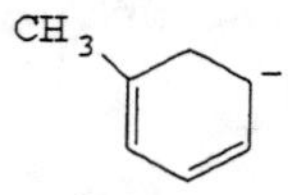

CH 3

Answer:

CH 3 ⟷ CH 3 ⟷ CH 3

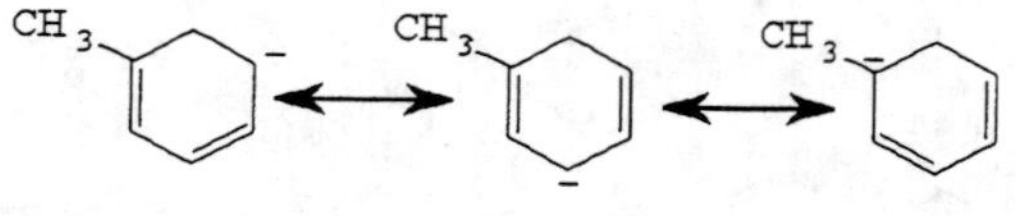

Section: 4

42. Draw the important resonance contributing forms for the structure shown below.

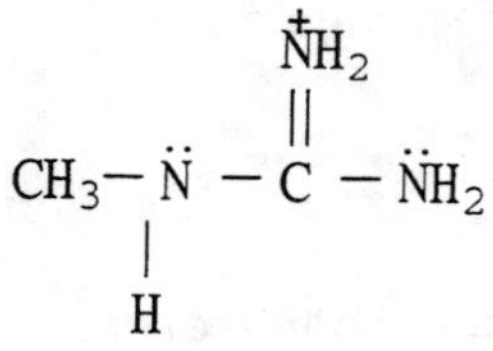

$CH_3-\ddot{N}(H)-C(=\overset{+}{N}H_2)-\ddot{N}H_2$

Answer:

$CH_3-\ddot{N}(H)-C(=\overset{+}{N}H_2)-\ddot{N}H_2 \longleftrightarrow CH_3-\overset{+}{N}(H)=C(\ddot{N}H_2)-\ddot{N}H_2 \longleftrightarrow CH_3-\ddot{N}(H)-C(\ddot{N}H_2)=\overset{+}{N}H_2$

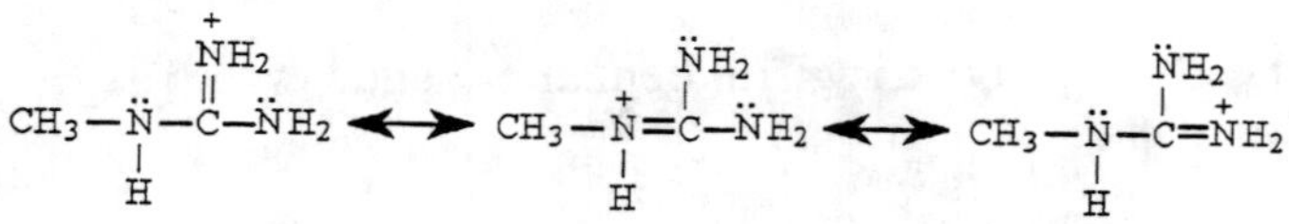

Section: 4

43. Draw the important resonance contributing forms for the structure shown below.

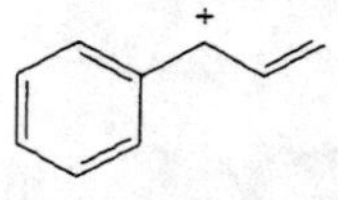

Answer:

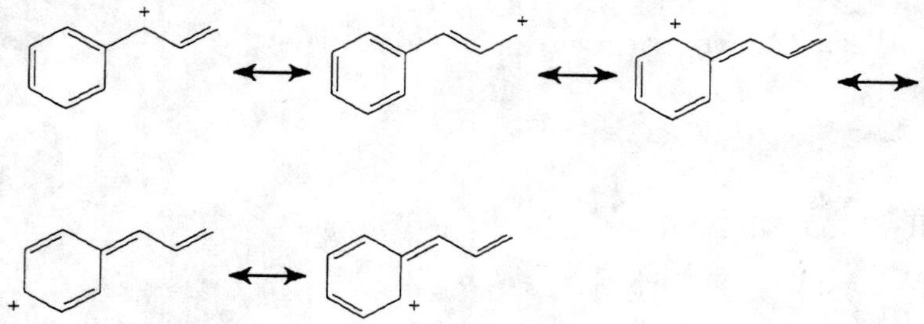

Section: 4

44. Draw the important resonance contributing forms for the structure shown below.

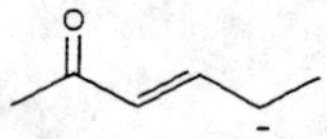

Answer:

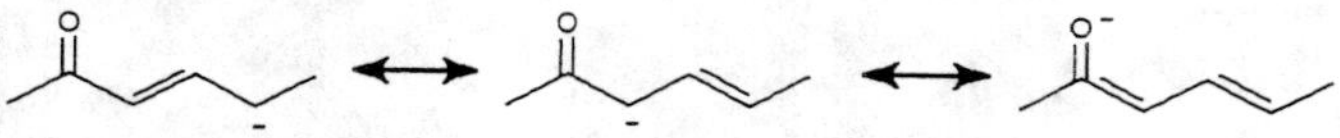

Section: 4

45. Stabilization of a charged species usually results when this species can be more accurately depicted as a hybrid of several resonance contributing forms. Why is this the case?

Answer: Stabilization results from delocalization of charge. Section: 5

46. What is suggested by the fact that benzene's molar heat of hydrogenation is 36 kcal less than three times the molar heat of hydrogenation of cyclohexene?

Answer:
This suggests that the type of π bonding in benzene lends special stability to the molecule.

Section: 6

47. Cyclic hydrocarbons which can be represented as structures containing alternating single and double bonds are called ________.

Answer: annulenes Section: 12

MULTIPLE CHOICE

48. In the molecular orbital representation of benzene, how many π molecular orbitals are present?

a. 1
b. 2
c. 4
d. 6
e. 8

Answer: d Section: 15

SHORT ANSWER

49. Provide a diagram which depicts the relative energies of the π molecular orbitals of benzene. Show which molecular orbitals are filled in benezene's ground state.

Answer:

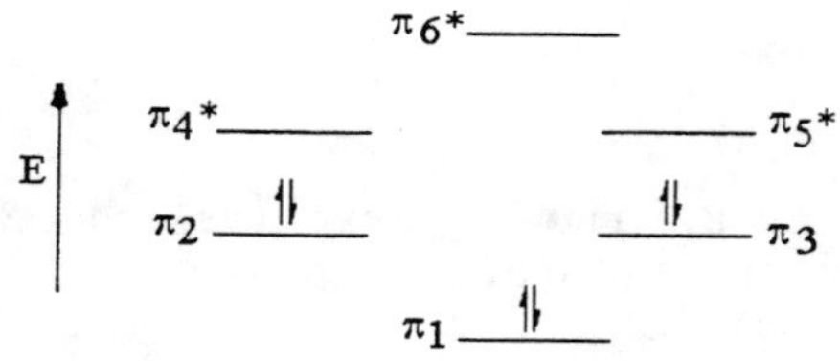

Section: 15

50. List the criteria which compounds must meet in order to be considered aromatic.

Answer:
1. The structure must be cyclic.
2. Each atom in the ring must have an unhybridized p orbital.
3. The structure must be planar or nearly planar so that overlap of these p orbitals is effective.
4. The π network must contain 4n + 2 electrons (where n is a whole number), so that delocalization of the π electrons results in a lowering of the molecule's electronic energy.

Section: 11

Chapter 6: Electron Delocalization, Resonance, and Aromaticity

MULTIPLE CHOICE

51. Aromatic molecules contain _____ π electrons.

    a. no
    b. 4n + 2 (with n an interger)
    c. 4n + 2 (where n = 0.5)
    d. 4n (with n an integer)
    e. unpaired

    Answer: b Section: 11

SHORT ANSWER

52. Provide the major resonance structures of the ion which results when the most acidic hydrogen of cyclopentadiene is lost.

    Answer:

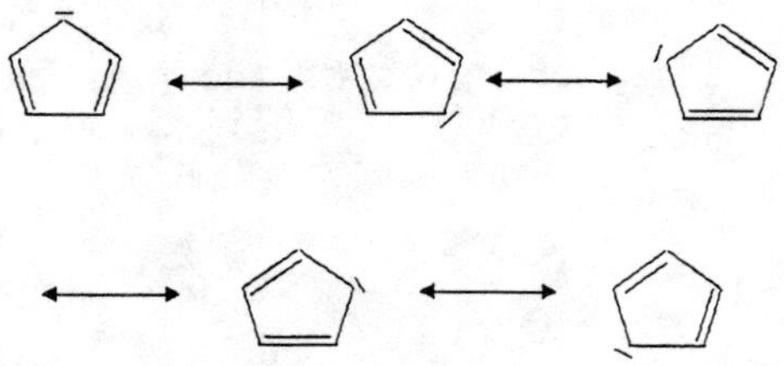

    Section: 13

53. When cycloheptatriene is deprotonated, an anion with seven resonance forms of equal energy can be drqwn. Given this fact, explain why cycloheptatriene is only slightly more acidic than propene.

    Answer: Each resonance form of the cycloheptatrienyl anion is antiaromatic.

    Section: 13

54. Classify cyclopentadienyl cation as aromatic, antiaromatic, or nonaromatic. Assume planarity of the π netowrk.

    Answer: antiaromatic Section: 15

55. Classify cyclopheptatrienyl cation as aromatic, antiaromatic, or nonaromatic. Assume planarity of the π netowrk.

    Answer: aromatic Section: 15

56. Classify the compound below as aromatic, antiaromatic, or nonaromatic. Assume planarity of the π netowrk.

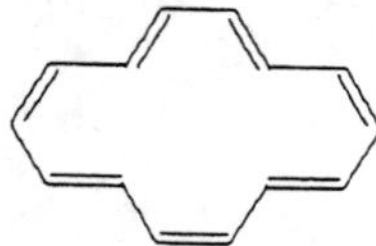

Answer: aromatic Section: 12

57. Classify the compound below as aromatic, antiaromatic, or nonaromatic. Assume planarity of the π netowrk.

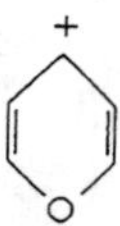

Answer: aromatic Section: 12

58. Is cyclooctatetraene a planar molecule? Explain.

Answer:
No. In addition to increasing other forms of strain, a planar conformation would make this molecule antiaromatic.
Section: 14

MULTIPLE CHOICE

59. Which of the structures below would be aromatic?

a. I and IV
b. I, III, and IV
c. III and IV
d. II

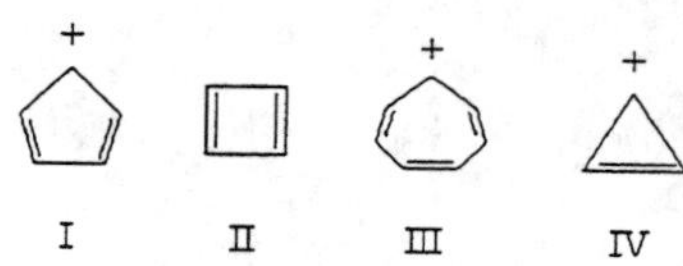

Answer: c Section: 12

SHORT ANSWER

60. Indicate whether the following molecule is aromatic, antiaromatic, or nonaromatic.

Answer: aromatic Section: 12

Chapter 7: Reactions of Dienes

MULTIPLE CHOICE

1. Which of the following is a cumulated diene?

a.

b.

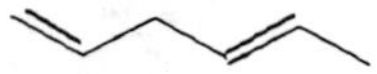

c.

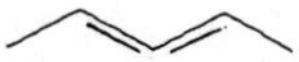

d.

e.

Answer: c

2. Which of the following is an isolated diene?

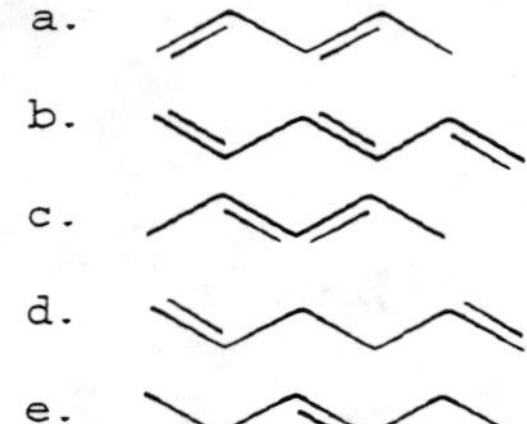

Answer: d

3. Which of the following is a conjugated diene?

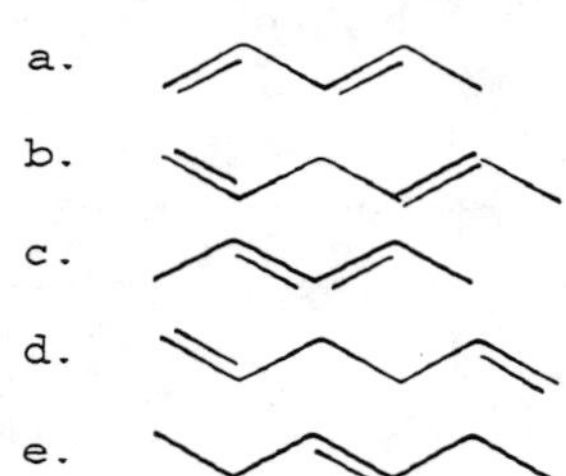

Answer: a

4. Give the IUPAC name for the following compound.

   a. Isoprene
   b. 3-Methyl-1,3-butadiene
   c. 2-Methyl-1,3-butadiene
   d. 2-Methyl-2,3-butadiene
   e. Isobutadiene

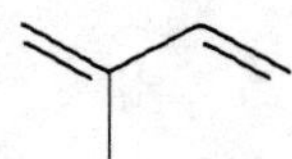

Answer: c Section: 1

5. What is the IUPAC name for the following compound?

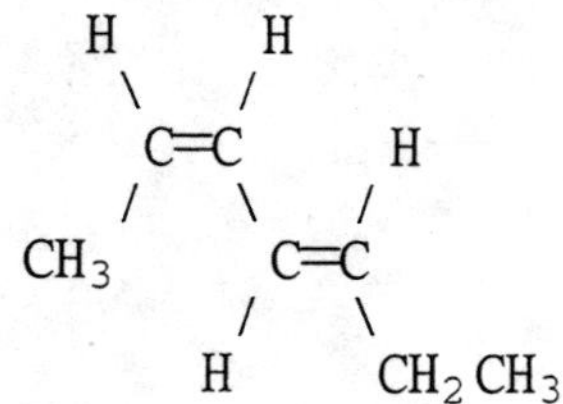

   a. cis,trans-2,4-heptadiene
   b. 2Z, 4Z-2,4-heptadiene
   c. cis, cis-2,4-heptadiene
   d. trans,trans-2,4-heptadiene
   e. 2E, 4E-2,4-heptadiene

Answer: a Section: 1

6. Which of the following is the most stable diene?

a. 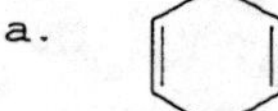

b.

c.

d.

e.

Answer: c Section: 3

7. How many nodes would you expect in the molecular orbitals alignment of overlapping p orbitals in 1,3-butadiene?

   a. zero
   b. one
   c. two
   d. three
   e. four

Answer: c Section: 3

8. What is/are the product(s) from the following reaction?

   a. I. only
   b. II. only
   c. I. is minor, II. is major
   d. I. is major, II. is minor
   e. Equal amounts of I. and II.

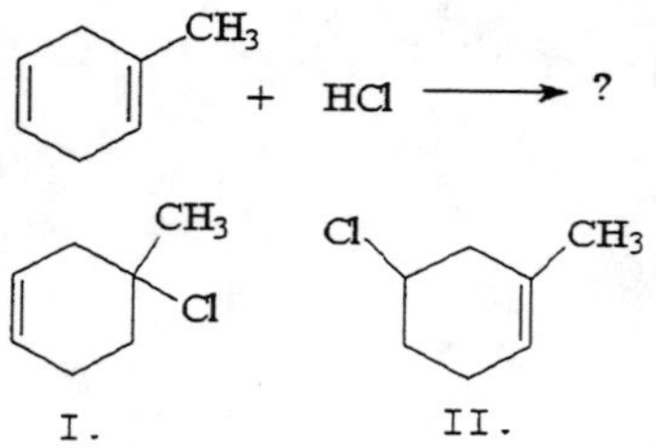

CH_3 + HCl ⟶ ?

CH_3 Cl   Cl CH_3

I.   II.

Answer: d Section: 5

9. What is/are the product(s) of the following reaction?

$CH_2{=}CHCH{=}CH_2 + HBr \longrightarrow ?$

I. $CH_2{=}CHCH(Br)CH_3$

II. $CH_3CH(Br){-}CH(Br)CH_3$

III. $CH_2(Br){-}CH{=}CHCH_3$

a. I. only
b. II. only
c. III. only
d. I. and II.
e. I. and III.

Answer: e Section: 6

10. What is/are the product(s) from the following reaction?

a. I. and II.
b. I. and III.
c. I. and IV.
d. II. and III.
e. II. and IV.

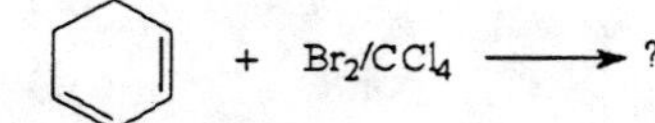

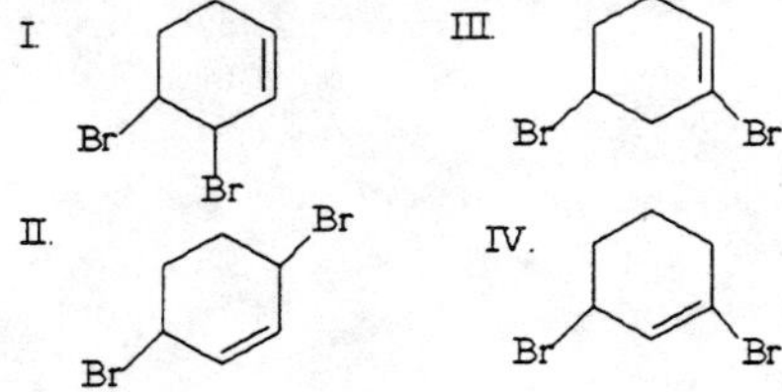

Answer: a Section: 6

11. What is/are the product(s) from the following reaction?

$CH_2{=}CHCH=CH_2$ + HBr $\xrightarrow{-80°C}$?

I. $CH_2CHCH=CH_2$ (Br on C1)

II. $CH_3CHCH=CH_2$ (Br on C2)

III. $CH_3CH=CHCH_2$ (Br on C4)

a. II. Minor, III. Major
b. II. Major, III. Minor
c. III. only
d. I. and II.
e. II. and III.

Answer:
b

Section: 6

12. What is/are the product(s) from the following reaction?

$CH_2{=}CHCH=CH_2$ + HBr $\xrightarrow{45°}$?

I. $CH_2CH_2CH=CH_2$ (Br on C1)

II. $CH_3CHCH=CH_2$ (Br on C2)

III. $CH_3CH=CHCH_2$ (Br on C4)

a. I. only
b. II. only
c. III. only
d. I. and II.
e. II. and III.

Answer: c Section: 6

13. What is the thermodynamic product for the following reaction?

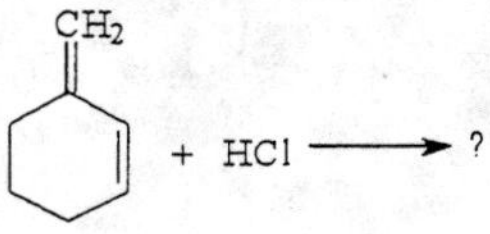

a. 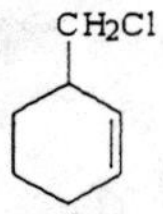

b. 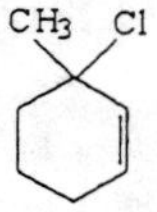

c. 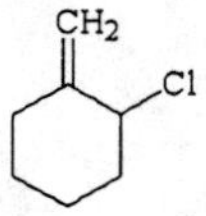

d. 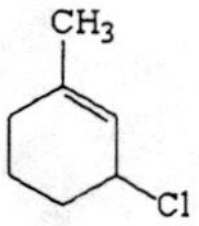

e. 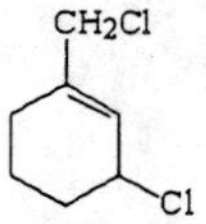

Answer: d Section: 6

14. What is the kinetic product for the following reaction?

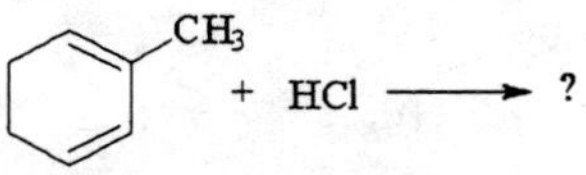

a. 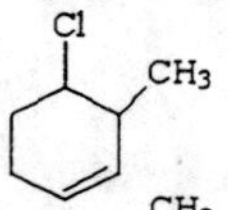

b. 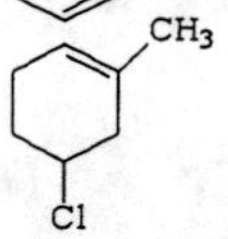

c. 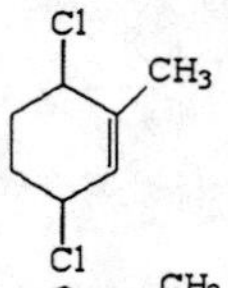

d. 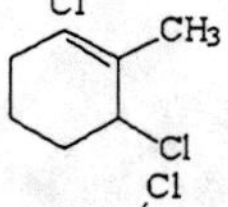

e. 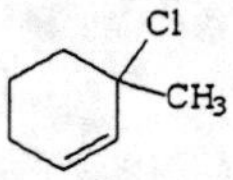

Answer: e Section: 6

15. What is the product of the following reaction?

heat ?

a.

b.

c.

d.

e.

Answer: e Section: 8

16. What is the product from the following reaction?

H COOH H COOH heat ?

a. COOH COOH

b. COOH COOH

c. COOH COOH

d. COOH COOH

e. COOH COOH

Answer: b Section: 8

17. Which of the following conjugated dienes would <u>not</u> react with a dieneophile in a Diels-Alder reaction?

a. 

b. 

c.

d.

e. a and c

Answer: e Section: 8

18. Which of the following dienes is the most reactive in a Diels-Alder reaction?

a.

b.

c.

d.

e.

Answer: c Section: 8

19. What diene and what dieneophile would be used to synthesize the following compound?

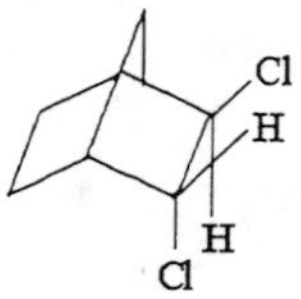

a.

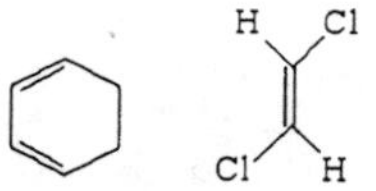

b.

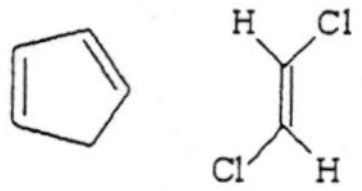

c.

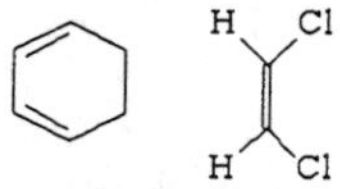

d.

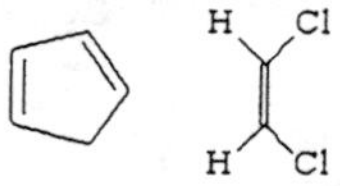

e.

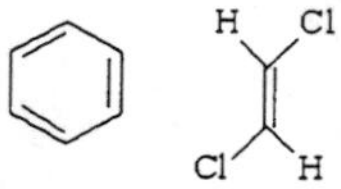

Answer: b Section: 8

20. What is the name of the following compound?

a. Spiro [2.2.2] octane
b. Bicyclo [2.2.0] octane
c. Bicyclo [2.2] octane
d. Bicyclo [2.2.2] octane
e. Tricyclo [2.2.2] octane

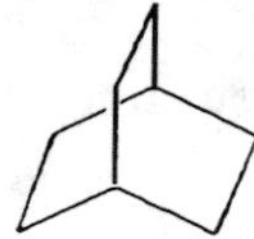

Answer: d Section: 9

21. What is the name of the following compound?

    a. Spiro [5.2] octane
    b. Spiro [5.2.0] octane
    c. Bicyclo [5.2.0] octane
    d. Bicyclo [5.2] octane
    e. Spiro [6.4] octane

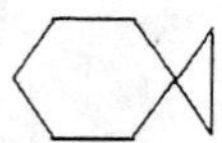

Answer: a Section: 9

SHORT ANSWER

22. Arrange the following unsaturated hydrocarbons in order of decreasing reactivity toward HBr addition:

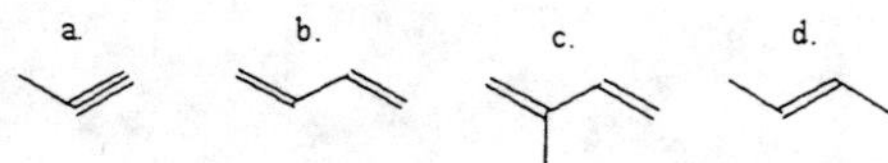

Answer: c>b>d>a Section: 5

23. Name the following compound:

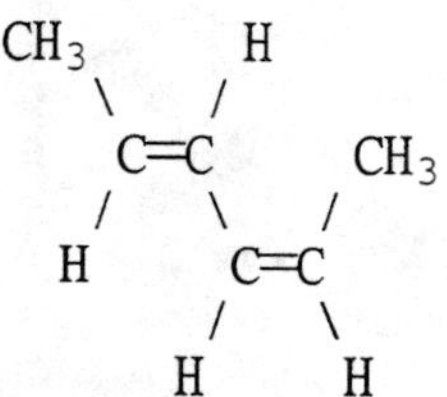

Answer: (2E, 4Z)-2,4-hexadiene or trans,cis-2,4-hexadiene Section: 1

24. What is the hybridization of the carbons labeled A, B, and C?

$$CH_3-\underset{A}{\overset{\overset{CH_3}{|}}{C}}=CH-\underset{B}{CH_2}-C\equiv\underset{C}{CH}$$

Answer:

sp^2, sp^3, sp

25. What is the major product obtained from the addition of one mole of HCl at high temperature to 1,3-butadiene?

Answer: 1-chloro-2-butene (thermodynamic product) Section: 7

26. What is the product of the following Diels-Alder reaction?

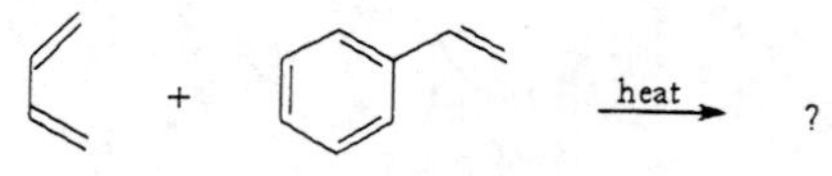

heat → ?

Answer:

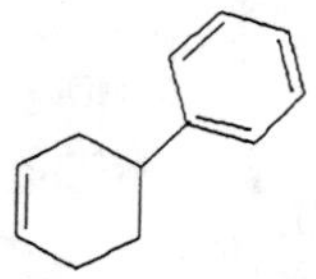

Section: 8

27. What diene and what dieneophile could be used to prepare the following?

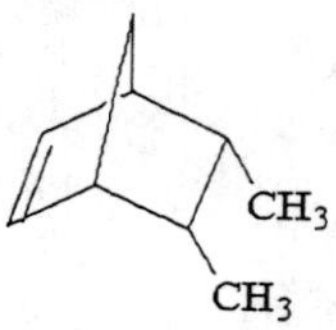

CH_3

CH_3

Answer:

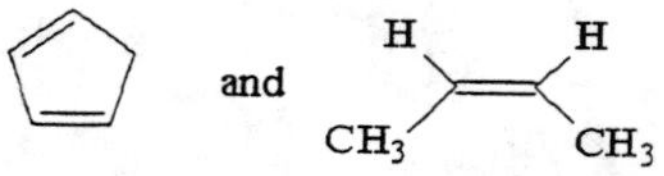

and

H H

CH_3 CH_3

Section: 8

28. How could the following compound be synthesized using a Diels-Alder reaction?

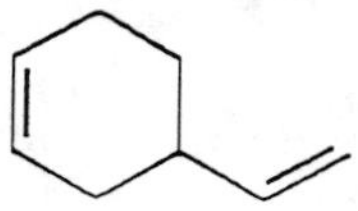

Answer:

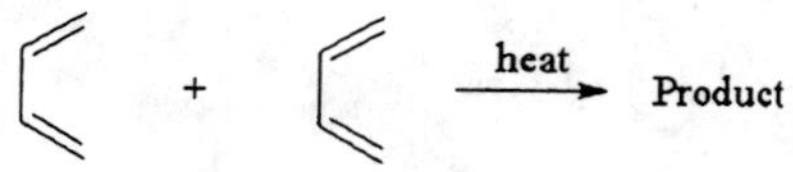

\+ heat → Product

Section: 8

Chapter 7: Reactions of Dienes

29. Cyclopentadiene can react with itself in a Diels-Alder reaction. Draw the endo and exo products.

    Answer:

    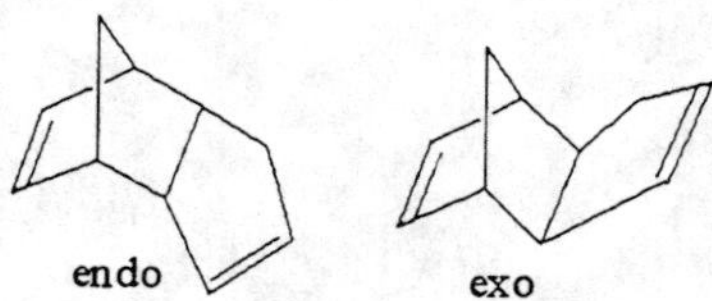

    Section: 8

30. Rank the following dienes in order of increasing stability: trans-1,3-pentadiene, cis-1,3-pentadiene, 1,4-pentadiene, and 1,2-pentadiene.

    Answer:
    1, 2-pentadiene < 1, 4-pentadiene < cis-1,3-pentadiene < trans-1,3-pentadiene

    Section: 3

31. Among a series of isomeric trienes, the more negative the ΔH° of the hydrogenation reaction of a given triene, the _______ stable it is relative to the others in the isomeric series.

    Answer: less Section: 3

32. Consider the hydrogenation reaction of each compound listed and rank the compounds in order of increasing ΔH° of this reaction. The most negative ΔH° should be listed first.

    cis-2-pentene, 2,3-pentadiene, and trans-1,3-pentadiene

    Answer: 2,3-pentadiene < trans-1,3-pentadiene < cis-2-pentene Section: 3

33. Consider the hydrogenation reaction of each compound listed and rank the compounds in order of increasing ΔH° of this reaction. The most negative ΔH° should be listed first.

    2,5-dimethyl-1,3-cycloheptadiene,
    1,4-dimethyl-1,3-cycloheptadiene, and
    3,6-dimethyl-1,4-cycloheptadiene

    Answer:
    3,6-dimethyl-1,4-cycloheptadiene <
    2,5-dimethyl-1,3-cycloheptadiene <
    1,4-dimethyl-1,3-cycloheptadiene

    Section: 3

MULTIPLE CHOICE

34. What is the hybridization of the central carbon of allene (1,2-propadiene)?

    a. sp
    b. sp^2
    c. sp^3
    d. p
    e. none of the above

    Answer: a Section: 3

35. What descriptive term is applied to the type of diene represented by 2,4-hexadiene?

    a. conjugated diene
    b. cumulated diene
    c. isolated diene
    d. alkynyl diene
    e. none of the above

    Answer: a

36. What descriptive term is applied to the type of diene represented by 1,5-octadiene?

    a. conjugated diene
    b. cumulated diene
    c. isolated diene
    d. alkynyl diene
    e. none of the above

    Answer: c

SHORT ANSWER

37. Give a representation of the highest occupied π MO of 1,3-butadiene in its ground state.

    Answer:

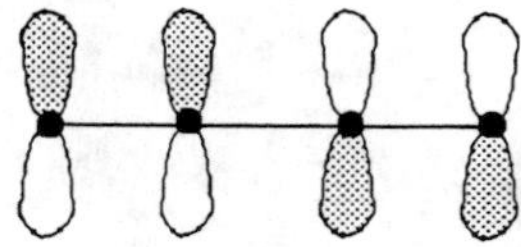

    Section: 3

38. Give a representation of the lowest unoccupied π MO of 1,3-butadiene in its ground state.

Answer:

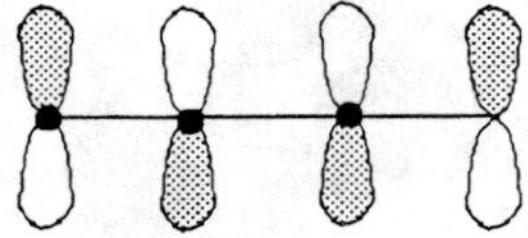

Section: 3

MULTIPLE CHOICE

39. How many nodes, other than the node coincident with the molecular plane, are found in the highest energy π MO of 1,3-butadiene?

a. 0
b. 1
c. 2
d. 3
e. none of the above

Answer: d Section: 3

SHORT ANSWER

40. Draw (Z)-1,3-hexadiene in its s-trans conformation.

Answer:

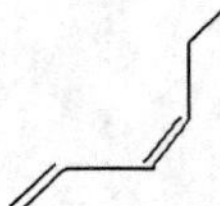

Section: 2

41. Provide the structure of the major product which results from 1,2-addition of HBr to the diene shown below.

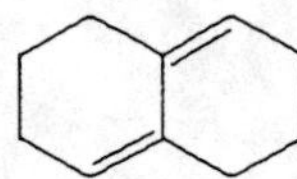

Answer:

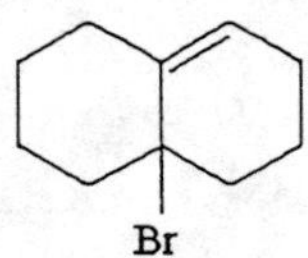

Section: 6

42. Provide the structure of the major product which results from 1,4-addition of Br_2 to the diene shown below.

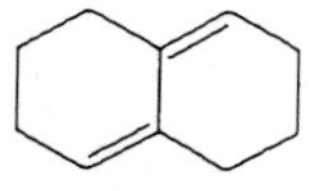

Answer:

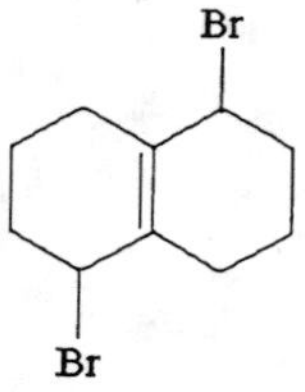

Br

Br

Section: 6

43. Draw the structure of the major product which results when the diene shown is treated with HBr at 40° C.

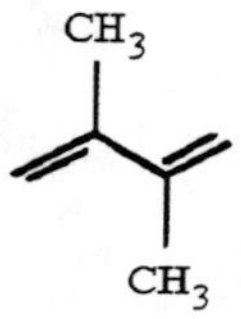

CH_3

CH_3

Answer:

CH_3

CH_2Br

CH_3

CH_3

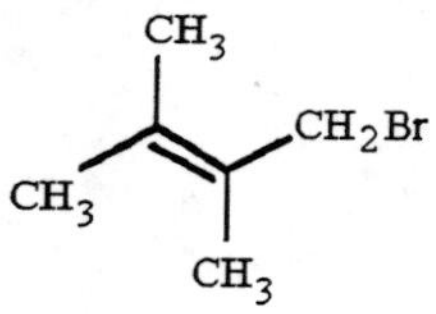

Section: 6

44. Draw the structure of the major product which results when the diene shown is treated with HBr at -80° C.

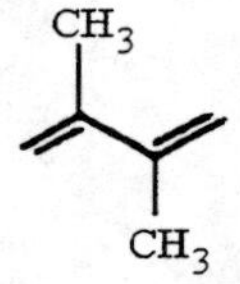

CH_3

CH_3

Answer:

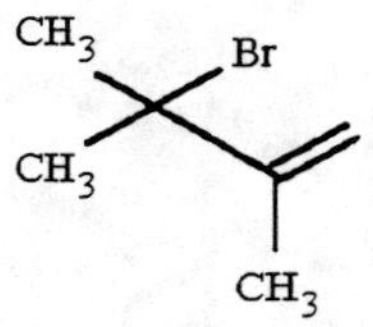

CH_3 Br

CH_3

CH_3

Section: 6

45. In the addition of HBr to conjugated dienes, is the product which results from 1,2-addition or that which results from 1,4-addition typically the product of kinetic control?

Answer: The product of kinetic control typically results from 1,2-addition.

Section: 7

MULTIPLE CHOICE

46. Which of the following terms best describes a Diels-Alder reaction?

a. a [4+2] cycloaddition
b. a [2+2] cycloaddition
c. a sigmatropic rearrangement
d. a 1,3-dipolar cycloaddition
e. a substitution reaction

Answer: a Section: 8

47. When 1,3-butadiene reacts with $CH_2{=}CHCN$, which of the terms below best describes the product mixture?

a. a mixture of two diastereomers
b. a single compound
c. a racemic mixture
d. optically active
e. a mixture of bicyclic compounds

Answer: c Section: 8

48. The Diels-Alder reaction is a concerted reaction; this means:

a. a mixture of endo and exo products is formed.
b. all bond making and bond breaking occurs simultaneously.
c. the products contain rings.
d. the reaction follows Markovnikov's rule.
e. the reaction is highly endothermic.

Answer: b Section: 8

SHORT ANSWER

49. Provide the structure of the major organic product in the following reaction.

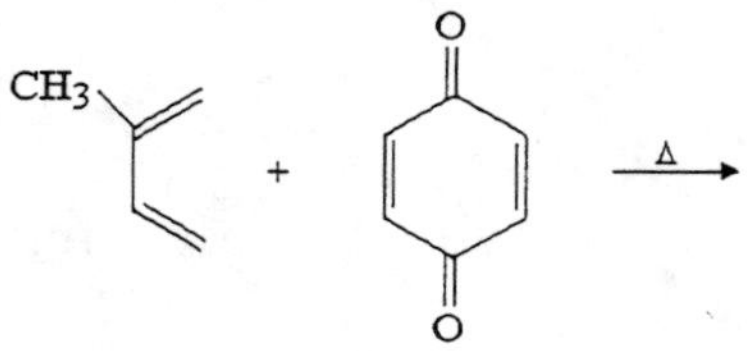

CH_3 + O O Δ

Answer:

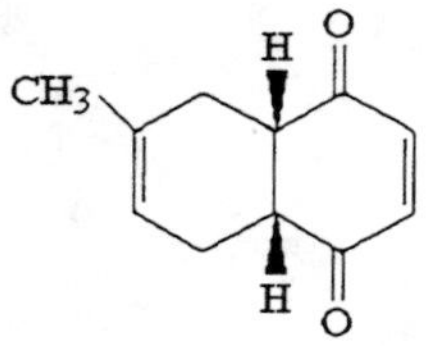

CH_3 H O H O

Section: 8

50. Provide the structure of the major organic product in the following reaction.

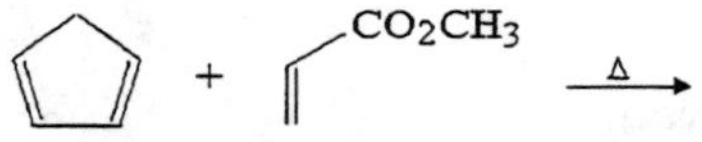

+ CO_2CH_3 Δ

Answer:

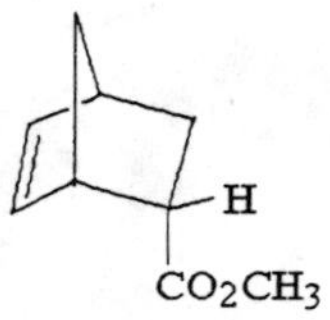

H CO_2CH_3

Section: 8

51. Provide the structure of the major organic product in the following reaction.

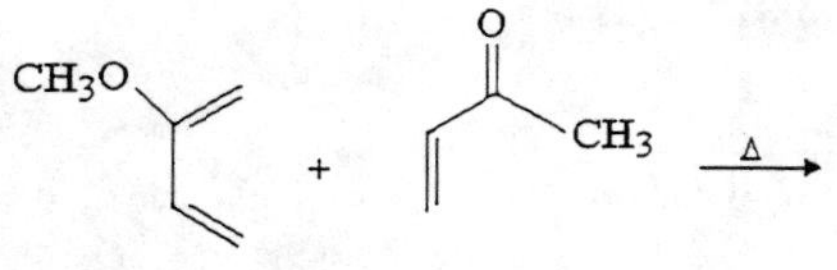

Answer:

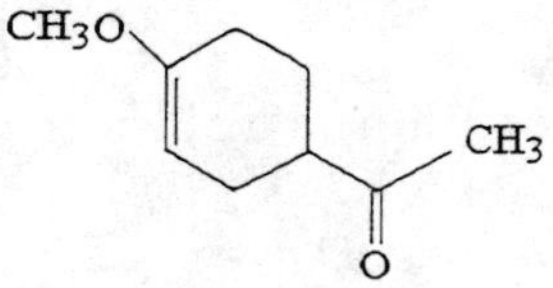

Section: 8

52. Provide the structure of the major organic product in the following reaction.

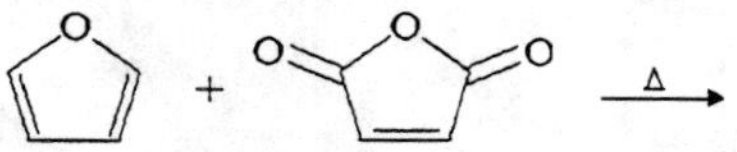

Answer:

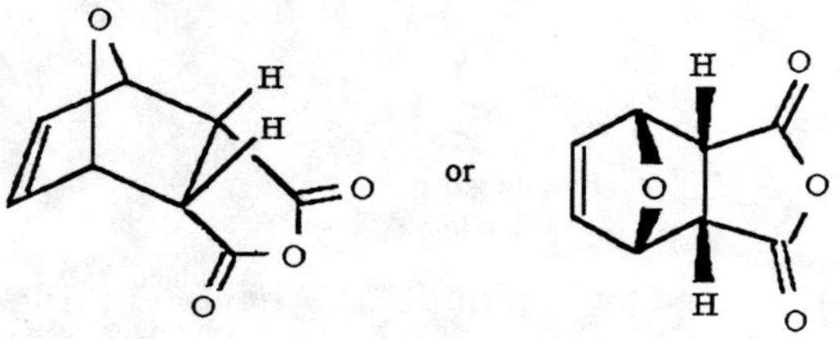

Section: 8

53. Provide the structure of the major organic product in the following reaction.

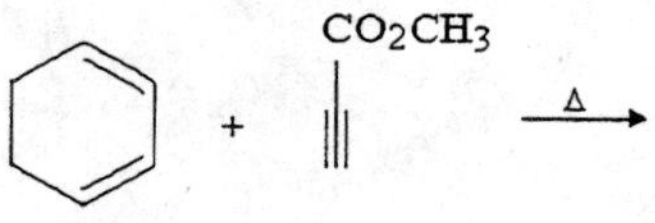

Answer:

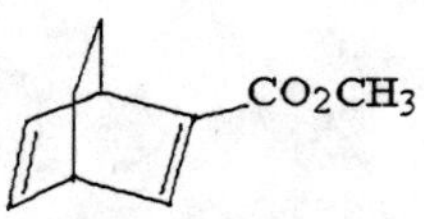

Section: 8

54. Provide the structure of the major organic product in the following reaction.

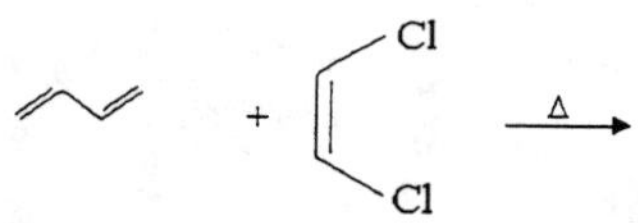

Cl
+
Δ
Cl

Answer:

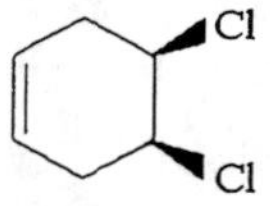

Cl
Cl

Section: 8

55. What diene and dienophile would react to give the product below?

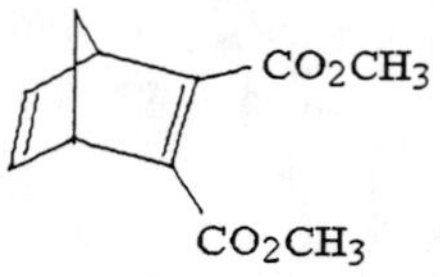

CO_2CH_3
CO_2CH_3

Answer:

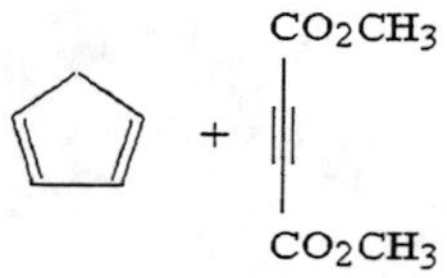

CO_2CH_3
+
CO_2CH_3

Section: 8

56. What diene and dienophile would react to give the product below?

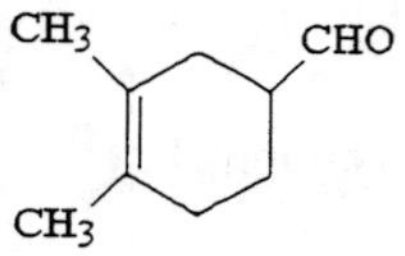

CH_3 CHO
CH_3

Answer:

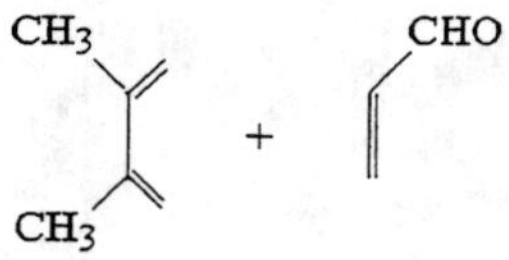

CH_3 CHO
+
CH_3

Section: 8

57. What diene and dienophile would react to give the product below?

Answer:

Section: 8

58. What diene and dienophile would react to give the product below?

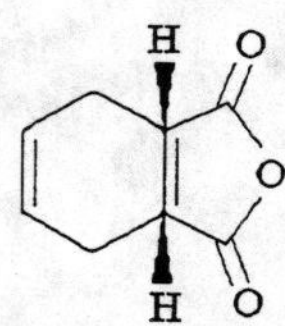

Answer:

Section: 8

MULTIPLE CHOICE

59. Which of the following compounds is the most reactive dienophile in a Diels-Alder reaction with 1,3-butadiene?

a. $CH_2{=}CHOCH_3$
b. $CH_2{=}CHCHO$
c. $CH_3CH{=}CHCH_3$
d. $(CH_3)_2C{=}CH_2$
e. $CH_2{=}CH_2$

Answer: b Section: 8

SHORT ANSWER

60. Why does the (E,E)-isomer of 2,4-hexadiene react much more rapidly with dienophiles than does the (Z,Z)-isomer?

    Answer:
    The s-cis conformation, which is required for participation in a Diels-Alder reaction, is hindered by a severe CH_3/CH_3 steric interaction in the (Z,Z)-isomer.

    Section: 8

Chapter 8: Reactions of Alkanes, Radicals

MULTIPLE CHOICE

1. The major type of reactions that alkanes undergo is:

   a. electrophilic substitution reactions.
   b. electrophilic addition reactions.
   c. free radical substitution reactions.
   d. free radical addition reactions.
   e. nucleophilic substitution reactions.

   Answer: c Section: 2

2. Which of the following is the best description of propane, $CH_3CH_2CH_3$, at room temperature?

   a. Liquid, soluble in H_2O
   b. Gas, soluble in gasoline
   c. Liquid, soluble in gasoline
   d. Gas, soluble in water
   e. Solid, soluble in gasoline

   Answer: b

3. Which of the following is the <u>most</u> stable radical?

   a. $CH_2{=}\dot{C}H$
   b. $CH_2{=}CHCH_2\dot{C}H_2$
   c. $CH_3\dot{C}HCH_3$
   d. $CH_3\dot{C}H_2$
   e. $CH_3\dot{C}H{-}C_6H_5$ (benzene ring)

   Answer: e Section: 5

4. Which of the following is the initiation step for the monobromination of cyclohexane?

a. 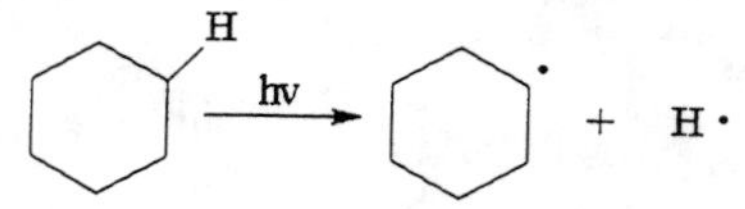 H hv → · + H·

b. 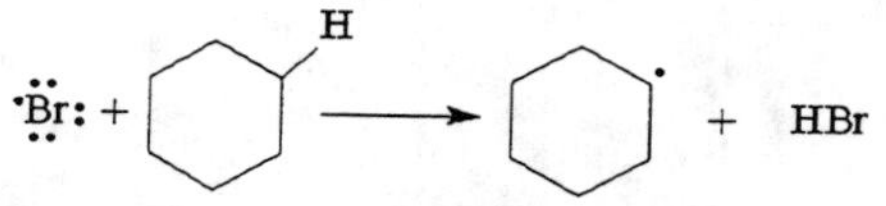 ·Br: + H → · + HBr

c. 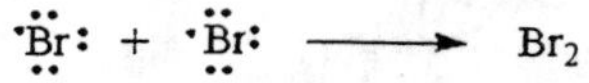 ·Br: + ·Br: → Br_2

d. 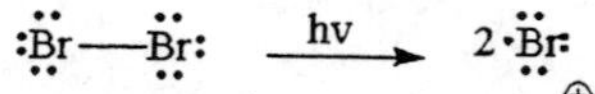 :Br—Br: hv → 2·Br:

e. 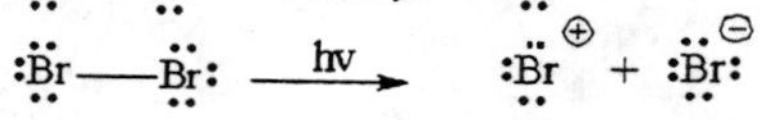 :Br—Br: hv → :Br⊕ + :Br:⊖

Answer: d Section: 2

5. Which of the following is the rate-determining step for the monobromination of cyclohexane?

a. 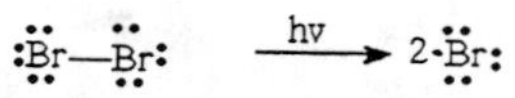 :Br—Br: hv → 2·Br:

b. 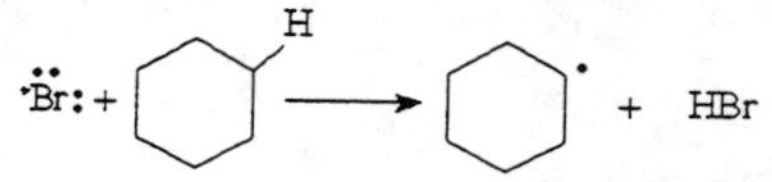 ·Br: + H → · + HBr

c.  · + :Br· → Br

d. 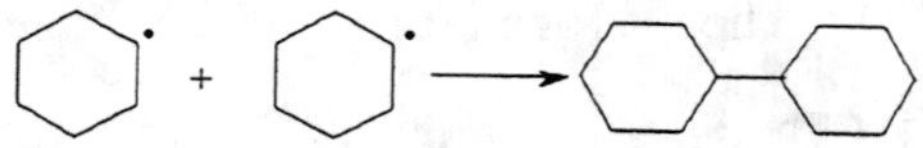 · + · →

e. 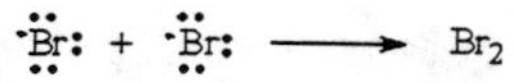 ·Br: + ·Br: → Br_2

Answer: b Section: 2

6. Calculate the overall ΔH° for the reaction given below:

$$(CH_3)_3\text{-C-H} + \text{Cl-Cl} \rightarrow (CH_3)_3\text{-Cl} + \text{H-Cl}$$

(ΔH° =91) (ΔH° =58) (ΔH° =78.5) (ΔH° =103)

a. + 181.5 kcal/mole
b. + 58.0 kcal/mole
c. + 32.5 kcal/mole
d. -32.5 kcal/mole
e. -57.5 kcal/mole

Answer: d Section: 2

7. How many products are formed from the monochlorination of ethylcyclohexane? Ignore stereoisomers.

   a. 6
   b. 8
   c. 5
   d. 9
   e. 11

   CH_2CH_3

   Answer: a Section: 2

8. Which of the following are the products obtained from the disproportionation of the propyl radical?

   $CH_3CH_2\dot{C}H_2$

   a. $CH_3\dot{C}HCH_3$ + $CH_2{=}CHCH_3$
   b. $CH_3CH_2CH_3$ + $CH_3\dot{C}HCH_3$
   c. $CH_3CH_2CH_2CH_2CH_2CH_3$
   d. $CH_3CH_2CH_3$ + $CH_2{=}CHCH_3$
   e. $CH_2{=}C{=}CH_2$ + $CH_3CH_2CH_3$

   Answer: d Section: 2

9. Which of the following is true about the halogenation of methane, CH_4?

   a. The activation energy for F• + CH_4 → •CH_3 + HF is larger than any other halogen.
   b. O_2 slows these reactions by inhibiting the termination step.
   c. The activation energy for I• + CH_4 → •CH_3 + HI is relatively small.
   d. The reaction must take place in the liquid state.
   e. I_2 is unreactive because the activation energy for I• + CH_4 → •CH_3 + HI is relatively large.

   Answer: e Section: 4

10. When (R)-2-bromobutane reacts with Cl_2/hv, which of the following is true?

I: CH_3 (top); Br— —H; Cl— —H; CH_3 (bottom)

II: CH_3 (top); Br— —H; H— — Cl; CH_3 (bottom)

a. Only I is formed.
b. Only II is formed.
c. I and II are formed in equal amounts.
d. I and II are formed in unequal amounts.
e. I and II could never form under these conditions.

Answer: d Section: 6

11. When (R)-2-bromobutane reacts with Cl_2/hv, which of the following is true?

I: CH_2Cl (top); Br— —H; CH_2CH_3 (bottom)

II: CH_2Cl (top); H— —Br; CH_2CH_3 (bottom)

a. I is formed.
b. II is formed.
c. I and II are formed in equal amounts.
d. I and II are formed in unequal amounts.
e. I and II could never form under these conditions.

Answer: a Section: 6

12. When (R)-2-chlorobutane reacts with Br_2/hv, which of the following is true?

CH_3 | CH_3
Br—|—Cl | Cl—|—Br
CH_2CH_3 | CH_2CH_3

I II

a. Only I is formed.
b. Only II is formed.
c. Both I and II are formed in equal amounts.
d. Both I and II are formed in unequal amounts.
e. Both I and II could never form under these conditions.

Answer: d Section: 6

13. How many products can be isolated when (S)-2-chlorobutane reacts with Cl_2/hv?

CH_3CH_2
Cl—|—H
CH_3

(S)-2-chlorobutane

a. 1
b. 2
c. 3
d. 4
e. 5

Answer: e Section: 2

14. What is the major product of the following reaction?

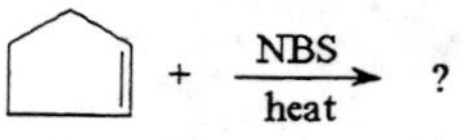
+
NBS
heat
?

a. 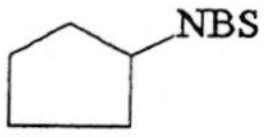
NBS

b. 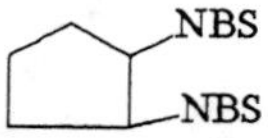
NBS
NBS

c. 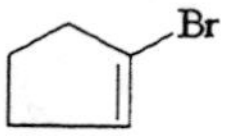
Br

d. 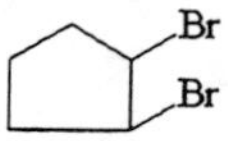
Br
Br

e. 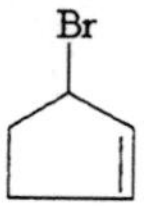
Br

Answer: e Section: 5

15. What is the major product of the following reaction?

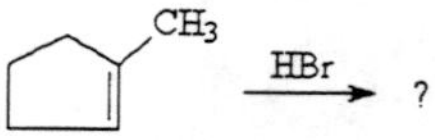
CH3
HBr
?

a. 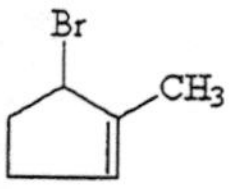
Br
CH3

b. 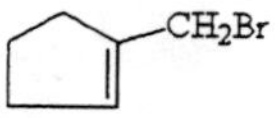
CH2Br

c. 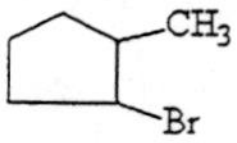
CH3
Br

d. 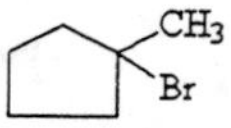
CH3
Br

e. 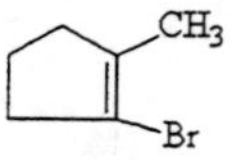
CH3
Br

Answer: d Section: 5

16. What is the major product of the following reaction?

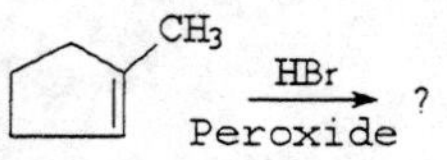

1-methylcyclopentene (CH_3) $\xrightarrow[\text{Peroxide}]{\text{HBr}}$?

a. 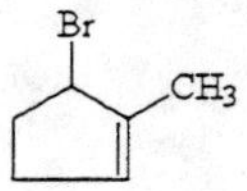 Br, CH_3

b. 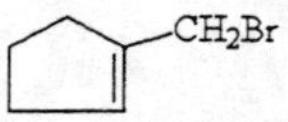 CH_2Br

c. 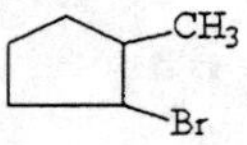 CH_3, Br

d. 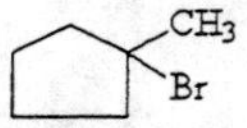 CH_3, Br

e. 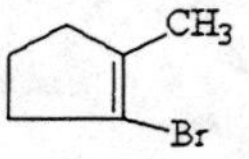 CH_3, Br

Answer: c Section: 5

17. What is the major product of the following reaction?

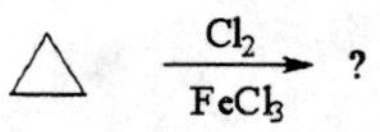

cyclopropane $\xrightarrow[\text{FeCl}_3]{\text{Cl}_2}$?

a. 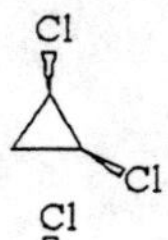 Cl, Cl

b. 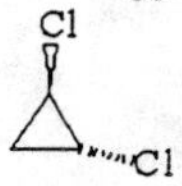 Cl, Cl

c. 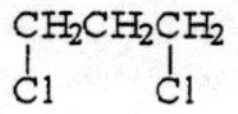 $\underset{\text{Cl}}{\text{CH}_2}\text{CH}_2\underset{\text{Cl}}{\text{CH}_2}$

d.  Cl

e. 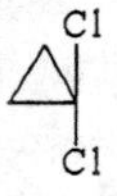 Cl, Cl

Answer: c Section: 7

18. Which of the following statement(s) for dichlorocyclohexane is/are correct?

I. cis 1,2- is more stable than trans 1,2.
II. cis 1,3- is more stable than trans 1,3.
III. cis 1,4- is more stable than trans 1,4.

a. I
b. II
c. III
d. I and II
e. I and III

Answer: b

19. Which of the following is not an intermediate in the reaction of

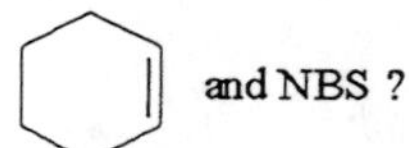

and NBS ?

a. Br_2
b. Br·
c. 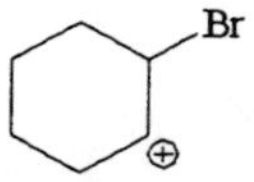 Br ⊕
d. 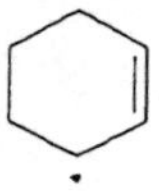
e. HBr

Answer: c Section: 5

SHORT ANSWER

20. Complete the following tree of reactions by supplying the missing reagents:

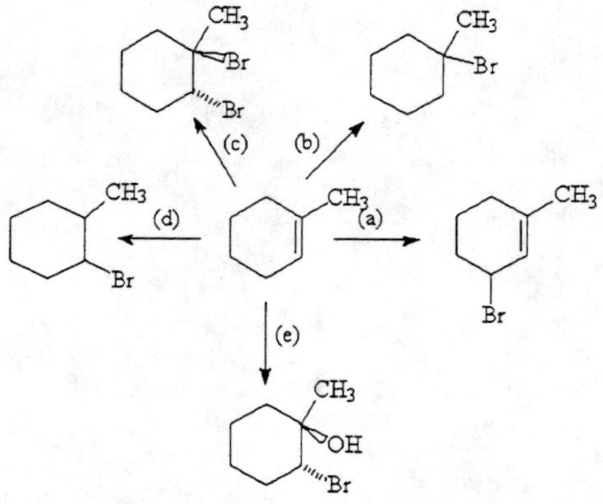

Answer:
a. NBS/heat
b. HBr
c. Br_2/CCl_4
d. HBr/Peroxide
e. Br_2/H_2O

Section: 5

21. An alkane with the molecular formula C_5H_{10} forms only one monochlorinated product when heated with Cl_2/hv. Give the structure and the IUPAC name for this alkane.

Answer:

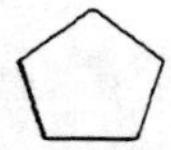

cyclopentane

Section: 2

22. How many monochlorinated products would be obtained from 2-methylbutane? Show the structures and give their IUPAC names.

Answer:

2-chloro-2-methylbutane 1-chloro-2-methylbutane

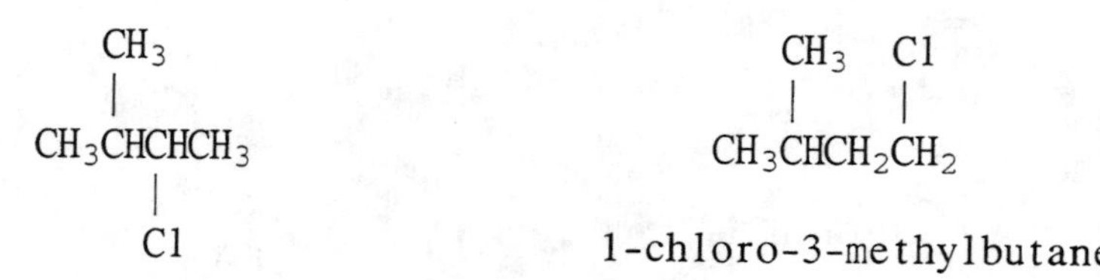

1-chloro-3-methylbutane

2-chloro-3-methylbutane

Section: 2

23. Calculate the percent yield obtained for each product in problem 22. Assume that the relative ease of hydrogen abstraction in the chlorination process is 5 for 3° ; 3.8 for 2° ; and 1 for 1° hydrogens.

Answer:

2-chloro-2-methylbutane
1 x 5 = 5 ; %yield= 5 x 100/ 21.6 = 23.1%

1-chloro-2-methylbutane
6 x 1 = 6 ; %yield= 6 x 100/ 21.6 = 27.8%

2-chloro-3-methylbutane
2 x 3.8 = 7.6 ; %yield= 7.6 x 100/ 21.6 = 35.2%

1-chloro-3-methylbutane
3 x 1 = 3 ; %yield= 3 x 100/ 21.6 = 13.9%

Section: 3

24. Give the structure of the free radical intermediate for each product in problem 22.

Answer:

$CH_3C(CH_3)(Cl)CH_2CH_3 \quad = \quad CH_3\dot{C}(CH_3)CH_2CH_3$

$ClCH_2CH(CH_3)CH_2CH_3 \quad = \quad \dot{C}H_2CH(CH_3)CH_2CH_3$

$CH_3CH(CH_3)CH(Cl)CH_3 \quad = \quad CH_3CH(CH_3)\dot{C}HCH_3$

$CH_3CH(CH_3)CH_2CH_2Cl \quad = \quad CH_3CH(CH_3)CH_2\dot{C}H_2$

Section: 3

25. Consider the following monobromination reaction, then answer the following questions.

$$CH_3CH_2CH_2CH_3 + \xrightarrow[hv]{Br_2} \text{Products}$$

a. Give the structures and the IUPAC names for the products.
b. Give the common names for the products.
c. Calculate $\Delta H°$ for the overall reaction using the following data:

$CH_3CH_2CH(H)CH_3$ (H on C-2)	$\Delta H°$ = 100 kcal/mole
$CH_3CH_2CH_2CH_2$–H	$\Delta H°$ = 105 kcal/mole
Br—Br	$\Delta H°$ = 46 kcal/mole
$CH_3CH_2CH(Br)CH_3$ (Br on C-2)	$\Delta H°$ = 70 kcal/mole
$CH_3CH_2CH_2CH_2Br$	$\Delta H°$ = 74 kcal/mole
H—Br	$\Delta H°$ = 87.5 kcal/mole

d. Calculate the percent yield for each product. (relative rate of abstraction of 3° hydrogen is 1600; 2° is 82; and 1° is 1.)

e. Propose a step-by-step mechanism for the <u>major</u> product only.

f. Draw a schematic potential energy diagram for the rate-determining step (RDS) only.

g. Does the transition state of the RDS resemble more closely the reactants or the products?

h. Would the value of the activation energy be different for different alkanes? Explain.

i. Would the reaction slow down or speed up if I_2 is used instead of Br_2? Explain.

Answer:

a. $CH_3CHCH_2CH_3$ with Br on C-2 (2-bromobutane); $CH_3CH_2CH_2CH_2$ with Br on C-1 (1-bromobutane)

```
CH3CHCH2CH3          CH3CH2CH2CH2
   |                             |
   Br                            Br
```

2-bromobutane   1-bromobutane

b.

```
CH3CHCH2CH3          CH3CH2CH2CH2
   |                             |
   Br                            Br
```

sec-butyl bromide   butyl bromide

c. $\Delta H°$ for 2-bromobutane formation = (100 + 46) - (70 + 87.5) = -11.5 kcal/mole
$\Delta H°$ for 1-bromobutane formation = (105 + 46) - (74 + 87.5) = -10.5 kcal/mole

d. 2-bromobutane = 4 x 82 = 328 %yield = 328 x 100/334 = 98.2%
1-bromobutane = 6 x 1 = 6 %yield = 6 x 100/334 = 1.8%

e. See graph.

f. See graph.

g. Product, since the transition state is closer to the products than it is to the reactants (Hammond Principle).

h. Yes, since each alkane has its own bond dissociation energy.

i. The reaction would slow down if I_2 is used. This is true because the activation energy for $I\cdot + CH_3CH_2CH_2CH_3$ is larger than in $Br\cdot + CH_3CH_2CH_2CH_3$.

e.

i. Br—Br $\xrightarrow{h\nu}$ 2•Br

ii. •Br + $CH_3CHCH_2CH_3$ (H) ⟶ HBr + $CH_3\dot{C}HCH_2CH_3$

iii. $CH_3\dot{C}HCH_2CH_3$ + Br—Br ⟶ $CH_3CH(Br)CH_2CH_3$ + •Br

iv. •Br + •Br ⟶ Br_2

(One of many possibilities)

f.

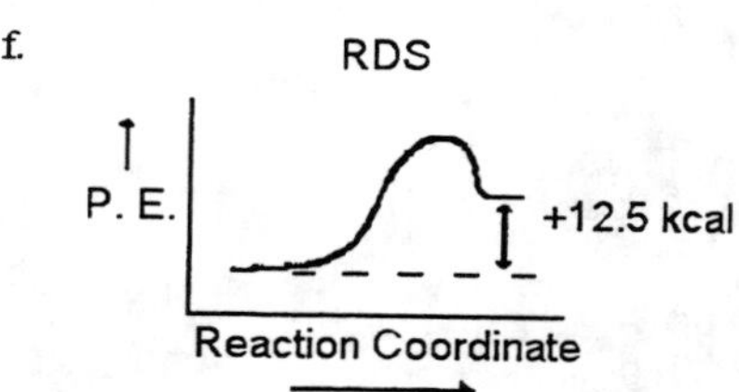

Section: 2

26. Write an equation to describe the initiation step in the chlorination of methane.

Answer: Cl-Cl + photon ($h\upsilon$) ⟶ 2 Cl• Section: 2

27. When light is shown on a mixture of chlorine and chloromethane, carbon tetrachloride is one of the components of the final reaction mixture. Propose a series of mechanistic steps which explain this observation.

Answer:

CH_3Cl + Cl• ⟶ $\cdot CH_2Cl$ + HCl
$\cdot CH_2Cl$ + Cl-Cl ⟶ Ch_2Cl_2 + Cl•
CH_2Cl_2 + Cl• ⟶ $\cdot CHCl_2$ + HCl
$\cdot CHCl_2$ + Cl-Cl ⟶ $CHCl_3$ + Cl•
$CHCl_3$ + Cl• ⟶ $\cdot CCl_3$ + HCl
$\cdot CCl_3$ + Cl-Cl ⟶ CCl_4 + Cl•

Section: 2

28. Species with unpaired electrons are called ________.

Answer: radicals <u>or</u> free radicals Section: 2

29. Chlorination of methane can result in a mixture of chlorinated products. What experimental conditions should be used to favor the production of chloromethane over the other chlorinated products?

Answer: Make sure the molar ratio of methane to chlorine is relatively large.

Section: 2

30. Write a detailed, stepwise mechanism for the following reaction.

$$C_6H_5-CH_3 + Cl_2 \xrightarrow{\Delta} C_6H_5-CH_2Cl + HCl$$

Answer:

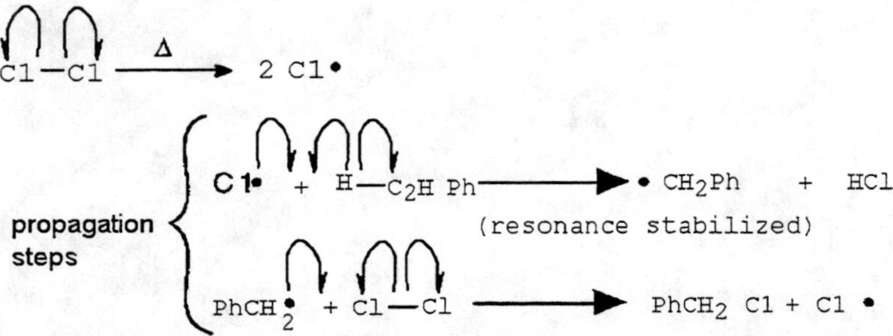

Section: 2

MULTIPLE CHOICE

31. Consider the bond dissociation energies listed below in kcal/mol.

CH_3-Br	70
CH_3CH_2-Br	68
$(CH_3)_2$CH-Br	68
$(CH_3)_3$C-Br	65

These data show that the carbon-bromine bond is <u>weakest</u> when bromine is bound to a __________.

a. methyl carbon
b. preimary carbon
c. secondary carbon
d. tertiary carbon
e. quanternary carbon

Answer: d Section: 2

32. Given the bond dissociation energies below (in kcal/mol), estimate the $\Delta H°$ for the propagation step $(CH_3)_2CH\cdot + Cl_2 \longrightarrow (CH_3)_2CHCl + Cl\cdot$.

$CH_3CH_2CH_2$-H	98
$(CH_3)_2CH$-H	95
Cl-Cl	58
H-Cl	103
$CH_3CH_2CH_2$-Cl	81
$(CH_3)_2CH$-Cl	80

a. -22 kcal/mol
b. +22 kcal/mol
c. -40 kcal/mol
d. +45 kcal/mol
e. -45 kcal/mol

Answer: a Section: 2

SHORT ANSWER

33. Given the bond dissociation energies below (in kcal/mol), calculate the overall $\Delta H°$ for the following reaction:

$(CH_3)_3CH + Br_2 \longrightarrow (CH_3)_3CBr + HBr$

$(CH_3)_3C$-H	91
$(CH_3)_3C$-Br	65
Br-Br	46
H-Br	88
CH_3-Br	70

Answer: -16 kcal/mol Section: 2

34. Consider the reaction: $CH_3CH_2\cdot + Br_2 \longrightarrow Ch_3CH_2Br + Br\cdot$. Given that this reaction has an activation energy of +6 kcal/mol and a $\Delta H°$ of -22 kcal/mol, sketch a reaction-energy profile for this reaction. Label the axes and show E_a and $\Delta H°$ on your drawing.

Answer:

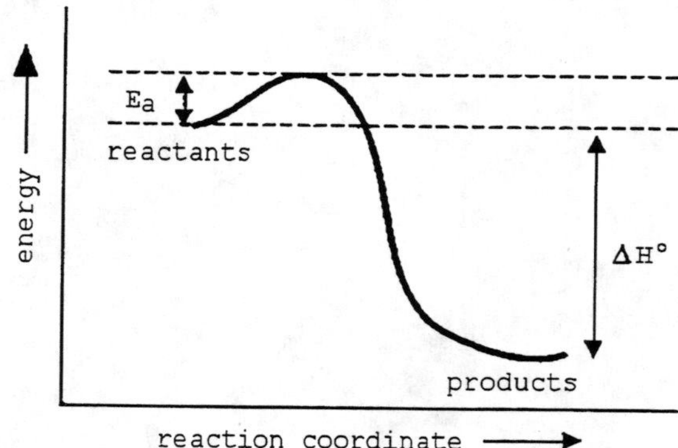

Section: 2

Chapter 8: Reactions of Alkanes, Radicals

MULTIPLE CHOICE

35. The major monobominated product which results when methylcyclohexane is subjected to free radical bromination is:

a. a promary bromide
b. a secondary bromide
c. a tertiary bromide
d. a quaternary bromide
e. bromomethane

Answer: c Section: 4

36. Which of the halogens below undergoes free radical halogenation with ethane most rapidly?

a. fluroine
b. chlorine
c. iodine
d. bromine
e. pyridine

Answer: a Section: 4

37. Rank the free radicals (I-III) shown below in order of <u>decreasing</u> stability (ie, from most stable to least stable).

$\cdot CH_2CH_2CH(CH_3)_2$ I

$CH_3CH_2\dot{C}(CH_3)_2$ II

$CH_3\dot{C}HCH(CH_3)_2$ III

a. I > III > II
b. II > III > I
c. I > II > III
d. II > I > III
e. III > II > I

Answer: b Section: 3

SHORT ANSWER

38. Write the structures of all of the monobromination products of 1,1,3,3-tetramethylcyclobutane.

Answer:

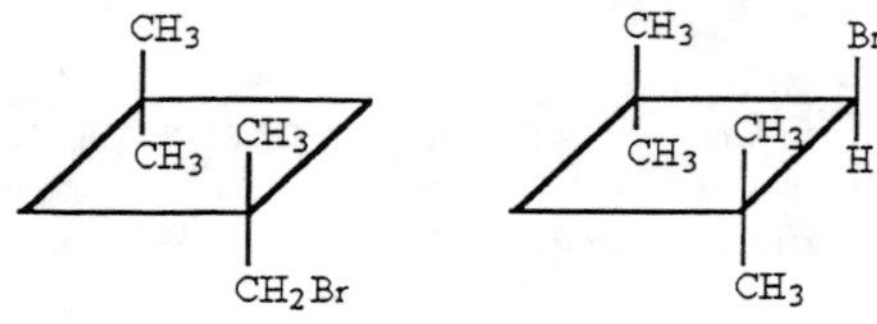

Section: 7

39. When 1,1,3,3-tetramethylcyclorbutane is brominated at 125° C, the relative reactivity of the 1° :2° :3° hydrogens is approximately 1:82:1600. Estimate the amount of each monobromination product.

Answer: primary bromide, 3.5%; secondary bromide, 96.5% Section: 3

40. What is the relative reactivity of 2° vs 1° hydrogens in the free radical bromination of n-butane if the ratio of 1-bromobutane to 2-bromobutane formed is 7:93?

Answer: The 2° hydrogens are 20 times more reactive than the 1° ones.

Section: 3

41. What C_5H_{12} isomer will give only a single monochloration product?

Answer: $(CH_3)_4C$ <u>or</u> neopentane <u>or</u> 2, 2-dimethylpropane Section: 2

42. Use the Hammond Postulate to explain why free radical brominations are more selective than free radical chlorinations.

Answer:
The first propagation step in free radical bromination is endothermic while the analogous step in free radical chlorination is exothermic. From the Hammond Postulate, this means that the transition state for the bromination is product-like (ie, radical-like) while the transition state for the chlorination is reactant-like. The product-like transition state for bromination has the C-H bond nearly broken and a great deal of radical character on the carbon atom. The energy of this transition state reflects most of the energy difference of the radical products. This is not true in the chlorination case where the transition state possesses little radical character.

Section: 4

43. List the following radicals in order of increasing stability (ie, from least stable to most stable).

$(CH_3)_3C\cdot$, $CH_2{=}CHCH_2\cdot$, $CH_3CH_2\cdot$, $CH_3\cdot$, $(CH_3)_2CH\cdot$

Answer: $CH_3\cdot < CH_3CH_2\cdot < (CH_3)_2CH\cdot < (CH_3)_3C\cdot < CH_2{=}CHCH_2\cdot$

Section: 5

44. When Br• reacts with 1-butene ($CH_3CH_2CH{=}CH_2$), the hydrogen atom which is preferentially abstracted is the one which produces a resonance stabilized radical. Draw the major resonance contributing forms of this radical.

Answer:

$CH_3\dot{C}HCH=CH_2 \longleftrightarrow CH_3CH=CH\dot{C}H_2$

Section: 5

MULTIPLE CHOICE

45. During the free radical chlorination of methane, which of the following reactions has the <u>lowest</u> collision frequency?

a. $Cl\cdot + \cdot CH_3 \longrightarrow CH_3Cl$
b. $Cl\cdot + Cl_2 \longrightarrow Cl_2 + Cl\cdot$
c. $\cdot CH_3 + CH_4 \longrightarrow CH_4 + \cdot CH_3$
d. $\cdot CH_3 + Cl_2 \longrightarrow CH_3Cl + Cl\cdot$

Answer: a Section: 2

46. The reaction $Br_2 + CH_3Br \longrightarrow CH_2Br_2 + HBr$ was carried out. Which of the following mechanism steps is both productive and relatively likely to occur?

a. $Br\cdot + \cdot CH_2Br \longrightarrow CH_2Br_2$
b. $Br\cdot + \cdot CH_3 \longrightarrow CH_3Br$
c. $Br\cdot + Br_2 \longrightarrow Br_2 + Br\cdot$
d. $Br\cdot + CH_3Br \longrightarrow HBr + \cdot CH_2Br$

Answer: d Section: 2

47. The reation Br_2 + CH_3Br ⟶ CH_2Br_2 + HBr was carried out. Which of the following mechanism steps is productive, but relatively unlikely to occur?

a. Br• + CH_3Br ⟶ HBr + •CH_2Br
b. Br• + •CH_2Br ⟶ CH_2Br_2
c. Br• + Br_2 ⟶ Br_2 + Br•
d. Br• + •CH_3 ⟶ CH_3Br

Answer: b Section: 2

48. Which of the following reactions is a termination step in the free radical chlorination of methane?

a. Cl_2 + Cl• ⟶ Cl• + Cl_2
b. Cl_2 ⟶ 2 Cl•
c. •CH_3 + Cl• ⟶ CH_3Cl
d. CH_4 + Cl• ⟶ HCl + •CH_3

Answer: c Section: 2

49. Which of the following is a chain propagation step in the free radical chlorination of methane?

a. CH_4 + Cl• ⟶ •CH_3 + HCl
b. Cl_2 ⟶ 2 Cl•
c. Cl• + •CH_3 ⟶ CH_3Cl
d. •CH_3 + CH_4 ⟶ CH_4 + •CH_3

Answer: a Section: 2

50. Which of the following most nearly describes the geometry of the methyl radical?

a. Trigonal pyramid, bond angle 109.5° .
b. Trigonal pyramid, bond angle 120° .
c. Trigonal planar, bond angle 109.5° .
d. Trigonal planar, bond angle 120° .

Answer: d Section: 6

51. How many electrons are contained in p orbital of the methyl radical?

a. Zero
b. One
c. Two
d. Three

Answer: b Section: 6

Chapter 8: Reactions of Alkanes, Radicals

SHORT ANSWER

52. Predict the major monobromination product in the following reaction.

$$(CH_3)_3CCH_2CH_3 + Br_2 \xrightarrow{h\upsilon}$$

Answer: $(CH_3)_3CCHBrCH_3$ Section: 4

53. The following molecule contains how many 1°, 2°, and 3° hydrogens?

$(CH_3)_2CHCH_2CHCH_2CH_3$ (with CH_2CH_3 on the second CH)

Answer: 12 1°, 6 2°, 2 3° Section: 3

54. Predict the major monobromination product in the following reaction.

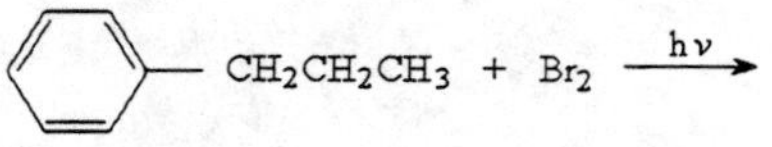

$$C_6H_5\text{—}CH_2CH_2CH_3 + Br_2 \xrightarrow{h\nu}$$

Answer:

$C_6H_5\text{—}CH(Br)CH_2CH_3$

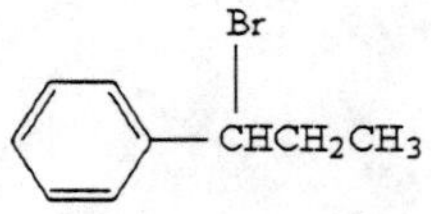

Section: 5

MULTIPLE CHOICE

55. How many distinct dichlorination products can result when isobutane is subjected to free radical chlorination?

a. 1
b. 2
c. 3
d. 4
e. 6

Answer: c Section: 2

56. What is the <u>major</u> product of the following reaction?

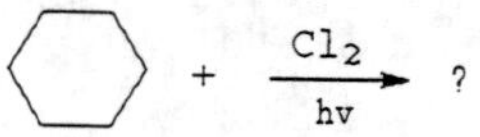

a. 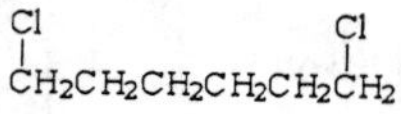

b. 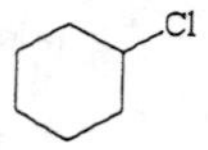

c. 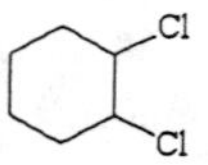

d. 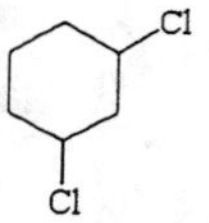

e. 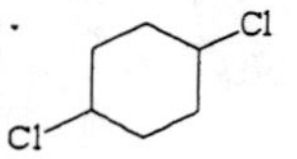

Answer: b Section: 7

SHORT ANSWER

57. An unknown sample is suspected of being either ethane or isobutane. How would you distinguish between the two alkanes?

Answer:
A knetic study of the rate of reaction with Br_2/hv between the two alkanes should reveal that the ease of abstraction of a 3° hydrogen is about a thousand times that of a 1° hydrogen.

Section: 3

Chapter 9: Reactions at an sp3 Hybridized Carbon I

MULTIPLE CHOICE

1. Which of the following are the substitution products of the reaction shown below?
$CH_3CH_2Br + {}^-OH \longrightarrow ?$

a. $CH_3CH_2Br^+H + O^-$
b. $HOCH_2CH_2Br$
c. $CH_3CH_2OH + Br^-$
d. $CH_2{=}CH_2 + Br^- + H_2O$
e. $CH_2{=}CHBr + H_2O$

Answer: c Section: 2

2. What is the nucleophile in the reaction shown below?

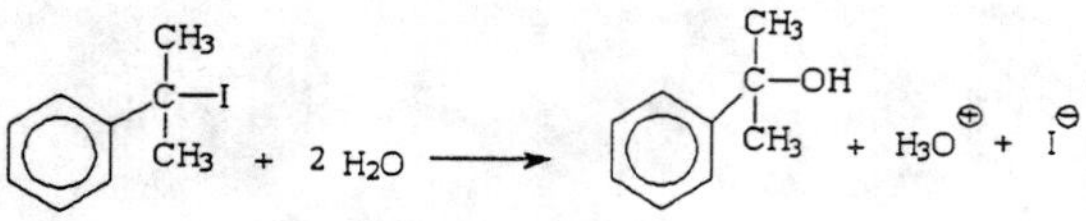

$C_6H_5C(CH_3)_2I + 2\ H_2O \longrightarrow C_6H_5C(CH_3)_2OH + H_3O^{\oplus} + I^{\ominus}$

a. 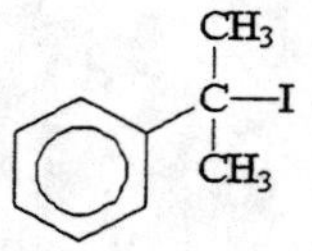 $C_6H_5C(CH_3)_2I$

b. H_2O
c. $H_3O^{\oplus}$
d. $I^{\ominus}$
e. $HO^{\ominus}$

Answer: b Section: 1

3. Which of the following is not a nucleophile?

a. CH_3NH_2
b. PH_3
c. ${}^+CH_3$
d. $CH_3O^{\ominus}$
e. $HC \equiv C:^{\ominus}$

Answer: c Section: 1

4. Which of the following alkyl halides gives the <u>slowest</u> S_N2 reaction?

a. CH_3CH_2Cl

b. 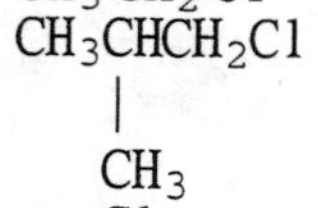 CH_3CHCH_2Cl with CH_3 on the CH (i.e., $CH_3CH(CH_3)CH_2Cl$)

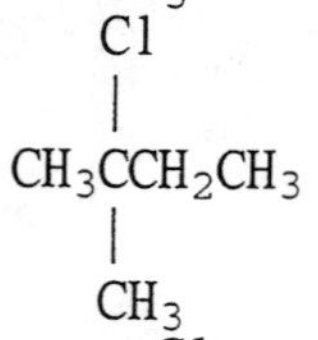

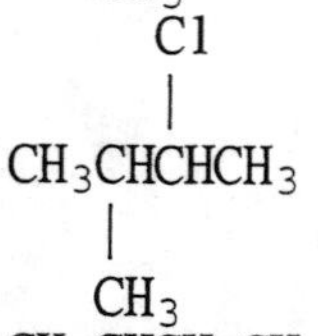

c. $CH_3CCH_2CH_3$ with Cl above and CH_3 below the central C (i.e., $CH_3CCl(CH_3)CH_2CH_3$)

d. $CH_3CHCHCH_3$ with Cl on one CH and CH_3 on the other (i.e., $CH_3CH(CH_3)CHClCH_3$)

e. $CH_3CHCH_2CH_3$ with CH_2–Cl on the CH (i.e., $CH_3CH(CH_2Cl)CH_2CH_3$)

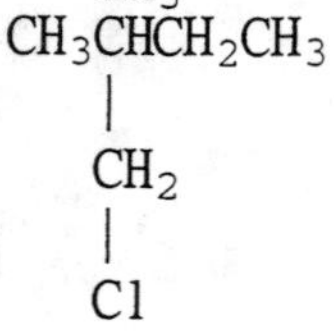

Answer: c Section: 2

5. Assuming no other changes, what is the effect of <u>doubling</u> <u>only</u> the concentration of the alkyl halide in the following S_N2 reaction?

$$CH_3Br + HO^{\ominus} \longrightarrow CH_3OH + Br^{\ominus}$$

a. No change
b. Doubles the rate
c. Triples the rate
d. Quadruples the rate
e. Rate is halved

Answer: b Section: 2

6. Assuming no other changes, what is the effect of <u>doubling</u> <u>both</u> the alkyl halide and the nucleophile concentrations in the reaction given in problem #5?

a. No change
b. Doubles the rate
c. Triples the rate
d. Quadruples the rate
e. Rate is halved

Answer: d Section: 2

7. Which of the following is the strongest nucleophile in an aqueous solution?

a. HO^-
b. F^-
c. Cl^-
d. Br^-
e. I^-

Answer: e Section: 3

8. Which of the following is the best leaving group?

a. HO^-
b. F^-
c. Cl^-
d. Br^-
e. I^-

Answer: e Section: 3

9. Which of the following species is most reactive in an S_N2 reaction?

a. $CH_3CH_2{-}Cl$
b. $CH_3CH_2{-}Br$
c. $CH_3CH_2{-}I$
d. $CH_3CH_2{-}F$
e. $CH_3CH_2{-}OH$

Answer: c Section: 3

10. Which of the following S_N2 reactions is the <u>fastest</u>?

a. CH_3CHCH_3 + HO^- ⟶ CH_3CHCH_3 + Br^-
   | |
   Br   OH
b. $CH_3CH_2CH_2I$ + HO^- ⟶ $CH_3CH_2CH_2OH$ + I^-
c. CH_3CHCH_3 + HO^- ⟶ CH_3CHCH_3 + I^-
   | |
   I   OH
d. $CH_3CH_2CH_3Br$ + HO^- ⟶ $CH_3CH_2CH_3OH$ + Br^-
e. $CH_3CH_2CH_2I$ + H_2O ⟶ $CH_3CH_2CH_2OH$ + HI

Answer: b Section: 2

11. Which of the following S_N2 reactions is the <u>slowest</u>?

a. $CH_3CH_2CH_3Br + HO^- \longrightarrow CH_3CH_2CH_3OH + Br^-$
b. $CH_3CH_2CH_2Cl + HO^- \longrightarrow CH_3CH_2CH_2OH + Cl^-$
c. $CH_3CH_2CH_2I + HO^- \longrightarrow CH_3CH_2CH_2OH + I^-$
d. $CH_3CH_2CH_3F + HO^- \longrightarrow CH_3CH_2CH_2OH + F^-$
e. $CH_3CH(F)CH_3 + HO^- \longrightarrow CH_3CH(OH)CH_3 + F^-$

Answer: e Section: 2

12. When (S)-2-bromobutane undergoes an S_N2 reaction with CH_3O^- the product is the compound shown below. What is/are the configuration(s) of the product(s) obtained from this reaction?

$$CH_3CH_2CH(OCH_3)CH_3$$

a. S only
b. R only
c. A mixture of enantiomers with more R than S
d. A mixture of enantiomers with more S than R
e. Equal mixture of R and S

Answer: b Section: 7

13. The specific rotation of optically pure (R)-sec-butyl alcohol is -13.52° . An optically pure sample of (R)-sec-butyl bromide was converted into the corresponding sec-butyl alcohol via an S_N2 reaction. What is the specific rotation of the product, assuming 100% yield?

a. -13.52°
b. Between 0° and -13.52°
c. Between 0° and +13.52°
d. +13.52°
e. zero

Answer: d Section: 7

14. Which of the following is the best method for preparing THF,

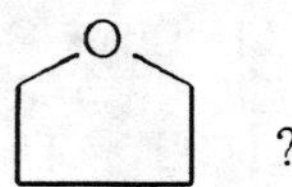

?

a. $CH_3CH_2O^{\ominus} + CH_3CH_2Br$
b. $CH_3CH_2CH_2O^{\ominus} + CH_3Br$
c. $BrCH_2CH_2CH_2CH_2O^{\ominus}$
d. $CH_3CH(O^{\ominus})CH_2CH_2CH_2Br$
e. $CH_3CH_2CH_2Br + CH_3O^{\ominus}$

Answer: c Section: 4

15. Assuming no other changes, what is the effect of <u>doubling</u> <u>only</u> the concentration of the alkyl halide in the following S_N1 reaction?

$$(CH_3)_3CBr + I^{\ominus} \longrightarrow (CH_3)_3CI + Br^{\ominus}$$

a. No change
b. Doubles the rate
c. Triples the rate
d. Quadruples the rate
e. Rate is halved

Answer: b Section: 5

16. Assuming no other changes, what is the effect of <u>doubling</u> <u>only</u> the concentration of the nucleophile in the reaction given in problem #15?

a. No change
b. Doubles the rate
c. Triples the rate
d. Quadruples the rate
e. Rate is halved

Answer: a Section: 5

17. Assuming no other changes, what is the effect of <u>doubling</u> <u>both</u> the concentration of the alkyl halide and the nucleophile in the reaction given in problem #15?

a. No change
b. Doubles the rate
c. Triples the rate
d. Quadruples the rate
e. Rate is halved

Answer: b Section: 5

18. Which of the following alkyl halides gives the fastest S_N1 reaction?

a. $CH_3CH_2CH_2Br$
b. CH_3CHCH_3
|
Br
c. CH_3CHCH_3
|
I
d. $CH_3CH_2CH_2I$
e. $CH_3CH_2CH_2Cl$

Answer: c Section: 5

19. Which of the following alkyl halides gives the slowest S_N1 reaction?

Br
|
a. $CH_3CHCHCH_3$
|
CH_3
b. $CH_3CHCH_2CH_2Br$
|
CH_3
Br
|
c. $CH_3CCH_2CH_3$
|
CH_3
d. $ClCH_2CHCH_2CH_3$
|
CH_3
e. $CH_3CHCH_2CH_3$
|
CH_2Cl

Answer: b Section: 5

20. Which of the following factors has <u>no</u> effect on the rate of S_N1 reactions?

a. The nature of the alkyl halide
b. The nature of the leaving group
c. The concentration of the alkyl halide
d. The concentration of the nucleophile
e. The value of the rate constant

Answer: d Section: 5

21. Which of the following carbocations is the most stable?

a. $CH_3\overset{\oplus}{C}H_2$

b. $CH_3\overset{\oplus}{C}HCH_3$

c. $CH_3\overset{\oplus}{C}(CH_3)_2$ (CH₃ above and below C)

d. $CH_3C(CH_3)_2\overset{\oplus}{C}H_2$

e. $CH_3CH_2\overset{\oplus}{C}H_2$

Answer: c Section: 5

22. Which of the following carbocations is the least stable?

a. $CH_3\overset{\oplus}{C}H_2$

b. $CH_3\overset{\oplus}{C}HCH_3$

c. $CH_3\overset{\oplus}{C}(CH_3)_2$

d. $CH_3C(CH_3)_2\overset{\oplus}{C}H_2$

e. $CH_3CH_2\overset{\oplus}{C}H_2$

Answer: a Section: 5

23. Which of the following carbocations does not rearrange?

a. $CH_3\overset{\oplus}{C}H_2$

b. $CH_3\overset{\oplus}{C}HCH_3$

c. $CH_3\overset{\oplus}{C}(CH_3)_2$ (drawn as $CH_3C^{\oplus}$ with a CH_3 above and below)

d. $CH_3CH_2\overset{\oplus}{C}HCH_3$

e. All the above

Answer: e Section: 6

24. Which of the following can't undergo nucleophilic substitution reactions?

a. $CH_2{=}CHCH_2Br$

b. 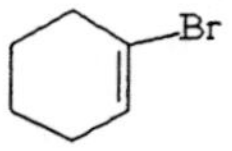 Br

c. 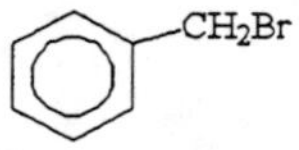 CH_2Br

d. 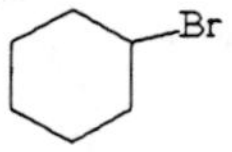 Br

e. 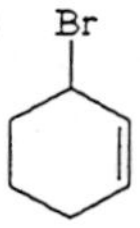 Br

Answer: b Section: 8

25. Which of the following is the rate law for S_N1 mechanisms?

a. Rate = k[Alkyl Halide] [Nucleophile]
b. Rate = k[Nucleophile]
c. Rate = k[Alkyl Halide]
d. Rate = k[Alkyl Halide] [Nucleophile] + k_2 [Alkyl Halide]
e. Rate = k_1 [Alkyl Halide] + k_2 [Nucleophile]

Answer: c Section: 5

26. Which of the following solvents is <u>aprotic</u>?

    a. CH_3CH_2OH
    b. $CH_3CH_2CH_2NH_2$
    c. $CH_3CH_2OCH_2CH_3$
    d. $CH_3CH_2NHCH_3$
    e. $CH_3\underset{\underset{O}{\|}}{C}OH$

Answer: c Section: 10

27. Which of the following solvents is <u>protic</u>?

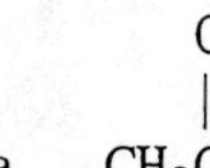

    a. $CH_3\overset{\overset{O}{\|}}{C}H$
    b. $CH_3\overset{\overset{O}{\|}}{C}CH_3$
    c. $CH_3CH_2OCH_3$
    d. CH_3CH_2OH
    e. $CH_3CH_2CH_2Cl$

Answer: d Section: 10

28. Protic and aprotic solvents are very similar as solvents except for their

    a. polarity.
    b. dielectric constant.
    c. ability to stabilize anions by hydrogen bonding.
    d. ability to stabilize cations by hydrogen bonding.
    e. ability to stabilize cations with unshared pairs of electrons.

Answer: c Section: 10

29. Which of the following best explains why S_N1 reactions involving a neutral reactant are <u>faster</u> in polar solvents?

    a. The substrate is more soluble in polar solvents.
    b. The substrate is less soluble in polar solvents.
    c. The nucleophile is solvated by polar solvents.
    d. Solvation by polar solvents stabilizes the carbocation.
    e. Solvation by polar solvents stabilizes the transition state.

Answer: e Section: 10

30. Which of the following statements is generally true for S_N1 reactions?

    a. Complete inversion of configuration occurs.
    b. These are favored by nonpolar solvents.
    c. These reactions are favored by polar solvents.
    d. Reaction rates depend only on the concentration of the nucleophile.
    e. The mechanism is a one-step back attack.

    Answer: c Section: 10

SHORT ANSWER

31. Explain why allyl chloride undergoes S_N1 reactions even though it is a 1° halide?

$$CH_2{=}CHCH_2Cl$$
allyl chloride

Answer:
Allyl chloride forms a relatively stable allylic carbocation with two contributing resonance structures.

Section: 8

32. Consider the following reaction mechanism:

What effect will the following have on the rate?
a. addition of Br^-
b. using a more polar solvent
c. using a stronger nucleophile
d. using a more concentrated nucleophile

I. $R{-}Br \rightleftharpoons R^{\oplus} + Br^{\ominus}$

II. $R^{\oplus} + Nu^{\ominus} \longrightarrow R{-}Nu$

Answer:
a. reaction will slow down because of competition with Nu^-
b. reaction will speed up because a more polar solvent assists in ionization and the formation of the carbocation
c. no effect on S_N1 reactions
d. no effect on S_N1 reactions

Section: 5

33. Explain why S_N2 reactions proceed faster in the solvent dimethylsulfoxide than in ethanol?

CH_3SOCH_3 CH_3CH_2OH

dimethlysolfoxide ethanol

Answer:
Dimethlysulfoxide is an aprotic solvent. Ethanol is a protic solvent which can hydrogen bond to the nucleophice, decreasing its nucleophilicity. Therefore, S_N2 reactions favor aprotic solvents.
Section: 10

MULTIPLE CHOICE

34. Which of the following best describes the carbon-chlorine bond of an alkyl chloride?

a. nonpolar; no dipole
b. polar; δ+ at carbon and δ- at chlorine
c. polar; δ- at carbon and δ+ at chlorine
d. ionic
e. none of the above

Answer: b Section: 1

35. What is the leaving group in the reaction shown below?

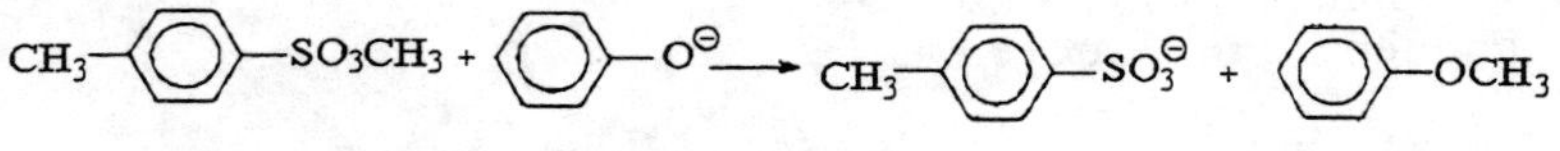

$CH_3-C_6H_4-SO_3CH_3 + C_6H_5-O^{\ominus} \longrightarrow CH_3-C_6H_4-SO_3^{\ominus} + C_6H_5-OCH_3$

a. 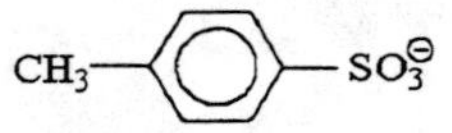 $CH_3-C_6H_4-SO_3^{\ominus}$

b. $C_6H_5-OCH_3$

c. $C_6H_5-O^{\ominus}$

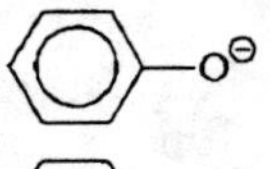

d. 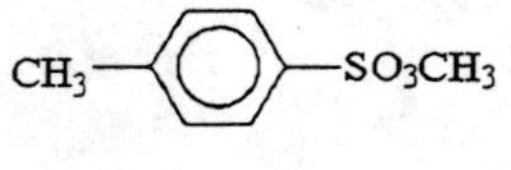 $CH_3-C_6H_4-SO_3CH_3$

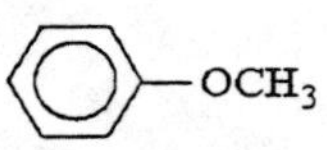

e. $CH_3O^{\ominus}$

Answer: a

36. Which of the following is classified as a vinyl halide?

    a. $CH_3CH{=}CHOH$
    b. $CH_3CH{=}CHCl$
    c. $CH_3CH{=}CHCH_2Cl$
    d. $CH_3CH_2CH_2CH_2Br$
    e. $BrCH_2CH{=}CH_2$

    Answer: b Section: 8

37. Which of the following species is the <u>least</u> nucleophilic?

    a. $(CH_3)_3CO^-$
    b. H_2O
    c. $(CH_3)_3N$
    d. BF_3
    e. CN^-

    Answer: d Section: 3

SHORT ANSWER

38. Provide a detailed, stepwise mechanism for the reaction below.

    $$(CH_3)_2CHCH_2CH_2CH_2I \; + \; CN^- \longrightarrow (CH_3)_2CHCH_2CH_2CH_2CN \; + \; I^-$$

    Answer:

    $$(CH_3)\ CHCH_2CH_2CH_2{-}I \; + \; CN^- \longrightarrow (CH_3)_2CHCH_2CH_2CH_2{-}CN$$

    Section: 2

39. Do all primary alkyl iodides undergo S_N2 reactions with sodium cyanide in DMSO at identical rates? Explain.

    Answer:
    No. All primary iodides are not equally accessible to attack by the CN^-. Staric hindrance varies among primary iodides.
    Section: 2

40. Rank the species below in order of increasing nucleophilicity in protic solvents: $CH_3CO_2^-$, CH_3S^-, HO^-, H_2O.

    Answer: $H_2O < CH_3CO_2^- < HO^- < CH_3S^-$ Section: 3

41. What type of solvent is best for S_N2 reactions which employ anionic nucleophiles: polar, protic solvents; polar, aprotic solvents; or nonpolar solvents? Explain.

Answer:
Polar, aprotic colvents are best. These solvents have strong dipole moments to enhance solubility of the anionic species but lack the ability to solvate the anion by hydrogen bonding.

Section: 10

42. Provide the structure of the major organic product which results when (S)-2-iodopentane is treated with KCN in DMF.

Answer:

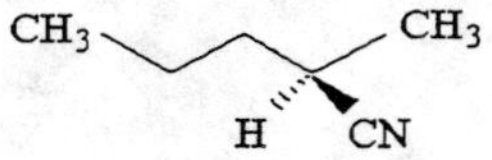

CH3 CH3 H CN

Section: 2

MULTIPLE CHOICE

43. Which of the following compounds will undergo an S_N2 reaction most readily?

a. $(CH_3)_3CCH_2I$
b. $(CH_3)_3CCl$
c. $(CH_3)_2CHI$
d. $(CH_3)_2CHCH_2CH_2CH_2Cl$
e. $(CH_3)_2CHCH_2CH_2CH_2I$

Answer: e Section: 2

SHORT ANSWER

44. Provide the structure of the major organic product in the following reaction.

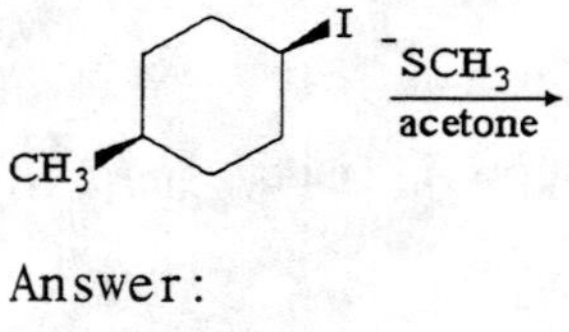

I $^-SCH_3$ acetone CH3

Answer:

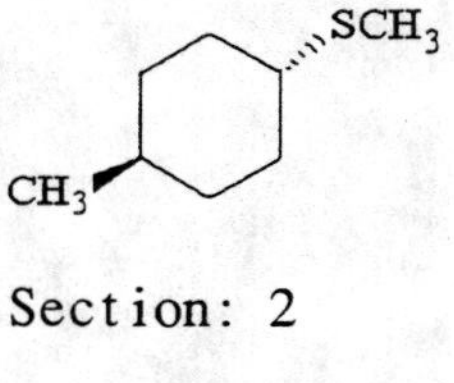

SCH3 CH3

Section: 2

45. Provide the structure of the major organic product in the following reaction.

$(CH_3)_3N$ + $CH_3CH_2CH_2I$ ⟶

Answer: $[(CH_3)_3NCH_2CH_2CH_3]^+$ I^- Section: 2

46. Provide the major organic product of the reaction below and a detailed, stepwise mechanism which accounts for its formation.

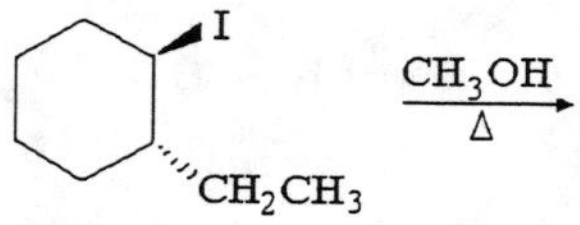

I
CH_2CH_3
CH_3OH
Δ

Answer:

I
CH_2CH_3
+
CH_2CH_3
+ I^-

+
H
CH_2CH_3
rearrangement
+
CH_2CH_3

+
CH_2CH_3
+ $CH_3\ddot{O}H$
H
+
$\ddot{O}CH_3$
CH_2CH_3

OCH_3
CH_2CH_3
+ H^+

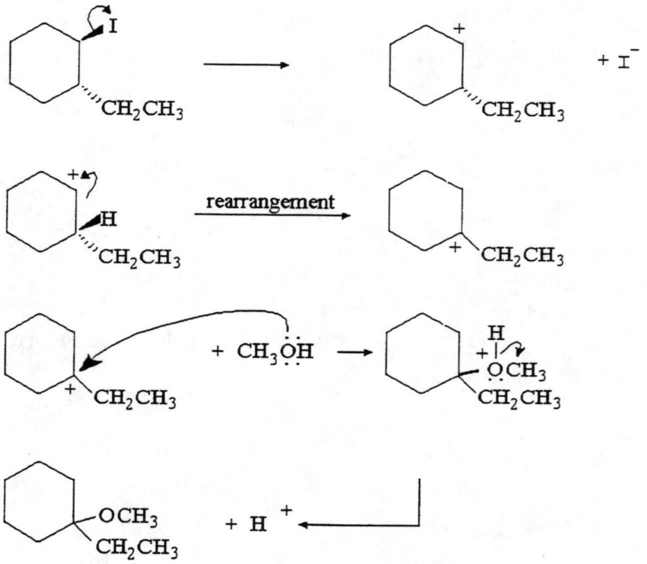

Section: 5

MULTIPLE CHOICE

47. S_N1 reactions usually proceed with:

    a. equal amounts of inversion and retention at the center undergoing substitution.
    b. slightly more inversion than retention at the center undergoing substitution.
    c. slightly more retention than inversion at the center undergoing substitution.
    d. complete inversion at the center undergoing substitution.
    e. complete retention at the center undergoing substitution.

    Answer: b Section: 7

48. Which of the compounds below undergoes solvolysis in aqueous ethanol most rapidly?

    a. cyclohexyl bromide
    b. methyl iodide
    c. isopropyl chloride
    d. 3-chloropentane
    e. 3-iodo-3-methylpentane

    Answer: e Section: 5

SHORT ANSWER

49. Provide the structure of the major organic products which result in the reaction below.

    CH_3 (wedge) cyclopentane ,,,ll Cl $\xrightarrow[\Delta]{CH_3CH_2OH}$

    Answer:

    CH_3 (wedge) cyclopentane (wedge) OCH_2CH_3 + CH_3 (wedge) cyclopentane ,,,ll OCH_2CH_3

    Section: 5

50. Provide the structure of the major organic product which results in the following reaction.

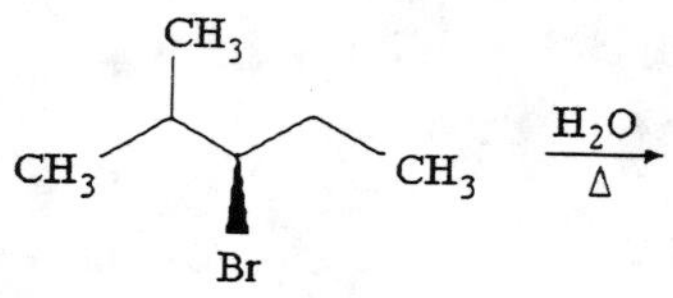

Answer:

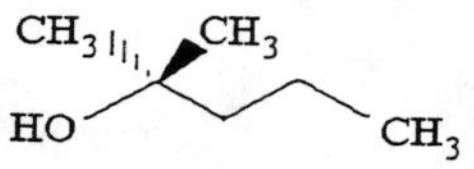

Section: 6

51. List the following compounds in order of increasing reactivity in an S_N1 reaction.

CH_3Br, $CH_3CH_2CH_2I$, $(CH_3)_3CI$, $CH_3CHBrCH_3$, CH_3CHICH_3

Answer: $CH_3Br < CH_3CH_2CH_2I < CH_3CHBrCH_3 < CH_3CHICH_3 < (CH_3)_3CI$

Section: 6

MULTIPLE CHOICE

52. Which of the following is a <u>secondary</u> alkyl halide?

a. CH_3Br
b. $(CH_3)_3CBr$
c. $(CH_3)_2CHBr$
d. $(CH_3)_2CHCH_2Br$

Answer: c Section: 2

53. Which of the following is <u>not</u> normally considered to be a nucleophile?

a. NH_3
b. NH_2CH_3
c. $HC\equiv C:^-$
d. $CH_3CH_2^+$

Answer: d Section: 1

54. Which of the following S_N2 reactions is the fastest?

a. $CH_3CH_2CH_2CH_2Br + OH^- \longrightarrow CH_3CH_2CH_2CH_2OH + Br^-$
b. $CH_3CH_2CH_2CH_2Br + H_2O \longrightarrow CH_3CH_2CH_2CH_2OH + HBr$
c. $CH_3CH_2CHBrCH_3 + OH^- \longrightarrow CH_3CH_2CHOHCH_3 + Br^-$
d. $CH_3CH_2CHBrCH_3 + H_2O \longrightarrow CH_3CH_2CHOHCH_3 + HBr$

Answer: a Section: 2

55. Which of the following S_N2 reactions is the slowest?

a. $CH_3CH_2CHBrCH_3 + OH^- \longrightarrow CH_3CH_2CHOHCH_3 + Br^-$
b. $CH_3CH_2CHBrCH_3 + H_2O \longrightarrow CH_3CH_2CHOHCH_3 + HBr$
c. $CH_3CH_2CH_2CH_2Br + OH^- \longrightarrow CH_3CH_2CH_2CH_2OH + Br^-$
d. $CH_3CH_2CH_2CH_2Br + H_2O \longrightarrow CH_3CH_2CH_2CH_2OH + HBr$

Answer: b Section: 2

56. The specific rotation of optically pure (R)-$C_6H_5CHOHCH_3$ is -42.3°. An optically pure sample of (R)-$C_6H_5CHClCH_3$ was converted into the corresponding alcohol via an S_N2 reaction. What is the specific rotation of the product?

a. -42.3°
b. Between 0° and -42.3°
c. Between +42.3° and 0°
d. +42.3°

Answer: d Section: 7

57. When (S)-1-bromo-1 phenylethane undergoes an S_N1 reaction with methanethiol (CH_3SH), the product is the compound shown. What is/are the configuration(s) of the product as obtained from this reaction?

```
C6H5CH-CH3
    |
   SCH3
```

a. S only.
b. R only.
c. A mixture of the enantiomers, with slightly more S than R.
d. A mixture of the enantiomers, with slightly more R than S.

Answer: d Section: 7

58. Which of the following does <u>not</u> provide evidence that there are two different mechanisms for nucleophilic substitution?

a. Reaction products when CH_3I is used as the substrate.
b. Reaction products when $(CH_3)_3CCH_2I$ is used as substrate.
c. The stereochemistry of nucleophilic substutions.
d. The effect of nuclepophile concentration on rate.

Answer: a Section: 2

59. Which of the following is the best nucleophile in water?

a. I^-
b. CH_3SCH_3
c. CH_3OCH_3
d. Cl^-

Answer: a Section: 3

SHORT ANSWER

60. In each of the pairs below, which is the best nucleophile in alcoholic solvents?

a. CH_3S^- or CH_3O^-
b. $(CH_3)_2NH$ or $(CH_3)_3N$
c. Cl^- or F^-
d. SCN^- or OCN^-

Answer:
a. CH_3S^-
b. $(CH_3)_2NH$
c. Cl^-
d. SCN^-

Section: 3

Chapter 10: Reactions at an sp3 Hybridized Carbon II

MULTIPLE CHOICE

1. Which of the following are the elimination products of the reaction shown below?

$$CH_3CH_2Br + {}^-OH \longrightarrow ?$$

a. $CH_3CH_2Br^+H + O^-$
b. $HOCH_2CH_2Br$
c. $CH_3CH_2OH + Br^-$
d. $CH_2{=}CH_2 + Br^- + H_2O$
e. $CH_2{=}CHBr + H_2O$

Answer: d Section: 1

2. Which of the following compounds undergoes E2 reactions with the <u>fastest</u> rate?

a. CH_3CHCH_3 (with Cl on the central carbon)
b. $CH_3CH_2CH_2Cl$
c. $CH_3CH_2CH_2I$
d. CH_3CHCH_3 (with I on the central carbon)
e. $CH_3CH_2CH_2Br$

Answer: d Section: 2

3. Which of the following bases gives the highest anti-Zaitsev product in E2 reactions when reacted with 2-bromo-2,3-dimethylbutane?

a. $CH_3O^{\ominus}$
b. $CH_3CH_2O^{\ominus}$
c. $CH_3CHO^{\ominus}$ (with CH_3 on the central carbon)
d. $CH_3CO^{\ominus}$ (with CH_3 above and CH_3 below the central carbon)
e. $CH_3CH_2CO^{\ominus}$ (with CH_3 above and CH_3 below the central carbon)

Answer: e Section: 2

4. Which of the following alkyl halides undergoes E1 reactions with the <u>fastest</u> rate?

   a. CH_3CHCH_3 with F on the middle carbon
   b. CH_3CHCH_3 with Cl on the middle carbon
   c. CH_3CHCH_3 with Br on the middle carbon
   d. CH_3CHCH_3 with I on the middle carbon
   e. CH_3I

   Answer: d Section: 3

5. Which of the following alkyl halides forms the most stable carbocation when it undergoes an E1 reaction?

   a. 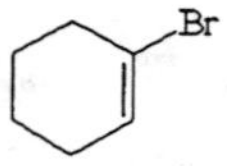 Br
   b. 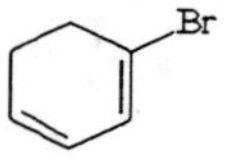 Br
   c. 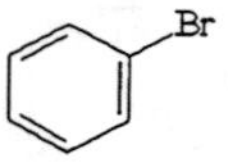 Br
   d. 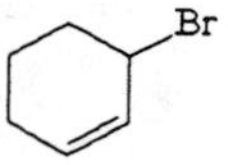 Br
   e. 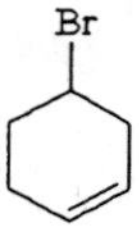 Br

   Answer: d Section: 3

6. What is the major product of the following E2 reaction?

$$CH_3CHCH_2CH_3 \text{ (Br on C-2)} + HO^- \xrightarrow{\text{heat}} \text{Product(s)}$$

a. $CH_3CH_2C(H)=CH_2$

b. $CH_3CH_2CH=CH_2$

c. $CH_3(H)C=C(H)CH_3$ (CH₃ groups trans)

d. $CH_3(H)C=C(CH_3)H$ (CH₃ groups cis)

e. $CH_3C(CH_3)=CH_2$

Answer: c Section: 2

7. Consider the following experimental data for the rate of the reaction given below:

$$(CH_3)_3CI + CH_3OH \xrightarrow{\text{heat}} (CH_3)_2C=CH_2 + CH_3\overset{+}{O}H_2 + I^-$$

Experiment #1	[Alkyl Halide]	[Base]	Rate
1	0.01	0.01	1
2	0.02	0.01	2
3	0.01	0.02	1

What is the mechanism for the reaction?

a. first order, S_N1
b. first order, S_N2
c. first order, E1
d. first order, E2
e. none of the above

Answer: c Section: 3

8. Which of the compounds shown below is/are the product(s) of this reaction:

a. I. only
b. II. only
c. I. and II. of equal yield
d. I. is major, II. is minor
e. I. is minor, II. is major

CH_3 H H Br + HO^- heat → ?

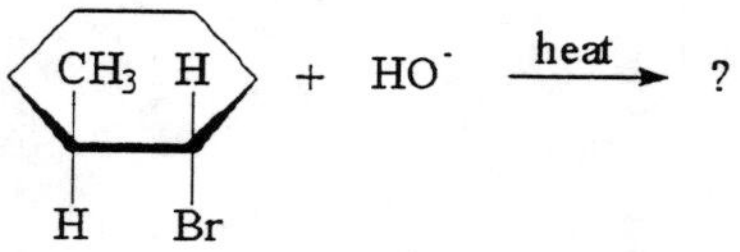

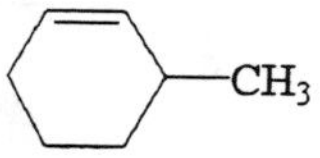

I. 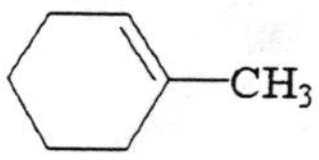 II.

Answer: a Section: 6

9. Which of the following compounds is/are the products of this reaction:

a. I. only
b. II. only
c. I. and II. of equal yield
d. I. is major, II. is minor
e. I. is minor, II. is major

H H CH_3 Br + HO^- heat → ?

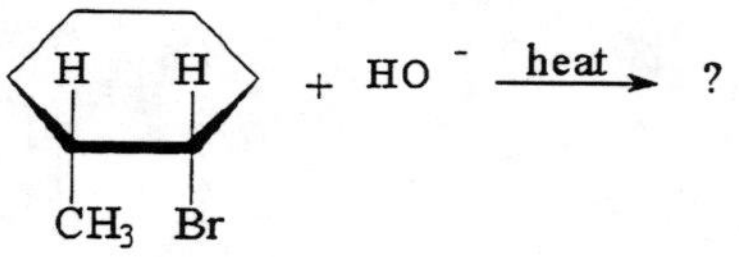

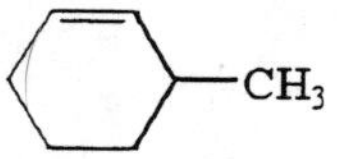

I. 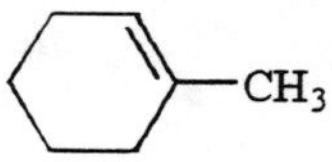 II.

Answer: e Section: 6

10. What is the major product of the following reaction?

$$\underset{\displaystyle Br}{CH_3\overset{}{C}HCH_2CH_3} + CH_3CH_2O^{\ominus} \xrightarrow[\text{heat}]{\text{alcohol}} ?$$

a. $CH_2{=}CHCH_2CH_3$
b. $CH_3CH{=}CHCH_3$
c. $CH_3\underset{\displaystyle OCH_2CH_3}{CH}CH_2CH_3$
d. $CH_3CH_2CH_2CH_2OCH_2CH_3$
e. $CH_3\underset{\displaystyle CH_3}{C}{=}CH_2$

Answer: b Section: 8

11. What is the major product of the following reaction?

$$\underset{\displaystyle Br}{CH_3\overset{}{C}HCH_2CH_3} + CH_3CH_2OH \xrightarrow[\text{temp.}]{\text{room}} ?$$

a. $CH_2{=}CHCH_2CH_3$
b. $CH_3CH{=}CHCH_3$
c. $CH_3\underset{\displaystyle OCH_2CH_3}{CH}CH_2CH_3$
d. $CH_3CH_2CH_2CH_2OCH_2CH_3$
e. $CH_3\underset{\displaystyle CH_3}{C}{=}CH_2$

Answer: c Section: 8

12. What is the major product of the following reaction?

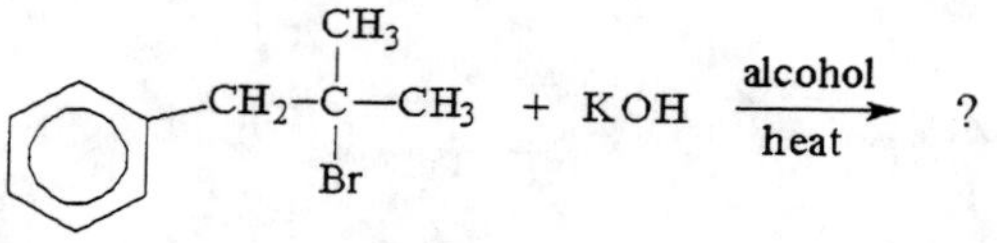

$C_6H_5-CH_2-C(CH_3)(Br)-CH_3 + KOH \xrightarrow[\text{heat}]{\text{alcohol}} ?$

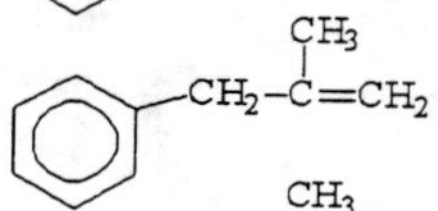

a. 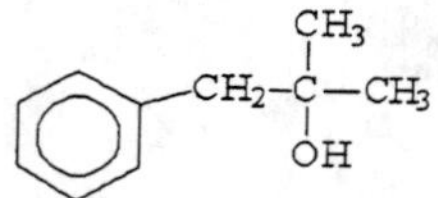 $C_6H_5-CH_2-C(CH_3)(OH)-CH_3$

b. $C_6H_5-CH_2-C(CH_3)=CH_2$

c. $C_6H_5-CH_2-CH(CH_3)-CH_2OH$

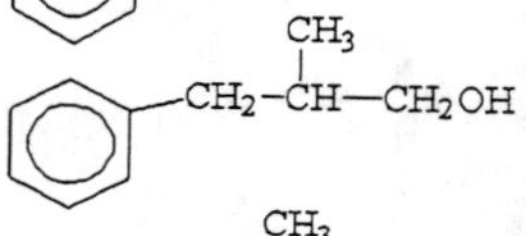

d. 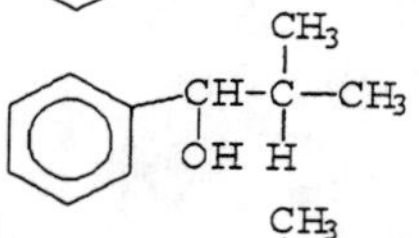 $C_6H_5-CH(OH)-C(CH_3)(H)-CH_3$

e. 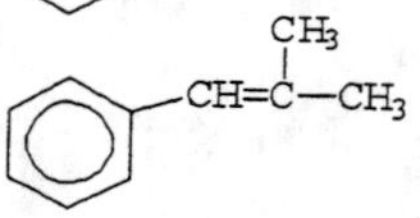 $C_6H_5-CH=C(CH_3)-CH_3$

Answer: e Section: 8

13. What is/are the product(s) of the following reaction?

$CH_3CH_2O^{\ominus} + CH_3C(CH_3)_2Br \longrightarrow ?$

a. $CH_3CH_2OC(CH_3)_2-CH_3$
b. $CH_2=CH_2$
c. $CH_3C(CH_3)=CH_2$
d. a. and b.
e. a. and c.

Answer: c Section: 8

14. What is/are the product(s) of the following reaction?

$$CH_3-\underset{\underset{CH_3}{|}}{\overset{\overset{CH_3}{|}}{C}}-O^{\ominus} + CH_3CH_2Br \longrightarrow ?$$

a. $CH_3CH_2O\underset{\underset{CH_3}{|}}{\overset{\overset{CH_3}{|}}{C}}-CH_3$
b. $CH_2{=}CH_2$
c. $CH_3\underset{\underset{CH_3}{|}}{C}=CH_2$
d. a. and b.
e. a. and c.

Answer: d Section: 8

15. Starting with 2-butene, which of the following is the best method for preparing 2-butyne?

a. HBr ; H_2 /Ni ; Zn/H^+
b. HBr ; Zn/H^+ ; H_2 /Ni
c. Br_2 /CCl_4 ; Zn/H^+ ; H_2 /Ni
d. Br_2 /CCl_4 ; $2NaNH_2$
e. HBr ; $NaNH_2$

Answer: d Section: 9

16. What is the major product of the following reaction?

$$CH_2{=}CHCH_2\underset{\underset{Br}{|}}{C}HCH_3 + KOH \xrightarrow[\text{heat}]{\text{alcohol}} ?$$

a. $CH_2{=}CHCH_2CH=CH_2$
b. $CH_3CH_2CH_2\underset{\underset{OH}{|}}{C}HCH_3$
c. $CH_2{=}CHCH=CHCH_3$
d. $CH_2{=}CH\underset{\underset{OH}{|}}{C}HCH_2CH_3$
e. $CH_2{=}CHCH_2CH{=}CH_2$

Answer: c Section: 10

SHORT ANSWER

17. Which of the following alkyl chlorides would undergo substitution most rapidly when treated with NaCCH: chloroethane, 2-chloropropane, or 1-chloro-2,2-dimethylpropane? Provide the structure of the substitution product.

Answer: $CH_3CH_2C{\equiv}CH$ Section: 9

18. Provide a series of synthetic steps by which $(CH_3)_2C{=}CH_2$ could be prepared from 2-methylpropane.

Answer:
1. Br_2, hυ
2. $NaOCH_3$, CH_3OH

Section: 11

19. When 1-bromo-2,2-dimethylcyclopentane is heated in ethanol, one of the products which results is shown below. Provide a detailed, stepwise mechanism for the production of this compound, and give the name of the mechanism by which it is produced.

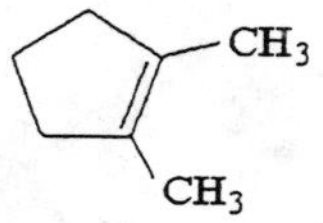

Answer:

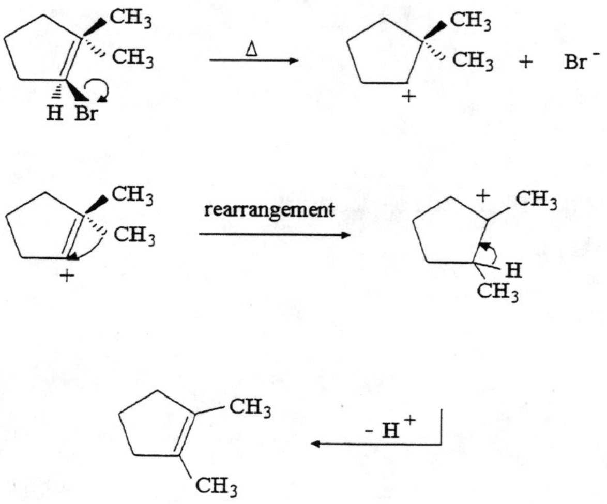

E1 mechanism.

Section: 3

20. Provide the structure of the major organic product which results in the following reaction.

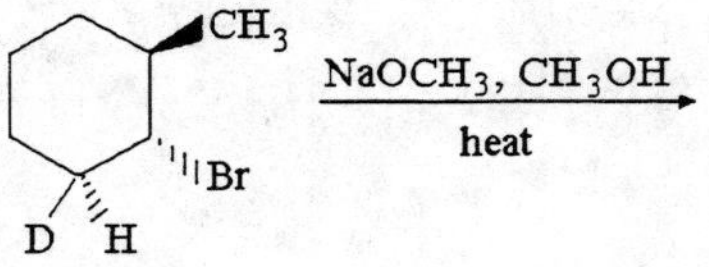

CH_3 ... Br ... D H
$\xrightarrow[\text{heat}]{NaOCH_3,\ CH_3OH}$

Answer:

CH_3 ... D

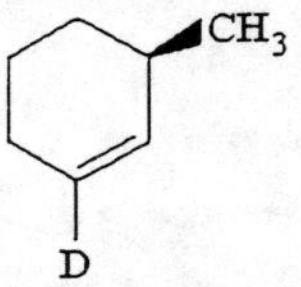

Section: 6

21. Provide the structure of the major organic product which results in the following reaction.

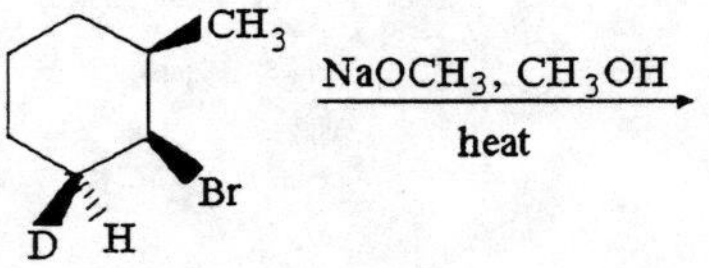

CH_3 ... Br ... D H
$\xrightarrow[\text{heat}]{NaOCH_3,\ CH_3OH}$

Answer:

CH_3 ... D H

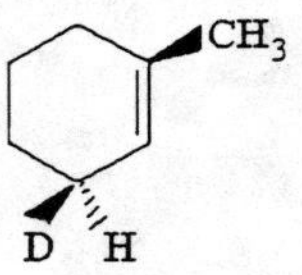

Section: 6

22. Provide the structure of the major organic product which results in the following reaction.

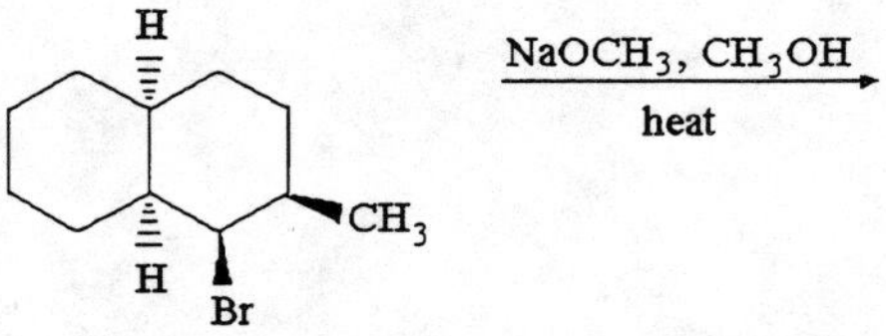

H ... CH_3 ... H ... Br
$\xrightarrow[\text{heat}]{NaOCH_3,\ CH_3OH}$

Answer:

H ... CH_3

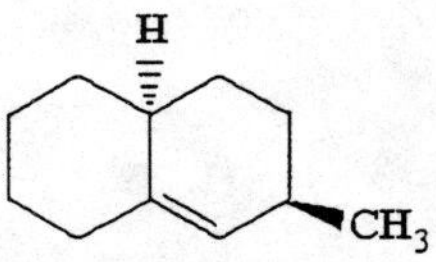

Section: 6

23. Which diastereomer of 1-bromo-4-t-butylcyclohexane, the cis or the trans, undergoes elimination more rapidly when treated with sodium ethoxide? Explain your answer.

Answer:
The cis isomer reacts more quickly. An axial orientation of the C-Br bond is required for E2 in six-membered ring systems. In order for this to occur in the trans isomer, the bulky t-butyl group must also assume an axial orientation; this requires substantial energy of activation.

Section: 6

24. When 1-iodo-1-methylcyclohexane is treated with $NaOCH_2CH_3$, the more highly substituted alkene product predominates. When $KOC(CH_3)_3$ is used instead, the less highly substituted alkene product predominates. Offer an explanation.

Answer:
The unhindered ethoxide produces the more stable alkene product (ie, the more highly substituted alkene possible). When the bulky t-butoxide is used, the most accessible hydrogen is removed. This results in the least highly substituted alkene possible.

Section: 2

MULTIPLE CHOICE

25. Which of the following statements correctly describe(s) E1 reactions of alkyl halides (RX)?

I. Rate = k[base]
II. Rate = k[base][RX]
III. Rate = k[RX]
IV. The reactions occur in two distinct steps.
V. Rearrangements are sometimes seen.

a. II and IV
b. III and V
c. I, IV, and V
d. I only
e. III, IV, and V

Answer: e Section: 3

Chapter 10: Reactions at an sp3 Hybridized Carbon II

SHORT ANSWER

26. Provide the structure of the major organic product which results when 2-bromo-2-methylbutane is treated with sodium ethoxide.

Answer:

```
 H       CH3
   \    /
    C=C
   /    \
CH3      CH3
```

Section: 2

MULTIPLE CHOICE

27. How many distinct alkene products are possible when the alkyl iodide below undergoes E2 elimination?

a. 1
b. 2
c. 3
d. 4
e. 5

CH_3 CH_3 I CH_3

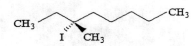

Answer: e Section: 2

SHORT ANSWER

28. Provide the structure of the major organic product in the following reaction.

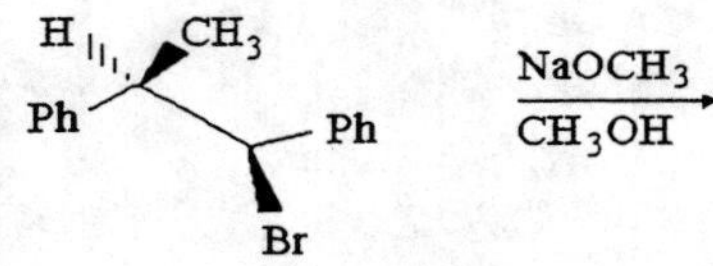

Answer:

Ph H CH_3 Ph

Section: 2

29. Show the best way to prepare $CH_3OCH(CH_3)_2$ by an S_N2 reaction.

Answer: CH_3I + $NaOCH(CH_3)_2$ Section: 9

MULTIPLE CHOICE

30. Predict the two most likely mechanisms for the reaction of 2-iodohexane with sodium ethoxide.

    a. S_N2 and S_N1
    b. E1 and E2
    c. S_N2 and E2
    d. E1 and S_N1
    e. E2 and S_N1

    Answer: c Section: 8

31. Predict the two most likely mechanisms which occur when 2-iodohexane is heated in ethanol.

    a. S_N2 and S_N1
    b. E1 and E2
    c. S_N2 and E2
    d. E1 and S_N1
    e. E2 and S_N1

    Answer: d Section: 8

SHORT ANSWER

32. Draw all likely alkene products in the following reaction and circle the product you expect to predominate.

    Br, CH_3 — NaOH / acetone →

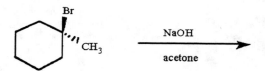

    Answer:

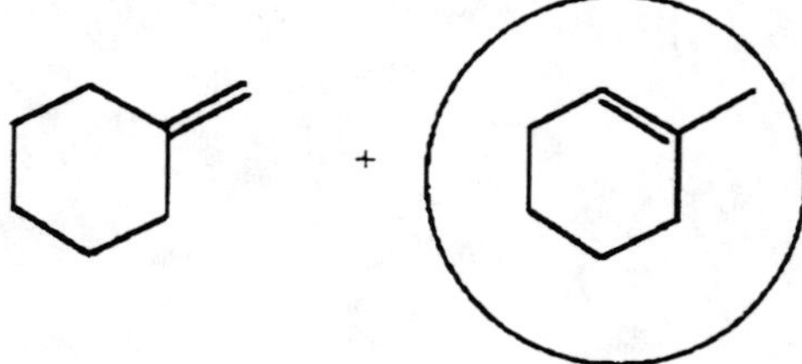

    Section: 2

33. Draw all likely alkene products in the following reaction and circle the product you expect to predominate.

Br

CH_3

$(CH_3)_3CO^- K^+$

$(CH_3)_3COH$

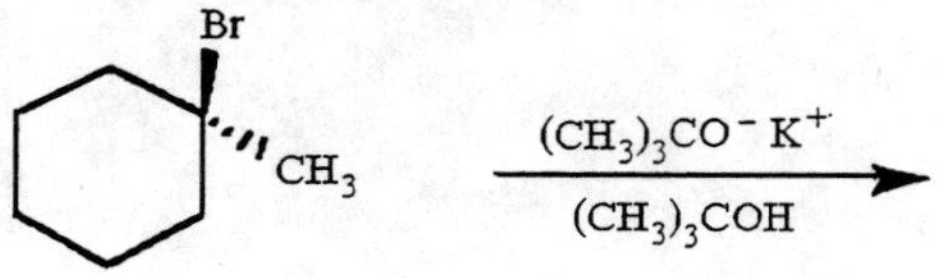

Answer:

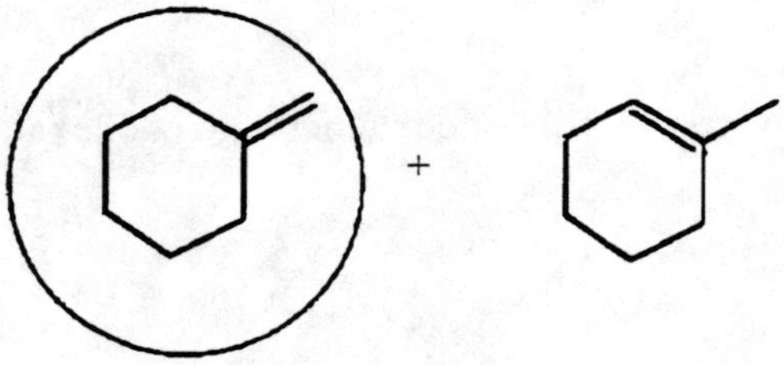

+

Section: 2

34. Draw all likely alkene products in the following reaction and circle the product you expect to predominate.

Br

NaOH

acetone

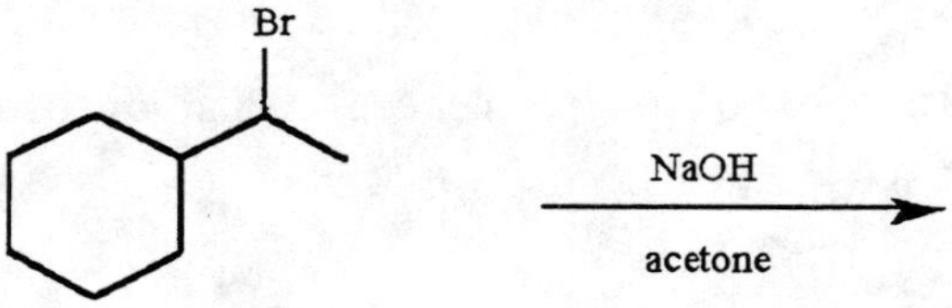

Answer:

+

Section: 2

35. Draw the alkene product which results when 1-bromopentane is heated in acetone containing NaOH. Give a detailed, step-by-step mechanism for the production of this compound.

Answer:

H H Br + -OH ⟶ H + H_2O + Br^-

Section: 1

MULTIPLE CHOICE

36. Which of the alkyl chlorides listed below undergoes dehydrohalogenation in the presence of a strong base to give 2-pentene as the <u>only</u> alkene product?

a. 1-chloropentane
b. 2-chloropentane
c. 3-chloropentane
d. 1-chloro-2-methylbutane
e. 1-chloro-3-methylbutane

Answer: c Section: 2

37. Dehydrohalogenation of 2-bromobutane in the presence of a strong base proceeds via which of the following mechanistic pathways?

a. S_N1
b. S_N2
c. E1
d. E2
e. none of the above

Answer: d Section: 4

SHORT ANSWER

38. Which base, ammonia (NH_3) or triethylamine [$(CH_3CH_2)_3N$], would be a better choice for use in converting 1-chlorohexane to 1-hexene? Explain briefly.

Answer:
Triethylamine. Amines can serve as both nucleophiles and as bases in reactions with alkyl halides. Increasing the steric bulk about the nitrogen diminishes the nucleophilicity while allowing the amine to continue to function effectively as a base.
Section: 8

39. Provide the reagents necessary for carrying out the transformation of cyclopentane to cyclopentene.

Answer:
1. Br_2, hυ
2. NaOH, acetone

Section: 11

40. Provide the reagents necessary for carrying out the transformation of 2-methylheptane to 2-methyl-1-heptene.

Answer:
1. Br_2, hυ
2. $(CH_3)_3CO^-K^+$, $(CH_3)_3COH$

Section: 11

41. Provide the reagents necessary for carrying out the transformation of 2-methylheptane to 2-methyl-2-heptene.

Answer:
1. Br_2, hυ
2. NaOH, acetone

Section: 11

42. Write the structures for the products of the following S_N2 reactions.

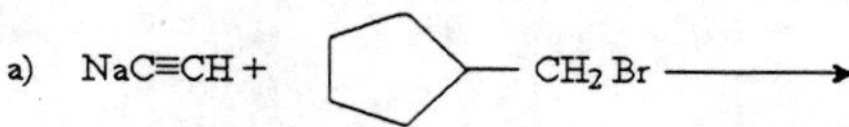

b) $(CH_3)_2CHCH_2CHCH_3$ + NaI ⟶ (Cl on the CH carbon)

c) $CH_2CH_2CH_2I$ + $NaOCH_3$

Answer:
a) 3-cyclopentyl-1-propyne
b) 2-iodo-4-methylpentane
c) 1-methoxypropane

Section: 9

43. List the following compounds in order of increasing reactivity in an E1 elimination.

$CH_3CH_2CHBrCH_3$, $CH_3CH_2CH_2CH_2Br$, $(CH_3)_3CBr$

Answer: $CH_3(CH_2)_3Br < CH_3CH_2CHBrCH_3 < (CH_3)_3CBr$ Section: 3

44. Supply the missing alkyl halide reactant in the elimination reactions shown below.

a) ? $\xrightarrow{EtO^-}$ (methylenecyclohexane, $=CH_2$)

b) ? $\xrightarrow{EtO^-}$ $CH_3CH = CHCH_2CH_3$

c) ? $\xrightarrow{EtO^-}$

Answer:
a. bromomethylcyclohexane
b. $CH_3CH_2CHBrCH_2CH_3$
c. $CH_3CH_2CH(CH_3)CH_2Br$

Section: 2

45. Write the product for the following reaction.

CH_3 (methylcyclohexane) $\xrightarrow[2)\ NaOEt,\ \Delta]{1)\ Br_2,\ hv}$

Answer: 1-methylcyclohexene Section: 11

46. Devise methods by which the following molecule could be synthesized from cyclohexane.

Answer: Cl_2, hv; NaOH, EtOH, heat Section: 11

47. Devise methods by which the following molecule could be synthesized. $(CH_3)_2C{=}CH_2$ from 2-methylpropane

Answer: Br_2, hv; NaOH, EtOH, heat Section: 11

Chapter 11: Reactions at an sp3 Hybridized Carbon III

MULTIPLE CHOICE

1. Which of the following is the best method for preparing CH_3Br?

   a. $CH_3OH + Br^-$
   b. $CH_3OH + HBr$
   c. $CH_3OH + Br_2$
   d. $CH_3OH + NaBr$
   e. $CH_3OH + Br^+$

   Answer: b Section: 1

2. What is the major product of the following reaction?

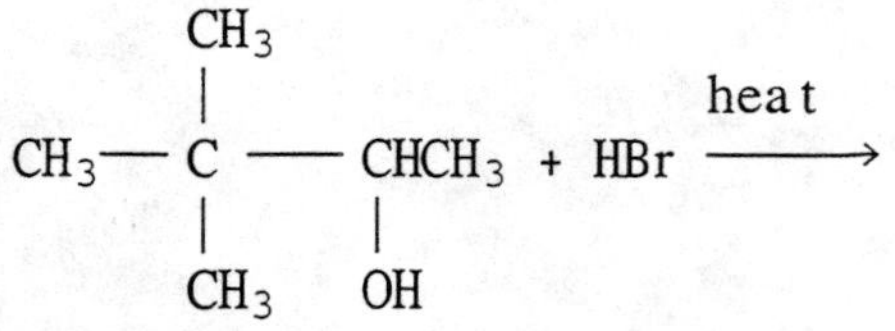

$$CH_3-\underset{CH_3}{\overset{CH_3}{\underset{|}{\overset{|}{C}}}}-\underset{OH}{\underset{|}{CH}}CH_3 + HBr \xrightarrow{heat}$$

   a. $CH_3-\underset{CH_3}{\overset{CH_3}{\underset{|}{\overset{|}{C}}}}-\overset{Br}{\overset{|}{CH}}CH_3$

   b. $CH_3-\underset{CH_3}{\overset{CH_3}{\underset{|}{\overset{|}{C}}}}-CH_2\underset{Br}{\underset{|}{CH_2}}$

   c. $CH_3-\underset{Br}{\overset{CH_3}{\underset{|}{\overset{|}{C}}}}-\underset{CH_3}{\underset{|}{CH}}CH_3$

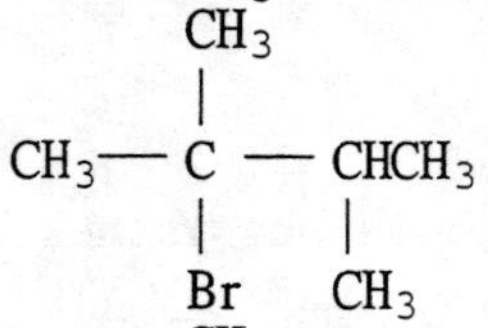

   d. 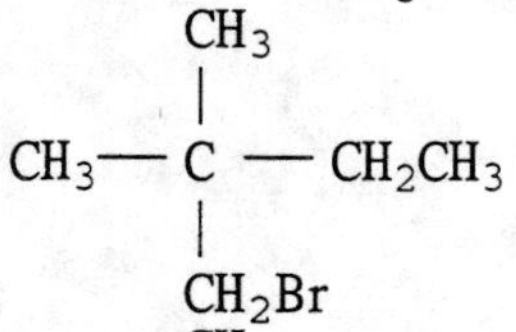 $CH_3-\underset{CH_2Br}{\overset{CH_3}{\underset{|}{\overset{|}{C}}}}-CH_2CH_3$

   e. 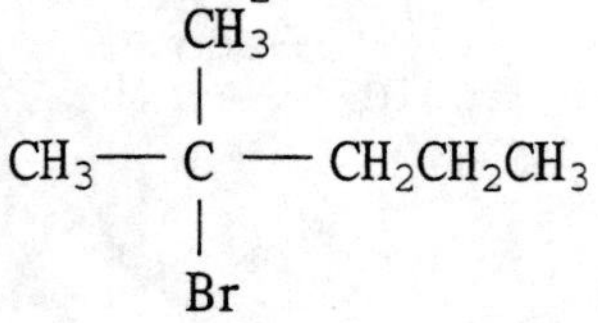 $CH_3-\underset{Br}{\overset{CH_3}{\underset{|}{\overset{|}{C}}}}-CH_2CH_2CH_3$

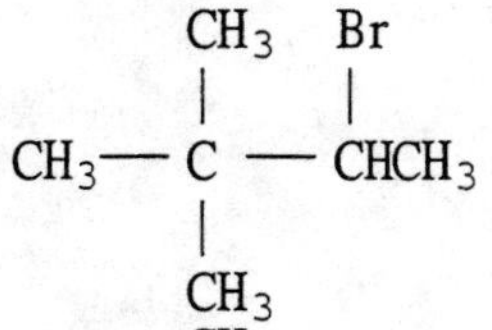

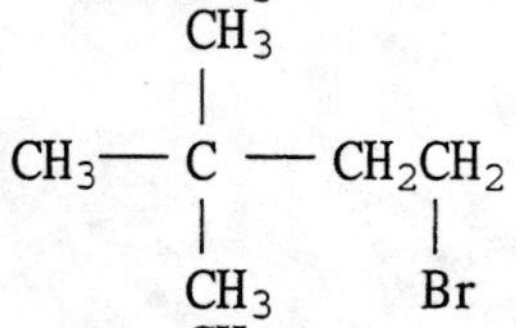

   Answer: c Section: 1

3. What is the major product of the following reaction?

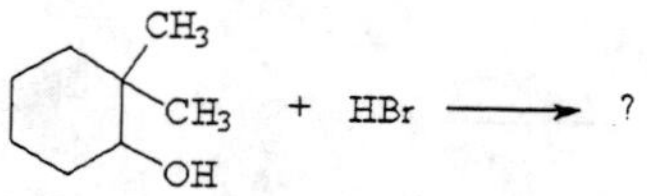

a. 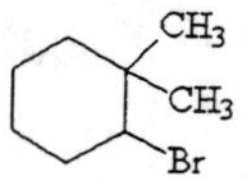

b. 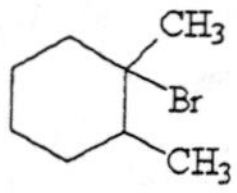

c. 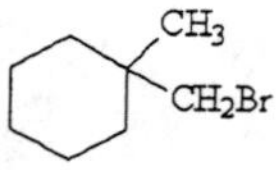

d. 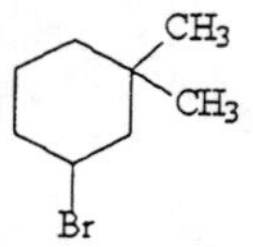

e. 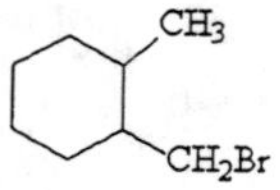

Answer: b Section: 1

4. Which of the folowing reagents is best used in the conversion of methyl alcohol to methyl chloride?

a. Cl_2/CCl_4
b. Cl_2/hv
c. Cl^-
d. $SOCl_2$
e. NaCl

Answer: d Section: 2

5. What is the product of the following reaction?

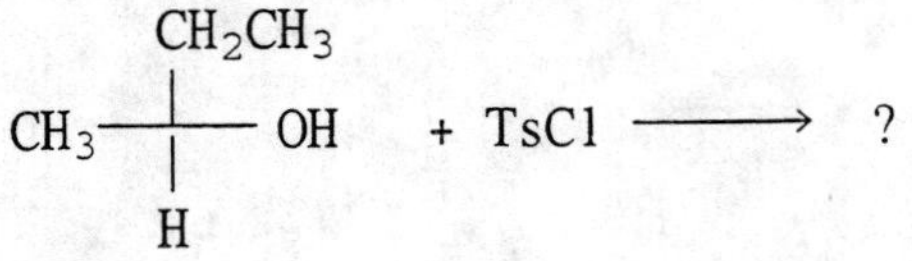

$$CH_3-\underset{\displaystyle H}{\overset{\displaystyle CH_2CH_3}{\overset{|}{\underset{|}{C}}}}-OH \quad + \ TsCl \longrightarrow \ ?$$

a. $CH_3-\underset{H}{\overset{CH_2CH_3}{C}}-Cl$

b. $CH_3-\underset{H}{\overset{CH_2CH_3}{C}}-OTs$

c. $Cl-\underset{H}{\overset{CH_2CH_3}{C}}-CH_3$.

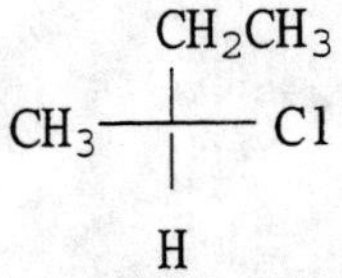

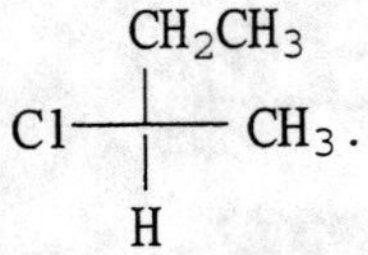

d. 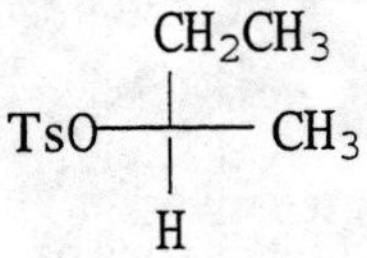 $TsO-\underset{H}{\overset{CH_2CH_3}{C}}-CH_3$

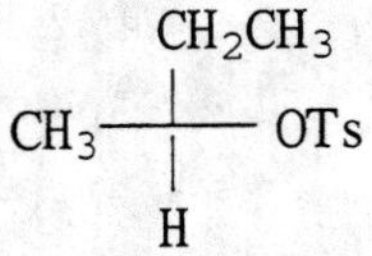

e. 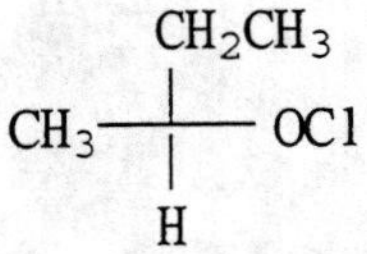 $CH_3-\underset{H}{\overset{CH_2CH_3}{C}}-OCl$

Answer: b Section: 3

6. What is the product of the following reaction?

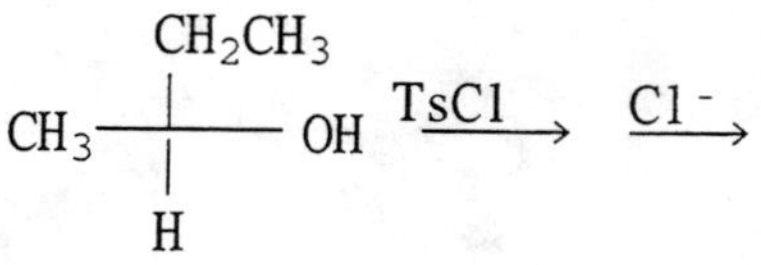

$$CH_3-\overset{CH_2CH_3}{\underset{H}{C}}-OH \xrightarrow{TsCl} \xrightarrow{Cl^-}$$

a. $CH_3-\overset{CH_2CH_3}{\underset{H}{C}}-Cl$

b. $CH_3-\overset{CH_2CH_3}{\underset{H}{C}}-OTs$

c. $Cl-\overset{CH_2CH_3}{\underset{H}{C}}-CH_3$

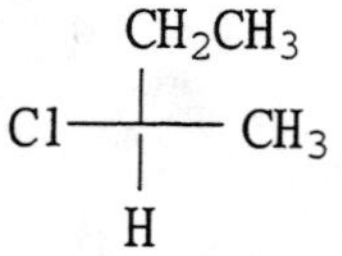

d. 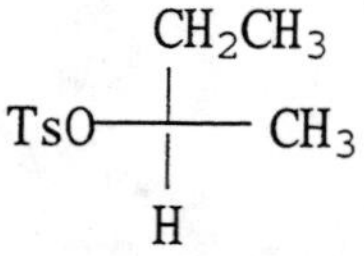 $TsO-\overset{CH_2CH_3}{\underset{H}{C}}-CH_3$

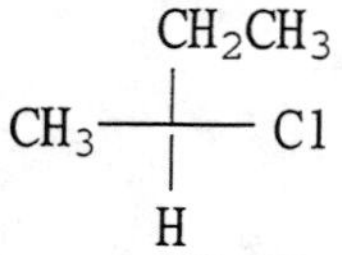

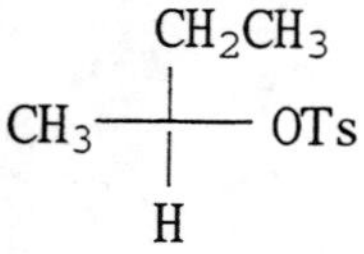

e. 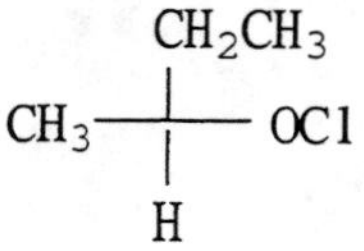 $CH_3-\overset{CH_2CH_3}{\underset{H}{C}}-OCl$

Answer: c Section: 3

7. Which of the following alcohols dehydrates with the <u>fastest</u> rate?

a. 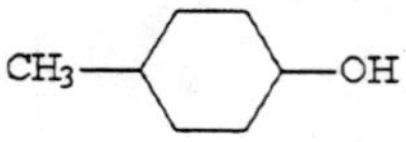 CH_3–(cyclohexane)–OH (1,4-substituted)

b. 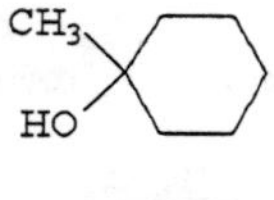 CH_3, HO (1,1-substituted cyclohexane)

c. 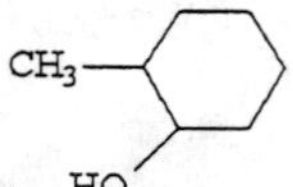 CH_3–(cyclohexane), HO (1,2-substituted)

d. 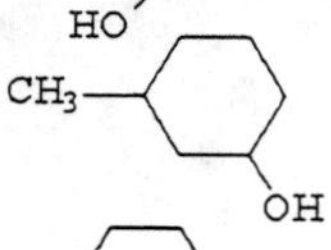 CH_3–(cyclohexane), OH (1,3-substituted)

e. 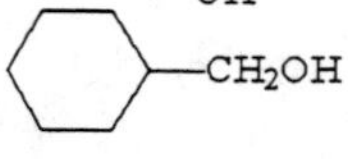 (cyclohexane)–CH_2OH

Answer: b Section: 4

8. Which of the following alcohols gives a rearranged carbocation when dehydrated?

a. 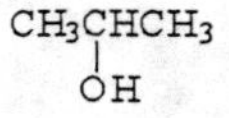 CH_3CHCH_3 (OH on C-2)

b. $CH_3CHCH_2CH_3$ (OH on C-2)

c. cyclohexane–OH

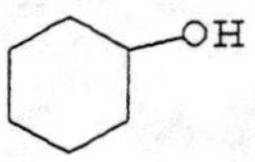

d. 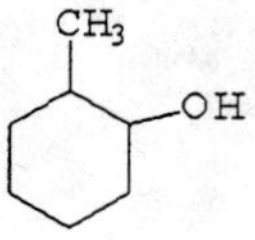 cyclohexane with CH_3 and OH on adjacent carbons

e. 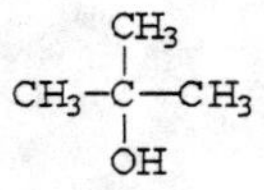 CH_3–C(CH_3)(OH)–CH_3

Answer: d Section: 4

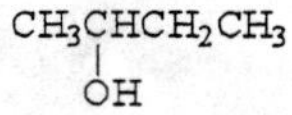

9. What is the product of the following reaction?

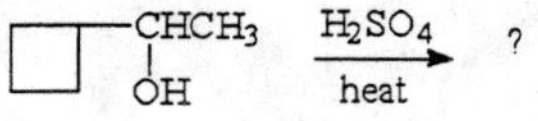

cyclobutyl–$CHCH_3$ (OH) $\xrightarrow[\text{heat}]{H_2SO_4}$?

a. 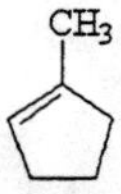 1-methylcyclopentene (CH_3)

b.  3-methylcyclopentene (CH_3)

c. 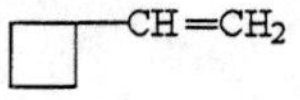 cyclobutyl–$CH{=}CH_2$

d. 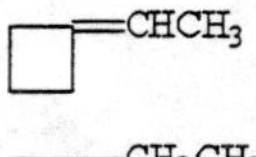 cyclobutylidene $=CHCH_3$

e. 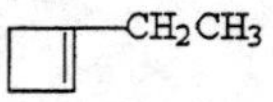 cyclobutene–CH_2CH_3

Answer: a Section: 4

10. What is the major product of the following reaction?

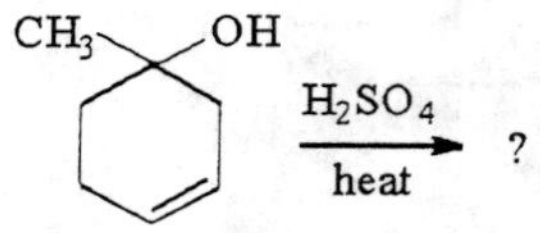

a.

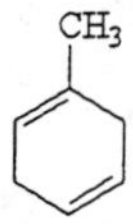

b.

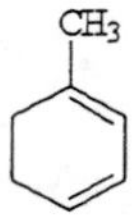

c.

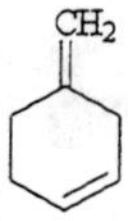

d.

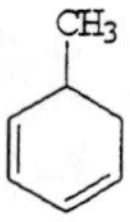

e.

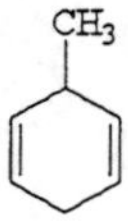

Answer: b Section: 4

11. What is the major product of the following reaction?

CH_3CH_2OH + $C_6H_5SO_2Cl$ → ?

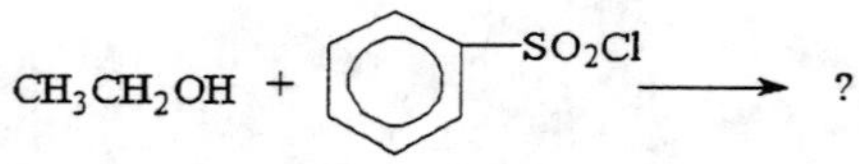

a. $CH_2{=}CH_2$

b. CH_3CH_2Cl

c. $CH_3CH_2SO_2Cl$

d. $CH_3CH_2SO_2-C_6H_5$

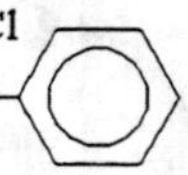

e. $CH_3CH_2OSO_2-C_6H_5$

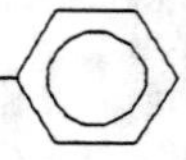

Answer: e Section: 3

12. What are the major products from the following reaction?

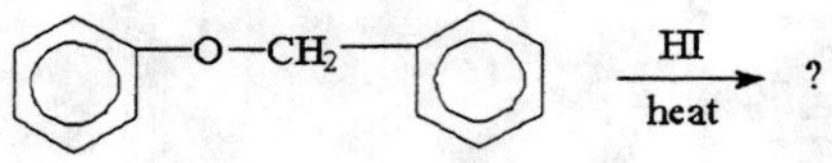

$C_6H_5-O-CH_2-C_6H_5 \xrightarrow[\text{heat}]{\text{HI}}$?

a. 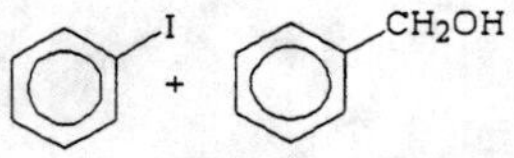 C_6H_5I + $C_6H_5CH_2OH$

b. C_6H_5OH + $C_6H_5CH_2I$

c. $C_6H_5OCH_2OH$ + C_6H_5I

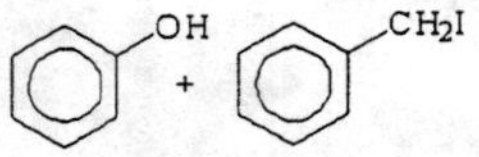

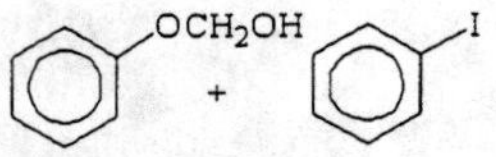

d. 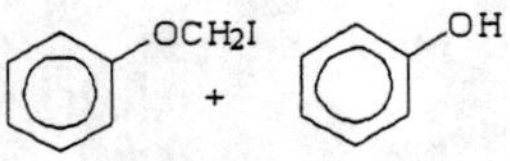 $C_6H_5OCH_2I$ + C_6H_5OH

e. 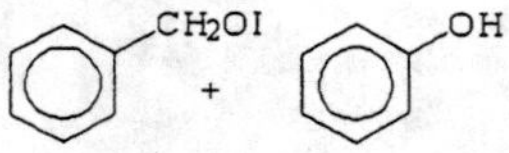 $C_6H_5CH_2OI$ + C_6H_5OH

Answer: b Section: 5

13. What is/are the product(s) from the following reaction?

ethylene oxide (epoxide) + HI ⟶ ?

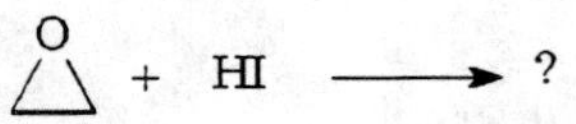

a. $HOCH_2CH_2I$

b. CH_3OH CH_3I

c. 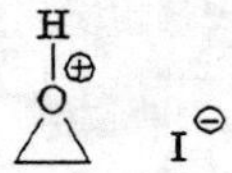 protonated epoxide ($O^{\oplus}$–H) $I^{\ominus}$

d. 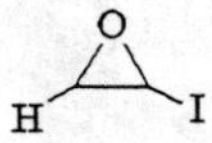 epoxide with H and I substituents

e. CH_3OCH_2I

Answer: a Section: 6

14. What is the major product from the following reaction?

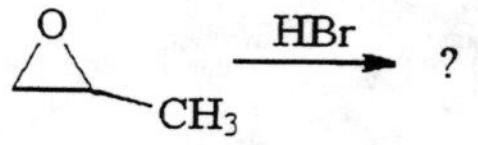

(epoxide with CH_3) $\xrightarrow{HBr}$?

a.  (epoxide)$-CH_2Br$

b. $HOCH_2CHCH_3$ with Br on the CH

c. $BrCH_2CHCH_3$ with OH on the CH

d. 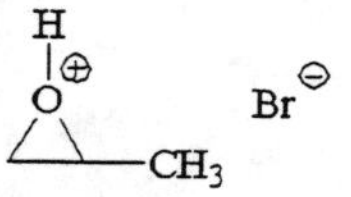 (protonated epoxide, $H-O^{\oplus}$)$-CH_3$ $Br^{\ominus}$

e. $HOCH_2CH_2CH_2Br$

Answer: b Section: 6

15. What is the major product for the following reaction?

(epoxide with two CH_3 on one carbon) $\xrightarrow[CH_3OH]{H^+}$?

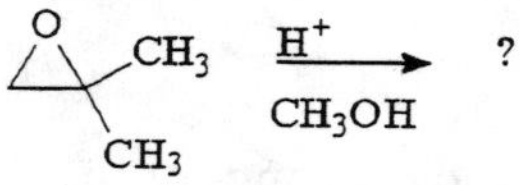

a. $CH_3OCH_2CH_2CH_2CH_2OH$

b. $HOCHCHOCH_3$ with CH_3 on each CH

c. 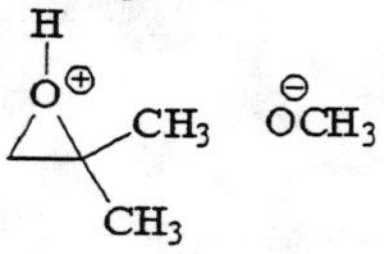 (protonated epoxide, $H-O^{\oplus}$, with two CH_3) $\overset{\ominus}{O}CH_3$

d. $HOCH_2CCH_3$ with CH_3 and OCH_3 on the C

e. $CH_3OCH_2CCH_3$ with CH_3 and OH on the C

Answer: d Section: 6

16. What is the major product for the following reaction?

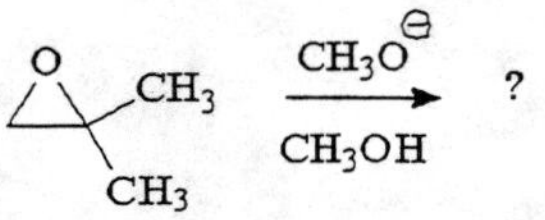

(2,2-dimethyloxirane) $\xrightarrow[CH_3OH]{CH_3O^{\ominus}}$?

a. $CH_3OCH_2CH_2CH_2CH_2OH$

b. 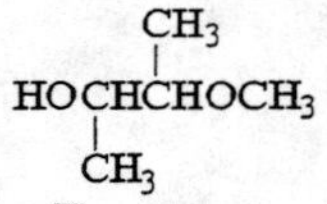 $HOCH(CH_3)CH(CH_3)OCH_3$

c. 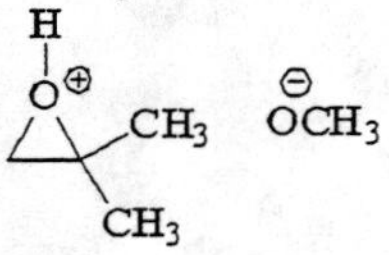 (protonated 2,2-dimethyloxirane, $O^{\oplus}$–H) $\; ^{\ominus}OCH_3$

d. 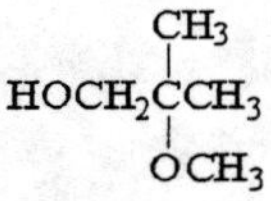 $HOCH_2C(CH_3)_2OCH_3$

e. 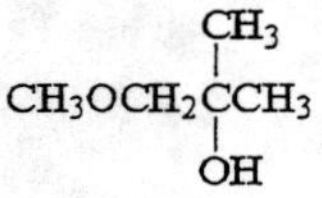 $CH_3OCH_2C(CH_3)_2OH$

Answer: e Section: 6

17. Which of the following is the strongest base?

a. HOMgBr
b. H_2O
c. CH_4
d. CH_3OH
e. CH_3MgBr

Answer: e Section: 8

18. What is the product of the following reaction?

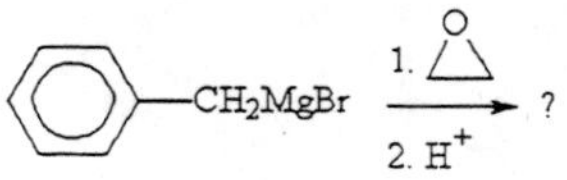

$C_6H_5CH_2MgBr \xrightarrow[\text{2. } H^+]{\text{1. ethylene oxide}} ?$

a. 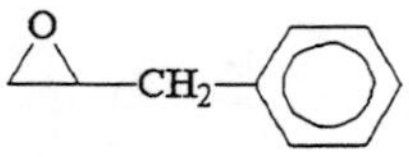 (epoxide)–CH_2–C_6H_5

b. $CH_3CH(OH)CH_2$–C_6H_5

c. $HOCH_2CH_2$–C_6H_5

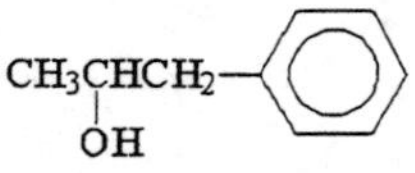

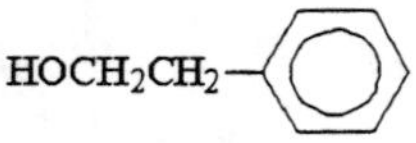

d. 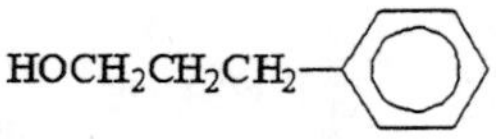 $HOCH_2CH_2CH_2$–C_6H_5

e. 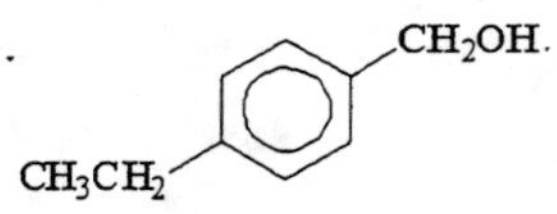 CH_3CH_2–C_6H_4–CH_2OH.

Answer: d Section: 8

19. What is the product of the following reaction sequence?

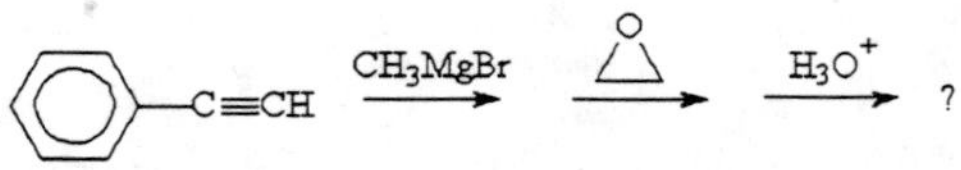

$C_6H_5C{\equiv}CH \xrightarrow{CH_3MgBr} \xrightarrow{\text{ethylene oxide}} \xrightarrow{H_3O^+} ?$

a. 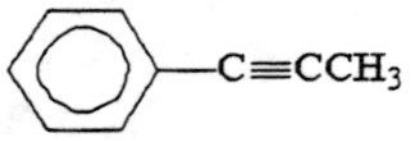 C_6H_5–$C{\equiv}CCH_3$

b. C_6H_5–$C{\equiv}CCH_2CH_2OCH_3$

c. C_6H_5–$C{\equiv}CCH_2OCH_2CH_3$

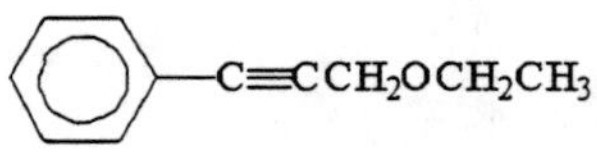

d. 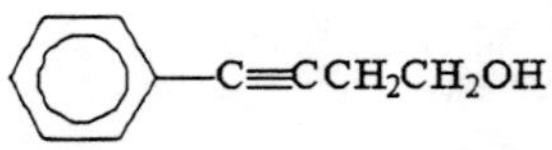 C_6H_5–$C{\equiv}CCH_2CH_2OH$

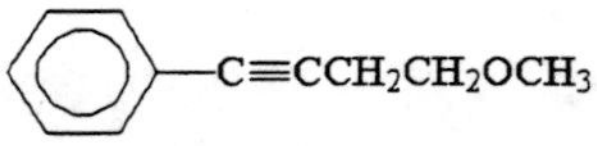

e. 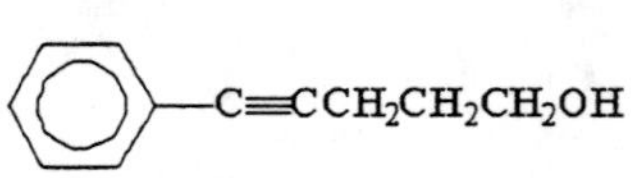 C_6H_5–$C{\equiv}CCH_2CH_2CH_2OH$

Answer: d Section: 8

20. What is the product of the following reaction?

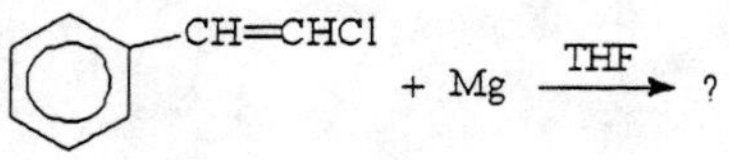

(phenyl)–CH=CHCl + Mg $\xrightarrow{THF}$?

a. 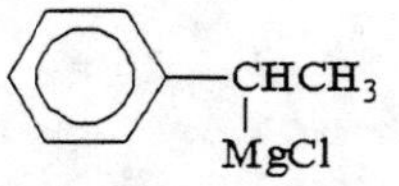 (phenyl)–CH($MgCl$)CH_3

b. 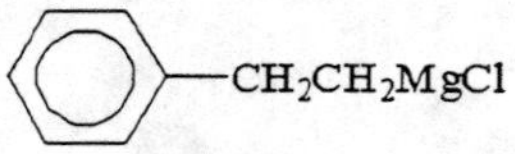 (phenyl)–CH_2CH_2MgCl

c. 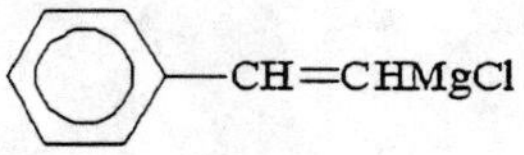 (phenyl)–CH=CHMgCl

d. 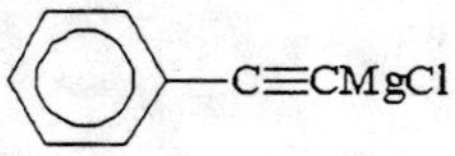 (phenyl)–C≡CMgCl

e. 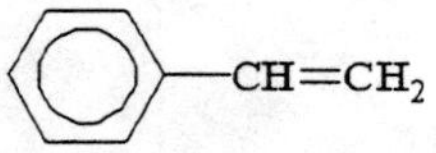 (phenyl)–CH=CH_2

Answer: c Section: 8

21. Which of the following is phenylthiol?

a. 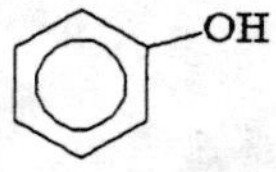 (phenyl)–OH

b. 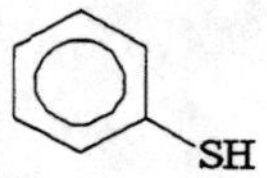 (phenyl)–SH

c. 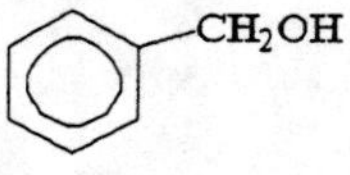 (phenyl)–CH_2OH

d. 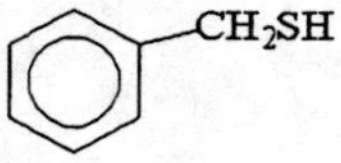 (phenyl)–CH_2SH

e. 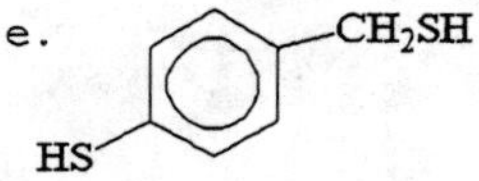 HS–(phenyl)–CH_2SH

Answer: b Section: 10

22. Which of the following is the strongest acid?

    a. CH_3NH
    b. CH_3OH
    c. CH_3SH
    d. CH_3OCH_3
    e. CH_3Cl

    Answer: c Section: 10

23. Which of the following is a quaternary ammonium ion?

```
            R
            |
a.   R  —  C  —  R

            ⊕

            R
            |
b.   R  —  C  —  R
            |
            R

            R
            |
c.   R  —  N  —  R
            ··
            R

            R
            |
d.   R  —  N  —  R
            |
            R

            R
            |⊕
e.   R  —  N  —  R
            |
            R
```

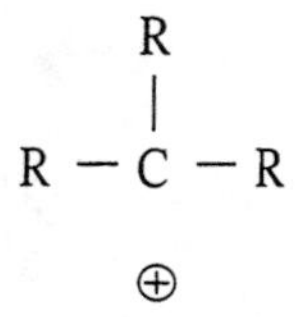

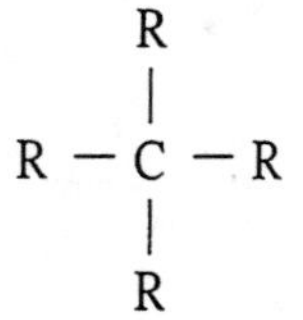

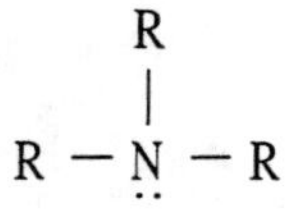

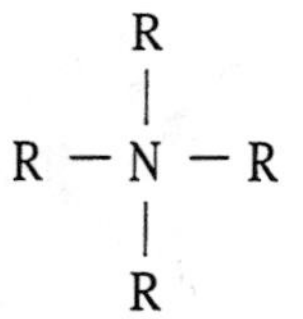

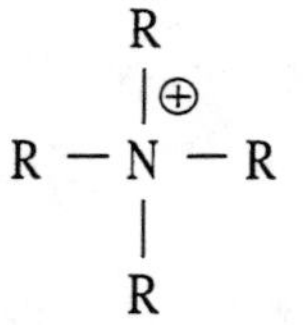

Answer: e Section: 11

24. What is the major alkene formed in the following Hofmann elimination?

$$CH_3CH_2CH_2\overset{\oplus}{N}(CH_3)_2-C(CH_3)_3 \;\; {}^{\ominus}OH \xrightarrow{\text{heat}} ?$$

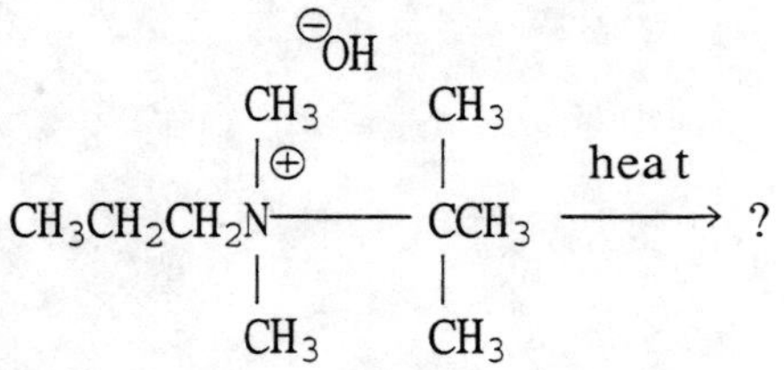

a. $CH_3CH=CH_2$
b. $CH_3-\underset{\|}{C}(=CH_2)-CH_3$
c. $CH_3CH_2CH=CH_2$
d. $CH_3CH(CH_3)CH=CH_2$
e. $CH_3CH_2C(CH_3)=CH_2$

Answer: a Section: 11

25. What is the major alkene formed in the following Hofmann elimination?

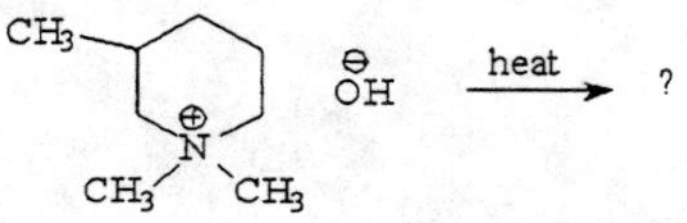

CH_3 / $\overset{\oplus}{N}$ / CH_3 CH_3 $\overset{\ominus}{O}H$ $\xrightarrow{\text{heat}}$?

a. 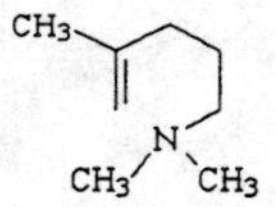 CH_3 / N / CH_3 CH_3

b. 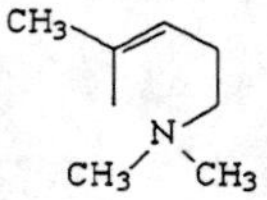 CH_3 / N / CH_3 CH_3

c. 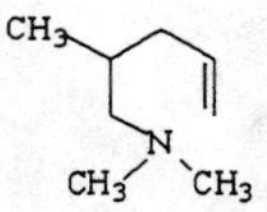 CH_3 / N / CH_3 CH_3

d. 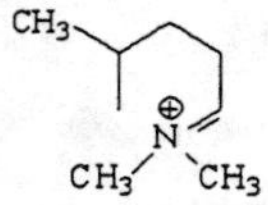 CH_3 / $\overset{\oplus}{N}$ / CH_3 CH_3

e. 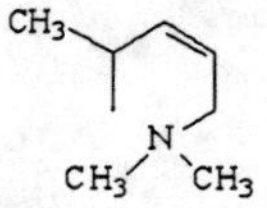 CH_3 / N / CH_3 CH_3

Answer: c Section: 11

26. What is the product of the following reaction?

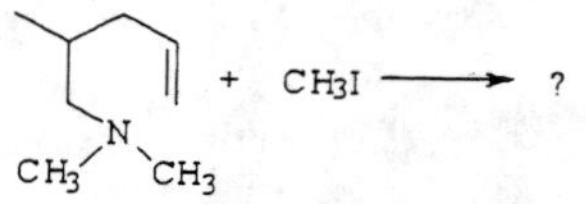

+ CH_3I ⟶ ?

CH3 N CH3

a.

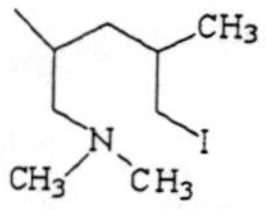

CH3 I N CH3 CH3

b.

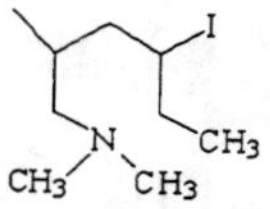

I N CH3 CH3 CH3

c.

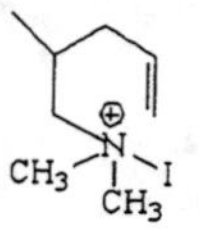

⊕ N CH3 I CH3

d.

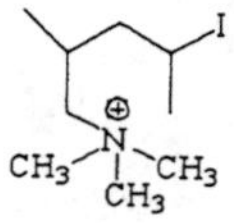

I ⊕ N CH3 CH3 CH3

e.

⊕ N CH3 CH3 CH3 I⊖

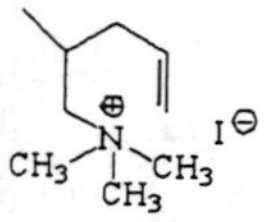

Answer: e Section: 11

27. What is the <u>major</u> alkene formed in the following reaction?

+ $\xrightarrow[H_2O]{Ag_2O}$ $\xrightarrow{heat}$?

⊕ N CH3 CH3 CH3 I⊖

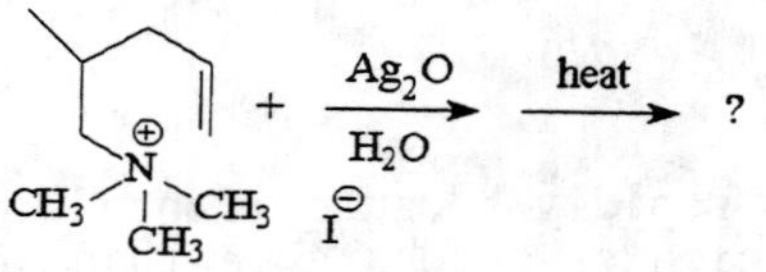

a.

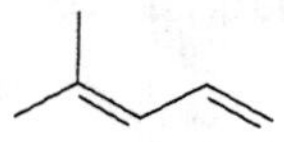

b.

c.

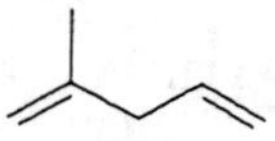

d.

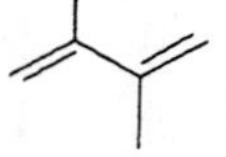

e.

Answer: c Section: 11

28. What is the major alkene produced in the following Cope elimination?

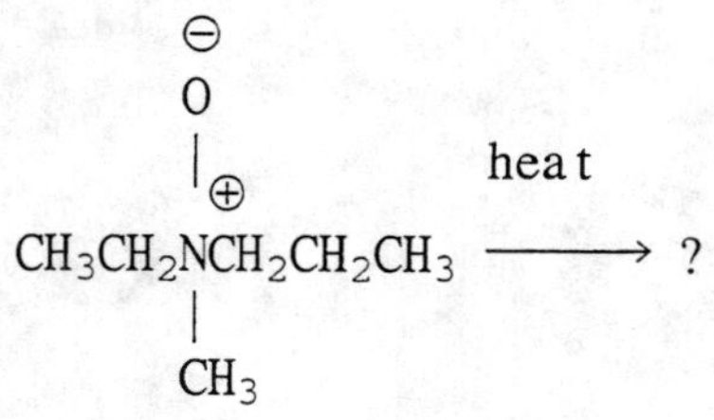

$$CH_3CH_2\overset{\overset{O^{\ominus}}{|}}{\underset{\underset{CH_3}{|}}{N^{\oplus}}}CH_2CH_2CH_3 \xrightarrow{\text{heat}} ?$$

a. $CH_2{=}CH_2$
b. $CH_3CH{=}CH_2$
c. $CH_3CH_2N{=}CH_2$
d. $CH_3CH_2N{=}O$
e. $CH_3CH_2CH_2N{=}O$

Answer: a Section: 11

SHORT ANSWER

29. Which of the following alcohols, when heated with H_2SO_4, will undergo dehydration more rapidly? Explain.

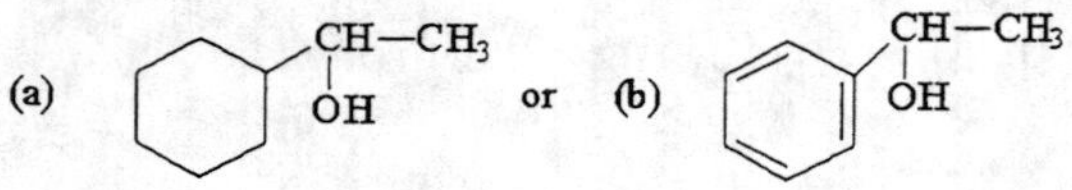

(a) cyclohexyl–CH(OH)–CH$_3$ or (b) phenyl–CH(OH)–CH$_3$

Answer:
(b) will undergo dehydration more rapidly because it produces a secondary benzylic carbocation, which is more stable than secondary carbocation.

Section: 4

30. A compound (A) $C_4H_{10}O$ is optically active. (A) reacts with HCl and $ZnCl_2$ (Lucas test) to form a cloudy product within 10 minutes. Deduce the structure of (A).

Answer:

One of the enantiomers of $CH_3CH_2\overset{\overset{H}{|}}{\underset{\underset{CH_3}{|}}{C}}—OH$

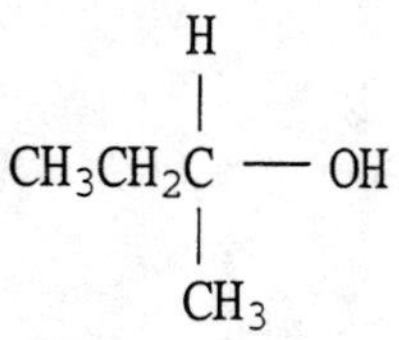

Section: 1

31. The following alkyl halide could not be used to form a Grignard reagent. Explain.

$$HO\text{-}CH_2CH_2\text{-}Br$$

Answer:
Since the alkyl halide contains an acidic proton on the oxygen, the Grignard reagent could not be formed.

Section: 8

32. Show how the following synthesis could be carried out.

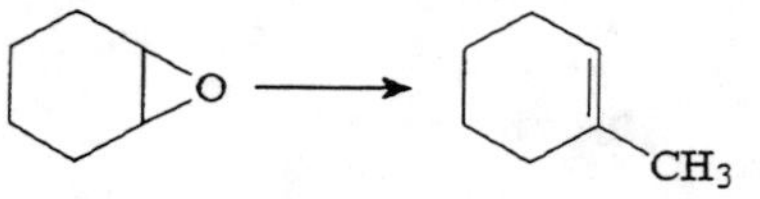

CH_3

Answer:

1. CH_3MgBr
2. H_2O

OH

CH_3

H^+
heat

CH_3

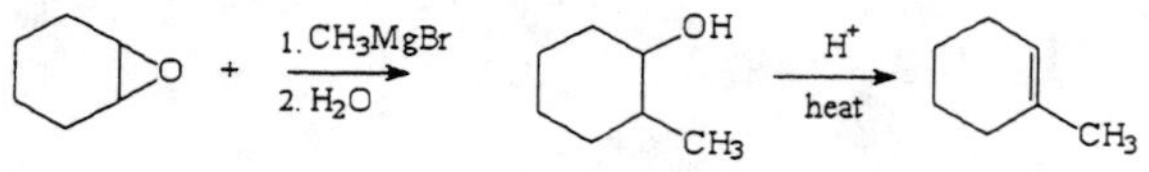

Section: 8

33. Starting with isopropyl alcohol as the only organic material, synthesize isopropyl propyl ether.

$$CH_3CH_2CH_2OCH(CH_3)CH_3$$

Answer:

$$CH_3CH(OH)CH_3 \xrightarrow[\text{heat}]{H_2SO_4} CH_3CH{=}CH_2 \xrightarrow[\text{Peroxide}]{HBr} CH_3CH_2CH_2Br$$

$$CH_3CH(OH)CH_3 \xrightarrow{Na} CH_3CH(O^-)CH_3$$

$$CH_3CH_2CH_2Br + CH_3CH(O^-)CH_3 \longrightarrow CH_3CH_2CH_2OCH(CH_3)_2$$

Section: 3

34. Propose a mechanism for the following reaction:

$$\underset{OH}{CH_2}CH_2CH_2\underset{OH}{CH_2} \xrightarrow{H^+} \text{(tetrahydrofuran)} + H_2O$$

Answer:

$$\underset{OH}{CH_2}CH_2CH_2\underset{:\ddot{O}H}{CH_2} + H^+ \longrightarrow \text{(protonated diol, intramolecular attack)} \longrightarrow H_2\ddot{O} + \text{(protonated cyclic ether)} \longrightarrow \text{(tetrahydrofuran)} + H_3O^+$$

Section: 1

35. Propose a mechanism for the following reaction:

$$\text{(tetrahydrofuran)} + HBr \xrightarrow{heat} \underset{Br}{CH_2}CH_2CH_2\underset{Br}{CH_2}$$

Answer:

$$\text{(tetrahydrofuran)} + H{-}Br \longrightarrow \text{(protonated cyclic ether)} + :\ddot{Br}:^{\ominus} \longrightarrow \underset{:\ddot{O}H}{CH_2}CH_2CH_2\underset{Br}{CH_2} + H{-}Br$$

$$\longrightarrow \underset{:\overset{\oplus}{O}H_2}{CH_2}CH_2CH_2\underset{Br}{CH_2} + :\ddot{Br}:^{\ominus} \longrightarrow H_2O + \underset{Br}{CH_2}CH_2CH_2\underset{Br}{CH_2}$$

Section: 5

36. Propose a detailed, step-by-step mechanism for the reaction pathway shown below.

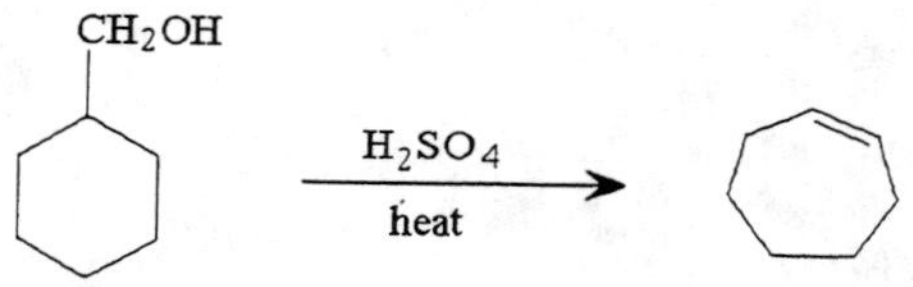

CH_2OH | H_2SO_4 / heat

Answer:

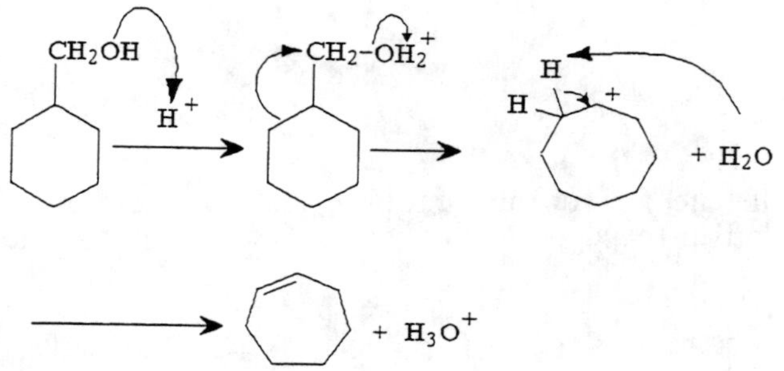

H^+ | $CH_2-OH_2^+$ | H | H | + | $+ H_2O$

$+ H_3O^+$

Section: 4

37. Propose a detailed, step-by-step mechanism for the reaction pathway shown below.

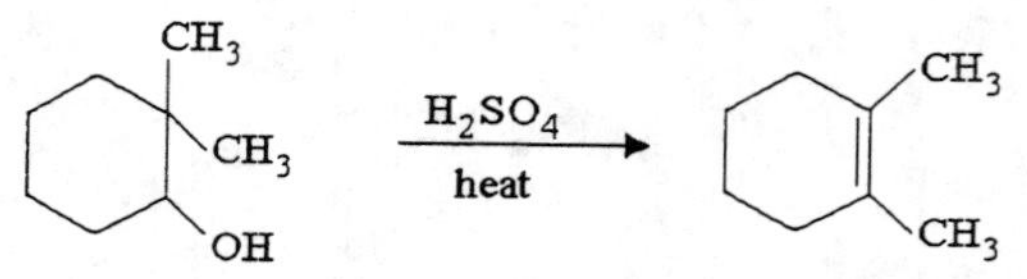

CH_3 | CH_3 | OH | H_2SO_4 / heat | CH_3 | CH_3

Answer:

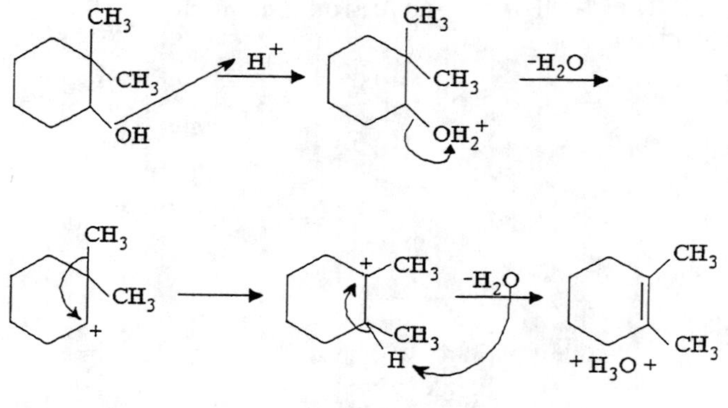

CH_3 | CH_3 | OH | H^+ | CH_3 | CH_3 | OH_2^+ | $-H_2O$

CH_3 | CH_3 | + | + | CH_3 | CH_3 | H | $-H_2O$ | CH_3 | CH_3 | $+ H_3O +$

Section: 4

38. Draw all likely products of the following reaction and circle the product you expect to predominate.

2-pentanol $\xrightarrow{H_2SO_4}$

Answer:

Section: 4

39. Draw all likely products of the following reaction and circle the product you expect to predominate.

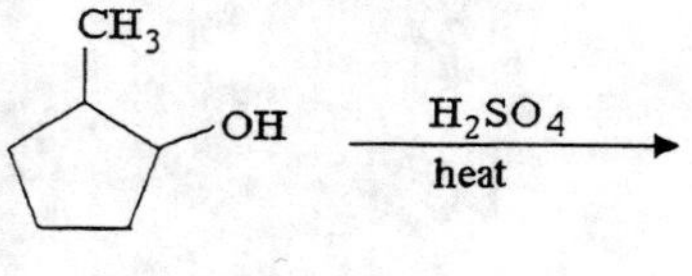

CH_3 OH $\xrightarrow[\text{heat}]{H_2SO_4}$

Answer:

CH_3 + CH_3 + CH_2

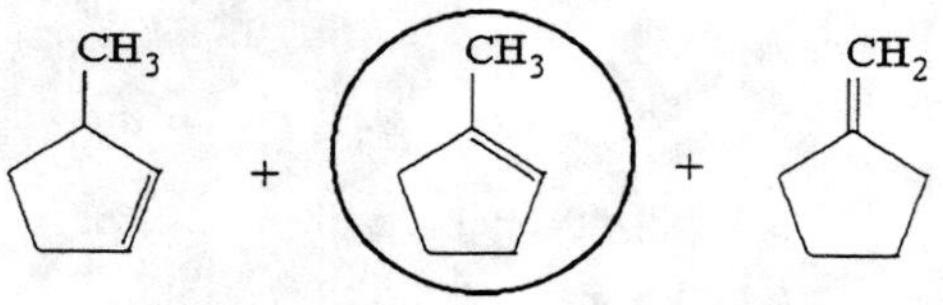

Section: 4

40. Draw all likely products of the following reaction and circle the product you expect to predominate.

OH $\xrightarrow[\text{heat}]{H_2SO_4}$

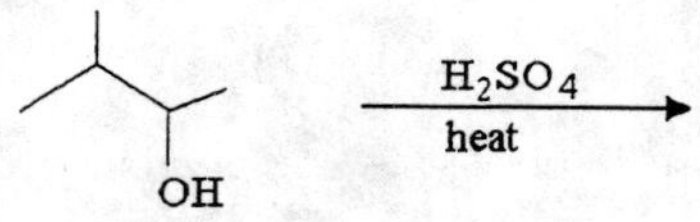

Answer:

Section: 4

41. Provide the structure of the major organic product in the reaction below.

I^- $^+N(CH_3)_2$

1. Ag_2O
2. Δ
3. CH_3I
4. Ag_2O
5. Δ

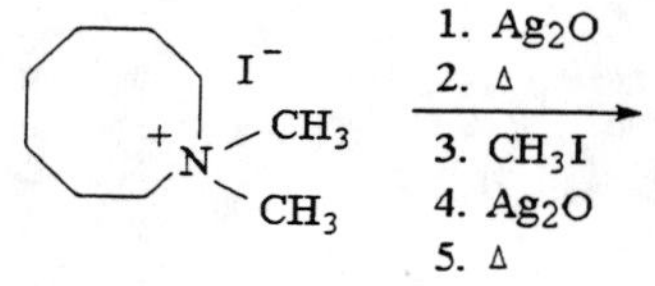

Answer:

Section: 11

42. Provide the structure of the major organic product in the reaction below.

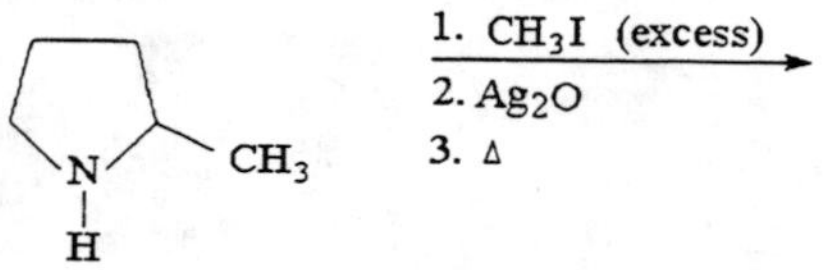

N–H CH_3

1. CH_3I (excess)
2. Ag_2O
3. Δ

Answer:

$N(CH_3)_2$ CH_3

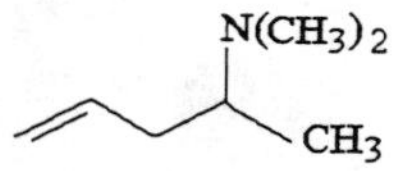

Section: 11

43. Provide the structure of the major organic product in the reaction below.

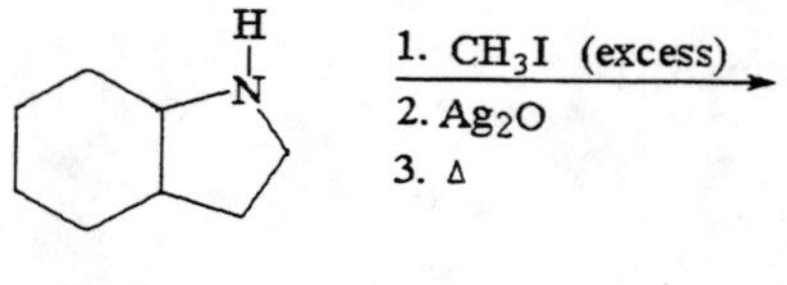

H–N

1. CH_3I (excess)
2. Ag_2O
3. Δ

Answer:

$N(CH_3)_2$

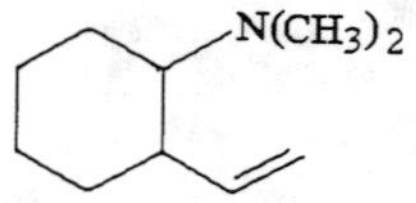

Section: 11

44. Provide the structure of the major organic product in the reaction below.

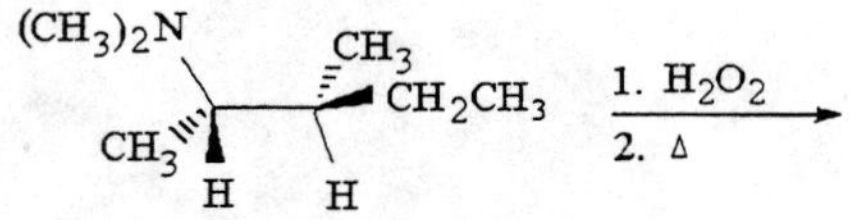

$(CH_3)_2N$, CH_3, H, H, CH_3, CH_2CH_3 — 1. H_2O_2 2. Δ →

Answer:

CH_3, CH_2CH_3, H

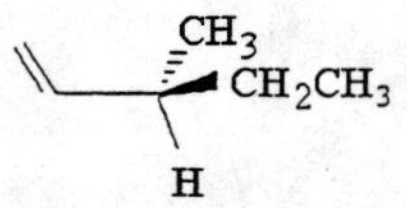

Section: 11

45. Provide the structure of the major organic product in the reaction below.

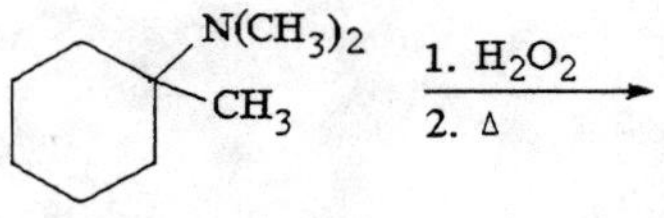

$N(CH_3)_2$, CH_3 — 1. H_2O_2 2. Δ →

Answer:

CH_2

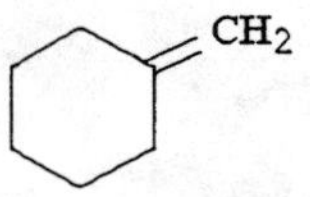

Section: 11

46. How would one use a Gringnard-based synthesis to accomplish the following transformation?

benzyl bromide ($PhCH_2Br$) to 3-phenyl-1-propanol

Answer:
1. Mg, Et_2O
2. ethylene oxide (oxirane)
3. H_3O^+

Section: 8

47. Provide the reagents necessary to accomplish the following transformation.

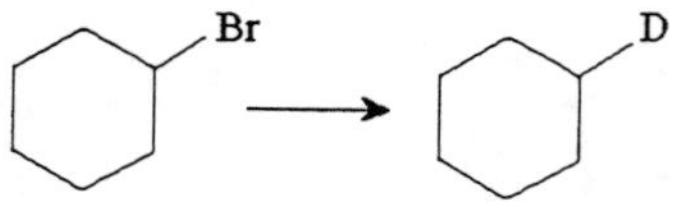

Answer:
1. Mg, Et_2O
2. D_2O

Section: 8

48. Draw the tosylate ion and explain why it is a particularly good leaving group.

Answer:
Extensive resonance stabilization of this ion makes it a particularly good leaving group.

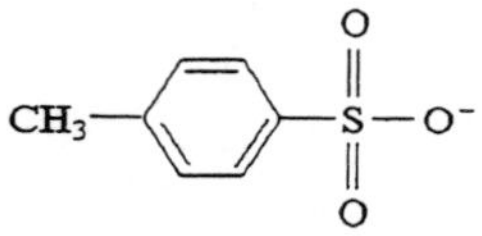

Section: 3

49. Why are alcohols typically poor electrophiles?

Answer:
The hydroxide ion, a relatively strong base, is a very poor leaving group.

Section: 1

50. Provide the structure of the major organic product in the reaction below.

$$PhCH_2CH_2OH \xrightarrow[\text{2. NaCN, acetone}]{\text{1. TsCl, pyridine}}$$

Answer: $PhCH_2CH_2CN$ Section: 3

51. Provide the structure of the major organic product in the reaction below.

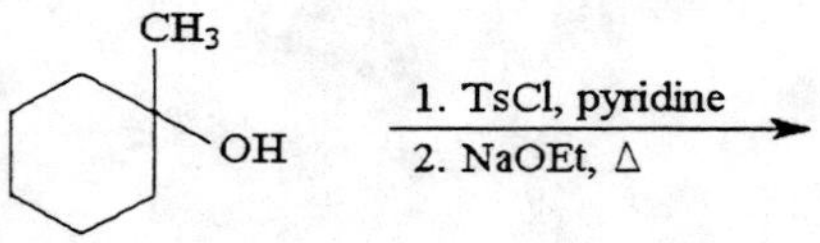

CH3 OH 1. TsCl, pyridine 2. NaOEt, Δ

Answer:

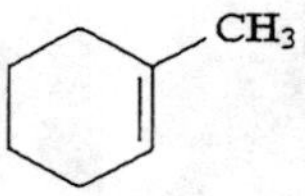

CH3

Section: 3

MULTIPLE CHOICE

52. When (R)-2-butanol is treated with TsCl in pyridine, the product formed is:

a. an achial compound.
b. a mixture of diastereomers.
c. a racemic mixture.
d. a single enantiomer.
e. none of the above.

Answer: d Section: 3

SHORT ANSWER

53. Predict the major product of the reaction below and provide a stepwise mechanism which accounts for its formation.

$CH_3(CH_2)_6CH_2OH$ + HBr ⟶

Answer:

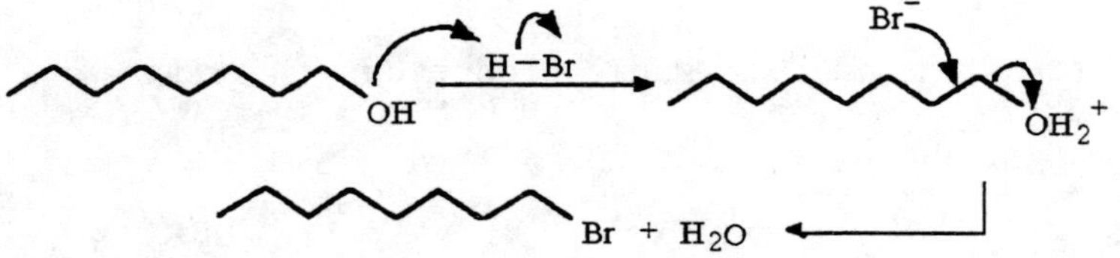

OH H−Br Br^- OH_2^+

Br + H_2O

Section: 1

MULTIPLE CHOICE

54. Which of the following alcohols will react most rapidly with the Lucas reagent (HCl, $ZnCl_2$)?

    a. $(CH_3)_3COH$
    b. $CH_3CH_2CH_2CH_2OH$
    c. $CH_3CHOHCH_2CH_3$
    d. $(CH_3)_2CHCH_2OH$

    Answer: a Section: 1

SHORT ANSWER

55. Provide the structure of the major organic product in the reaction below.

    CH_3 CH_3 CH_3 HO —HCl→

    Answer:

    CH_3 CH_3 CH_3 Cl

    Section: 1

56. Provide the structure of the major organic product in the reaction below.

    CH_3 OH —HCl, $ZnCl_2$→

    Answer:

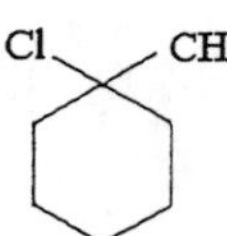

    Section: 1

57. Provide the structure of the major organic product in the reaction below.

$$(CH_3)_2CHCH_2OH \xrightarrow{PBr_3}$$

Answer: $(CH_3)_2CHCH_2Br$ Section: 2

58. Provide the major organic product in the reaction below.

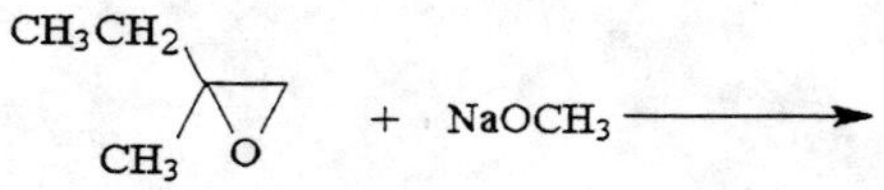

CH_3CH_2, CH_3, O + $NaOCH_3$ →

Answer:

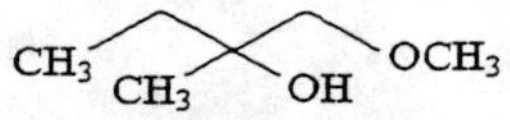

CH_3 CH_3 OH OCH_3

Section: 6

59. Provide the major organic product in the reaction below.

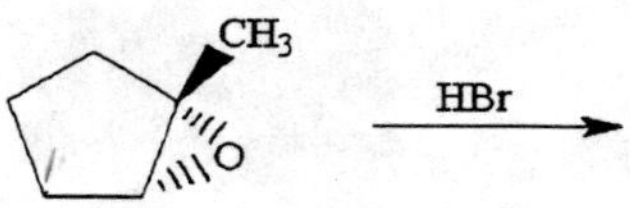

CH_3, O; HBr →

Answer:

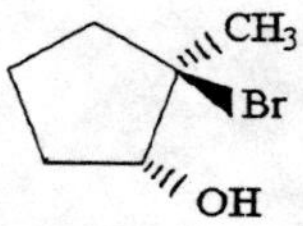

CH_3, Br, OH

Section: 6

60. Provide the major organic product in the reaction below.

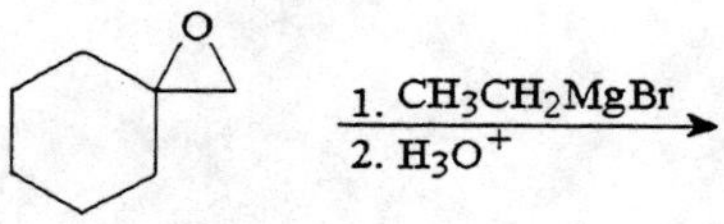

O; 1. CH_3CH_2MgBr 2. H_3O^+ →

Answer:

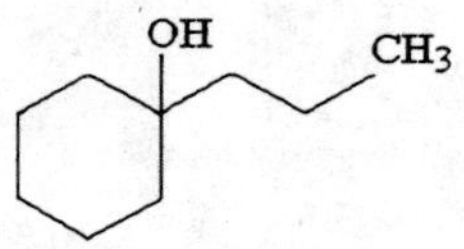

OH CH_3

Section: 6

Chapter 12: Mass Spectrometry and Infrared Spectroscopy

MULTIPLE CHOICE

1. Which of the following statements best explains the information we can gain from mass spectrometry?

   a. It allows us to determine the number of protons in a compound.
   b. It allows us to determine the kinds of functional groups in a compound.
   c. It allows us to determine the molecular weight and the mass of some fragments of a compound.
   d. It allows us to determine the presence and nature of a carbocation in the compound.
   e. It allows us to determine the presence and nature of a free radical in the compound.

   Answer: c

2. Which of the following statements best explains the information we can gain from infrared spectroscopy?

   a. It allows us to determine the number of protons in a compound.
   b. It allows us to determine the kinds of functional groups in a compound.
   c. It allows us to determine the molecular weight and the mass of some fragments of a compound.
   d. It allows us to determine the presence and nature of a carbocation in the compound.
   e. It allows us to determine the presence and nature of a free radical in the compound.

   Answer: b

3. Which of the following molecular changes is necessary for mass spectrometry to occur?

   a. Excitation of an electron from the ground state to higher energy state
   b. Change of alignment of an electron in a magnetic field
   c. Change of alignment of a proton in a magnetic field
   d. Loss of an electron
   e. Molecular vibration

   Answer: d Section: 1

4. Which of the following statements is true about what a molecular ion is?

   a. A compound that lost a pair of electrons
   b. A compound that gained a pair of electrons
   c. A compound that gained one electron
   d. A compound that lost one electron
   e. A compound that carries a free radical and a negative charge

   Answer: d Section: 1

5. Which of the following statements best describes the meaning of the following species:

$$[CH_3CH_2CH_3]^{\bullet +} ?$$

a. It is the molecular ion of propane.
b. It is the parent ion of propane.
c. It is the radical cation of propane.
d. The m/z value is 44.
e. All of the above

Answer: e Section: 1

6. Which of the following is true about the base peak in mass spectrometry?

a. The m/z value equals the molecular weight of the compound.
b. The m/z value corresponds to a very stable carbocation.
c. It has the largest peak height in the spectrum.
d. It has the highest m/z value of all the peaks in the spectrum.
e. The base peak is assigned a relative abundance equal to that of the parent ion.

Answer: c Section: 2

7. Which of the following is <u>not</u> true about the M+1 peak?

a. It is one m/z unit higher than the base peak.
b. It is one m/z unit higher than the molecular ion peak.
c. It is one m/z unit higher than the parent ion peak.
d. It occurs because there are two naturally occurring isotopes of carbon.
e. This means that the number of carbon atoms in a compound can be calculated if the relative abundance of both the M and M+1 peaks is known.

Answer: a Section: 3

8. Which of the following statements best explains how a hydrocarbon can show an M+2 peak in mass spectrometry?

a. From ^{13}C and ^{1}H
b. From ^{12}C and ^{2}H
c. From ^{13}C
d. From ^{2}H
e. From ^{13}C and ^{2}H

Answer: e Section: 3

9. Which of the following m/z values is the base peak for benzyl alcohol?

   a. 77
   b. 108
   c. 91
   d. 17
   e. 52

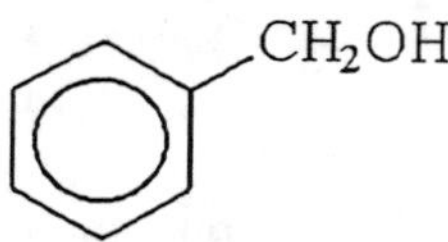

Answer: c Section: 5

10. Which of the following m/z values is the molecular ion for 2-butanone?

$$CH_3-\overset{\overset{\displaystyle O}{\|}}{C}-CH_2CH_3$$

   a. 15
   b. 29
   c. 43
   d. 57
   e. 72

Answer: e Section: 1

11. Which of the following compounds exhibits the pattern of m/z values shown below?

41, 43, 57, 87, 101, 116

   a. propylbromide
   b. isopropyl bromide
   c. sec-butyl isopropyl ether
   d. 2-hexanol
   e. 2-butanone

Answer: c Section: 5

12. Which of the following statements is <u>not</u> true about electromagnetic radiation?

   a. The velocity of light is directly proportional to the energy.
   b. All molecules absorb electromagnetic radiation at some frequency.
   c. Frequency is inversely proportional to wavelength.
   d. Energy is directly proportional to frequency.
   e. Energy is inversely proportional to wavelength.

Answer: a Section: 6

13. Which of the following regions of the electromagnetic spectrum has the greatest energy per photon?

a. Visible
b. Microwave
c. Radio
d. Ultraviolet
e. Infrared

Answer: d Section: 6

14. Which of the following normally occurs in a molecule when a photon of infrared light is absorbed?

a. An electron moves to an orbital of higher potential energy.
b. The vibration energy increases.
c. An electron changes alignment in a magnetic field.
d. The molecule gains an electron.
e. The molecule loses an electron.

Answer: b Section: 6

15. How many centimeters (cm) are there in one micrometer (μm)?

a. 10^{-6}
b. 10^{-3}
c. 10^{-5}
d. 10^{-8}
e. 10^{-4}

Answer: e Section: 6

16. Which of the following are considered to be bending (in-plane) vibrations?

a. scissoring and wagging
b. scissoring and twisting
c. rocking and wagging
d. rocking and twisting
e. scissoring and rocking

Answer: e Section: 7

17. Which of the following solvents is best used in infrared spectroscopy?

 a. Water, H_2O
 b. Carbon tetrachloride, CCl_4
 c. Methanol, CH_3OH
 d. Ethanol, CH_3CH_2OH
 e. Benzene,

Answer: b Section: 7

18. Which of the infrared regions is considered to be the <u>fingerprint</u> region?

 a. $4000 cm^{-1}$ - $1000 cm^{-1}$
 b. $4000 \mu m$ - $1000 \mu m$
 c. $2200 \mu m$ - $1000 \mu m$
 d. $1000 cm^{-1}$ - $400 cm^{-1}$
 e. $1000 \mu m$ - $400 \mu m$

Answer: d Section: 7

19. Which of the following wavenumbers corresponds to the bond shown below?

$C \equiv C$

 a. $1650\ cm^{-1}$
 b. $2100\ cm^{-1}$
 c. $1100\ cm^{-1}$
 d. $3300\ cm^{-1}$
 e. $2850\ cm^{-1}$

Answer: b Section: 8

20. Which of the following carbonyl groups exhibits the highest wavenumber in infrared spectroscopy?

a. 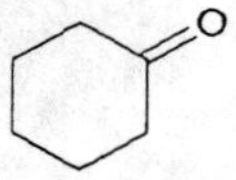 (cyclohexanone)

b. 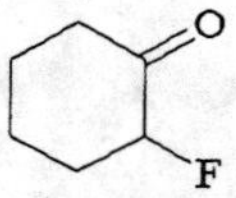 (2-fluorocyclohexanone)

c. 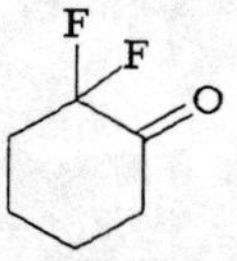 (2,2-difluorocyclohexanone)

d. 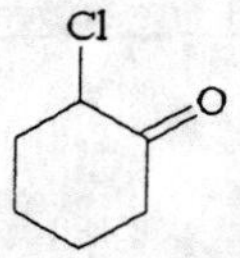 (2-chlorocyclohexanone)

e. 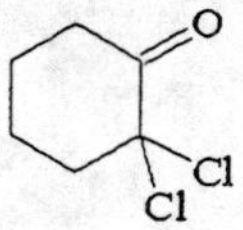 (2,2-dichlorocyclohexanone)

Answer: c Section: 8

21. Which of the following carbon-hydrogen bonds exhibits the lowest wavenumber for a C—H stretch in infrared spectroscopy?

a. 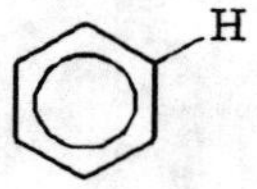 (benzene C—H)

b. O=C—H

c. C—C—H

d. C=C—H

e. C≡C—H

Answer: b Section: 8

22. Which of the following compounds has a vibration that is infrared inactive?

    a. $CH_3-C\equiv C-H$
    b. H_2O
    c. $CH_3-\overset{\overset{\displaystyle O}{\|}}{C}-CH_3$
    d. CO_2
    e. CO

    Answer: d Section: 9

23. Which of the following functional groups will exhibit no IR absorption at 1630 - 1780 cm^{-1} or at 3200 - 3550 cm^{-1}?

    a. An alcohol
    b. An amide
    c. A ketone
    d. An aldehyde
    e. An ether

    Answer: e Section: 8

24. In which of the following molecules is the presence of the IR peak for C=C not observed?

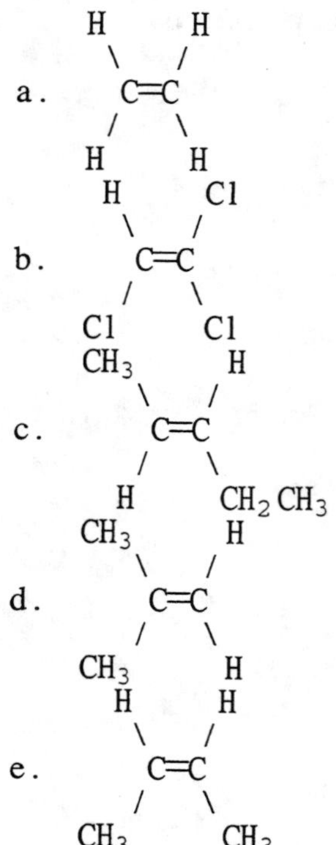

    Answer: a Section: 9

25. Predict the molecular formula of the compound represented below based on the MS data given.

m/z	Intensity
84	10.00
85	0.56
86	0.04

a. C_6H_{12}
b. C_5H_{24}
c. $C_4H_6O_2$
d. $C_3H_8O_2$
e. C_5H_8O

Answer: e Section: 3

26. Which of the following characterizes the unusually intense peak of alkyl halides in MS spectrometry?

a. M + 1 peak
b. M + 2 peak
c. Base peak
d. Parent peak
e. None of the above

Answer: b Section: 3

27. Which of the following is the base peak for the compound below?

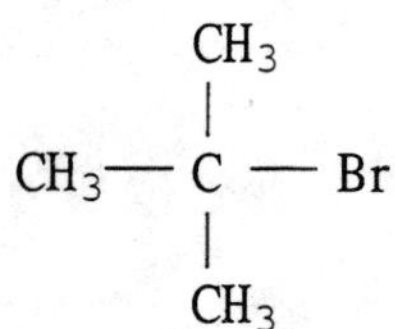

a. 77
b. 92
c. 15
d. 57
e. 43

Answer: d Section: 5

28. Which of the following structures will give a base peak of 43 in mass spectrometry?

    a. $CH_3CH_2CHCH_3$ with CH_2CH_3 on the CH
    b. $CH_3CH_2CH_2CH_2CH_2CH_3$
    c. $(CH_3)_2CH-CH(CH_3)_2$
    d. $CH_3C(CH_3)_2CH_2CH_3$
    e. None of the above

    Answer: c Section: 5

SHORT ANSWER

29. What does m/z stand for and what does it mean?

    Answer:
    The m stands for the mass and z for the charge. Since most fragments have a charge of +1, m/z is the molecular weight of the fragment.
    Section: 1

30. How could you distinguish the mass spectrum of 2,2-dimethyl propane from that of isopentane?

    Answer:
    The loss of a methyl from 2,2-dimethylpropane forms a stable tertiary carbocation, while loss of a methyl from isopentane forms a less stable secondary carbocation. So the M-15 peak will be greater for 2,2-dimethylpropane.
    Section: 2

31. Is it possible to have an M + 2 peak in mass spectrometry? Explain.

Answer:
A mass spectrum can show an M + 2 peak for a compound with a relatively abundant isotope two mass units greater than the most common isotope (such as oxygen, chlorine, and bromine). The presence of a relatively large M + 2 peak is evidence for a compound with chlorine or bromine. For instance, if M + 2 is 1/3 the height of the molecular ion, the compound contains chlorine. If M + 2 is about the same height as that of the molecular ion, the compound contains bromine.

Section: 3

32. The mass spectrum of an unknown compound has a molecular ion peak with a relative abundance of 43.27% and an M + 1 peak with a relative abundance of 3.81%. How many carbon atoms are in the compound?

Answer:

$$\text{\# of carbon atoms} = \frac{\text{relative abundance of M + 1}}{.011 \times (\text{relative abundance of M})}$$

$$\frac{3.81}{.011 \times 43.27} = 8 \text{ carbons}$$

Section: 3

33. Show the m/z values of the molecular ion and all the fragments for the compound ethyl ether,

$CH_3CH_2OCH_2CH_3$

Answer: 74, 59, 45, 44, 30, 29, 15 Section: 5

34. Does the position of the stretch of the oxygen-hydrogen bond in alcohols depend on the concentration of the alcohol?

Answer:
Yes, the more concentrated the alcohol, the more likely it is for the alcohol to form intermolecular hydrogen bonding. So the OH group of a concentrated (hydrogen-bonded) solution absorbs at a lower wavenumber than that of the diluted solution.

Section: 8

35. What are overtone bands and how useful are they in the interpretation of IR bands?

Answer:
Overtone bands arise at multiples of the fundamental absorption frequency. Weak overtones and combination bands in the 2000 - 1650 cm^{-1} region are useful in determining the number and relative position of the substituents on a benzene ring.
Section: 11

MULTIPLE CHOICE

36. Which compound would be expected to show intense IR absorption at 3300 cm^{-1}?

a. CH_3CCCH_3
b. butane
c. 1-butene
d. CH_3CH_2CCH

Answer: d Section: 8

37. Which compound would be expected to show intense IR absorption at 2710 and 1705 cm^{-1}?

a. $CH_3COCH_2CH_3$
b. $PhCOCH_3$
c. PhCHO
d. $CH_2{=}CHCOCH_3$

Answer: c Section: 8

38. Which compound would be expected to show intense IR absorption at 2250 cm^{-1}?

a. $CH_3CH_2CH_2CO_2H$
b. $CH_3CH_2CH_2CN$
c. $CH_3CH_2CH_2CH_2OH$
d. $CH_3CH_2CH_2CONH_2$

Answer: b Section: 8

SHORT ANSWER

39. Deduce a possible structure for the compound with the IR absorptions below.

C_3H_3Br: 3300, 2900, 2100 cm^{-1}

Answer: $HCCCH_2Br$ Section: 8

40. Deduce a possible structure for the compound with the IR absorptions below.

C_3H_5N: 3000, 2250 cm^{-1}

Answer: CH_3CH_2CN Section: 8

41. Deduce a possible structure for the compound with the IR absorptions below.

C_5H_8O: 3000, 1750 cm^{-1}

Answer: cyclopentanone Section: 8

42. Deduce a possible structure for the compound with the IR absorptions below.

C_4H_8O: 3000, 2715, 1715 cm^{-1}

Answer: $CH_3CH_2CH_2CHO$ Section: 8

43. How could IR spectroscopy be used to distinguish between the following pair of compounds?

$CH_3OCH_2CH_3$ and $CH_3CH_2CH_2OH$

Answer: O-H stretch at 3300 cm^{-1} Section: 8

44. How could IR spectroscopy be used to distinguish between the following pair of compounds?

$HOCH_2CH_2CHO$ and $CH_3CH_2CO_2H$

Answer:
Carboxylic acid will have a very broad O-H absorption; aldehyde has a characteristic C-H stretch
Section: 8

MULTIPLE CHOICE

45. Ethyne (HCCH) does not show IR absorption in the region 2000-2500 cm^{-1} because:

a. C-H stretches occur at lower energies.
b. CC stretches occur at about 1640 cm^{-1}.
c. there is no change in the dipole moment when the CC bond in ethyne stretches.
d. there is a change in the dipole moment when the CC bond in ethyne stretches.

Answer: c Section: 9

SHORT ANSWER

46. Describe the molecular ion region in the mass spectrum of CH_3CH_2Br.

Answer:
The natural abundance of the isotopes ^{79}Br and ^{81}Br are about the same; therefore, there will be peaks of equal intensity at m/z 108 and 110.
Section: 3

47. 2-Methylhexane shows an intense peak in the mass spectrum at m/z = 43. Propose a likely structure for this fragment.

Answer:
$(CH_3)_2CH^+$

Section: 2

MULTIPLE CHOICE

48. Which compound would be expected to show intense IR absorption at 1715 cm^{-1}?

a. $CH_3CH_2CO_2H$
b. 1-hexene
c. 2-methylhexane
d. $CH_3CH_2CH_2NH_2$

Answer: a Section: 8

49. Which compound would be expected to show intense IR absorption at 1746 cm^{-1}?

a. $CH_3CH_2OCH_2CH_3$
b. $CH_3CO_2CH_3$
c. CH_3CH_2CCH
d. $CH_3CH_2SCH_3$

Answer: b Section: 8

SHORT ANSWER

50. An infrared wavelength of 4.48μm is equivalent to a wavenumber of __________ cm^{-1}.

Answer: 2230 Section: 6

51. Which has the higher speed in a vacuum, ultraviolet or infrared light?

Answer: They have the same speed Section: 6

52. Which region of the electromagnetic spectrum, IR or UV, contains photons of the higher energy?

Answer: UV Section: 6

53. Which region of the electromagnetic spectrum, radio or visible, is characterized by waves of shorter wavelength?

Answer: visible Section: 6

54. Which region of the electromagnetic spectrum, IR or X-ray, is characterized by waves of lower frequency?

Answer: IR Section: 6

55. Which has a lower characteristic stretching frequency, the C-H or C-D bond? Explain briefly.

Answer: C-D; heavier atoms vibrate more slowly Section: 8

56. Which has a lower characteristic stretching frequency, the C=O bond or the C-O bond? Explain briefly.

Answer:
Stronger bonds are generally stiffer, thus requiring more force to stretch or compress them. The C-O bond is the weaker of the two and hence has the lower stretching frequency.

Section: 8

57. Describe the fate of a molecule from introduction to detection in a mass spectrometer.

Answer:
Upon introduction, sample molecules are ionized by an electron beam passing through a vacuum chamber. The resulting radical cation and fragment cations are accelerated into the flight tube of the magnet by a negatively charged plate. In the flight tube, the path of the ions is bent by the existing magnetic field. For a given magnetic field strength, only one m/z will be bent in such a way that its path matches the curvature of the tube and reaches the detector. The entire m/z range is scanned by varying the strength of the magnetic field.

Section: 1

58. What technique can be used to determine the molecular formula of a compound?

Answer: High resolution mass spectrometry Section: 4

MULTIPLE CHOICE

59. The mass spectra of alcohols often fail to exhibit detectable M peaks but instead show relatively large ______ peaks.

a. M+1
b. M+2
c. M-16
d. M-17
e. M-18

Answer: e Section: 5

SHORT ANSWER

60. How could IR spectroscopy be used to distinguish between the following pair of compounds?

$(CH_3)_3N$ and $CH_3NHCH_2CH_3$

Answer: N-H absorption near 3300 cm^{-1} Section: 8

Chapter 13: NMR Spectroscopy and Ultraviolet/Visible Spectroscopy

MULTIPLE CHOICE

1. Which of the following spectroscopic techniques uses the lowest energy of the electromagnetic radiation spectrum?

   a. UV
   b. visible
   c. IR
   d. X-ray
   e. NMR

   Answer: e Section: 1

2. Which of the following is used in nuclear magnetic resonance spectroscopy?

   a. X-ray
   b. infrared
   c. visible
   d. radio
   e. ultraviolet

   Answer: d Section: 1

3. If the frequency for flipping an ^{1}H nucleus at an applied field of 1.4092 Tesla is 60 MHz, what would be the applied magnetic field if the frequency is 360 MHz?

   a. 8.4552 Tesla
   b. 0.2349 Tesla
   c. 1.4092 Tesla
   d. 4.2577 Tesla
   e. 3.0439 Tesla

   Answer: a Section: 1

4. How many signals would you expect to see in the ^{1}H NMR spectrum of the following compound?

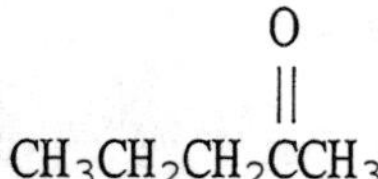

   a. 2
   b. 4
   c. 3
   d. 5
   e. 6

   Answer: b Section: 2

5. How many signals would you expect to see in the ^{1}H NMR spectrum of the following compound?

$$CH_3CH_2CH_2CH_3$$

a. 1
b. 3
c. 2
d. 4
e. 6

Answer: c Section: 2

6. How many signals would you expect to see in the ^{1}H NMR spectrum of the following compound?

a. 6
b. 3
c. 5
d. 4
e. 2

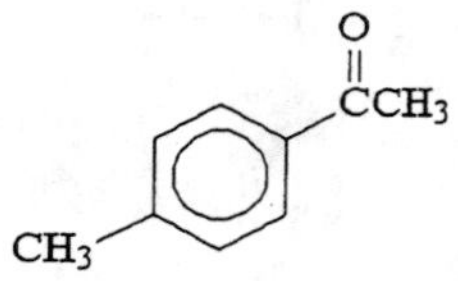

O
||
CCH3
CH3

Answer: d Section: 2

7. How many signals would you expect to see in the ^{1}H NMR spectrum of the following compound?

a. 6
b. 5
c. 4
d. 3
e. 2

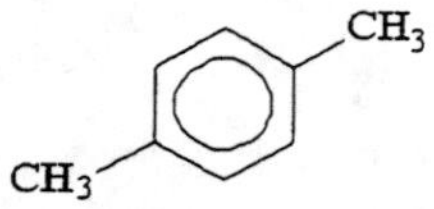

CH3
CH3

Answer: e Section: 2

8. How many signals would you expect to see in the ^{1}H NMR spectrum of the following compound?

   a. 1
   b. 2
   c. 3
   d. 4
   e. 5

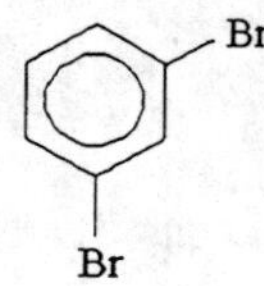

Answer: c Section: 2

9. How many signals would you expect to see in the ^{1}H NMR spectrum of the following compound?

   a. 5
   b. 4
   c. 2
   d. 3
   e. 1

OCH_3

OCH_3

Answer: d Section: 2

10. How many signals would you expect to see in the ^{1}H NMR spectrum of the following compound?

   a. 2
   b. 5
   c. 1
   d. 4
   e. 3

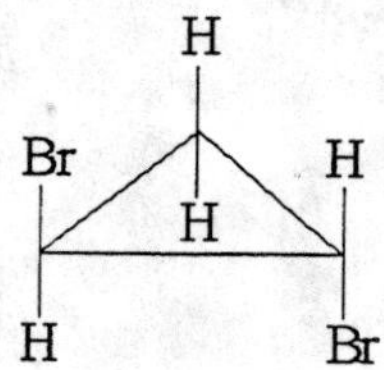

Answer: a Section: 2

11. How many signals would you expect to see in the ^{1}H NMR spectrum of the following compound?

    a. 1
    b. 2
    c. 4
    d. 5
    e. 3

Br

H H H

H H

Answer: e Section: 2

12. How many signals would you expect to see in the ^{1}H NMR spectrum of the following compound?

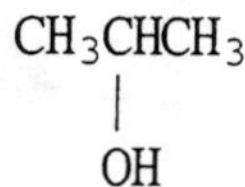

    a. 7
    b. 3
    c. 4
    d. 2
    e. 8

Answer: b Section: 2

13. How many signals would you expect to see in the ^{1}H NMR spectrum of the following compound?

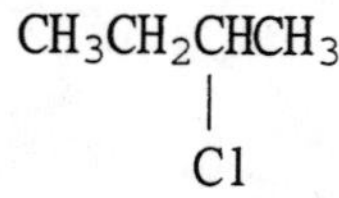

    a. 1
    b. 2
    c. 4
    d. 3
    e. 6

Answer: c Section: 2

14. How many signals would you expect to see in the ^1H NMR spectrum of the following compound?

$ClCH_2CH_2Cl$

a. 5
b. 4
c. 3
d. 2
e. 1

Answer: e Section: 2

15. If a chemical shift of an NMR signal is 7.2 ppm measured in a 60 MHz NMR spectrometer, how many Hz would this signal be from the TMS signal?

a. 8.3 Hz
b. 432 Hz
c. 0.12 Hz
d. 72 Hz
e. 60 Hz

Answer: b Section: 3

16. If two signals differ by 3 ppm, how do they differ in Hertz in a 60 MHz spectrometer?

a. 20 Hz
b. 300 Hz
c. 360 Hz
d. 180 Hz
e. 90 Hz

Answer: d Section: 3

17. What is the ratio of the protons in the following compound?

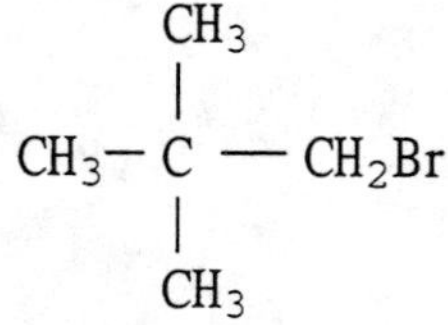

a. 3:3:3:2
b. 6:3:2
c. 9:2
d. 3:2
e. 6:2

Answer: c Section: 4

18. What is the ratio of the protons in the following compound?

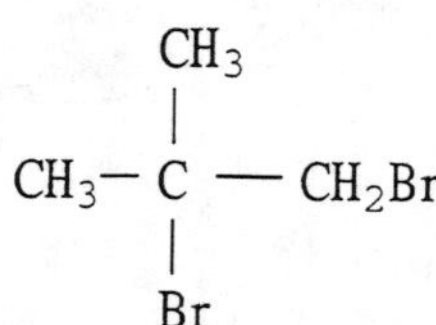

a. 3:3:2
b. 3:2
c. 6:2:1
d. 3:1
e. 3:2:1

Answer: d Section: 4

19. Which of the following protons gives an NMR signal with the highest chemical shift value (furthest downfield)?

$$F - CH_2CH_2CH_2CH_2CH_2 - Br$$
1 2 3 4 5

a. 1
b. 2
c. 3
d. 4
e. 5

Answer: a Section: 3

20. Which of the following protons gives an NMR signal with the lowest chemical shift value (furthest upfield)?

$$F - CH_2CH_2CH_2CH_2CH_2 - Br$$
1 2 3 4 5

a. 1
b. 2
c. 3
d. 4
e. 5

Answer: c Section: 3

21. Which of the following protons gives an NMR signal with the highest chemical shift value (furthest downfield)?

$$\underset{1}{(CH_3)_2}\underset{2}{CH} - O - \underset{3}{CH_2}\underset{4}{CH_2}\underset{5}{CH_3}$$

a. 1
b. 2
c. 3
d. 4
e. 5

Answer: b Section: 3

22. Which of the following compounds gives the highest chemical shift value (furthest downfield) in the NMR spectrum?

a. $(CH_3)_4C$

b. $(CH_3)_4Si$

c. $(CH_3)_3N$:

d. 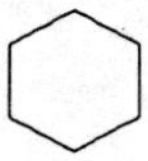 [cyclohexane structure]

e. 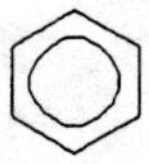 [benzene structure]

Answer: e Section: 5

23. Which of the following protons gives an NMR signal with the lowest chemical shift value (furthest upfield)?

a. 1
b. 2
c. 3
d. 4
e. 5

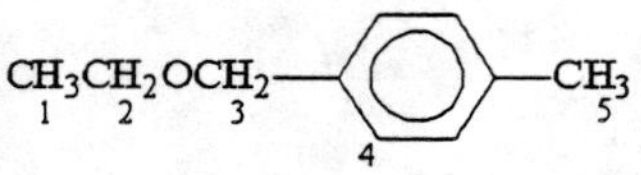

$\underset{1}{CH_3}\underset{2}{CH_2}O\underset{3}{CH_2}-C_6H_4(4)-\underset{5}{CH_3}$

Answer: a Section: 5

24. What splitting pattern is observed in the proton NMR spectrum for the indicated hydrogens?

 a. Singlet
 b. Doublet
 c. Triplet
 d. Quartet
 e. Septet

OCH_2CH_3

Answer: d Section: 6

25. What splitting pattern is observed in the proton NMR spectrum for the indicated hydrogens?

 a. Singlet
 b. Doublet
 c. Triplet
 d. Quartet
 e. Septet

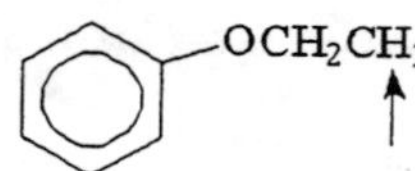

Answer: c Section: 6

26. What splitting pattern is observed in the proton NMR spectrum for the indicated hydrogens?

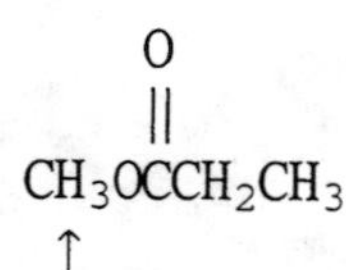

 a. Singlet
 b. Doublet
 c. Triplet
 d. Quartet
 e. Septet

Answer: a Section: 6

27. What splitting pattern is observed in the proton NMR spectrum for the indicated hydrogens?

$$CH_3OCH_2CH_2OCH_3$$
↑

a. Singlet
b. Doublet
c. Triplet
d. Quartet
e. Septet

Answer: a Section: 6

28. Which of the following methods is most suitable for studying conjugated systems?

a. Infrared
b. NMR
c. Mass spectrometry
d. X-ray
e. UV Visible

Answer: e Section: 16

29. Which of the following compounds exhibits the highest λ_{max} in UV spectra?

a. 
b. 
c. 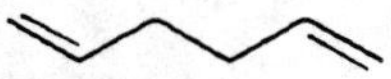
d. 
e.

Answer: d Section: 16

30. Which color is observed when a sample absorbs light at 650 nm?

a. yellow
b. orange
c. red
d. green
e. blue

Answer: e Section: 18

31. An unknown compound, C_9H_{12}, gave the following NMR spectrum:

Triplet at 1.21 ppm (3H)
Singlet at 2.30 ppm (3H)
Quartet at 2.60 ppm (2H)
Singlet at 7.04 ppm (4H)
What is the structure of the compound?

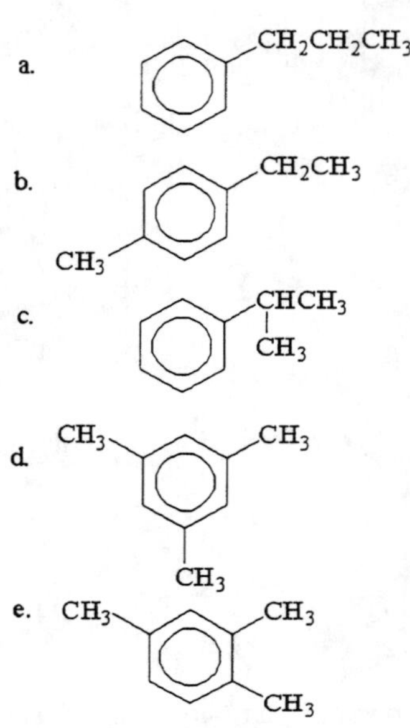

Answer: b Section: 1

32. Which of the following methyl groups will exhibit the most downfield (highest chemical shift) in ^1H NMR spectroscopy?

a. CH_3Cl
b. CH_3OH
c. $CH_3\overset{O}{\overset{\|}{C}}CH_3$
d. $CH_3CH_2CH_3$
e. CH_3 (on benzene ring)

Answer: b Section: 3

33. How many proton NMR singlets will 2-bromo-3-methyl-2-butene exhibit?

a. 1
b. 2
c. 3
d. 4
e. 5

Answer: c Section: 2

34. Which of the following technique(s) can readily distinguish between:

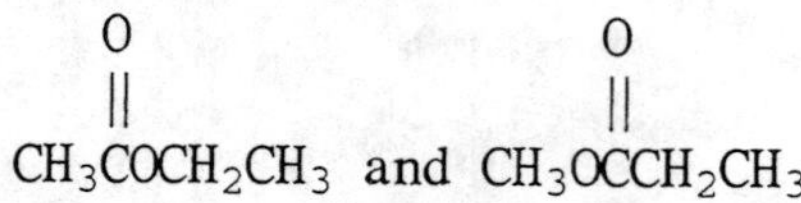

a. NMR
b. IR
c. MS
d. a and b
e. a and c

Answer: e

SHORT ANSWER

35. Which of the following compounds has a signal disappear when D_2O is added to it, ethyl alcohol or ethyl chloride?

Answer: The signal would disappear for ethyl alcohol and not for ethyl chloride.

Section: 9

36. An unknown compound, $C_4H_8Br_2$, gave the following proton NMR data:
Singlet at 1.97 ppm (6H)
Singlet at 3.89 ppm (2H)
What is the compound?

Answer:

singlet at 1.97 ppm { CH_3, CH_3 } — C(Br) — CH_2Br ← singlet at 3.89 ppm

Section: 1

37. An unknown compound, $C_3H_5Cl_3$, gave the following proton NMR data:
Doublet at 1.70 ppm (3H)
Multiplet at 4.32 ppm (1H)
Doublet at 5.85 ppm (1H)
What is the structure of the compound?

Answer:

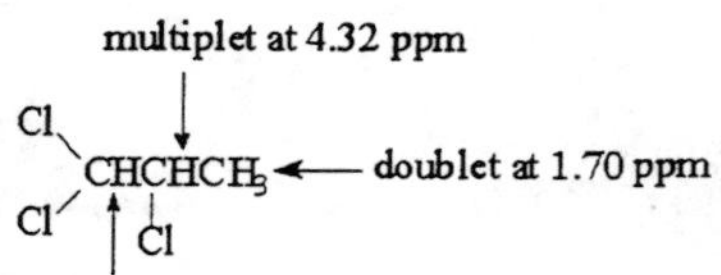

Section: 1

38. An unknown compound, $C_4H_{10}O$, gave the following proton NMR data:
Triplet at 1.13 ppm
Quartet at 3.38 ppm
What is the structure of the compound?

Answer:

quartet at 3.38 ppm

$CH_3CH_2OCH_2CH_3$

triplet at 1.13 ppm

Section: 1

MULTIPLE CHOICE

39. Absorption of UV-visible energy by a molecule results in:

a. vibrational transitions
b. electronic transitions
c. rotational transitions
d. nuclear transitions
e. none of the above

Answer: b Section: 16

40. In the UV-visible spectrum of (E)-1,3,5-hexatriene, the lowest energy absorption corresponds to:

a. a π to π^* transition.
b. a σ to π^* transition.
c. a π to σ^* transition.
d. a σ to σ^* transition.
e. a σ to π transition.

Answer: a Section: 16

41. Which of the following compounds absorbs the longest wavelength of UV-visible light?

a. (E)-2-butene
b. (Z)-2-butene
c. 1-hexene
d. (Z)-1,3-hexadiene
e. (E)-1,3,5-hexatriene

Answer: e Section: 16

SHORT ANSWER

42. Deduce the identity of the following compound from the spectral data given.

$C_9H_{10}O_2$: ^{13}C NMR, δ 18.06 (quartet), 45.40 (doublet), 127.32 (doublet), 127.55 (doublet), 128.61 (doublet), 139.70 (singlet), 180.98 (singlet); IR, broad 3500-2800, 1708 cm^{-1}

Answer: $PhCH(CH_3)CO_2H$

43. Deduce the identity of the following compound from the spectral data given.

$C_5H_{10}O$: ^1H NMR, δ 1.2 (6H, doublet), 2.1 (3H, singlet), 2.8 (1H, septet); IR, 2980, 1710 cm^{-1}; MS, m/z 71, 43

Answer: $(CH_3)_2CHCOCH_3$

44. Deduce the identity of the following compound from the spectral data given.

$C_4H_8O_2$: ^1H NMR, δ 1.23 (3H, triplet), 2.00 (3H, singlet), 4.02 (2H, quartet); IR, 2980, 1740 cm^{-1}

Answer: $CH_3CO_2CH_2CH_3$

45. Deduce the identity of the following compound from the spectral data given.

$C_7H_{10}O_2$: ^{1}H NMR, δ 1.16 (3H, singlet), 2.21 (2H, singlet); ^{13}C NMR, δ 216.25 (singlet), 52.57 (singlet), 34.51 (triplet), 20.22 (quartet)

Answer: 2,2-dimethyl-1, 3-cyclopentanedione Section: 1

46. Deduce the identity of the following compound from the ^{1}H NMR data given.

$C_5H_{10}O$: δ 1.1 (6H, doublet), 2.2 (3H, singlet), 2.5 (1H, septet)

Answer: $CH_3COCH(CH_3)_2$ Section: 1

47. Deduce the identity of the following compound from the ^{1}H NMR data given.

C_4H_7BrO: δ 2.2 (3H, singlet), 3.5 (2H, triplet), 4.5 (3H, triplet)

Answer: $CH_3COCH_2CH_2BR$ Section: 1

48. Deduce the identity of the following compound from the ^{1}H NMR data given.

$C_8H_{10}O$: δ 3.4 (3H, singlet), 4.5 (2H, singlet), 7.2 (5H, singlet)

Answer: $PhCH_2OCH_3$ Section: 1

49. Predict the number of signals expected, their splitting, and their relative area in the ^{1}H NMR spectrum of 1,2-dichloroethane ($ClCH_2CH_2Cl$).

Answer: 1 signal: singlet Section: 1

50. Predict the number of signals expected, their splitting, and their relative area in the ^{1}H NMR spectrum of $(CH_3)_3CCHO$.

Answer: 2 signals: (9H, singlet); (1H, singlet) Section: 1

51. Predict the number of signals expected, their splitting, and their relative area in the ^{1}H NMR spectrum of $CH_3CH_2OCH_3$.

Answer: 3 signals: (3H, triplet); (2H, quartet); (3H, singlet)

Section: 1

52. What multiplicities are observed in the spin coupled ^{13}C NMR spectrum of 2,3-dimethyl-2-butene?

Answer: a singlet and a quartet Section: 12

53. How might the two trimethylcyclohexane isomers shown below be most readily distinguished using NMR?

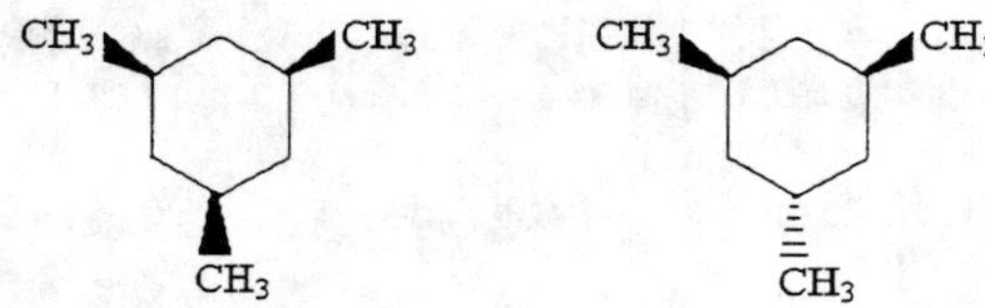

Answer: Number of signals in ^{13}C NMR, 3 signals versus 6. Section: 12

54. Why is Fourier transform NMR spectroscopy preferred over continuous wave as a technique for ^{13}C NMR?

Answer:
^{13}C nuclei have a low sensitivity which requires multiple acquisitions of the spectrum before data become useful. The time/acquisition is much lower using FT techniques so that ^{13}C spectra can be acquired in reasonable times using this method.

Section: 1

55. Give one reason why ^{13}C NMR is less sensitive than ^{1}H NMR.

Answer:
Natural isotopic abundance of ^{13}C is about 100 times less than that of ^{1}H <u>or</u> gyromagnetic ratio of ^{13}C is much smaller.

Section: 12

56. How might the proton spectrum of ultrapure dimethylamine, $(CH_3)_2NH$, differ from the spectrum of this compound to which D_2O has been added?

Answer:
N-H signal will broaden or disappear upon addition of D_2O as rapid hydrogen exchange occurs.

Section: 9

57. The chair form of cyclohexane has protons in two distinct environments, axial and equatorial. When the proton NMR of cyclohexane is run on a 100 MHz instrument at 23° C, only one signal for the compound is observed. Explain this apparent contradiction.

Answer:
Axial and equatorial positions are being rapidly interchanged by chair-chair conformational interconversions. This rapid exchange results in an average signal for these two positions. At very low temperatures, two distinct signals are observed as the conformational interconversion process is slowed.

Section: 8

MULTIPLE CHOICE

58. Using a 60 MHz spectrometer, the protons in dichloromethane appear at 5.30 ppm. When the same sample is placed in a 100 MHz instrument, where does the signal appear?

    a. 8.33
    b. 3.18
    c. 5.30
    d. cannot be determined from information given

    Answer: c Section: 3

SHORT ANSWER

59. ________________ is commonly used an internal reference in NMR spectroscopy; its signal is assigned δ=0 in ^1H and ^{13}C NMR spectroscopy.

    Answer: Tetramethylsilane, $(CH_3)_4Si$ Section: 3

MULTIPLE CHOICE

60. ^1H nuclei located near electronegative atoms tend to be ____________ relative to ^1H nuclei which are not.

    a. shielded
    b. deshielded
    c. resonanced
    d. split
    e. none of the above

    Answer: b Section: 3

Chapter 14: Reactions of Benzene and Substituted Benzenes

MULTIPLE CHOICE

1. Which of the following is an aromatic hydrocarbon?

a.

b.

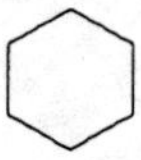

c. 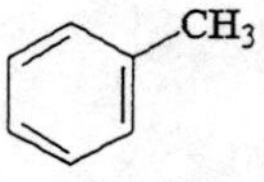 CH_3

d.

e. 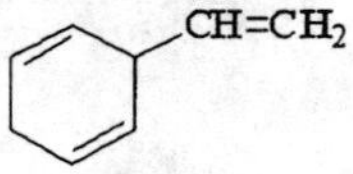 $CH{=}CH_2$

Answer: c

2. What is the structure of styrene?

a. 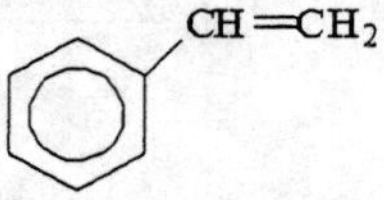 $CH{=}CH_2$

b. 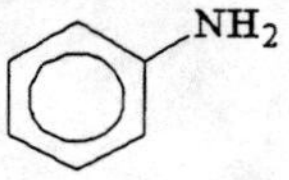 NH_2

c. 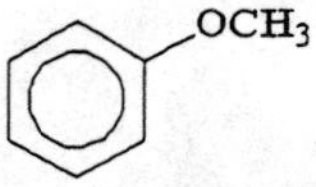 OCH_3

d. 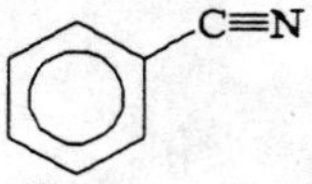 $C{\equiv}N$

e. 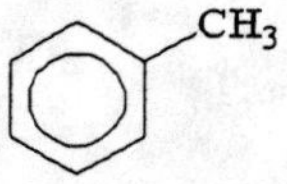 CH_3

Answer: a Section: 1

3. What is the structure of phenol?

a. CH_2OH

b. O
‖
CH

c. OCH_3

d. OH

e. SO_3H

Answer: d Section: 1

4. What is the structure of 3-phenylpentane?

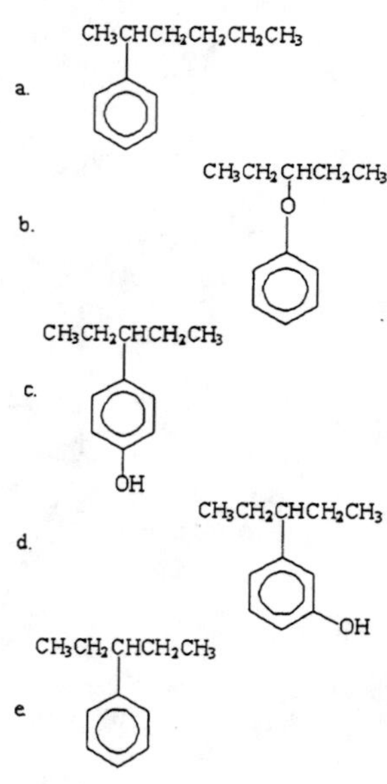

Answer: e Section: 1

5. What is the name of the following compound?

   a. p-Dichlorobenzene
   b. 1,4-Dichlorobenzene
   c. Phenyldichloride
   d. a and b
   e. b and c

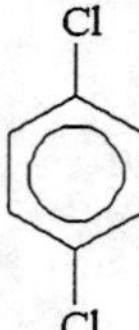

Answer: d Section: 1

6. What is the name of the following compound?

   a. m-Bromomethylbenzene
   b. m-Bromotoluene
   c. 3-Bromotoluene
   d. a and b
   e. b and c

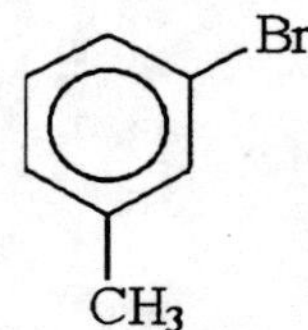

Answer: e Section: 1

7. What is the structure of p-toluidine?

a. 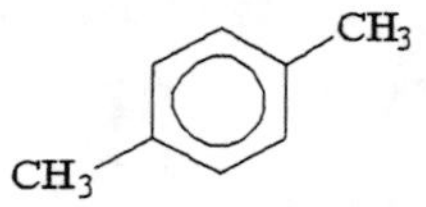 [structure: benzene ring with CH_3 and CH_3 para]

b. 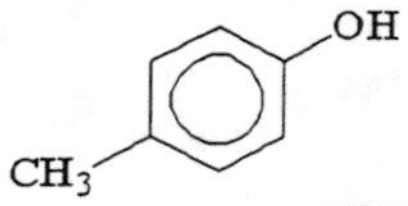 [structure: benzene ring with OH and CH_3 para]

c. 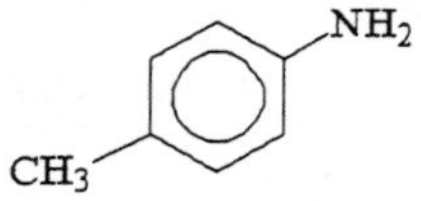 [structure: benzene ring with NH_2 and CH_3 para]

d. 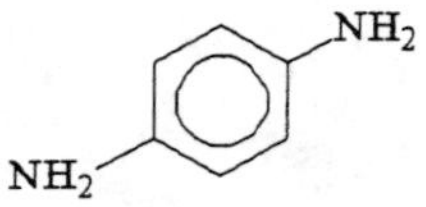 [structure: benzene ring with NH_2 and NH_2 para]

e. 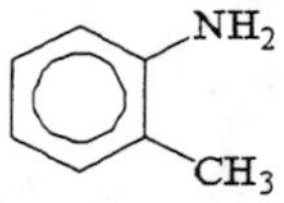 [structure: benzene ring with NH_2 and CH_3 ortho]

Answer: c Section: 1

8. What is the name of the following compound?

a. o-nitro-m-bromotoluene
b. 3-bromo-6-nitrotoluene
c. m-bromo-o-nitrotoluene
d. 5-bromo-2-nitrotoluene
e. 2-nitro-5-bromotoluene

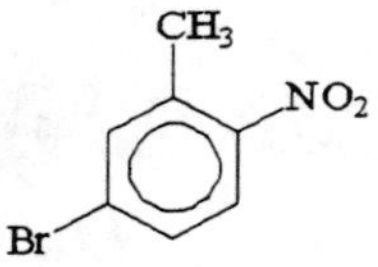

[structure: benzene ring with CH_3, NO_2 ortho to it, and Br para to the NO_2]

Answer: d Section: 1

9. The name 2,4,6-tribromobenzene is incorrect. Which of the following is the correct name?

a. tribromobenzene
b. m,m-dibromobromobenzene
c. 3,5-dibromobromobenzene
d. 1,3,5-tribromobenzene
e. m,m,m-tribromobenzene

Answer: d Section: 1

10. Which of the following is most likely to be the <u>first step</u> in the general mechanism for electrophilic substitution reactions?

a. + $E^{\oplus}$ ⟶ H, E, $\oplus$

b. + $E^{\oplus}$ ⟶ H, E, $\oplus$

c. + $E^{\oplus}$ ⟶ E, H + $H^{\oplus}$

d. + $E^{\oplus}$ ⟶ E, H + $H^{\oplus}$

e. H + $E^{\oplus}$ ⟶ $\oplus$ + HE

Answer: b Section: 3

11. Which of the following is the electrophile that attacks the aromatic ring during sulfonation?

a. $\overset{\oplus}{HSO_3}$
b. $SO_2^{\oplus}$
c. $HSO_3^{\ominus}$
d. H_2SO_4
e. $HSO_4^{\ominus}$

Answer: a Section: 6

12. Which of the following is the electrophile that attacks the aromatic ring during nitration?

a. NO_2
b. HNO_3
c. $NO_3^{\ominus}$
d. $NO_2^{\oplus}$
e. $NO_2^{\ominus}$

Answer: d Section: 5

13. Which of the following is the first step in the mechanism of bromination?

a. 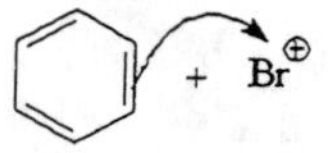 (benzene ring, arrow) + $Br^{\oplus}$

b.  $Br{-}Br \longrightarrow Br^{\oplus} + Br^{\ominus}$

c. 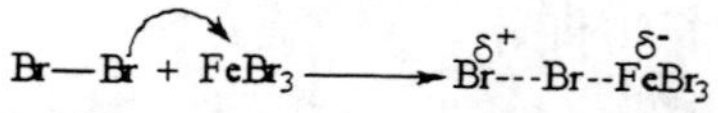 $Br{-}Br + FeBr_3 \longrightarrow \overset{\delta+}{Br}{-}{-}{-}Br{-}{-}\overset{\delta-}{FeBr_3}$

d. 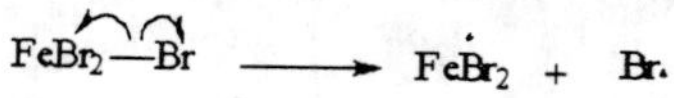 $FeBr_2{-}Br \longrightarrow \dot{F}eBr_2 + Br\cdot$

e.  $Br{-}Br \longrightarrow 2\,Br\cdot$

Answer: c Section: 4

14. Which of the following is the electrophile that attacks the aromatic ring during Friedel-Crafts acylation?

a. $R-\overset{O}{\overset{\|}{C}}-Cl$

b. $R-\underset{\oplus}{C}=O$

c. $R-\overset{O}{\overset{\|}{C}}-\overset{\oplus}{AlCl_3}$

d. $R-\overset{O}{\overset{\|}{C}}-\overset{\delta\ominus}{Cl}{-}{-}{-}\overset{\delta\oplus}{Cl}$

e. $AlCl_3$

Answer: b Section: 7

15. What is the major product of the following Friedel-Crafts alkylation?

Benzene + $CH_3CH(CH_3)CH_2Cl$ $\xrightarrow{AlCl_3}$?

a. $C_6H_5CH_2CH_2CH_2CH_3$

b. $C_6H_5CH_2CH(CH_3)CH_3$

c. $C_6H_5C(CH_3)_2CH_3$

d. $C_6H_5CH_2CH(CH_3)CH_3$

e. $C_6H_5CH(CH_3)CH_2CH_3$

Answer: c Section: 8

16. Which of the following is the best method for accomplishing this reaction?

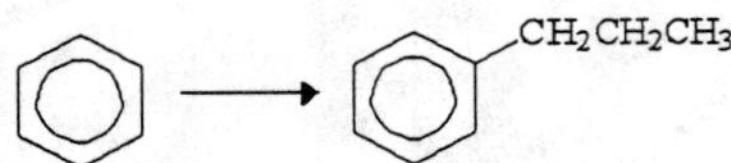

a. $CH_3CH_2CH_2Cl/AlCl_3$

b. $CH_3CH_2C(=O)Cl/AlCl_3$; Zn(Hg)/HCl/heat

c. $CH_3CH_2C(=O)Cl/AlCl_3$; $NH_2NH_2/{}^{\ominus}OH$, heat

d. a and b

e. b and c

Answer: e Section: 9

17. What is the <u>major</u> product of the following reaction?

CH_3CH_2–(benzene ring)–CH_3 $\xrightarrow[H^+,\ heat]{KMnO_4}$?

a. CH_3CH_2–(benzene ring)–COOH

b. HOOC–(benzene ring)–CH_3

c. $HOOCCH_2$–(benzene ring)–CH_3

d. HOOC–(benzene ring)–COOH

e. CH_3OOC–(benzene ring)–COOH

Answer: d Section: 10

18. What is the <u>major</u> product of the following reaction?

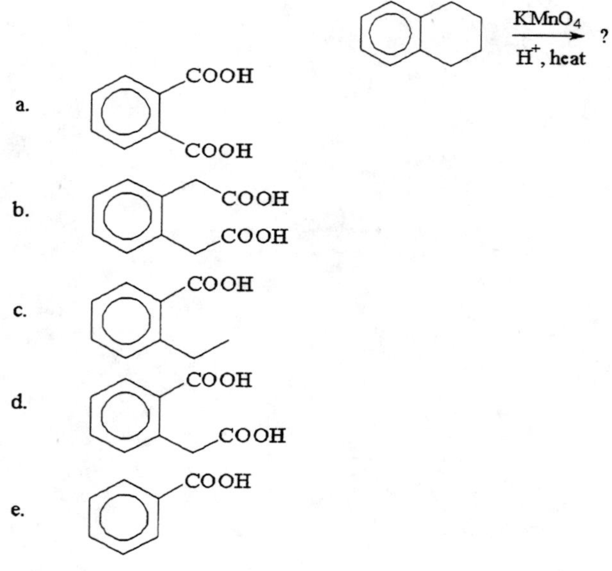

Answer: a Section: 10

19. What is the major product of the following reaction?

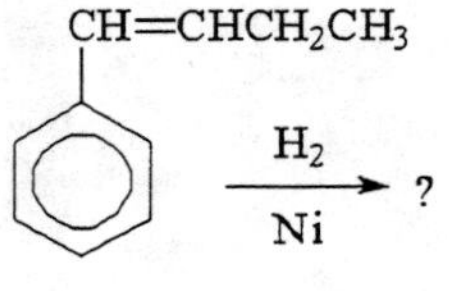

$C_6H_5CH{=}CHCH_2CH_3 \xrightarrow[Ni]{H_2} ?$

a. $C_6H_5C{\equiv}CCH_2CH_3$

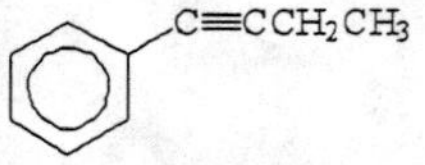

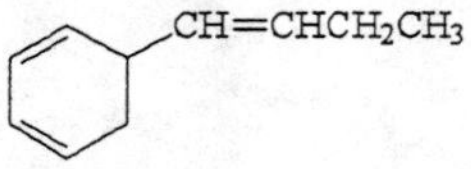

b. (cyclohexadienyl) $CH{=}CHCH_2CH_3$

c. (cyclohexenyl) $CH{=}CHCH_2CH_3$

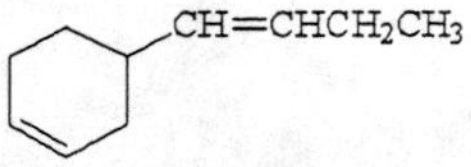

d. 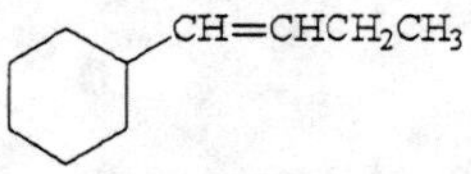 (cyclohexyl) $CH{=}CHCH_2CH_3$

e. 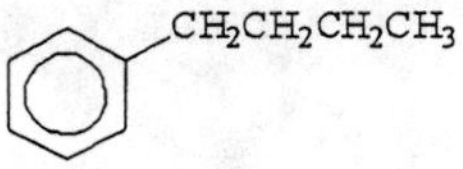 $C_6H_5CH_2CH_2CH_2CH_3$

Answer: e Section: 10

20. What is the best method for carrying out the following reaction?

$C_6H_6 \longrightarrow C_6H_5CH(OH)CH_3$

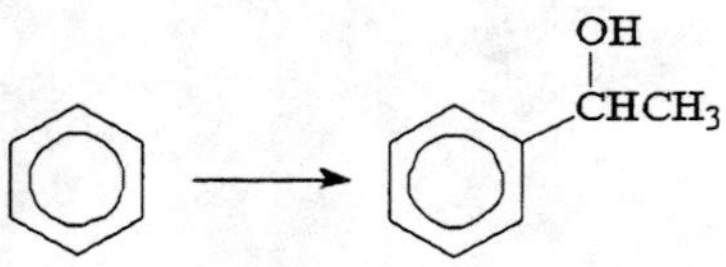

a. $CH_3CH_2OH/AlCl_3$

b. $CH_3CH_2Cl/AlCl_3$; NBS/heat ; $^{\ominus}OH$

c. $CH_3CH_2Cl/AlCl_3$; $^{\ominus}OH$

d. $CH_3CHCl(Cl)/AlCl_3$; $^{\ominus}OH$

e. $CH_3C(=O)Cl/AlCl_3$; $KMnO_4$, heat

Answer: b Section: 16

21. Which of the following compounds reacts most <u>slowly</u> during nitration?

a. $C_6H_5OCH_2CH_3$

b. C_6H_5OH

c. C_6H_6

d. $C_6H_5NHC(=O)CH_3$

e. $C_6H_5CH_3$

Answer: c Section: 11

22. Which of the following compounds reacts most rapidly during nitration?

a. $C_6H_5C{\equiv}N$

b. $C_6H_5NO_2$

c. $C_6H_5CF_3$

d. C_6H_5F

e. C_6H_5COOH

Answer: d Section: 11

23. Which of the following substrates is an electron donating group overall?

a. $-Br$
b. $-\underset{\underset{O}{\|}}{C}CH_3$
c. $-OCH_3$
d. $-\overset{\oplus}{N}(CH_3)_3$
e. $-CCl_3$

Answer: c Section: 11

24. Which of the following substrates is an electron withdrawing group overall?

a. $-CH_3$
b. $-OH$
c. $-C_6H_5$ (benzene ring)
d. $-OCH_3$
e. $-SO_3H$

Answer: e Section: 11

25. Which of the following structures is the most important contributor to the resonance hybrid formed when toluene undergoes para nitration?

a. 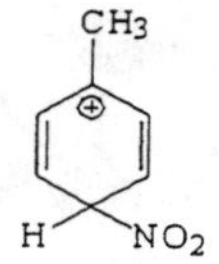 CH_3 ⊕ H NO_2

b. 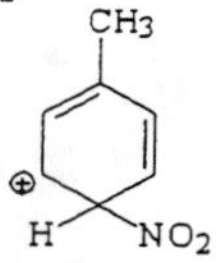 CH_3 ⊕ H NO_2

c. 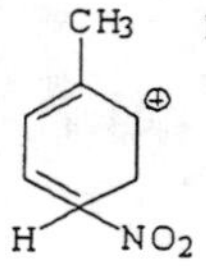 CH_3 ⊕ H NO_2

d. 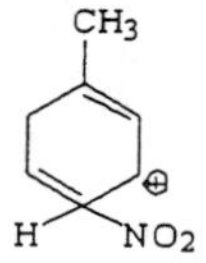 CH_3 ⊕ H NO_2

e. 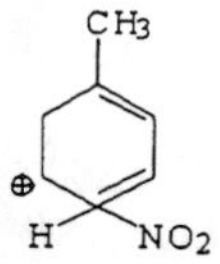 CH_3 ⊕ H NO_2

Answer: a Section: 12

26. Which of the following structures is the most important contributor to the resonance hybrid formed when anisole undergoes o-bromination?

a. 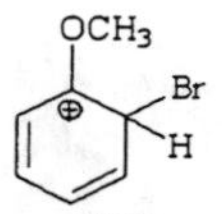 OCH_3 ⊕ Br H

b. 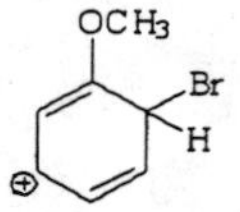 OCH_3 Br H ⊕

c. 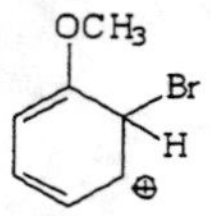 OCH_3 Br H ⊕

d. 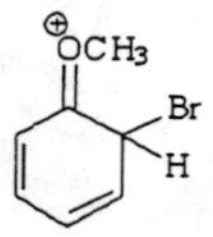 ⊕ OCH_3 Br H

e. 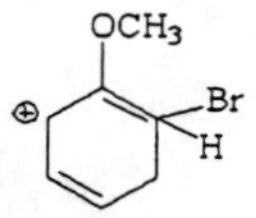 ⊕ OCH_3 Br H

Answer: d Section: 12

27. What is/are the product(s) from the following reaction?

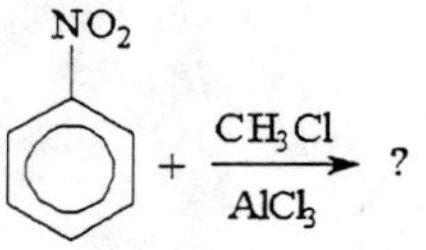

(Nitrobenzene) + CH_3Cl / $AlCl_3$ → ?

a. 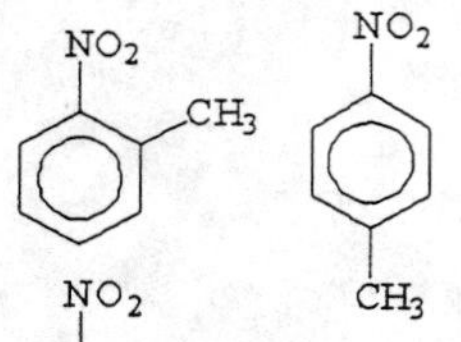 (o-nitrotoluene: NO_2, CH_3) (p-nitrotoluene: NO_2, CH_3)

b. 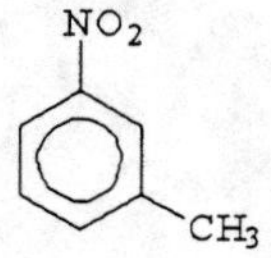 (m-nitrotoluene: NO_2, CH_3)

c. 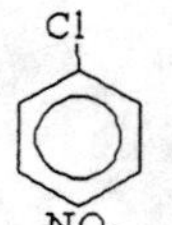 (chlorobenzene: Cl) CH_3NO_2

d. 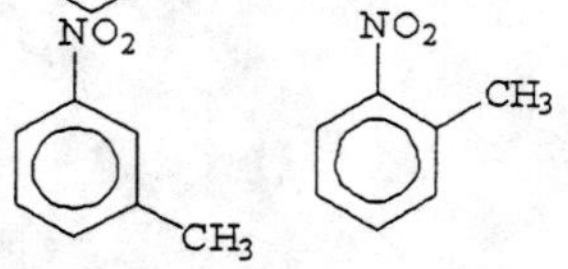 (m-nitrotoluene: NO_2, CH_3) (o-nitrotoluene: NO_2, CH_3)

e. No reaction

Answer: e Section: 15

28. Which is the best method for carrying out the following reaction?

a. $^{+}COOH$; HNO_3/H_2SO_4
b. $CH_3Cl/AlCl_3$; HNO_3/H_2SO_4 ; $KMnO_4/H^+$, heat
c. $CH_3Cl/AlCl_3$; $KMnO_4/H^+$, heat ; HNO_3/H_2SO_4
d. HNO_3/H_2SO_4 ; $CH_3Cl/AlCl_3$; $KMnO_4/H^+$, heat
e. HNO_3/H_2SO_4 ; $^{+}COOH$

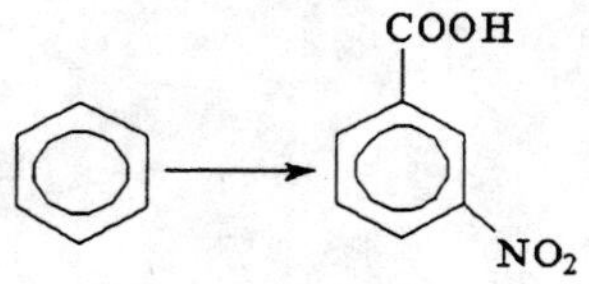

(Benzene) → (m-nitrobenzoic acid: COOH, NO_2)

Answer: c Section: 16

29. Which is the best method for carrying out the following reaction?

[Reaction scheme: benzene → 1-propyl-3-sulfobenzene ($CH_2CH_2CH_3$, SO_3H)]

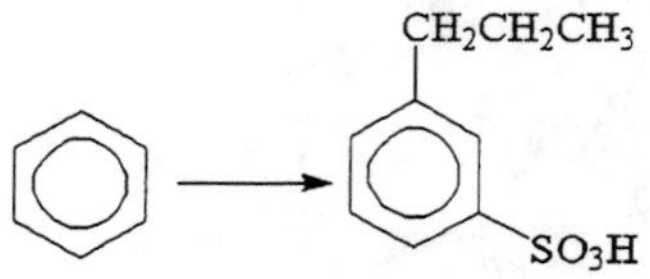

a. SO_3/H_2SO_4 ; $CH_3CH_2CH_2Cl/AlCl_3$

b. $CH_3CH_2CH_2Cl/AlCl_3$; SO_3/H_2SO_4

c. $CH_3CH_2C(=O)Cl/AlCl_3$; SO_3/H_2SO_4

d. $CH_3CH_2C(=O)Cl/AlCl_3$; Zn(Hg)/HCl/heat ; SO_3/H_2SO_4

e. $CH_3CH_2C(=O)Cl/AlCl_3$; SO_3/H_2SO_4 ; Zn(Hg)/HCl/heat

Answer: e Section: 16

30. What is the <u>major</u> product of the following reaction?

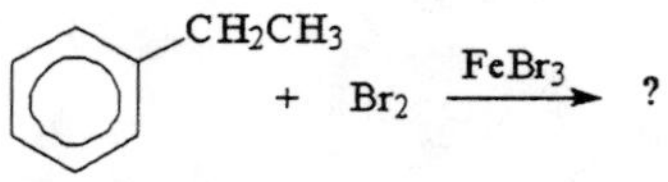

[Ethylbenzene (CH_2CH_3) + Br_2 $\xrightarrow{FeBr_3}$?]

a. 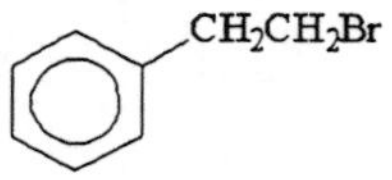 [benzene ring with CH_2CH_2Br]

b. 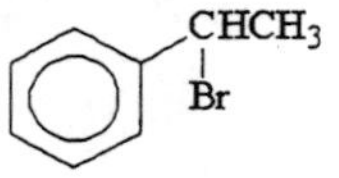 [benzene ring with $CHCH_3$ bearing Br]

c. 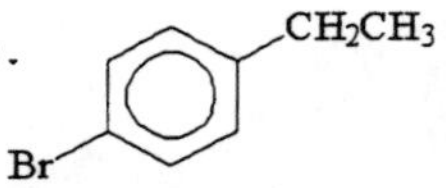 [benzene ring with CH_2CH_3 and Br para]

d. 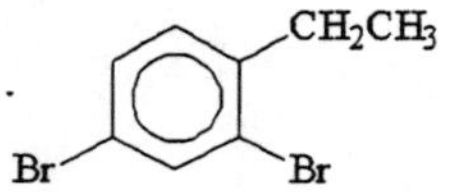 [benzene ring with CH_2CH_3, Br ortho and Br para]

e. 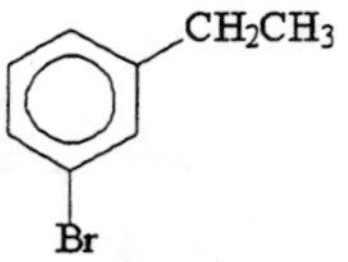 [benzene ring with CH_2CH_3 and Br meta]

Answer: c Section: 12

31. Which of the following is the best method for preparing m-chloroaniline?

a. NH_3 ;Cl_2 /$AlCl_3$
b. Cl_2 /$AlCl_3$; NH_3
c. Cl_2 /$AlCl_3$; HNO_3 /H_2SO_4 ; Sn/HCl, HO^-
d. HNO_3 /H_2SO_4 ; Cl_2 /$AlCl_3$; Sn/HCl, HO-
e. HNO_3 /H_2SO_4 ; Sn/HCl ; HO^- ; Cl_2 /$AlCl_3$

Answer: d Section: 16

32. What is the <u>major</u> product of the following reaction?

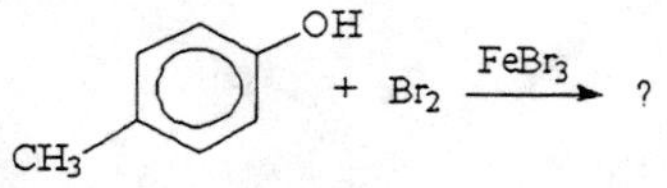

CH_3–C_6H_4–OH (para) + Br_2 $\xrightarrow{FeBr_3}$?

a. CH_3, OH, Br substituted benzene (Br ortho to CH_3)

b. $ClCH_2$–C_6H_4–OH

c. CH_3–C_6H_4–OCl

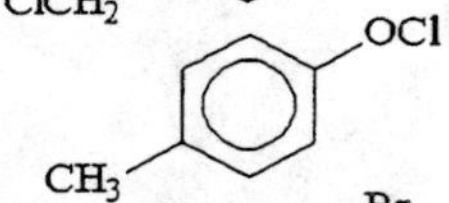

d. CH_3–C_6H_4–Br

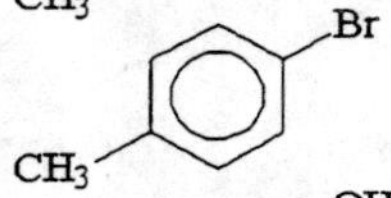

e. CH_3, OH, Br substituted benzene (Br ortho to OH)

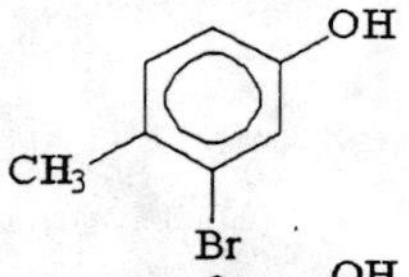

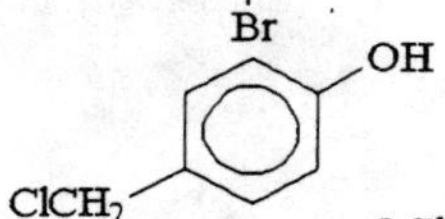

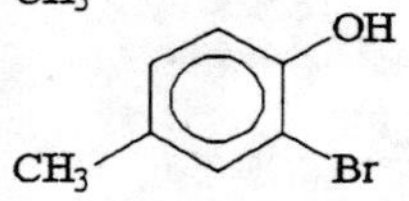

Answer: e Section: 17

33. What is the major product of the following reaction?

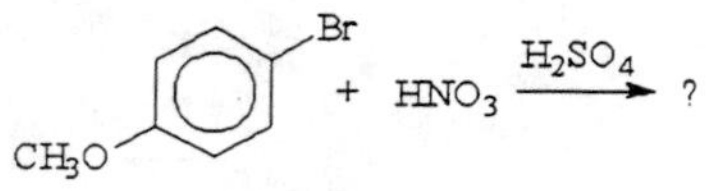

CH_3O–C$_6$H$_4$–Br (para) + HNO_3 $\xrightarrow{H_2SO_4}$?

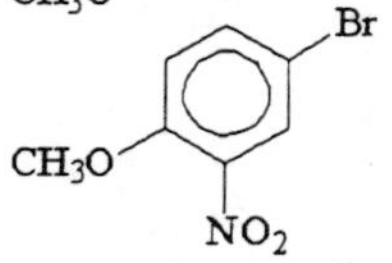

a. 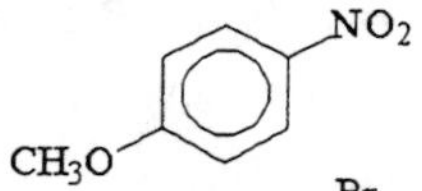 CH_3O, NO_2 (para)

b. CH_3O, Br, NO_2

c. CH_3O, Br, NO_2

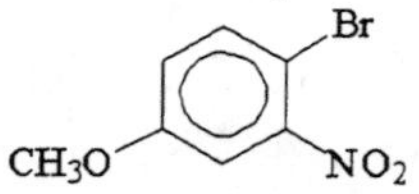

d. 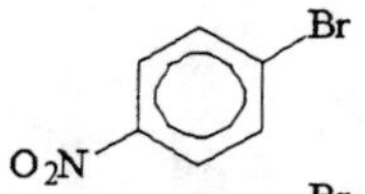 O_2N, Br (para)

e. 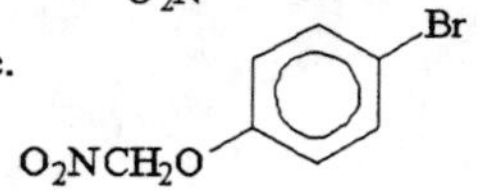 O_2NCH_2O, Br (para)

Answer: b Section: 17

34. What is the best method for the preparation of p-chlorotoluene in high yield?

a. Start with benzene; methylate; chlorinate
b. Start with benzene; chlorinate; methylate
c. Start with toluene; chlorinate
d. Start with chlorobenzene; methylate
e. Start with p-aminotoluene; $NaNO_2$/HCl, 0° C; CuCl

Answer: e Section: 17

35. What is the best method for the preparation of m-dibromobenzene from benzene?

a. Nitrate; Sn/HCl; $NaNO_2$/HCl, 0° C; brominate twice
b. Nitrate; Sn/HCl; $NaNO_2$/HCl, 0° C; brominate twice; H_3PO_2
c. Nitrate; Sn/HCl; $NaNO_2$/HCl, 0° C; H_3PO_2; brominate twice
d. Nitrate; brominate; Sn/HCl; $NaNO_2$/HCl, 0° C; CuBr
e. Brominate twice

Answer: d Section: 17

36. What is the major product of the following reaction?

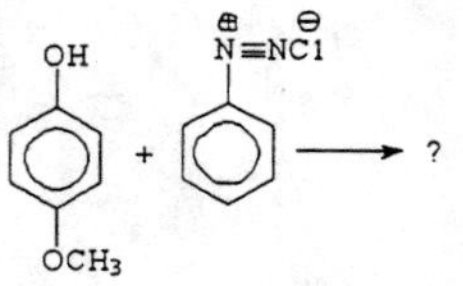

a.

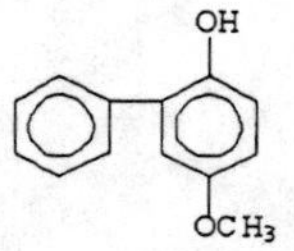

b.

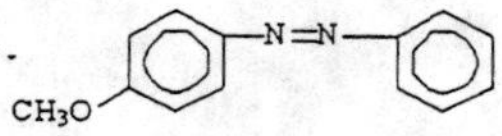

c.

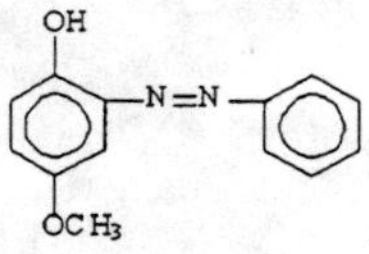

d.

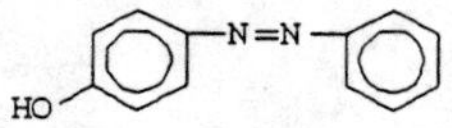

e.

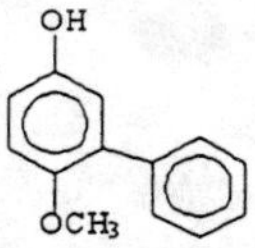

Answer: c Section: 19

37. Which of the following is <u>most</u> reactive toward nucleophilic aromatic substitution?

a.

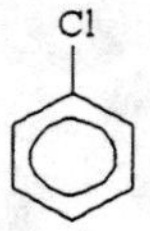

b.

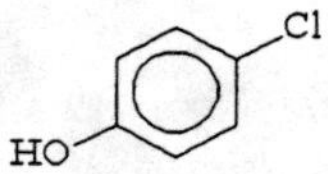

c.

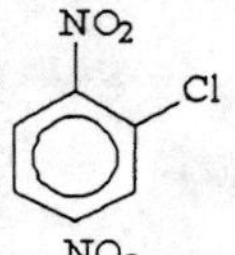

d.

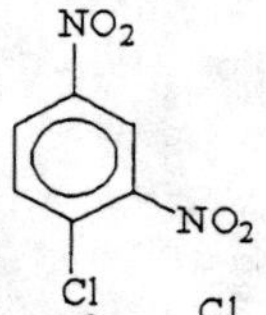

e.

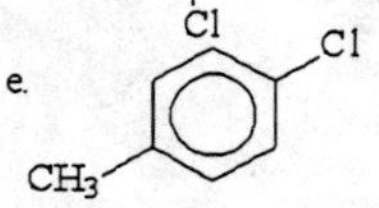

Answer: d Section: 21

38. Which of the following is known as benzyne?

a.

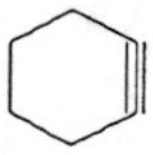

b.

c. 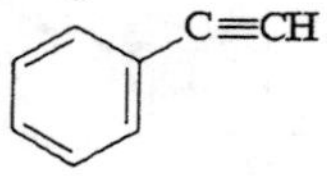 C≡CH

d.

e.

Answer: d Section: 22

39. What are the products of the following reaction?

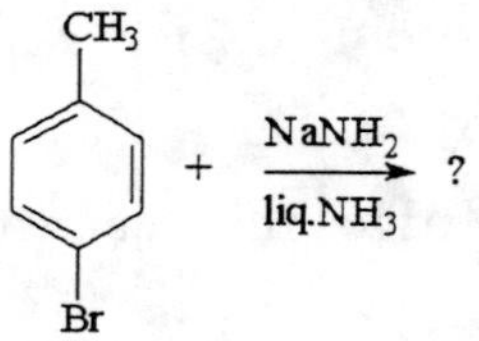

CH_3 / Br + $NaNH_2$, liq.NH_3 → ?

a. 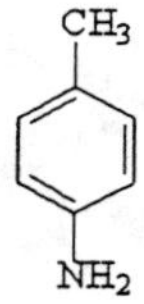 CH_3 / NH_2

b. 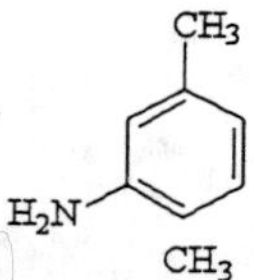 CH_3 / H_2N

c. 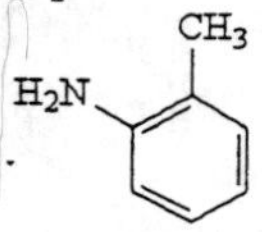 CH_3 / H_2N

d. a and b
e. b and c

Answer: d Section: 22

40. What is the <u>major</u> product from the following reaction?

$\xrightarrow[H_2SO_4]{HNO_3}$?

a. NO_2

b. NO_2

c. NO_2

d. a and b equally

e. a and c equally

Answer: a Section: 23

41. What is the <u>major</u> product of the following reaction?

CH_3 $\xrightarrow[H_2SO_4]{HNO_3}$?

a. CH_3 NO_2

b.

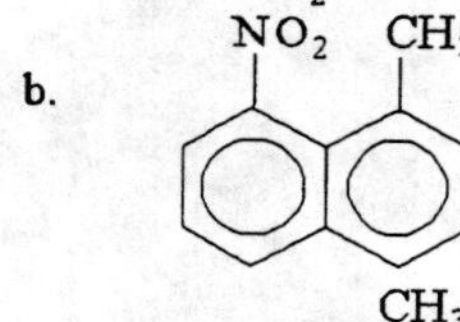

c. CH_3 NO_2

d. a and b equally

e. a and c equally

Answer: c Section: 23

SHORT ANSWER

42. What is the name of the following compound?

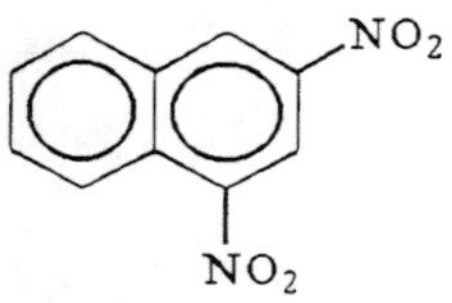

Answer: 1,3-Dinitronaphthalene Section: 23

43. Explain why nitrobenzene can be used as a solvent for Friedel-Crafts alkylation.

Answer:
Because of the deactivating nature of the NO_2 group, it will not undergo a F.C. reaction.
Section: 15

44. Explain why direct nitration of aniline yields, among other products, m-nitroaniline.

Answer:
In acidic media, the basic amino group forms a protonated ammonium ion which is a meta director.
Section: 15

45. Provide the major resonance structures of the intermediate sigma complex in the reaction of benzene with the generic electrophile E^+.

Answer:

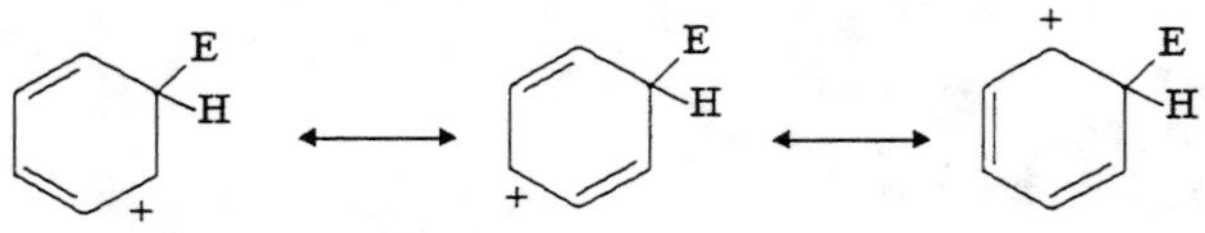

Section: 3

46. In the bromination of benzene using Br_2 and $FeBr_3$, is the intermediate carbocation aromatic? Explain.

Answer:
No, the carbocation is not aromatic since the ring contains an sp^3 center.
Section: 3

47. Provide a detailed, stepwise mechanism for the reaction of benzene with Br_2 and $FeBr_3$. Make sure to include the activating reaction between Br_2 and $FeBr_3$ in your mechanism.

Answer:

$$Br-Br: + FeB_3 \rightleftharpoons Br-\overset{+}{Br}-\overset{-}{Fe}Br_3$$

$$C_6H_6 + Br-\overset{+}{Br}-\overset{-}{Fe}Br_3 \longrightarrow C_6H_6Br^+ + FeBr_4^-$$

$$C_6H_6Br^+ \longrightarrow C_6H_5Br + H^+$$

Section: 4

48. Name the major organic product which results when 4-ethylbenzenesulfonic acid is heated in aqueous acid.

Answer: ethylbenzene Section: 6

49. Rank the following groups in order of increasing activating power in electrophilic aromatic substitution reactions: $-OCH_3$, $-OCOCH_2CH_3$, $-CH_2CH_3$, -Br.

Answer: $-Br < -CH_2CH_3 < -OCOCH_2CH_3 < -OCH_3$ Section: 11

MULTIPLE CHOICE

50. In electrophilic aromatic substitution reactions a chlorine substituent:

a. is a deactivator and a m-director.
b. is a deactivator and an o,p-director.
c. is an activator and a m-director.
d. is an activator and an o,p-director.
e. none of the above

Answer: b Section: 12

SHORT ANSWER

51. Draw the four major resonance structures of the carbocation intermediate in the reaction of anisole with HNO_3/H_2SO_4 to yield p-nitroanisole.

Answer:

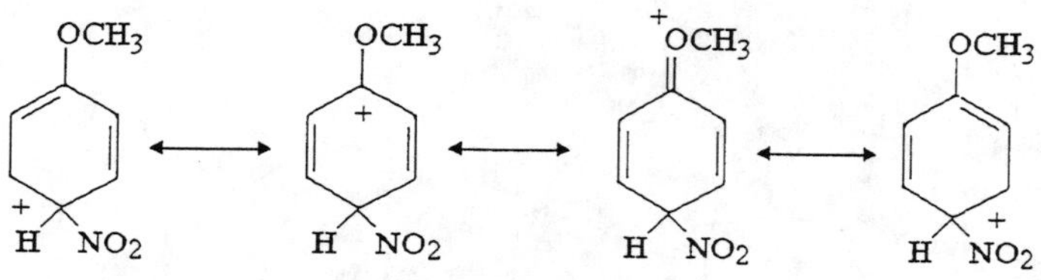

Section: 12

52. Draw the three major resonance structures of the carbocation intermediate in the reaction of acetophenone with HNO_3/H_2SO_4 to yield o-nitroacetophenone. Circle the resonance form which is less stable than the other two.

Answer:

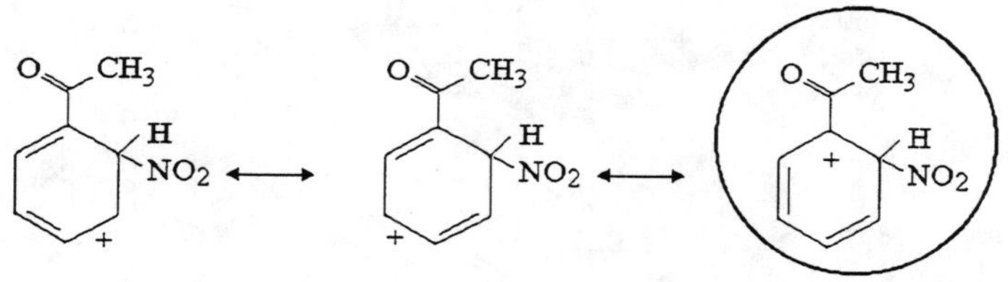

Section: 12

53. Provide a series of synthetic steps by which 2-bromo-4-nitrobenzoic acid can be prepared from toluene.

Answer:
1. HNO_3, H_2SO_4
2. Br_2, $FeBr_3$
3. $KMnO_4$, H^+

Section: 16

54. Provide the structure of the major mononitration product of the compound below.

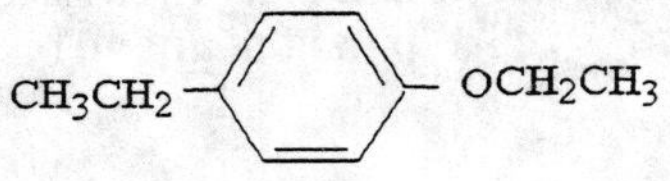

CH_3CH_2–C_6H_4–OCH_2CH_3

Answer:

NO_2

CH_3CH_2–C_6H_3–OCH_2CH_3

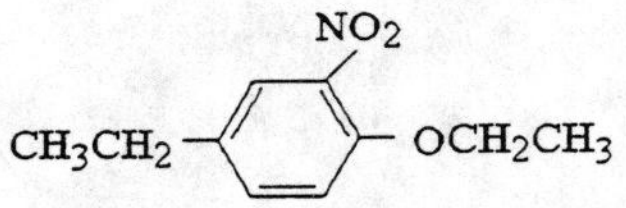

Section: 17

55. Provide the structure of the major mononitration product(s) of the compound below.

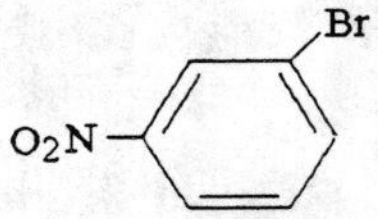

Br

O_2N–C_6H_4

Answer:

Br Br

O_2N–C_6H_3–NO_2 + O_2N–C_6H_3

O_2N

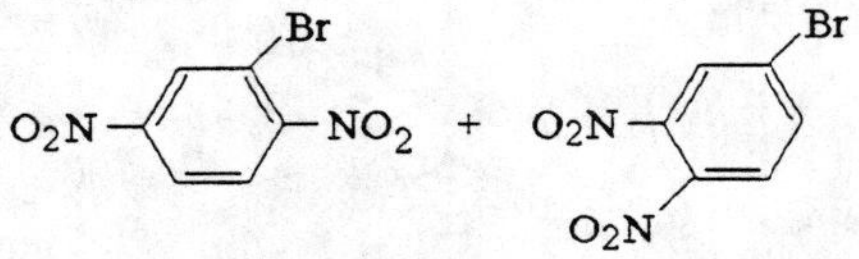

Section: 17

56. Provide the structure of the major organic product of the following reaction.

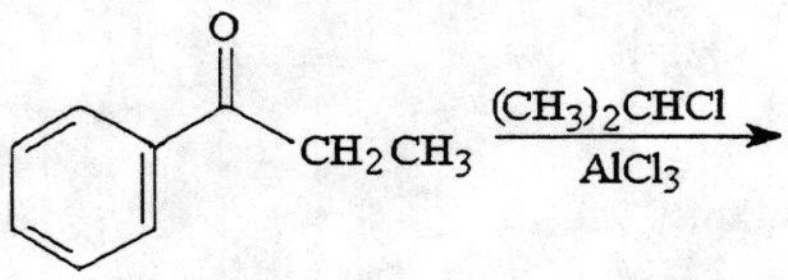

O

C_6H_5–C–CH_2CH_3 $\xrightarrow[AlCl_3]{(CH_3)_2CHCl}$

Answer:
No reaction. The ring is too deactivated by the acyl substituent to undergo Friedel-Crafts aklylation.

Section: 15

57. Provide the structure of the major organic product of the following reaction.

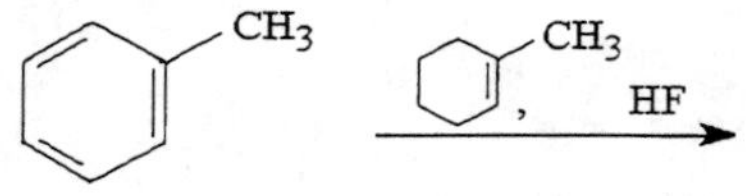

CH3 — CH3, HF

Answer:

H3C — CH3

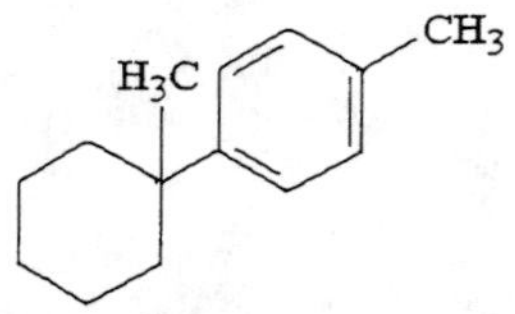

Section: 8

58. Provide a detailed, stepwise mechanism for the following reaction.

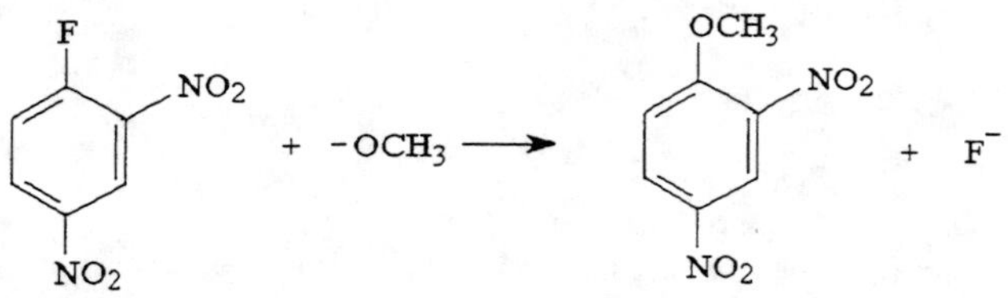

F, NO2, NO2 + -OCH3 → OCH3, NO2, NO2 + F^-

Answer:

F, NO2, NO2 + $^-OCH_3$ → F, OCH3, NO2, NO2 → OCH3, NO2, NO2

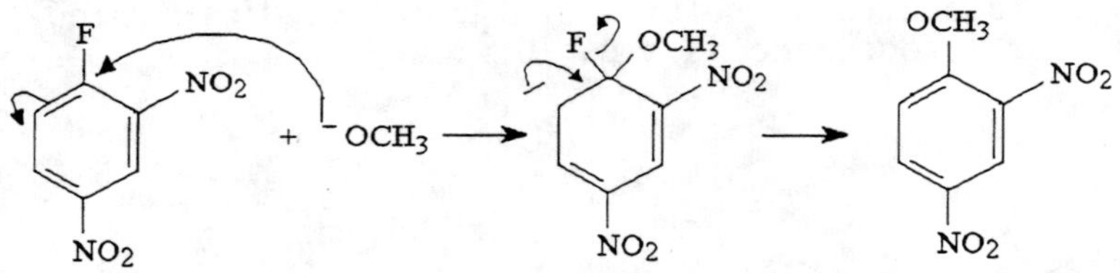

Section: 21

59. Provide a detailed, stepwise mechanism for the following reaction.

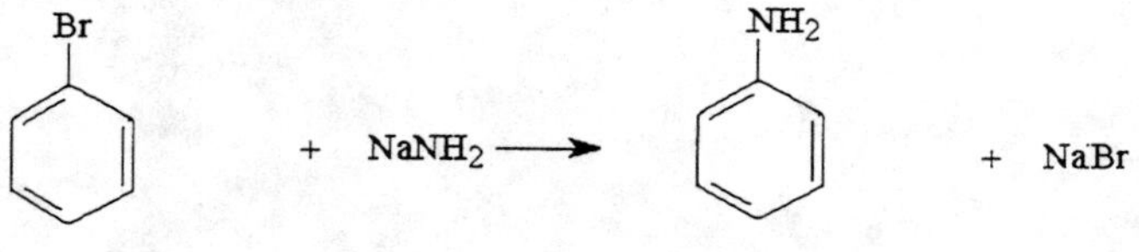

Br (bromobenzene) + $NaNH_2$ → NH_2 (aniline) + NaBr

Answer:

Br, H + $\bar{N}H_2$ → (benzyne) + NH_3 + Br^-

(benzyne) + $\bar{N}H_2$ → NH_2 (carbanion)

(carbanion) + H–NH_2 → NH_2 (aniline) + $^-NH_2$

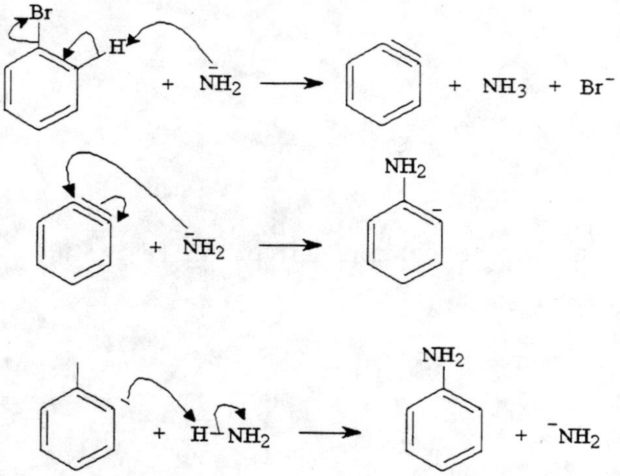

Section: 22

60. Draw the <u>four</u> major resonance structures of the intermediate which results when o-nitrochlorobenzene is treated with NaOH.

Answer:

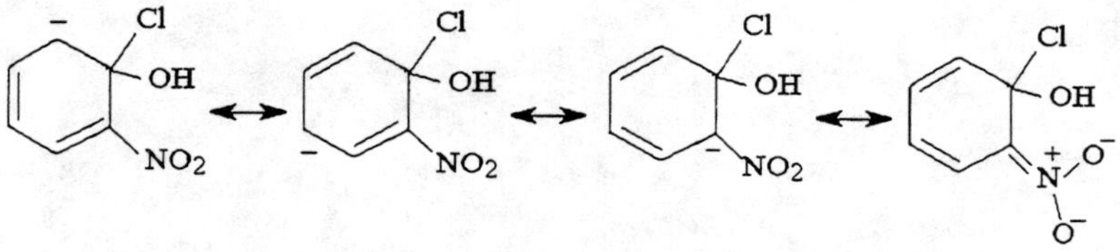

Cl, OH, NO_2 ↔ Cl, OH, NO_2 ↔ Cl, OH, NO_2 ↔ Cl, OH, $\overset{+}{N}$, O^-, O^-

Section: 21

Chapter 15: Carbonyl Compounds I

MULTIPLE CHOICE

1. Which of the following compounds is an amide?

a. $CH_3CH_2NH_2$

b. CH_3NHNH_2

c. $CH_3C(=O)NH_2$

d. $HOC(=O)CH_2NH_2$

e. 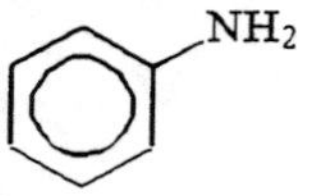 $C_6H_5NH_2$ (benzene ring with NH_2)

Answer: c

2. Which of the following compounds is an acyl chloride?

a. 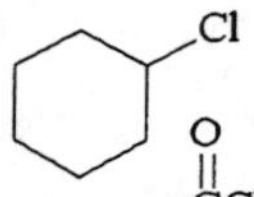 cyclohexane ring with Cl

b. 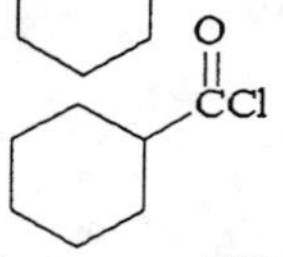 cyclohexane ring with $C(=O)Cl$

c. cyclohexane ring with CH_2Cl

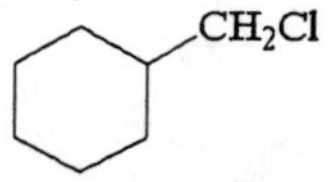

d. 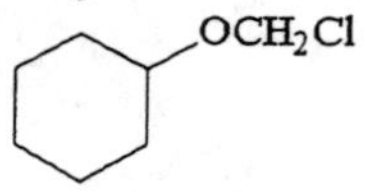 cyclohexane ring with OCH_2Cl

e. 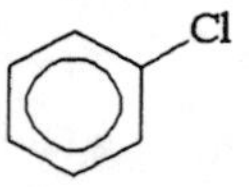 benzene ring with Cl

Answer: b

3. Which of the following compounds is an ester?

a. $CH_3\overset{\overset{O}{\|}}{C}O-C_6H_5$

b. $CH_3\overset{\overset{O}{\|}}{C}-C_6H_5$

c. $CH_3\overset{\overset{O}{\|}}{C}OH$

d. $CH_3\overset{\overset{O}{\|}}{C}OOH$

e. $CH_3\overset{\overset{O}{\|}}{C}O\overset{\overset{O}{\|}}{C}CH_3$

Answer: a

4. Which of the following compounds is an anhydride?

a. $CH_3\overset{\overset{O}{\|}}{C}O-C_6H_5$

b. $CH_3\overset{\overset{O}{\|}}{C}-C_6H_5$

c. $CH_3\overset{\overset{O}{\|}}{C}OH$

d. $CH_3\overset{\overset{O}{\|}}{C}OOH$

e. $CH_3\overset{\overset{O}{\|}}{C}O\overset{\overset{O}{\|}}{C}CH_3$

Answer: e

5. What is the common name for the following compound?

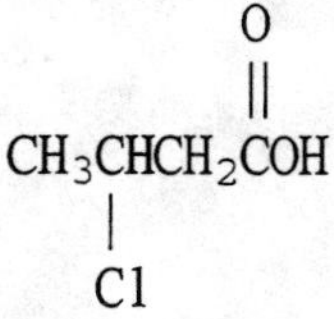

a. γ-Chlorobutanoic acid
b. β-Chlorobutanoic acid
c. β-Chlorobutyric acid
d. γ-Chlorobutyric acid
e. 3-Chlorobutyric acid

Answer: c Section: 1

6. What is the IUPAC name for the following compound?

```
         CH3
          |
CH3CH2CHCH2COOH
```

a. 3-Methylpentanoic acid
b. Isohexanoic acid
c. β-Methylvaleric acid
d. 2-Methylpentanoic acid
e. 3-Methylvaleric acid

Answer: a Section: 1

7. Which of the following compounds is benzoyl chloride?

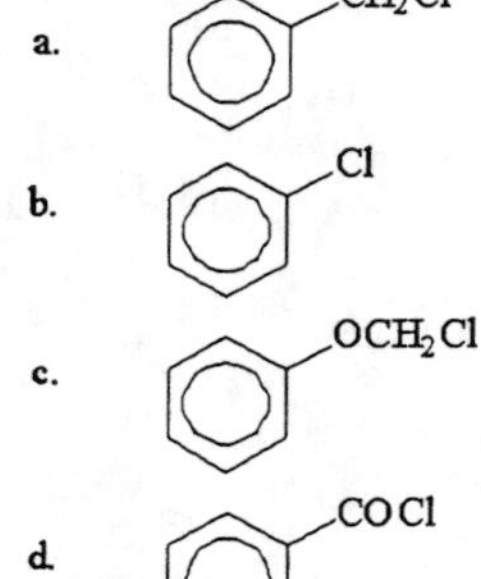

e. COOCl

Answer: d Section: 1

8. What is the common name for the following compound?

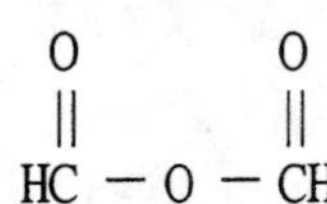

a. Methanoic anhydride
b. Formic anhydride
c. Formyl formate
d. Ethanoic anhydride
e. Diformyl ether

Answer: b Section: 1

9. What is the IUPAC name for the following compound?

   a. Phenyl acetate
   b. Methyl benzoate
   c. Methyl phenylacetate
   d. Benzyl acetate
   e. Benzyl methylacetate

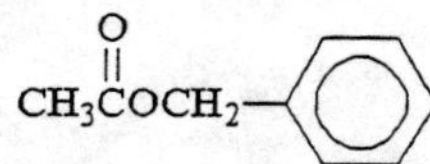

   Answer: d Section: 1

10. Which of the following compounds is methyl formate?

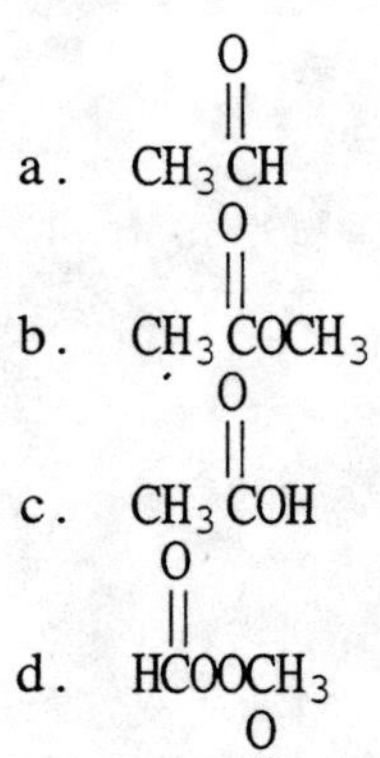

   Answer: e Section: 1

11. Which of the following compounds is γ-butyrolactone?

a.

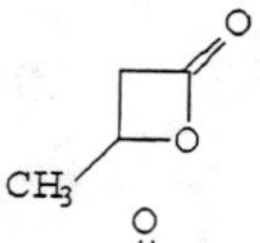

CH_3

b.

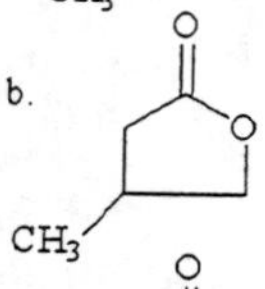

CH_3

c.

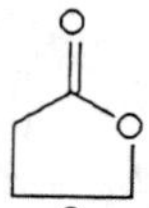

d.

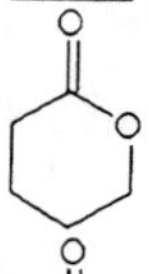

e.

CH_3

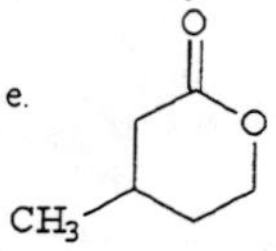

Answer: c Section: 1

12. What is the common name for the following compound?

a. Phenylcarbamide
b. Benzamide
c. Phenylmethanamide
d. Benzeneamide
e. Benzylamide

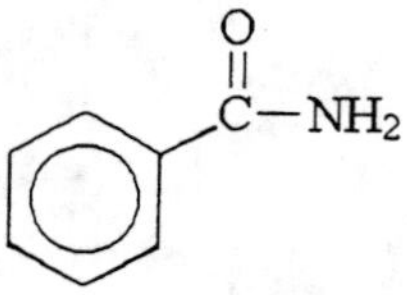

O
||
C—NH_2

Answer: b Section: 1

13. Which of the following compounds is N,N-dimethylbenzamide?

a. 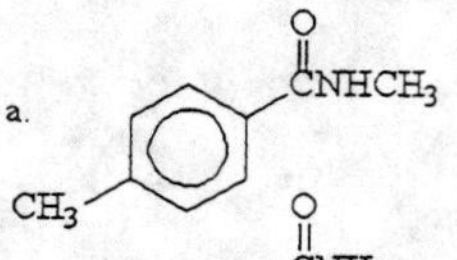 O ‖ CNHCH$_3$; CH$_3$

b. 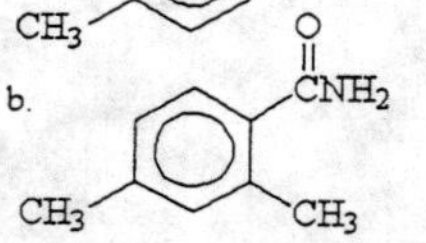 O ‖ CNH$_2$; CH$_3$ CH$_3$

c. 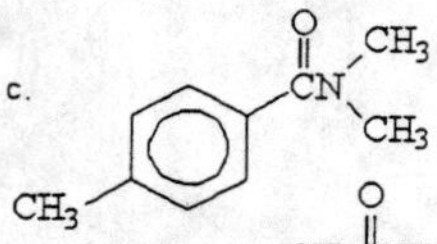 O ‖ CN(CH$_3$)CH$_3$; CH$_3$

d. 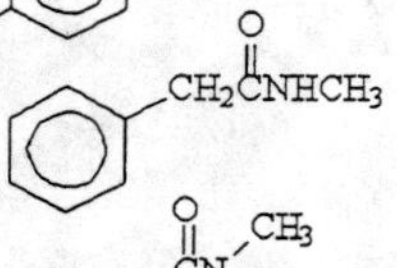 O ‖ CH$_2$CNHCH$_3$

e. 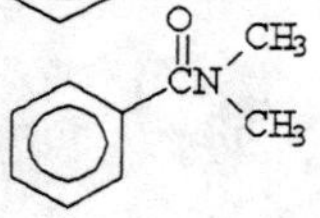 O ‖ CN(CH$_3$)CH$_3$

Answer: e Section: 1

14. Which of the following compounds is γ-valerolactam?

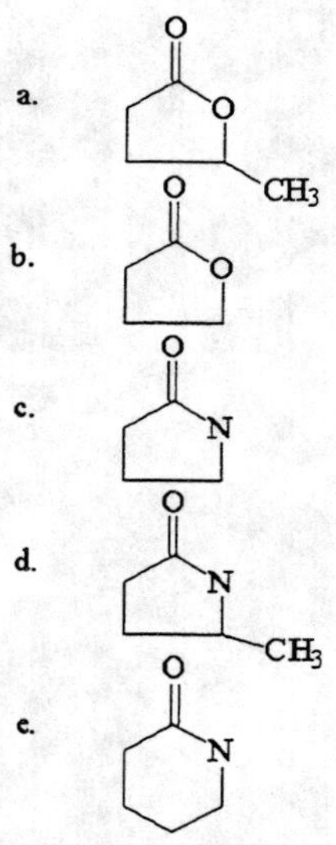

Answer: d Section: 1

15. What is the common name for the following compound?

   a. Benzonitrile
   b. Benzenecyanide
   c. Cyanobenzene
   d. Cyanophenyl
   e. Benzocyanide

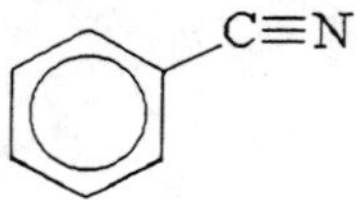

$C_6H_5-C\equiv N$

Answer: a Section: 1

16. What is the hybridization and geometry of the carbonyl carbon in carboxylic acids and their derivatives?

   a. sp^3 , tetrahedral
   b. sp^2 , trigonal planar
   c. sp^2 , tetrahedral
   d. sp^3 , trigonal planar
   e. sp, trigonal planar

Answer: b Section: 2

17. Which of the following compounds is most reactive toward nucleophilic acyl substitution?

   a. $R-C(=O)-NH_2$
   b. $R-C(=O)-OCH_3$
   c. $R-C(=O)-Cl$
   d. $R-C(=O)-O-C(=O)-R$
   e. $R-C(=O)-OH$

Answer: c Section: 4

18. Which of the following compounds is the least reactive toward nucleophilic acyl substitution?

a. $R-C(=O)-NH_2$

b. $R-C(=O)-OCH_3$

c. $R-C(=O)-Cl$

d. $R-C(=O)-O-C(=O)-R$

e. $R-C(=O)-OH$

Answer: a Section: 4

19. Which of the following compounds is hydrolyzed most rapidly in aqueous NaOH?

a. $R-C(=O)-NH_2$

b. $R-C(=O)-OCH_3$

c. $R-C(=O)-Cl$

d. $R-C(=O)-O-C(=O)-R$

e. $R-C(=O)-OH$

Answer: c Section: 4

20. Which of the following compounds is hydrolyzed most slowly in aqueous NaOH?

a. 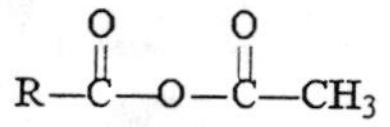 $R-\overset{O}{\overset{||}{C}}-O-\overset{O}{\overset{||}{C}}-CH_3$

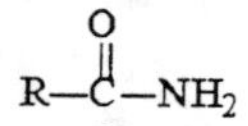

b. $R-\overset{O}{\overset{||}{C}}-NH_2$

c. $R-\overset{O}{\overset{||}{C}}-OCH_2-C_6H_5$

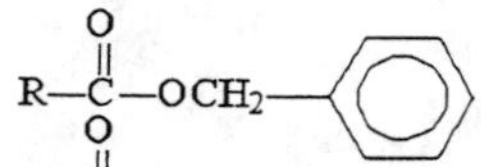

d. 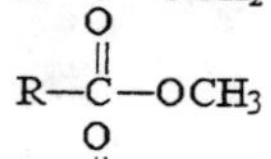 $R-\overset{O}{\overset{||}{C}}-OCH_3$

e. 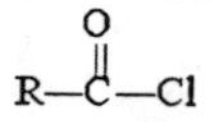 $R-\overset{O}{\overset{||}{C}}-Cl$

Answer: b Section: 4

21. Acetyl chloride undergoes nucleophilic substitution at a faster rate than methyl acetate because:

a. the ester is more sterically hindered than the acid chloride.
b. the acid chloride is more sterically hindered than the ester.
c. the methoxide is a better leaving group than chloride.
d. esters hydrolyze faster than acid chlorides.
e. chloride is a better leaving group than methoxide.

Answer: e Section: 4

22. Which of the following is the best method for preparing

$$CH_3-\overset{O}{\overset{||}{C}}-O-CH_3?$$

a. $CH_3-\overset{O}{\overset{||}{C}}-H + CH_3OH \xrightarrow{H^{\oplus}}$

b. $CH_3-\overset{O}{\overset{||}{C}}-Cl + CH_3\overset{\ominus}{O}\overset{\oplus}{Na} \longrightarrow$

c. $CH_3-\overset{O}{\overset{||}{C}}-H + CH_3\overset{\ominus}{O}\overset{\oplus}{Na} \longrightarrow$

d. $CH_3-\overset{O}{\overset{||}{C}}-OH + CH_3OH \xrightarrow{H^{\oplus}}$

e. $CH_3-\overset{O}{\overset{||}{C}}-NH_2 + CH_3\overset{\ominus}{O}\overset{\oplus}{Na} \longrightarrow$

Answer: b Section: 6

23. Which of the following compounds would yield acetic acid when hydrolyzed under heat?

a. $CH_3\overset{O}{\overset{\|}{C}}-O-\overset{O}{\overset{\|}{C}}CH_3$

b. $CH_3C\equiv N$

c. $CH_3-\overset{O}{\overset{\|}{C}}-Cl$

d. $CH_3-\overset{O}{\overset{\|}{C}}-OCH_3$

e. All the above

Answer: e Section: 6

24. Which of the following is the best method for preparing aspirin,

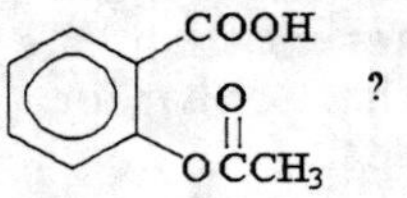

a.

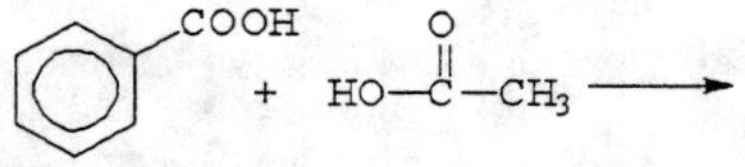

b.

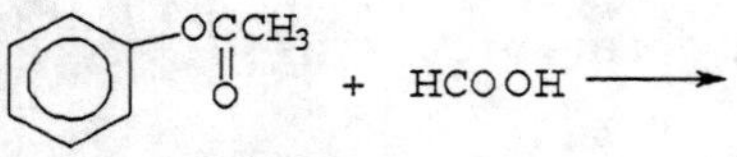

c.

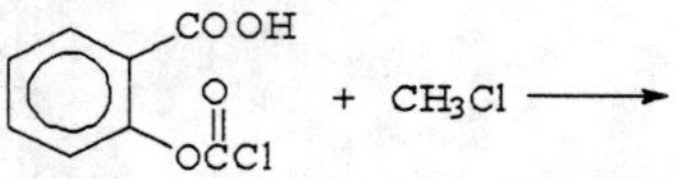

d.

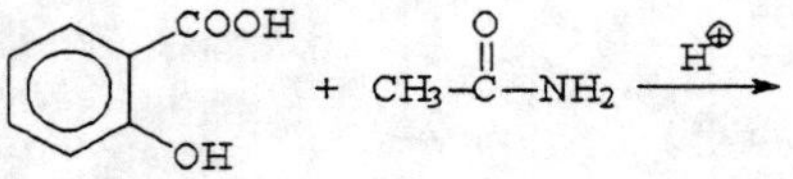

e.

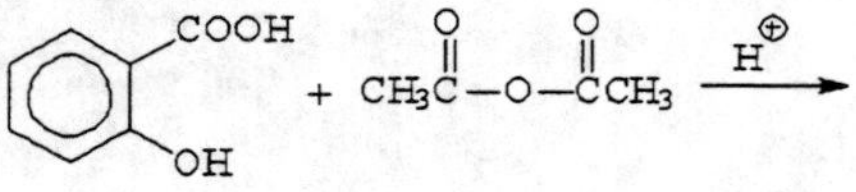

Answer: e Section: 7

25. Which of the following methods can be used to prepare

$$CH_3-C\equiv N?$$

a. $CH_3Br + {}^{-}C\equiv N \longrightarrow$

b. $CH_3\overset{O}{\overset{\|}{C}}NH_2 \xrightarrow[\text{heat}]{P_2O_5}$

c. $CH_3CH_2Br + NH_2 \longrightarrow$

d. a and b

e. b and c

Answer: d Section: 12

26. Which of the following is one of the steps in the mechanism of the following reaction?

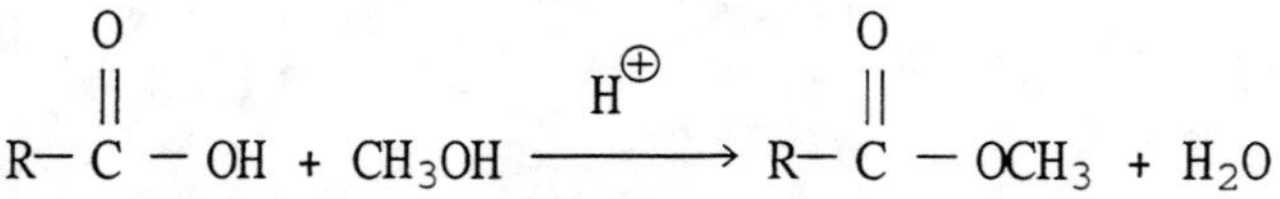

$$R-\overset{O}{\overset{\|}{C}}-OH + CH_3OH \xrightarrow{H^{\oplus}} R-\overset{O}{\overset{\|}{C}}-OCH_3 + H_2O$$

a. $R-\overset{O}{\overset{\|}{C}}-OH + CH_3OH \longrightarrow R-\overset{O^{\ominus}}{\overset{|}{\underset{\underset{HOCH_3}{|}}{C}}}-OH$

b. $R-\overset{O^{\oplus}H}{\overset{\|}{C}}-OH + CH_3O^{\ominus} \longrightarrow R-\overset{{}^{\oplus}OH}{\overset{|}{\underset{\underset{OCH_3}{|}}{C}}}-OH$

c. $R-\overset{O^{\oplus}H}{\overset{\|}{C}}-OH + CH_3OH \longrightarrow R-\overset{OH}{\overset{|}{\underset{\underset{H-\overset{\oplus}{O}-CH_3}{|}}{C}}}-OH$

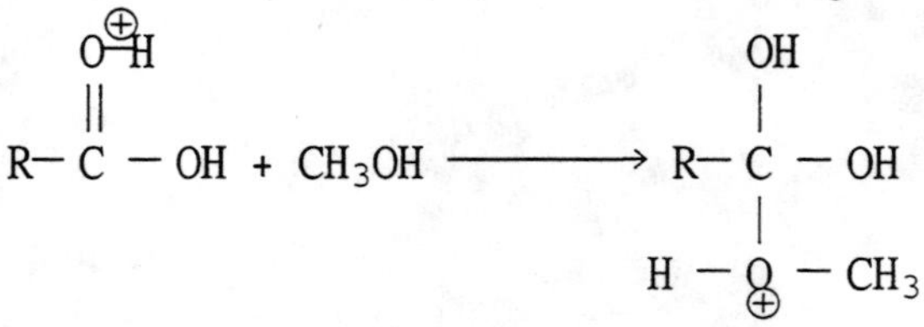

d. $R-\overset{O}{\overset{\|}{C}}-OH + CH_3O^{\ominus} \longrightarrow R-\overset{O^{\ominus}}{\overset{|}{\underset{\underset{OCH_3}{|}}{C}}}-OH$

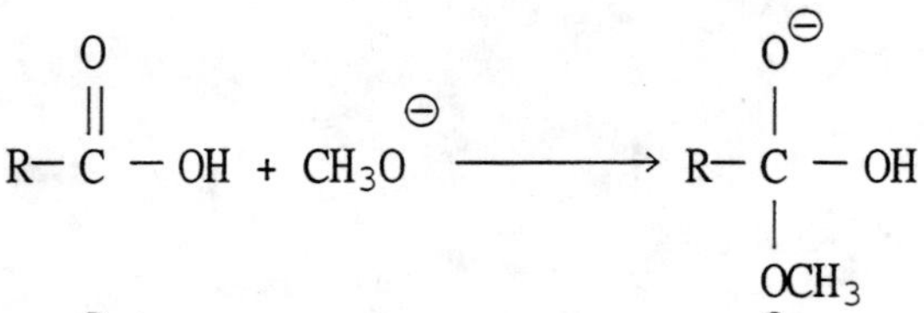

e. $R-\overset{O}{\overset{\|}{C}}-OH + CH_3OH \longrightarrow R-\overset{O}{\overset{\|}{C}}_{\oplus} + {}^{\ominus}OH$

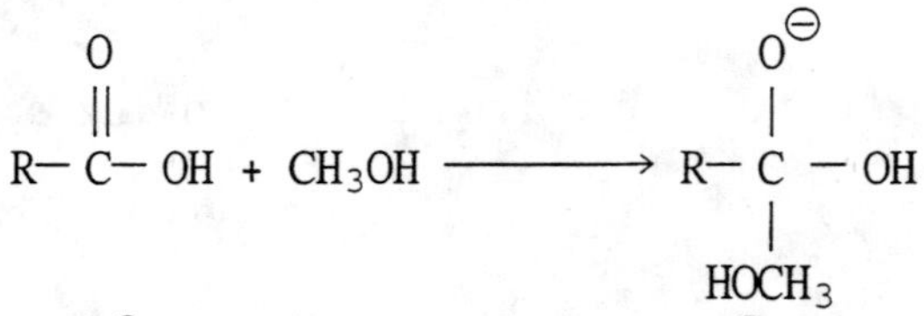

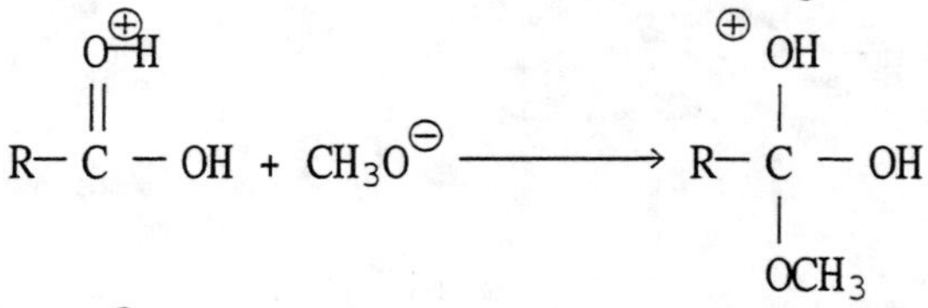

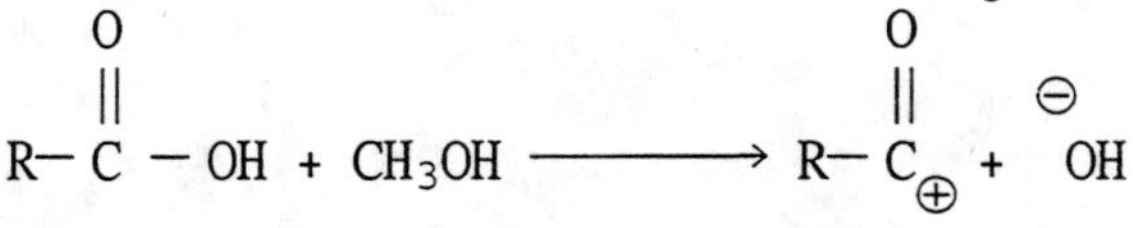

Answer: c Section: 9

27. What is the product obtained from the following reaction?

$$C_6H_5-\overset{O}{\overset{\|}{C}}-OH + CH_3\overset{18}{O}H \xrightarrow{H^{\oplus}} ?$$

a. $C_6H_5-\overset{O}{\overset{\|}{C}}-O-CH_3 + H_2\overset{18}{O}$

b. $C_6H_5-\overset{O}{\overset{\|}{C}}-\overset{18}{O}-CH_3 + H_2\overset{18}{O}$

c. $C_6H_5-\overset{\overset{18}{O}}{\overset{\|}{C}}-O-CH_3 + H_2\overset{18}{O}$

d. $C_6H_5-\overset{O}{\overset{\|}{C}}-\overset{18}{O}-CH_3 + H_2O$

e. $C_6H_5-\overset{\overset{18}{O}}{\overset{\|}{C}}-\overset{18}{O}-CH_3 + H_2\overset{18}{O}$

Answer: d Section: 11

28. What is the major organic product obtained from the following sequence of reactions?

$$CH_3-\overset{O}{\overset{\|}{C}}-OH \xrightarrow{SoCl_2} \xrightarrow{CH_3OH} ?$$

a. $CH_3-\overset{O}{\overset{\|}{C}}-O-CH_3$

b. $\underset{\underset{Cl}{|}}{CH_2}-\overset{O}{\overset{\|}{C}}-O-CH_3$

c. $CH_3-O-\overset{O}{\overset{\|}{C}}-H$

d. $CH_3-O-\overset{O}{\overset{\|}{C}}-H$

e. $CH_3-\overset{O}{\overset{\|}{C}}-O-Cl$

Answer: a Section: 17

29. Which of the following reagents is used in the following reaction?

a. Cl_2/P
b. Cl_2/CCl_4
c. Cl_2/hv
d. HCl
e. PCl_3

?

Answer: e Section: 17

30. Which of the following is phthalic acid?

a. COOH, COOH

b. COOH, HOOC

c. COOH, COOH

d.

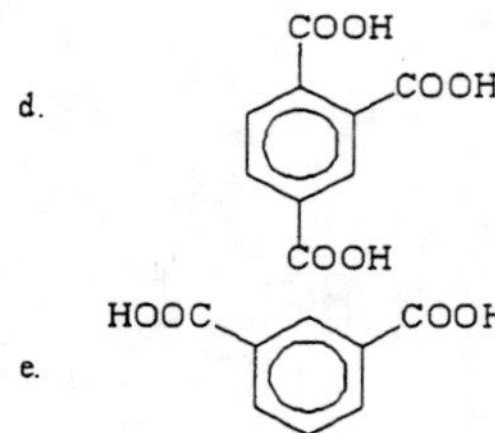

e. HOOC, COOH, COOH

Answer: c Section: 18

31. What is the <u>major</u> product of the following reaction?

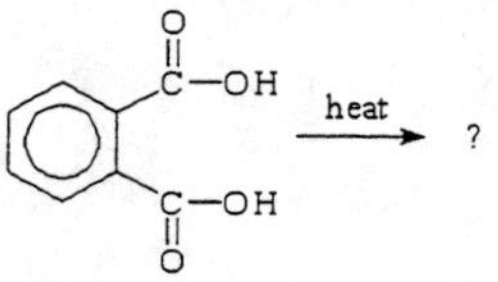

a.

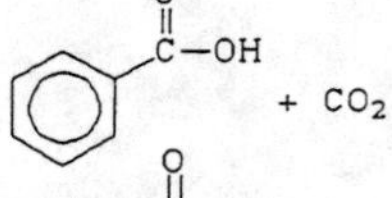

b.

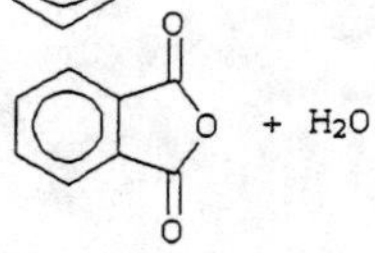

c.

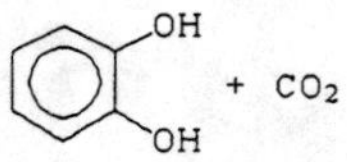

d.

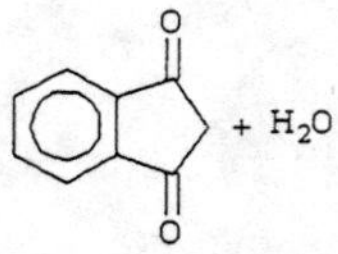

e.

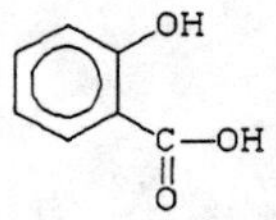

Answer: b Section: 18

32. What is the <u>major</u> organic product of the following reaction?

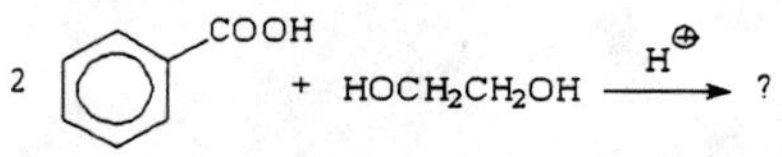

a.

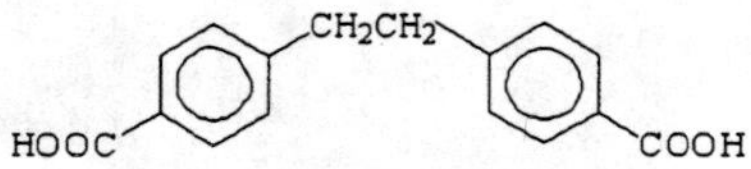

b.

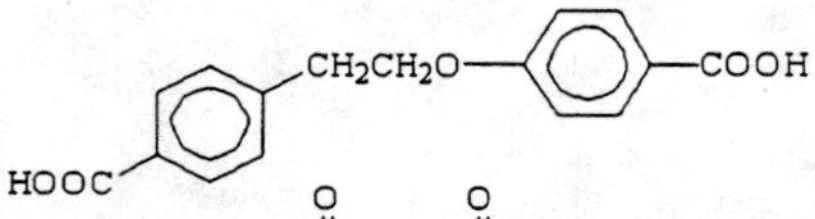

c.

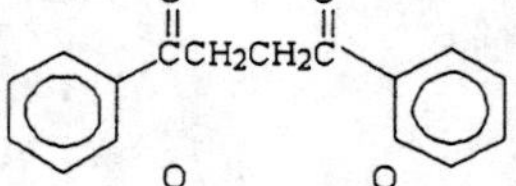

d.

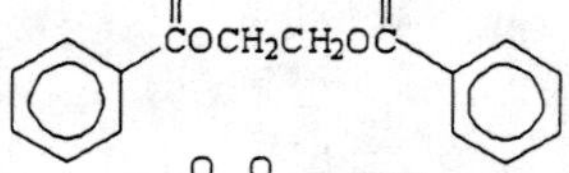

e.

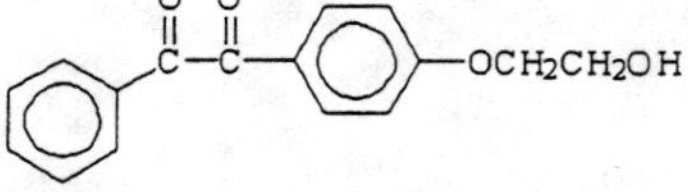

Answer: d Section: 11

SHORT ANSWER

33. Do you expect the following reaction to go to completion? Why or why not?

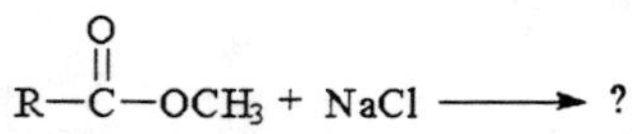

$$R{-}\overset{O}{\overset{\|}{C}}{-}OCH_3 + NaCl \longrightarrow ?$$

Answer:
No. The equilibrium will favor the reactants because the nucleophile, Cl^-, is a much weaker base than the leaving group CH_3O^-.

Section: 3

34. What are the products from the following reaction?

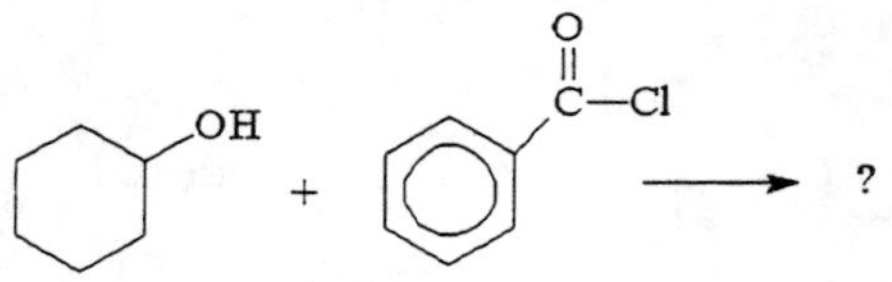

Cyclohexanol (cyclohexyl–OH) + benzoyl chloride ($C_6H_5{-}\overset{O}{\overset{\|}{C}}{-}Cl$) $\longrightarrow$?

Answer:

Cyclohexyl benzoate ($C_6H_5{-}\overset{O}{\overset{\|}{C}}{-}O{-}C_6H_{11}$) + HCl

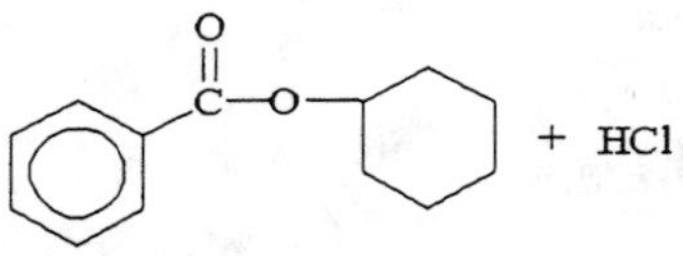

Section: 6

35. An unknown compound, $C_5H_{10}O_2$, gave the following proton NMR data:
(a) doublet, at 1.23 ppm (6H)
(b) singlet, at 2.10 ppm (3H)
(c) septet, at 4.98 ppm (1H)
Propose a structure for the compound.

Answer:

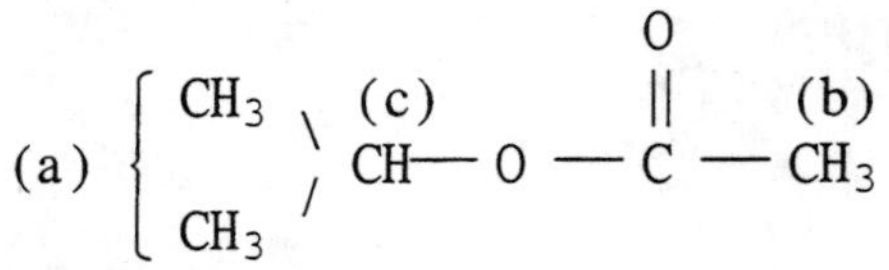

$$\text{(a)}\left\{\begin{array}{l}CH_3\\CH_3\end{array}\right\rangle \overset{(c)}{CH}{-}O{-}\overset{O}{\overset{\|}{C}}{-}\overset{(b)}{CH_3}$$

36. What is the product of the following reaction?

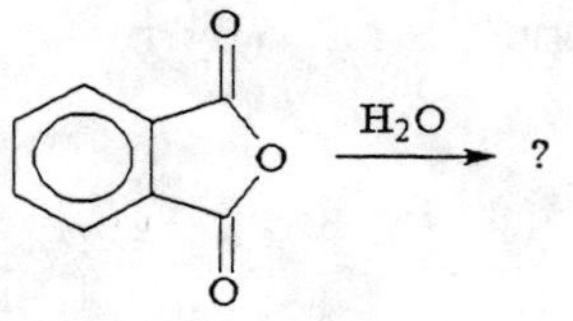

$$\text{phthalic anhydride} \xrightarrow{H_2O} ?$$

Answer:

Phthalic acid (benzene ring bearing two adjacent C(=O)–OH groups)

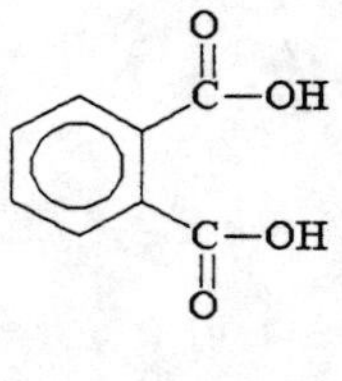

This is the hydrolysis of an anhydride to an acid.

Section: 7

37. Provide a detailed, stepwise mechanism for the reaction of acetyl chloride with methanol to produce methyl acetate and HCl.

Answer:

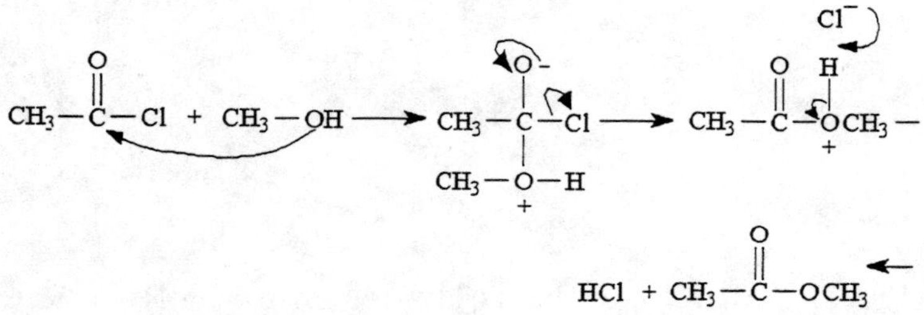

$$CH_3-\overset{O}{\overset{\|}{C}}-Cl + CH_3-OH \longrightarrow CH_3-\underset{CH_3-\overset{+}{O}-H}{\overset{O^-}{C}}-Cl \longrightarrow CH_3-\overset{O}{\overset{\|}{C}}-\overset{+}{\overset{H}{O}}CH_3 \;(Cl^-) \longrightarrow HCl + CH_3-\overset{O}{\overset{\|}{C}}-OCH_3$$

Section: 6

38. What are the products of the reaction of benzoic acid with thionyl chloride?

Answer: benzoyl chloride, SO_2 (g), and HCl (g) Section: 17

39. Give a detailed, stepwise mechanism for the Fischer esterification of acetic acid with methanol.

Answer:

$$CH_3-\overset{O}{\overset{\|}{C}}-OH + H^+ \rightleftharpoons CH_3-\overset{\overset{+}{O}-H}{\overset{\|}{C}}-OH + CH_3OH \rightleftharpoons CH_3-\overset{O-H}{\overset{|}{C}}-OH \text{ with } CH_3-\overset{+}{O}-H \text{ on C}$$

$$\downarrow -H^+$$

$$CH_3-\overset{O-H}{\overset{|}{C}}-OH \text{ with } CH_3-O \text{ on C} \xrightleftharpoons{+H^+} CH_3-\overset{H-\overset{+}{O}-H}{\overset{|}{C}}-OH \text{ with } CH_3-O \text{ on C} \rightleftharpoons H_2O + CH_3-C=\overset{+}{O}-H \text{ with } OCH_3 \text{ on C}$$

$$\rightleftharpoons H_3O^+ + CH_3-\overset{O}{\overset{\|}{C}}-OCH_3$$

Section: 11

MULTIPLE CHOICE

40. Which of the following conditions will drive the equilibrium of the Fischer esterification towards ester formation?

a. addition of water
b. removal of water as it is formed
c. addition of an inorganic acid as a catalyst
d. addition of alcohol
e. both b and d

Answer: e Section: 11

41. Esters and amides are most easily made by nucleophilic acyl substitution reactions on:

a. alcohols
b. acid anhydrides
c. carboxylates
d. carboxylic acids
e. acid chlorides

Answer: e Section: 6

42. In nucleophilic acyl substitution,:

a. protonation of the carbonyl is followed immediately by loss of the leaving group.
b. loss of the leaving group is followed by rearrangement of the carbocation.
c. addition to the carbonyl by a nucleophile is followed by loss of the leaving group.
d. ester hydrolysis is followed by deprotonation.
e. an S_N2 reaction occurs.

Answer: c Section: 3

SHORT ANSWER

43. Provide a detailed, stepwise mechanism for the reaction of propanoyl chloride with ammonia.

Answer:

$CH_3CH_2-C(=O)-Cl + NH_3 \longrightarrow CH_3CH_2-C(O^-)(^+NH_3)-Cl \longrightarrow CH_3CH_2-C(=O)-\overset{+}{N}H_2-H$ (Cl^-) $\longrightarrow HCl + CH_3CH_2-C(=O)-NH_2$

Section: 6

44. Provide a detailed, stepwise mechanism for the reaction of propanoic anhydride with phenol.

Answer:

$CH_3CH_2-C(=O)-O-C(=O)-CH_2CH_3 + HO-C_6H_5 \rightleftharpoons CH_3CH_2\,C(O^-)(O-C(=O)CH_2CH_3)-\overset{+}{O}(H)-C_6H_5$

$\longrightarrow {}^-O-C(=O)CH_2CH_3 + C_6H_5-\overset{+}{O}(H)-C(=O)CH_2CH_3$

$\longrightarrow CH_3CH_2CO_2H + CH_3CH_2CO_2Ph$

Section: 7

45. Provide a detailed, stepwise mechanism for the acid-catalyzed transesterification of ethyl acetate with n-propanol.

Answer:

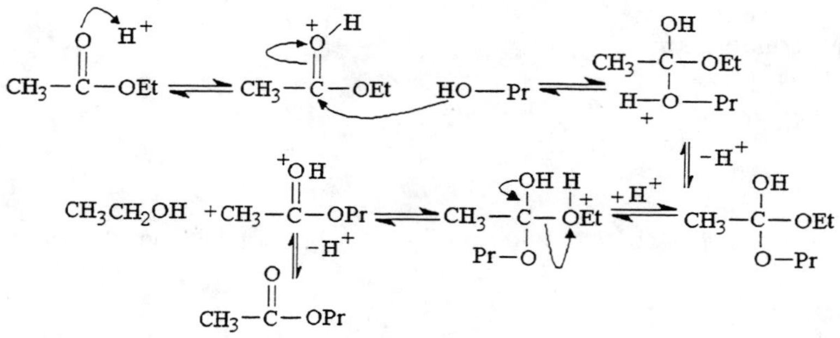

$CH_3-C(=O)-OEt \rightleftharpoons CH_3-C(=O^+H)-OEt$; $HO-Pr \rightleftharpoons CH_3-C(OH)(OEt)-O^+(H)-Pr$; $-H^+$ $\rightleftharpoons CH_3-C(OH)(OEt)-O-Pr$; $+H^+ \rightleftharpoons CH_3-C(OH)(O-Pr)-O^+(H)Et \rightleftharpoons CH_3CH_2OH + CH_3-C(=O^+H)-OPr$; $-H^+$ $\rightleftharpoons CH_3-C(=O)-OPr$

Section: 9

46. Provide a detailed, stepwise mechanism for the base-mediated transesterification of methyl benzoate with sodium ethoxide.

Answer:

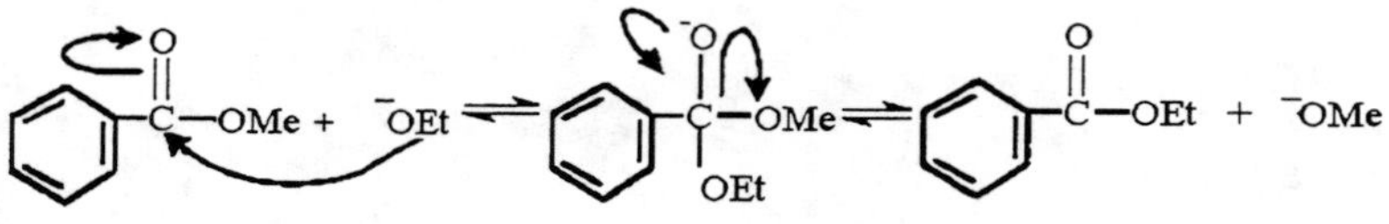

$C_6H_5-C(=O)-OMe + {}^-OEt \rightleftharpoons C_6H_5-C(O^-)(OEt)-OMe \rightleftharpoons C_6H_5-C(=O)-OEt + {}^-OMe$

Section: 10

47. Provide a detailed, stepwise mechanism for the reaction of butyrolactone with ammonia.

Answer:

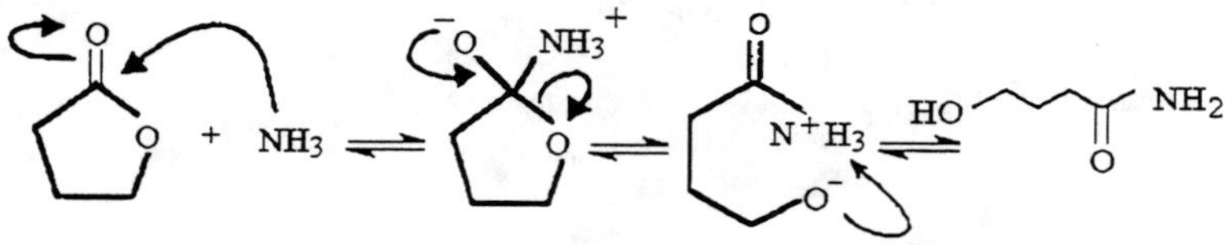

butyrolactone + NH_3 $\rightleftharpoons$ tetrahedral intermediate (^-O, NH_3^+) $\rightleftharpoons$ ring-opened $C(=O)-N^+H_3$ / O^- $\rightleftharpoons$ $HO-CH_2CH_2CH_2-C(=O)-NH_2$

Section: 8

MULTIPLE CHOICE

48. The hydrolysis of esters in base is called:

a. the Fischer esterification.
b. the Hunsdiecker reaction.
c. the Dieckmann condensation.
d. transesterification.
e. saponification.

Answer: e Section: 16

49. Typically, amides will hydrolyze under _________ conditions than esters.

a. milder
b. more dilute
c. stronger
d. less vigorous
e. more saline

Answer: c Section: 4

50. Phthalic acid produces what acid derivative upon heating?

a. an ester
b. a carboxylate
c. an acid chloride
d. an anhydride
e. an amide

Answer: d Section: 18

SHORT ANSWER

51. What are the major organic products when 1 equivalent of acetic formic anhydride reacts with 1 equivalent of methanol?

Answer: acetic acid and methyl formate Section: 8

MULTIPLE CHOICE

52. A nitrile can be made by dehydrating an amide. However, for this reaction to occur, the amide must be:

a. primary.
b. secondary.
c. tertiary.
d. N-methylated.
e. part of a lactam.

Answer: a Section: 12

SHORT ANSWER

53. What reagents are used to dehydrate amides to nitriles?

    Answer: P_2O_5 or $POCl_3$ Section: 12

54. Why are β-lactams, such as penicillins and cephalosporins, unusually reactive amides?

    Answer:
    Their enhanced reactivity is due to the considerable ring strain in the four-membered ring.
    Section: 13

55. Propose a synthesis of benzonitrile from benzene and any other necessary reagents.

    Answer:
    1. HNO_3, H_2SO_4
    2. Fe, HCl
    3. $NaNO_2$, HCl, cold
    4. CuCN

56. Carbonated beverages typically have pH's below 7.0. What acid is present which lowers the pH? Show the reaction by which it is formed.

    Answer: carbonic acid, $CO_2 + H_2O \longrightarrow H_2CO_3$ Section: 18

57. Provide the proper IUPAC name for the compound below.

    $CO_2CH(CH_3)_2$

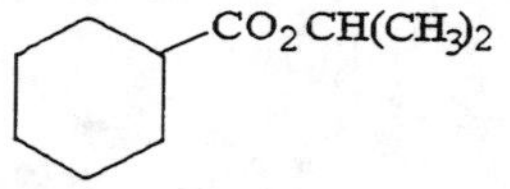

    Answer: isopropyl cyclohexanecarboxylate Section: 1

58. Provide the proper IUPAC name for $ClCH_2CH_2CONHCH_3$.

    Answer: N-methyl 3-chloropropanamide Section: 1

59. Provide the proper IUPAC name for the compound below.

    CH_3CH_2 H H $CH_2CH_2CH_2COCl$

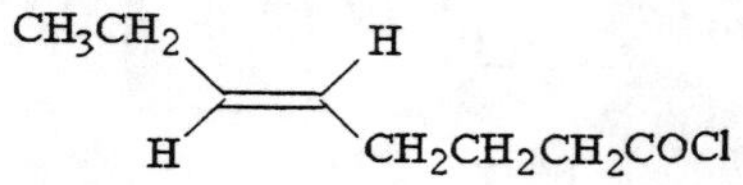

    Answer: trans-5-octenoyl chloride Section: 1

60. Provide the structure of propanoic anhydride.

    Answer: $(CH_3CH_2CO)_2O$ Section: 1

Chapter 16: Carbonyl Compounds II

MULTIPLE CHOICE

1. Which of the following compounds is an aldehyde?

a. 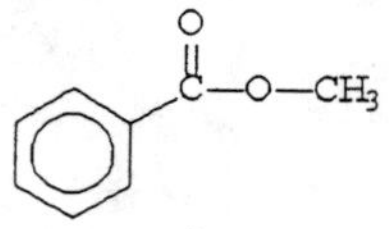 $C_6H_5C(=O)-O-CH_3$

b. $C_6H_5C(=O)-CH_3$

c. $C_6H_5C(=O)-OH$

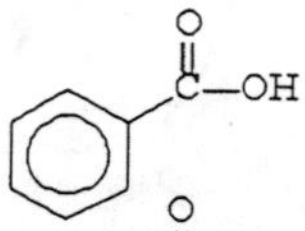

d. 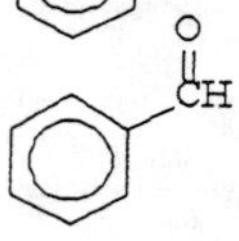 $C_6H_5C(=O)H$

e. 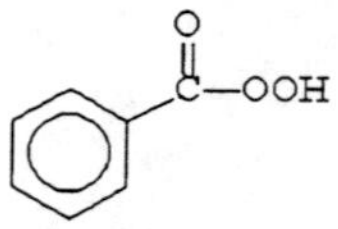 $C_6H_5C(=O)-OOH$

Answer: d

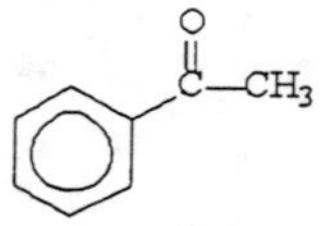

2. Which of the following compounds is a ketone?

a. 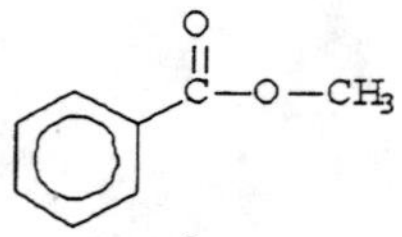 $C_6H_5C(=O)-O-CH_3$

b. $C_6H_5C(=O)-CH_3$

c. $C_6H_5C(=O)-OH$

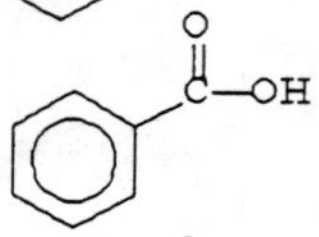

d. 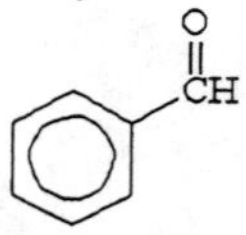 $C_6H_5C(=O)H$

e. 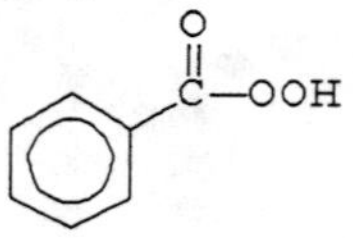 $C_6H_5C(=O)-OOH$

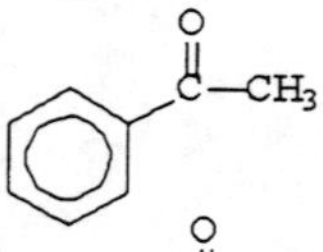

Answer: b

3. What is the common name for the following compound?

$$\underset{\underset{Cl}{|}}{CH_2}-\overset{\overset{O}{\|}}{C}-H$$

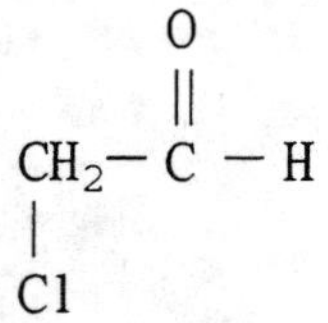

a. Chloroaldehyde
b. α-Chloroacetaldehyde
c. β-Chloroacetaldehyde
d. 2-Chloroethanal
e. α-Chloroethanal

Answer: b Section: 1

4. Which of the following compounds is acetone?

a. 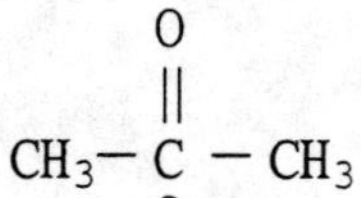 $CH_3-\overset{\overset{O}{\|}}{C}-CH_3$

b. $CH_3-\overset{\overset{O}{\|}}{C}-H$

c. $CH_3-O-\overset{\overset{O}{\|}}{C}-CH_3$

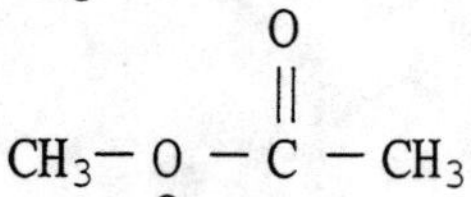

d. 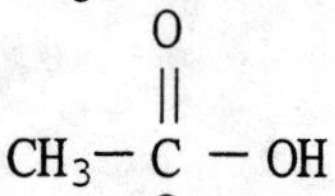 $CH_3-\overset{\overset{O}{\|}}{C}-OH$

e. 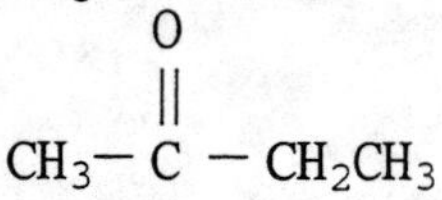 $CH_3-\overset{\overset{O}{\|}}{C}-CH_2CH_3$

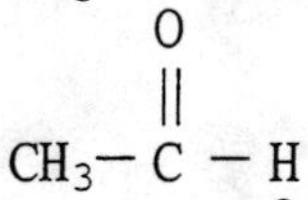

Answer: a Section: 1

5. What is the IUPAC name for the following compound?

$$CH_3CH(Br)CH_2-\overset{\overset{O}{\|}}{C}-H$$

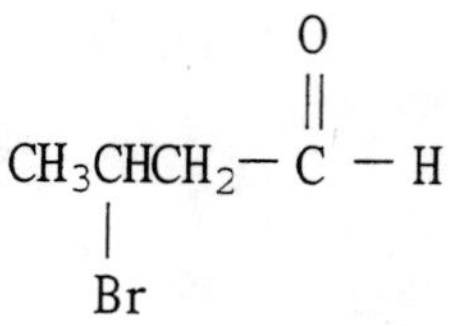

a. 2-Bromobutanal
b. α-Bromobutanal
c. 3-Bromobutanal
d. β-Bromobutyraldehyde
e. 3-Bromobutanone

Answer: c Section: 1

6. Which of the following compounds is benzaldehyde?

a. $C_6H_5-\overset{O}{\overset{\|}{C}}-OOH$

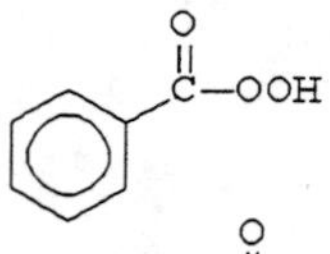

b. $C_6H_5-CH_2\overset{O}{\overset{\|}{C}}H$

c. $C_6H_5-\overset{O}{\overset{\|}{C}}CH_3$

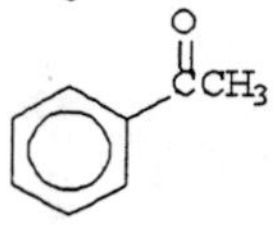

d. $C_6H_5-\overset{O}{\overset{\|}{C}}H$

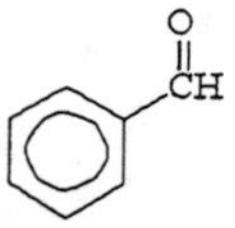

e. $C_6H_5-\overset{O}{\overset{\|}{C}}OH$

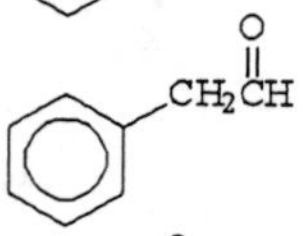

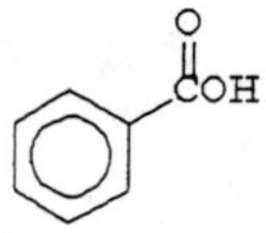

Answer: d Section: 1

7. Which of the following is cyclohexanone?

a.

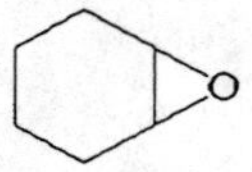

b.

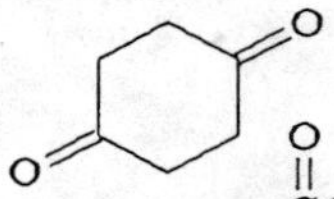

c.

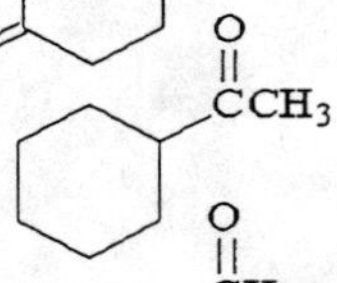

d.

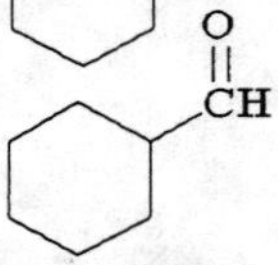

e.

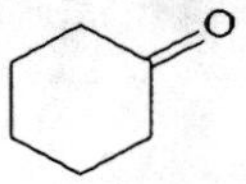

Answer: e Section: 1

8. What is the IUPAC name for the following compound?

$$CH_3\overset{O}{\overset{\|}{C}}\underset{CH_3}{\underset{|}{CH}}CH_2\underset{CH_3}{\underset{|}{CH}}Cl$$

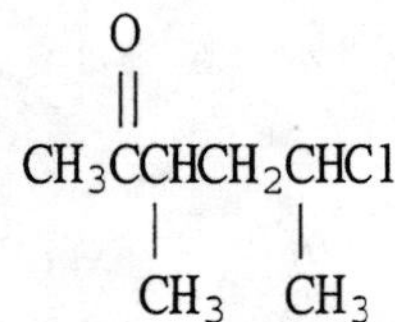

a. 5-Chloro-3-methylhexanone
b. 1-Chloro-1,3-dimethyl-4-pentanone
c. 5-Chloro-3,5-dimethyl-2-pentanone
d. 5-Chloro-3-methyl-2-hexanone
e. 2-Chloro-4-methyl-5-hexanone

Answer: d Section: 1

9. Which of the following characterizes the reactions of aldehydes and ketones?

a. Electrophilic addition
b. Electrophilic substitution
c. Nucleophilic acyl substitution
d. Nucleophilic addition
e. Free radical addition

Answer: d Section: 3

10. Which of the following reactions can be used to prepare 3-methyl-3-hexanol,

$$CH_3CH_2\overset{\overset{OH}{|}}{\underset{\underset{CH_3}{|}}{C}}CH_2CH_2CH_3 \quad ?$$

a. $CH_3CH_2\overset{\overset{O}{||}}{C}CH_2CH_2CH_3 \xrightarrow[2.\ H_3O^{\oplus}]{1.\ CH_3MgBr}$

b. $CH_3\overset{\overset{O}{||}}{C}CH_2CH_2CH_3 \xrightarrow[2.\ H_3O^{\oplus}]{1.\ CH_3CH_2MgBr}$

c. $CH_3\overset{\overset{O}{||}}{C}CH_2CH_3 \xrightarrow[2.\ H_3O^{\oplus}]{1.\ CH_3CH_2CH_2MgBr}$

d. a and b

e. a, b, and c

Answer: e Section: 4

11. What is/are the <u>major</u> product(s) of the following reaction?

$$CH_3MgBr + HOCH_2\overset{\overset{O}{||}}{C}H \longrightarrow ?$$

a. $HO{-}CH_2\overset{\overset{O}{||}}{C}CH_3 + HMgBr$

b. $HO{-}CH_2\overset{\overset{OMgBr}{|}}{\underset{\underset{CH_3}{|}}{C}}H$

c. $CH_4 + BrMgOCH_2\overset{\overset{O}{||}}{C}H$

d. $BrMgCH_2\overset{\overset{O}{||}}{C}H + CH_3OH$

e. $HOCH_2C(=O)CH_2MgBr + H_2$

Answer: c Section: 4

12. Which of the following is the best method for preparing lactic acid from acetaldehyde?

$CH_3CH(OH)COOH$ — lactic acid

$CH_3C(=O)H$ — acetaldehyde

a. Cl_2 ; ^-OH ; CH_3OH
b. CH_3MgBr ; ^-OH
c. $KMnO_4$; ^-OH
d. $HC{\equiv}N$; H_3O^+/heat
e. Cl_2 ; Mg/ether ; ^-OH

Answer: d Section: 4

13. Which of the following reagents can be used to reduce acetaldehyde to ethyl alcohol?

a. 1. $LiAlH_4$ /2. H_3O^+
b. 1. $NaBH_4$ /2. H_3O^+
c. H_2 /Pt
d. a and d
e. a, b, and c

Answer: e Section: 4

14. Which of the following reactions can be used to prepare:

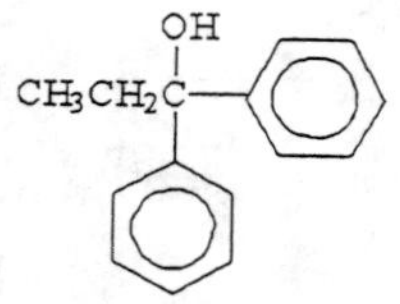

$$CH_3CH_2C(OH)(C_6H_5)_2 \; ?$$

a. $CH_3CH_2\overset{O}{\overset{\|}{C}}{-}OCH_3 \xrightarrow[2.\ H_3O^{\oplus}]{1.\ C_6H_5MgBr}$

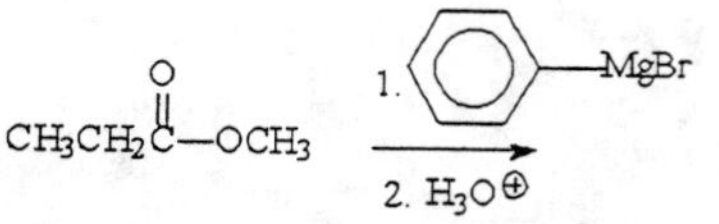

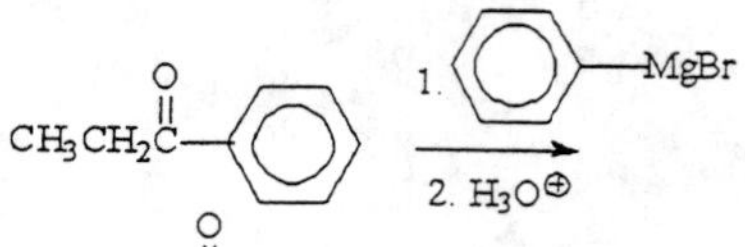

b. $CH_3CH_2\overset{O}{\overset{\|}{C}}{-}C_6H_5 \xrightarrow[2.\ H_3O^{\oplus}]{1.\ C_6H_5MgBr}$

c. $(C_6H_5)_2C{=}O \xrightarrow[2.\ H_3O^{\oplus}]{1.\ CH_3CH_2MgBr}$

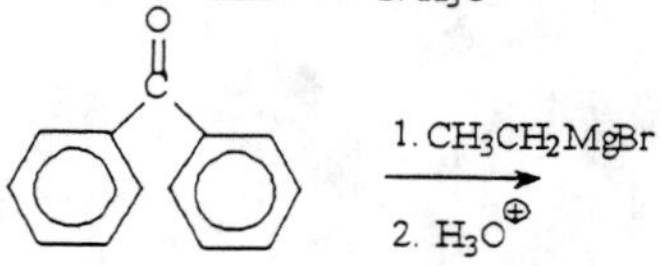

d. b and c

e. a, b and c

Answer: e Section: 4

15. Which of the following reactions is the best method for preparing acetaldehyde?

$$CH_3\overset{O}{\overset{\|}{C}}{-}H$$

a. $CH_3\overset{O}{\overset{\|}{C}}Cl \xrightarrow[2.\ H_3O^{\oplus}]{1.\ LiAlH[OC(CH_3)_3]_3}$

b. $CH_3\overset{O}{\overset{\|}{C}}Cl \xrightarrow[2.\ H_3O^{\oplus}]{1.\ LiAlH_4}$

c. $CH_3\overset{O}{\overset{\|}{C}}OH \xrightarrow[2.\ H_3O^{\oplus}]{1.\ LiAlH_4}$

d. $CH_3\overset{O}{\overset{\|}{C}}OCH_3 \xrightarrow[2.\ H_3O^{\oplus}]{1.\ NaBH_4}$

e. $CH_3\overset{O}{\overset{\|}{C}}NH_2 \xrightarrow[2.\ H_3O^{\oplus}]{1.\ LiAlH_4}$

Answer: a Section: 6

16. What is the major product of the following reaction?

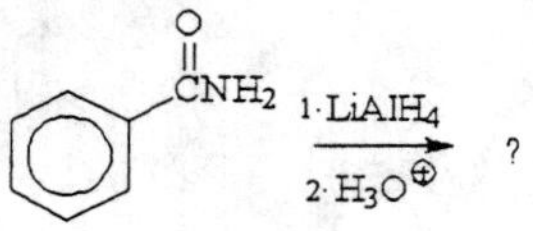

$C_6H_5CONH_2 \xrightarrow[2.\ H_3O^{\oplus}]{1.\ LiAlH_4} ?$

a. $C_6H_5\overset{\oplus}{N}H_3$

b. $C_6H_5CH_2\overset{\oplus}{N}H_3$

c. $C_6H_5CH(OH)\overset{\oplus}{N}H_3$

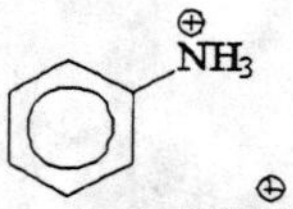

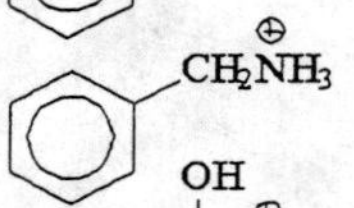

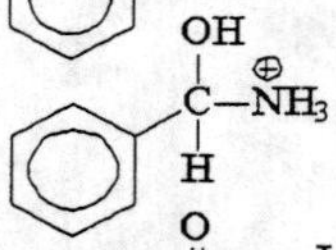

d. $C_6H_5C(=O)N(Li)AlH_2$

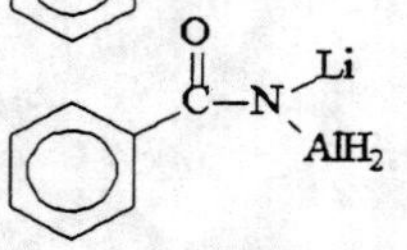

e. $C_6H_5CH_2OH$

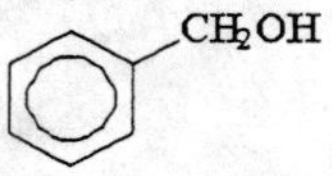

Answer: b Section: 6

17. Which of the following compounds is a hydrazone?

a. $C_6H_{11}NHNH_2$

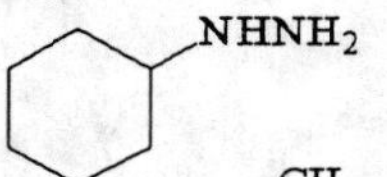

b. $C_6H_{11}N(CH_3)_2$

c. $C_6H_{11}-NH-NH-C_6H_5$

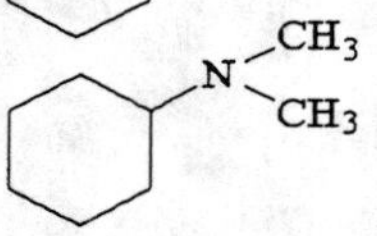

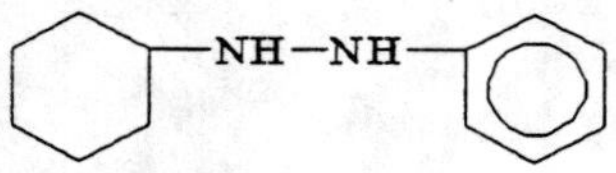

d. $C_6H_{10}=N-NH_2$

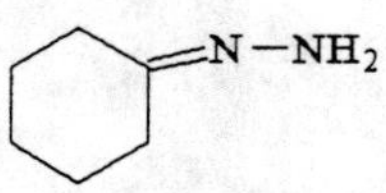

e. $C_6H_{11}NH_2$

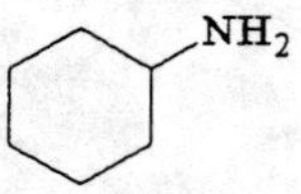

Answer: d Section: 7

18. What is the major organic product of the following reaction?

$$CH_3-\overset{\overset{\displaystyle O}{\|}}{C}-H + HONH_2 \longrightarrow ?$$

a. CH_3CHOH
   |
   NHOH

b. $CH_3C=O$
   |
   NHOH

c. $CH_3C=NOH$
   |
   H

d. $CH_3CHNHOCHCH_3$
   |　　　|
   OH　　OH

e. CH_3CHOH
   |
   ONH_2

Answer: c Section: 7

19. Which of the following is an enamine?

a.
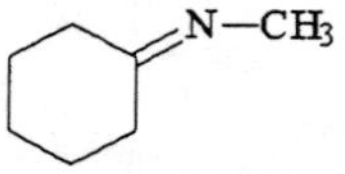

b.
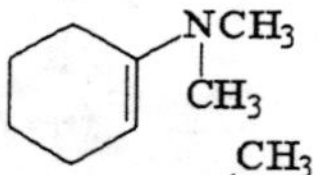

c.
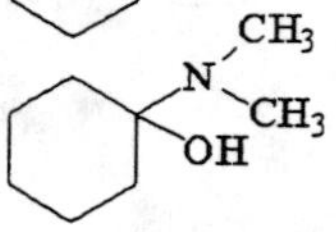

d.
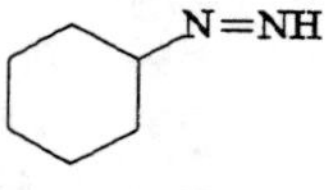

e.
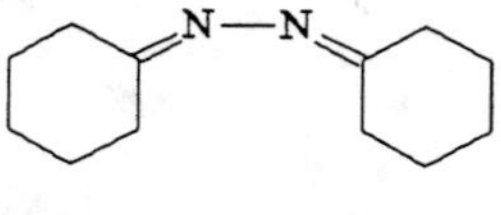

Answer: b Section: 7

20. What is the major organic product of the following reaction?

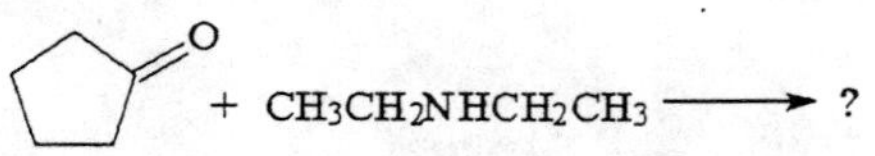

cyclopentanone + $CH_3CH_2NHCH_2CH_3$ → ?

a. 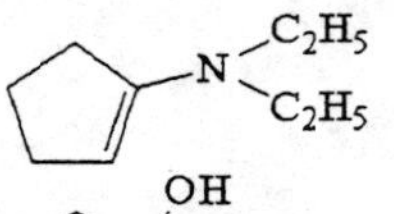 cyclopentene with $N(C_2H_5)_2$ on the ring double bond

b. cyclopentane with OH and $N(C_2H_5)_2$ on the same carbon

c. cyclopentylidene $=\overset{\oplus}{N}(C_2H_5)_2$

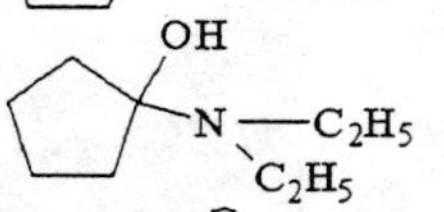

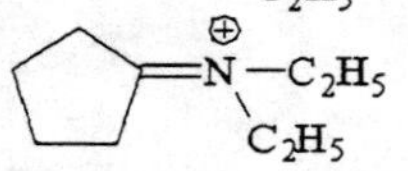

d. 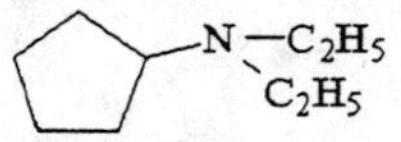 cyclopentane with $N(C_2H_5)_2$

e. 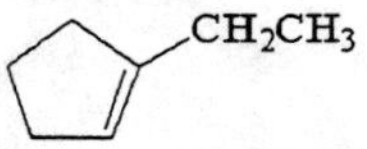 cyclopentene with CH_2CH_3

Answer: a Section: 7

21. What is the major organic product of the following reaction?

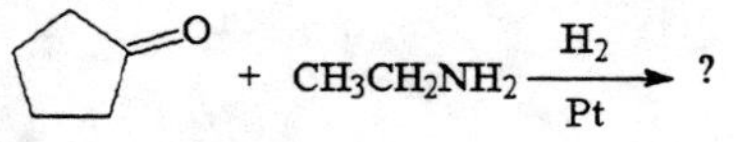

cyclopentanone + $CH_3CH_2NH_2$ $\xrightarrow[Pt]{H_2}$?

a. 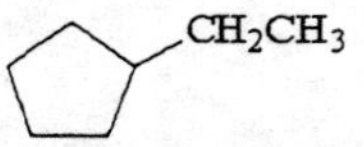 cyclopentane with CH_2CH_3

b.  cyclopentylidene $=NCH_2CH_3$

c. 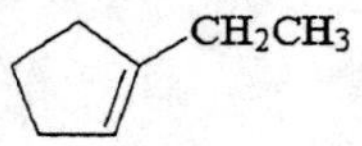 cyclopentene with CH_2CH_3

d. cyclopentane with OH and $NHCH_2CH_3$ on the same carbon

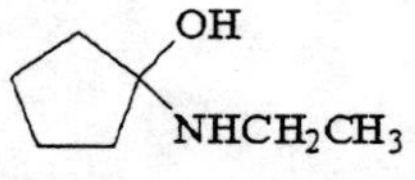

e.  cyclopentane with $NHCH_2CH_3$

Answer: e Section: 8

22. The compound:

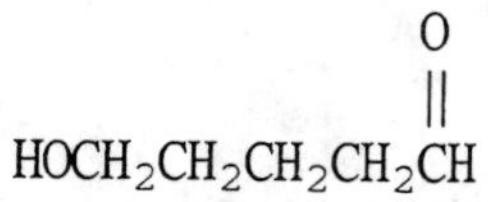

$$HOCH_2CH_2CH_2CH_2\overset{O}{\overset{\|}{C}}H$$

can form a hemiacetal by reacting with itself in solution. What is the structure of this hemiacetal?

a. 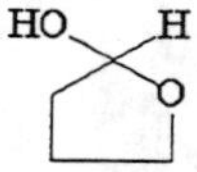 (five-membered ring structure with HO, H and ring O)

b. 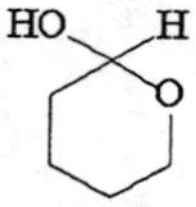 (six-membered ring structure with HO, H and ring O)

c. 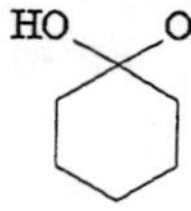 (cyclohexane ring with HO and OH on the same carbon)

d. 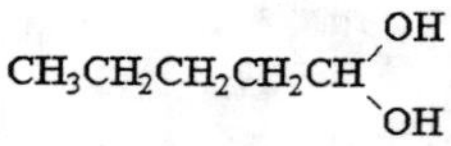 $CH_3CH_2CH_2CH_2CH(OH)_2$

e. 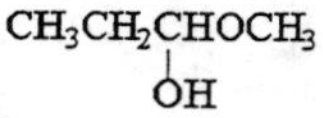 $CH_3CH_2\underset{OH}{\underset{|}{C}}HOCH_3$

Answer: b Section: 8

23. What sequence of reactions would best accomplish the following conversion?

a. $NaBH_4$ /H_3O^+
b. $HOCH_2CH_2OH/H^+$; $NaBH_4$; H^+ /H_2O
c. $KMnO_4$ / HO^- ; heat
d. $LiAlH_4$ /H_3O^+
e. H_2 /Pt ; PCC

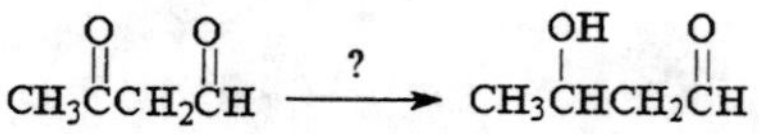

$$CH_3\overset{O}{\overset{\|}{C}}CH_2\overset{O}{\overset{\|}{C}}H \xrightarrow{?} CH_3\overset{OH}{\overset{|}{C}}HCH_2\overset{O}{\overset{\|}{C}}H$$

Answer: b Section: 9

24. What is the major product of the following reaction?

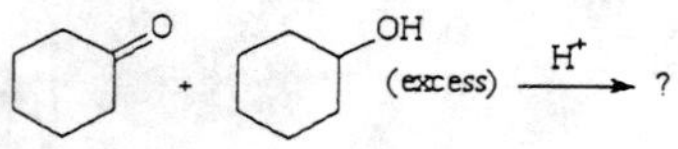

cyclohexanone + cyclohexanol (excess) $\xrightarrow{H^+}$?

a.

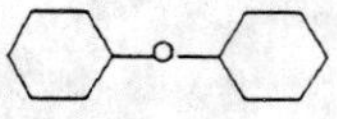

b. 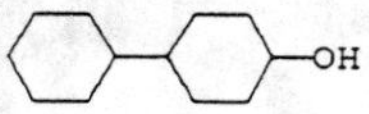 OH

c. 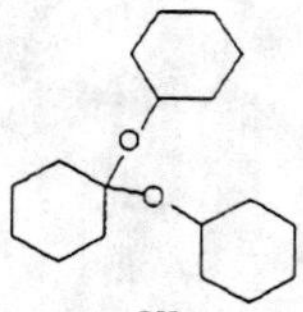 O O

d. 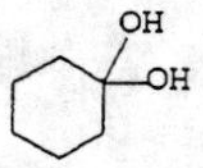 OH OH

e. 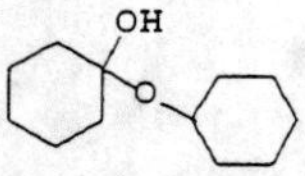 OH O

Answer: c Section: 8

25. What carbonyl compound and what phosphonium ylide are required for the synthesis of the following alkene?

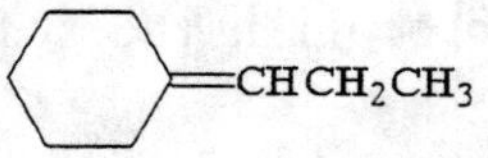

$=CHCH_2CH_3$

a. 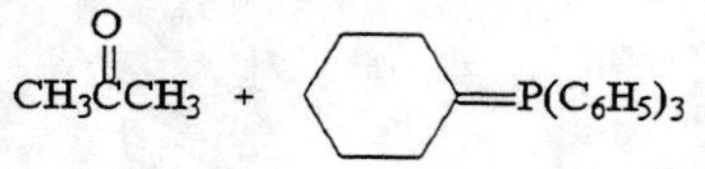 $CH_3\overset{O}{\overset{\|}{C}}CH_3$ + $=P(C_6H_5)_3$

b. 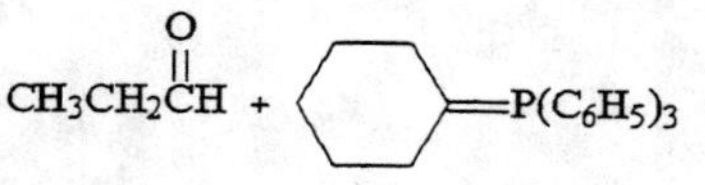 $CH_3CH_2\overset{O}{\overset{\|}{C}}H$ + $=P(C_6H_5)_3$

c. $=O$ + $CH_3CH_2CH{=}P(C_6H_5)_3$

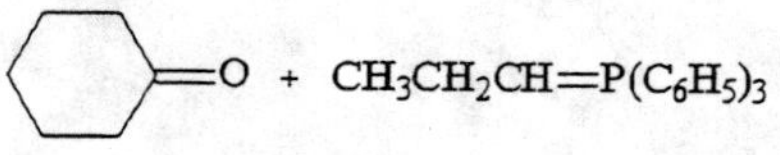

d. b or c

e. a, b or c

Answer: d Section: 11

26. Which enantiomer is formed from attack of a methyl Grignard reagent on the si face of benzaldehyde:

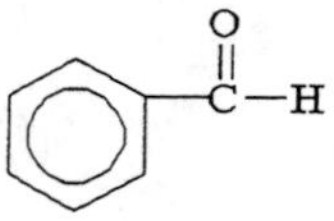

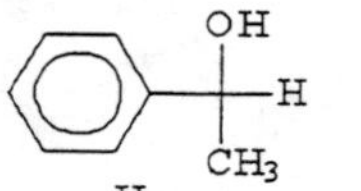

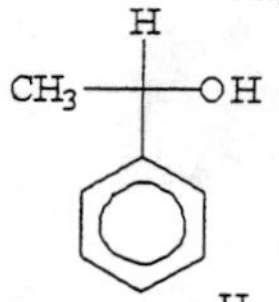

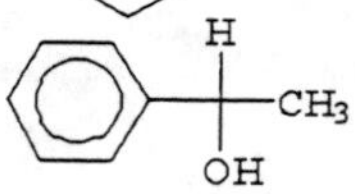

Answer: a Section: 12

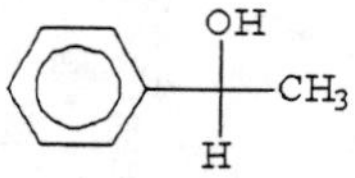

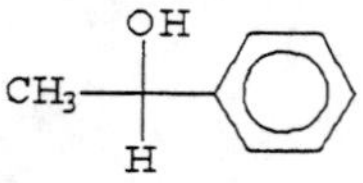

27. What reagent(s) would you use to accomplish the following conversion?

a. CH_3Br; H_3O^+
b. CH_3MgBr; H_3O^+
c. $(CH_3)_2CuLi$; H_3O^+
d. CH_3Br; $LiAlH_4$; H_3O^+
e. $LiAlH_4$; CH_3MgBr; H_3O^+

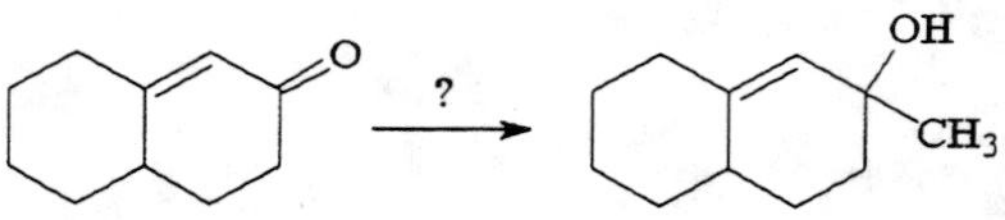

Answer: b Section: 14

28. What reagent(s) would you use to accomplish the following conversion?

a. CH_3Br; H_3O^+
b. CH_3MgBr; H_3O^+
c. $(CH_3)_2CuLi$; H_3O^+
d. CH_3Br; $LiAlH_4$; H_3O^+
e. $LiAlH_4$; CH_3MgBr; H_3O^+

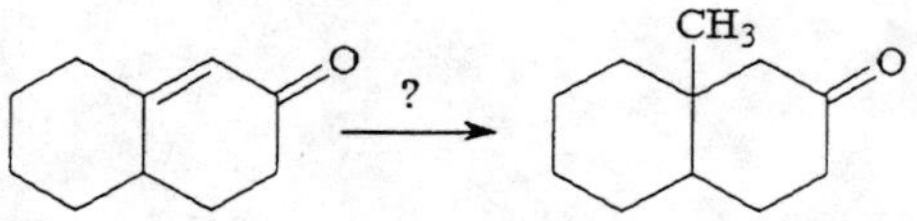

CH_3
?

Answer: c Section: 14

SHORT ANSWER

29. Propose an IR frequency that would distinguish between the following compounds:

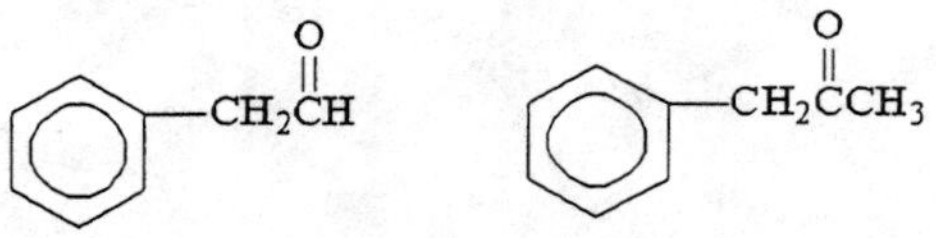

O
CH_2CH
I

O
CH_2CCH_3
II

Answer:
Compound I shows a C-H peak for the aldehyde at ~2700 cm^{-1}, while compound II shows a peak at ~1380 cm^{-1} for CH_3.

30. Identify the compound whose molecular formula and proton NMR spectra are given below:

$C_{15}H_{14}O$; (a) triplet, at 2.1 ppm (2H)
(b) triplet, at 2.4 ppm (2H)
(c) multiplet, at 7.9 - 7.2 ppm (10H)

Answer:

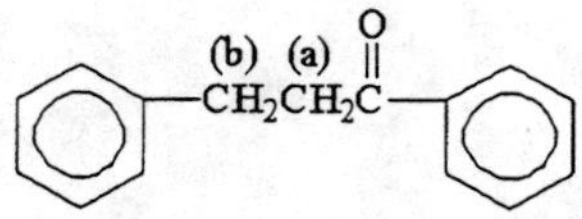

(b) (a) O
CH_2CH_2C
(c) (c)

31. An unknown compound (A), $C_{10}H_{12}O$, gave the following proton NMR data:
   (a) triplet, at 1.0 ppm (3H)
   (b) quartet, at 2.4 ppm (2H)
   (c) singlet, at 3.7 ppm (2H)
   (d) multiplet, at 7.2 ppm (5H)
   Propose a structure for (A)

Answer:

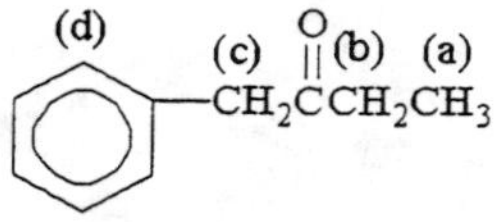

(d) (c) (b) (a)

C_6H_5–$CH_2\overset{O}{\overset{\|}{C}}CH_2CH_3$

32. Complete the following reaction sequence by supplying the missing information:

$CH_3OH \xrightarrow{?} CH_3Br \xrightarrow{?} ? \xrightarrow[2.?]{1.?} CH_3CH_2OH$

Answer:

$CH_3OH \xrightarrow{PBr_3} CH_3Br \xrightarrow[ether]{Mg} CH_3MgBr \xrightarrow[2.\ H_3O^+]{1.\ H-\overset{O}{\overset{\|}{C}}-H} CH_3CH_2OH$

Section: 4

33. Give the products obtained from the following Wittig reaction. Show both stereoisomers.

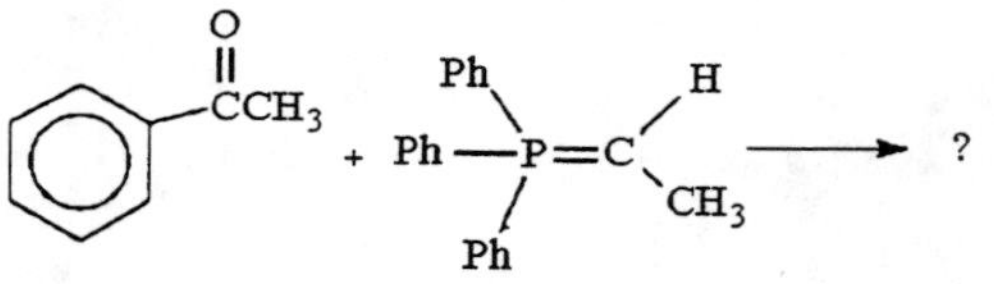

$C_6H_5\overset{O}{\overset{\|}{C}}CH_3$ + $Ph_3P{=}C(H)CH_3$ ⟶ ?

Answer:

$C_6H_5(CH_3)C{=}C(H)CH_3$ (CH₃ and H on the second carbon as drawn: H up, CH₃ down) and $C_6H_5(CH_3)C{=}C(CH_3)H$ (CH₃ up, H down)

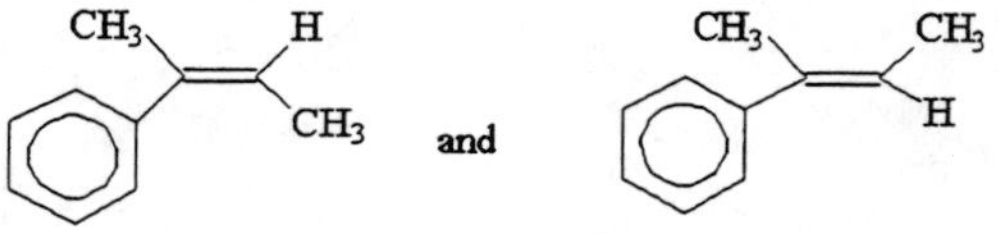

Section: 11

34. Give three different sets of reagents (a carbonyl compound and a Grignard reagent) that could be used to prepare the following compound.

$$CH_3-\overset{\displaystyle OH}{\underset{\displaystyle CH_3}{\overset{|}{\underset{|}{C}}}}-CH_2CH_3$$

Answer:

a. $CH_3CH_2\overset{\displaystyle O}{\overset{\|}{C}}OCH_3$ and 2 CH_3MgBr

b. $CH_3\overset{\displaystyle O}{\overset{\|}{C}}CH_2CH_3$ and CH_3MgBr

c. $CH_3\overset{\displaystyle O}{\overset{\|}{C}}CH_3$ and CH_3CH_2MgBr

Section: 4

35. Propose a plausible mechanism that would explain the following conversions:

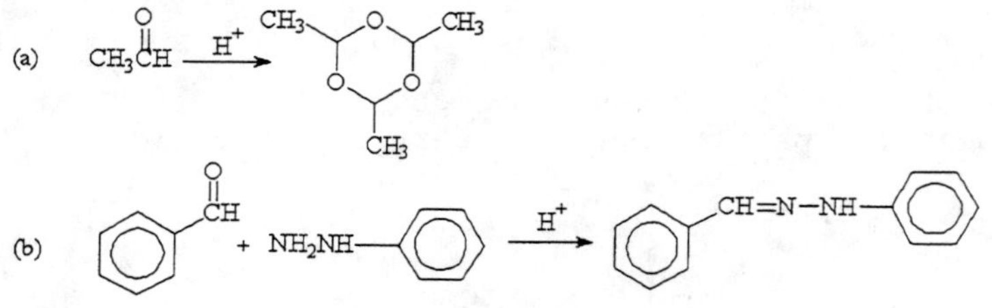

(a) $CH_3CHO \xrightarrow{H^+}$ 2,4,6-trimethyl-1,3,5-trioxane

(b) $C_6H_5CHO + NH_2NH-C_6H_5 \xrightarrow{H^+} C_6H_5CH{=}N{-}NH{-}C_6H_5$

Answer:

(a) $CH_3CHO + H^+ \rightleftharpoons CH_3CH{=}\overset{\oplus}{O}H$; attack by a second CH_3CHO, then a third CH_3CHO; ring closure and loss of H^+ give the trioxane.

(b) $C_6H_5CHO + H^+ \rightleftharpoons C_6H_5CH{=}\overset{\oplus}{O}H$; attack by $NH_2NH-C_6H_5$; $-H^+$ / $+H^+$ proton transfers give $C_6H_5CH(\overset{\oplus}{O}H_2)NHNHC_6H_5$; loss of water gives $C_6H_5CH{=}\overset{\oplus}{N}H{-}NH{-}C_6H_5$; $-H^+$ gives $C_6H_5CH{=}N{-}NH{-}C_6H_5$.

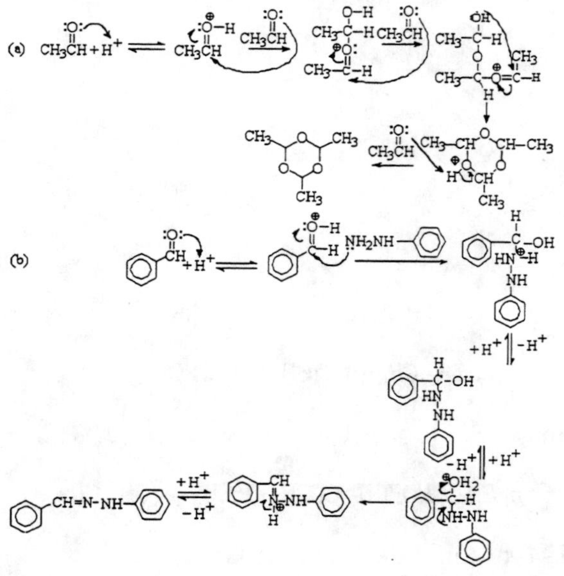

Section: 7

36. Would you expect the carbonyl carbon of benzaldehyde to be more or less electrophilic than that of acetaldehyde? Explain using resonance structures.

Answer:
The carbonyl carbon of benzldehyde would be less electrophilic. Delocalization reduces its partial positive charge as seen in the resonance forms below.

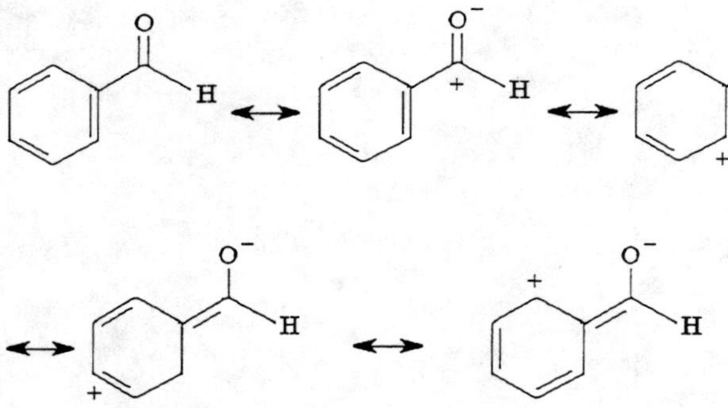

Section: 2

37. Provide the proper IUPAC name for $PhCH_2CH(CH_3)CH_2CH_2CHO$.

Answer: 4-methyl-5-phenylpentanal Section: 1

38. Provide the proper IUPAC name for $CH_3CHOHCH_2COCH_2C(CH_3)_2CH_2CH_3$.

Answer: 2-hydroxy-6,6-dimethyl-4-octanone Section: 1

39. Provide the structure of cyclohexanone oxime.

Answer:

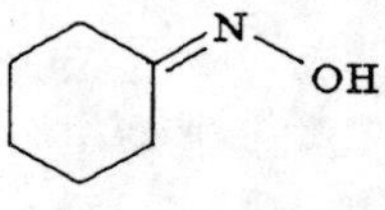

Section: 7

40. Provide the structure of the diethyl acetal of butanal.

Answer: $CH_3CH_2CH_2CH(OCH_2CH_3)_2$ Section: 8

MULTIPLE CHOICE

41. In carbon NMR, the carbon atom of the carbonyl group in aldehydes and ketones has a chemical shift of about:

a. 20 ppm.
b. 40 ppm.
c. 60 ppm.
d. 120 ppm.
e. 200 ppm.

Answer: e

42. In the proton NMR spectra of aldehydes and ketones, the protons bonded to carbons adjacent to the carbonyl group typically fall into which of the chemical shift ranges below?

a. 1.0-2.0 ppm
b. 2.0-3.0 ppm
c. 4.0-4.5 ppm
d. 7.0-8.0 ppm
e. 9.0-10.0 ppm

Answer: b

SHORT ANSWER

43. By which single-step process can benzene be readily converted into acetophenone?

Answer: CH_3COCl, $AlCl_3$ - a Friedel-Crafts acylation reaction

44. What reagent can be used to convert benzophenone into triphenylmethanol?

Answer: PhMgBr followed by hydrolysis Section: 4

45. Beginning with sodium acetylide (NaCCH), propose a three-step synthesis of hexanal.

Answer:
1. $CH_3CH_2CH_2CH_2Br$
2. Sia_2BH
3. H_2O_2, NaOH

46. Provide the major organic product which results when $(CH_3)_2CHCH_2CH_2COCl$ is treated with $LiAlH[OC(CH_3)_3]_3$.

Answer: $(CH_3)_2CHCH_2CH_2CHO$ Section: 6

47. Provide a detailed, stepwise mechanism for the acid-catalyzed reaction of 2-butanone with ethylene glycol ($HOCH_2CH_2OH$) to produce an acetal.

Answer:

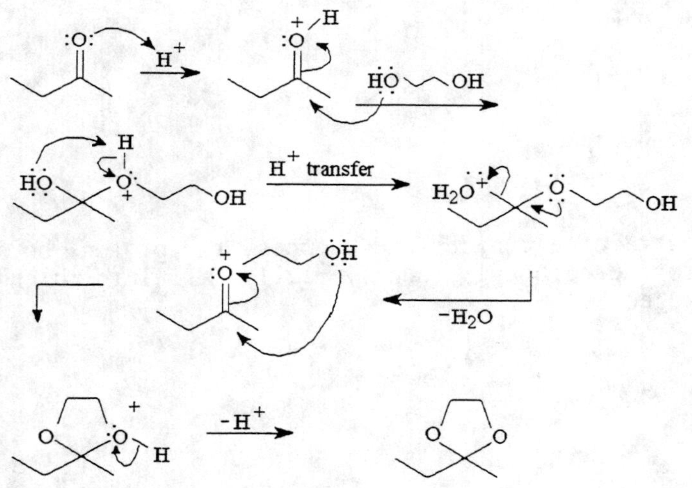

Section: 8

48. Provide a detailed, stepwise mechanism for the acid-catalyzed condensation reaction between benzaldehyde and methylamine.

Answer:

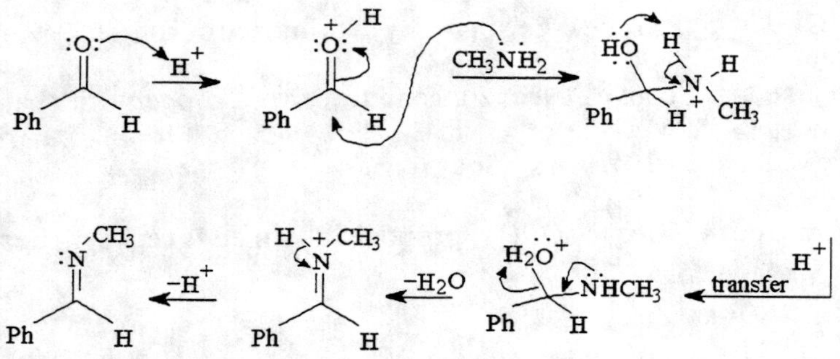

Section: 7

49. Provide a detailed, stepwise mechanism for the acid-catalyzed condensation reaction between cyclohexanone and H_2NOH.

Answer:

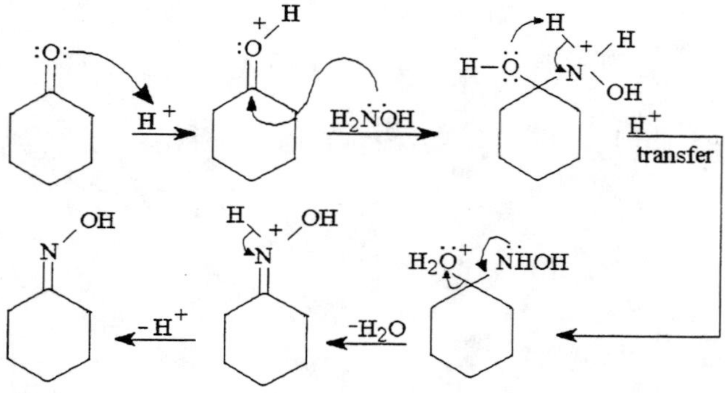

H^+ $H_2\ddot{N}OH$ H^+ transfer $-H_2O$ $-H^+$

Section: 7

MULTIPLE CHOICE

50. When the carbonyl group of a neutral ketone is protonated,

a. the resulting species becomes more electrophilic.
b. the resulting species is activated toward nucleophilic attack.
c. subsequent nucleophilic attack on the resulting species is said to occur under acid-catalyzed conditions.
d. the resulting species has a positive charge.
e. all of the above.

Answer: e Section: 7

SHORT ANSWER

51. Draw the two major resonance forms of the cation which results when cyclohexanone's carbonyl group is protonated.

Answer:

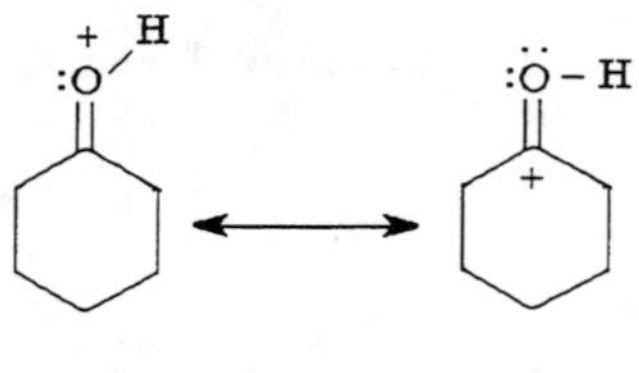

Section: 7

52. Propose a synthesis of 4-phenyl-2-butanol from 3-phenylpropanal.

Answer:
1. CH_3MgBr
2. H_3O^+

Section: 4

53. Propose a synthesis of 3-methyl-4-heptyn-3-ol from 1-butyne.

Answer:
1. $NaNH_2$
2. $CH_3COCH_2CH_3$
3. H_3O^+

Section: 4

54. Provide the structure of the hydrate of cyclopentanone.

Answer:

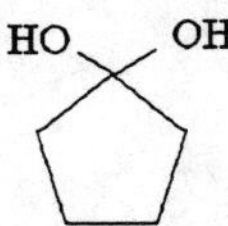

Section: 8

55. Conversion of aldehydes and ketones to imines is an acid-catalyzed process. Explain why this conversion is actually hindered by the presence of too much acid.

Answer:
At very low pH, much more of the amine is protonated and can no longer function as a nuclepohile.

Section: 7

56. Provide the major organic product of the reaction of aniline with 3-pentanone.

Answer:

Ph
N
CH_3 CH_3

Section: 7

57. Propose a sequence of steps to carry out the following conversion.

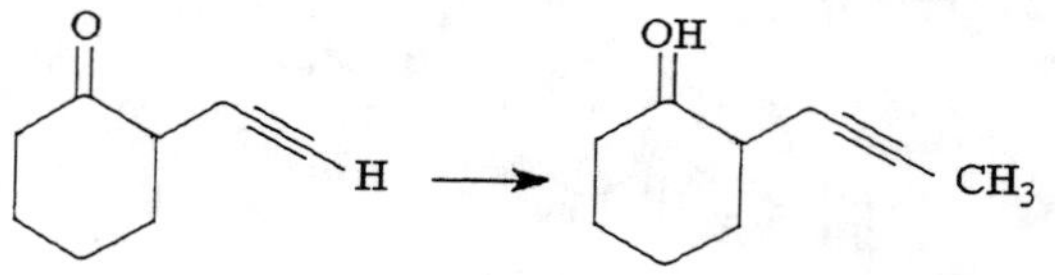

Answer:
1. $HOCH_2CH_2OH$, H^+
2. $NaNH_2$
3. CH_3I
4. H_3O^+
5. $NaBH_4$ or $LiAlH_4$

Section: 9

58. When $HOCH_2CH_2CH_2CH_2COCH_2CH_2CH_2CH_2OH$ is heated in the presence of an acid catalyst, a reaction occurs. The product has the formula $C_9H_{16}O_2$. Provide the structure of this product.

Answer:

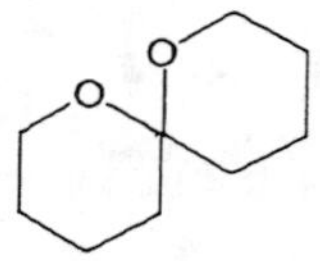

Section: 8

59. Propose a sequence of steps to carry out the following conversion.

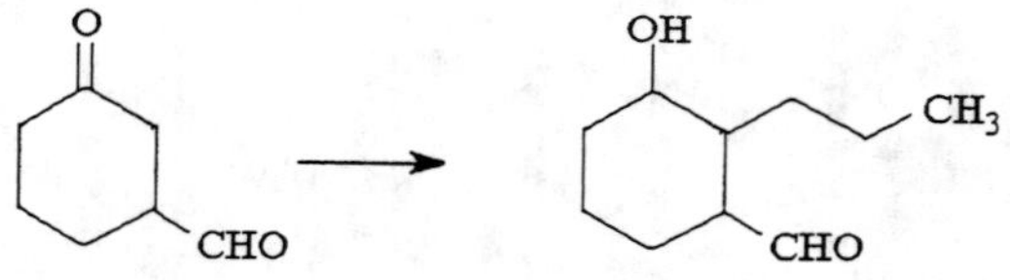

Answer:
1. $HOCH_2CH_2OH$, H^+
2. $CH_3CH_2CH_2MgBr$
3. H_3O^+

Section: 9

60. Which would be more appropriate as the reduction in the following sequence, a Clemmensen or a Wolff-Kishner? Explain.

1. $HOCH_2CH_2OH$ (1 eq), H^+
2. reduction
3. H_3O^+

Answer:
Wolff-Kishner. The acidic conditions of the Clemmensen would remove the protecting group causing reduction of the aldehyde as well.

Section: 7

Chapter 17: Oxidation-Reduction Reactions

MULTIPLE CHOICE

1. Which of the following is <u>not</u> a reduction reaction?

a. 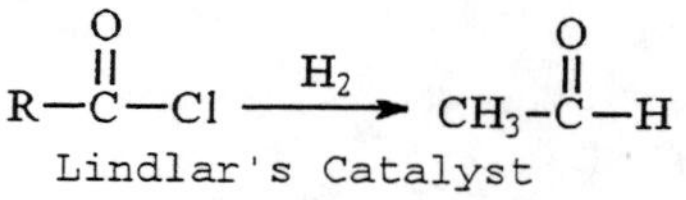 $R-\overset{O}{\overset{\|}{C}}-Cl \xrightarrow[\text{Lindlar's Catalyst}]{H_2} CH_3-\overset{O}{\overset{\|}{C}}-H$

b. $CH_2{=}CH_2 \xrightarrow[CCl_4]{Br_2} \underset{Br}{CH_2}-\underset{Br}{CH_2}$

c. $CH_3-\overset{O}{\overset{\|}{C}}-CH_3 \xrightarrow[{}^{-}OH,\ heat]{NH_2NH_2} CH_3CH_2CH_3$

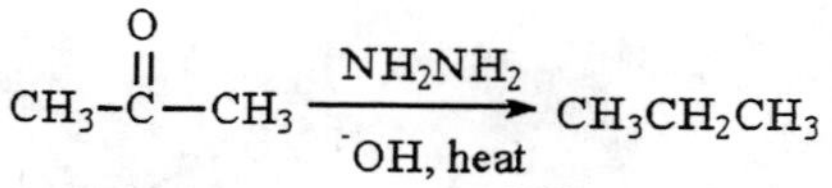

d. 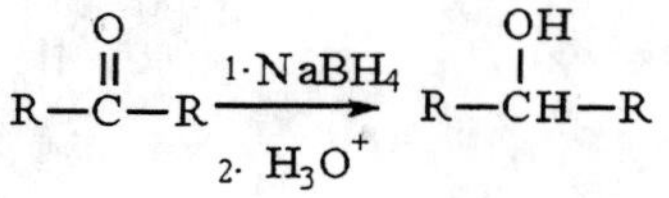 $R-\overset{O}{\overset{\|}{C}}-R \xrightarrow[2.\ H_3O^+]{1.\ NaBH_4} R-\overset{OH}{\overset{|}{C}H}-R$

e. 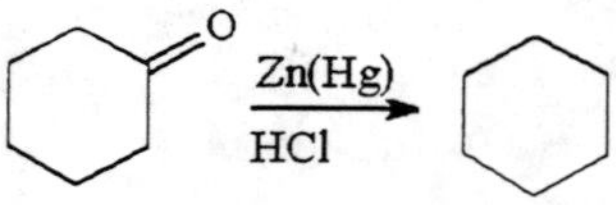 cyclohexanone $\xrightarrow[HCl]{Zn(Hg)}$ cyclohexane

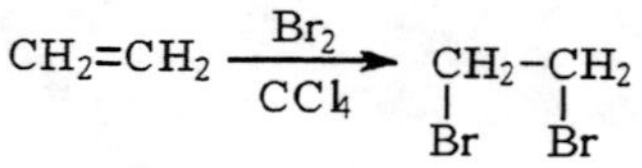

Answer: b Section: 1

2. What is the major organic product of the following reaction?

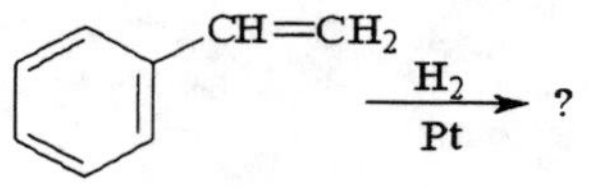

$C_6H_5CH{=}CH_2 \xrightarrow[Pt]{H_2}$?

a. (cyclohexadiene ring)–$CH{=}CH_2$

b. (cyclohexane ring)–CH_2CH_3

c. (benzene ring)–CH_2CH_3

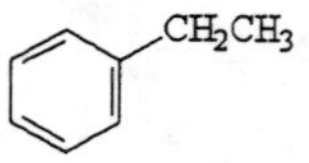

d. 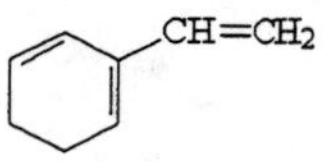 (cyclohexene ring)–$CH{=}CH_2$

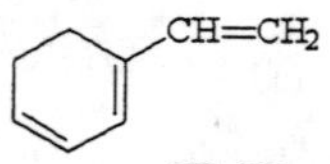

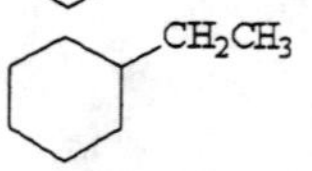

e. 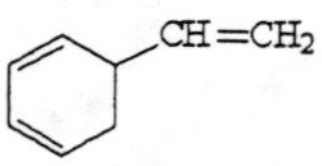 (cyclohexadiene ring)–$CH{=}CH_2$

Answer: c Section: 1

3. Which of the following reducing agents can cause the formation of cis-2-butene as a major product of the reaction shown below?

   a. H_2 /Lindlar's Catalyst
   b. Na/NH_3
   c. Li/NH_3
   d. 1.$LiAlH_4$ /2.H_3O^+
   e. 1.$NaBH_4$ /2.H_3O^+

   $$CH_3C\equiv CCH_3 \xrightarrow{?} \underset{CH_3}{\overset{H}{}}\!\!\!\diagdown C=C \diagup\!\!\!\underset{CH_3}{\overset{H}{}}$$

   Answer: a Section: 1

4. Which of the following carbonyl compounds can be reduced by H_2 /Raney Ni?

   a. $CH_3-\overset{O}{\overset{\|}{C}}-OH$

   b. $CH_3-\overset{O}{\overset{\|}{C}}-OCH_3$

   c. $CH_3-\overset{O}{\overset{\|}{C}}-NHCH_3$

   d. $CH_3-\overset{O}{\overset{\|}{C}}-Cl$

   e. $CH_3-\overset{O}{\overset{\|}{C}}-NH_2$

   Answer: d Section: 1

5. Which of the following reagents gives the reaction shown?

   a. H_2/H_2SO_4
   b. H_2O/Ni
   c. H_2O/H_2SO_4
   d. H^+/H_2O
   e. H_2/Ni

   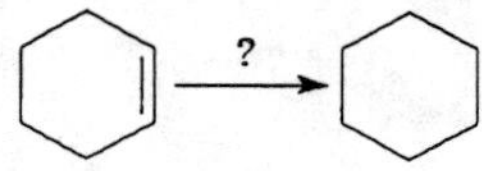

   Answer: e Section: 1

6. Which of the following reagents would reduce carboxylic acids and esters into alcohols?

a. H_2 /Raney Ni
b. 1.$LiAlH_4$ /2.H_3O^+
c. Na/NH_3
d. 1.$NaBH_4$ /2.H_3O^+
e. Zn(Hg)/H^+

Answer: b Section: 1

7. Which of the following reducing agents is best used in the reaction shown below?

$$CH_3-\overset{\overset{\displaystyle O}{\|}}{C}-NH_2 \xrightarrow{?} CH_3CH_2NH_2$$

a. H_2 /Raney Ni
b. 1.$LiAlH_4$ /2.H_3O^+
c. Na/NH_3
d. 1.$NaBH_4$ /2.H_3O^+
e. Zn(Hg)/H^+

Answer: b Section: 1

8. Which of the following reducing agents is best used in the reaction shown below?

$$CH_2=CH-\overset{\overset{\displaystyle O}{\|}}{C}-CH_3 \xrightarrow{?} CH_2=CH-\overset{\overset{\displaystyle OH}{|}}{CH}-CH_3$$

a. H_2 /Raney Ni
b. 1.$LiAlH_4$ /2.H_3O^+
c. Na/NH_3
d. 1.$NaBH_4$ /2.H_3O^+
e. Zn(Hg)/H^+

Answer: d Section: 1

9. Which of the following alcohols is oxidized to a ketone by chromic acid?

a. 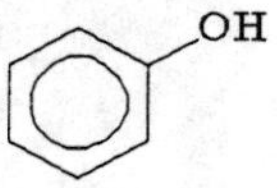 (phenol: benzene ring with OH)

b. 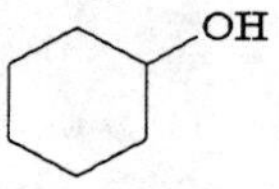 (cyclohexanol: cyclohexane ring with OH)

c. 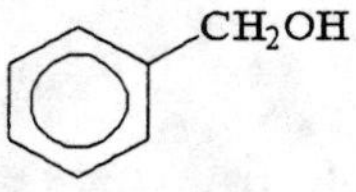 (benzyl alcohol: benzene ring with CH_2OH)

d. 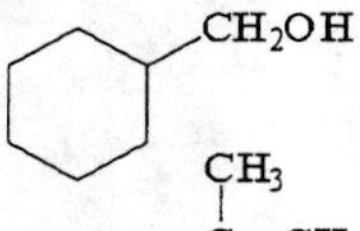 (cyclohexylmethanol: cyclohexane ring with CH_2OH)

e. 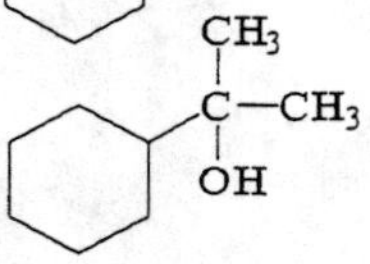 (cyclohexane ring with $C(CH_3)_2OH$)

Answer: b Section: 2

10. Which of the following reagents does <u>not</u> form an alcohol from an aldehyde as shown below?

a. H_2 /Pt
b. 1.$NaBH_4$ /2.H_3O^+
c. 1.$LiAlH_4$ /2.H_3O^+
d. 1.Ag_2O,NH_3 /2.H_3O^+
e. H_2 /Raney Ni

$$R-\overset{\overset{\large O}{||}}{C}-H \xrightarrow{?} R-CH_2-OH$$

Answer: d Section: 1

11. Which of the following reagents <u>does</u> form an alcohol from a ketone as shown below?

a. NH_2NH_2 / ^-OH
b. Zn(Hg)/H^+
c. $Ag(NH_3)_2NO_3$
d. $K_2Cr_2O_7$ /H^+
e. H_2 /Raney Ni

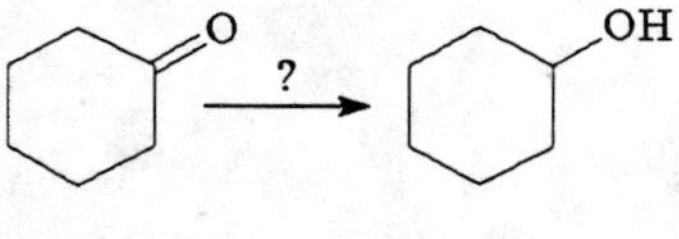

(cyclohexanone $\xrightarrow{?}$ cyclohexanol)

Answer: e Section: 1

12. Which of the following oxidizing agents is used in a Baeyer-Villiger oxidation?

    a. NH_2NH_2 / ^{-}OH
    b. Zn(Hg)/H^+
    c. $Ag(NH_3)_2NO_3$
    d. $K_2Cr_2O_7$/H^+
    e. RCOOH with =O on the C (peroxy acid, $RC(=O)OOH$)

    Answer: e Section: 3

13. What is the major organic product of the following reaction?

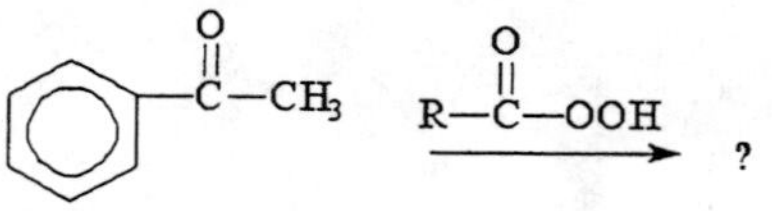

$$C_6H_5-C(=O)-CH_3 \xrightarrow{R-C(=O)-OOH} ?$$

a. 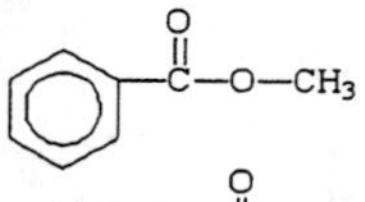 $C_6H_5-C(=O)-O-CH_3$

b. $C_6H_5-O-C(=O)-CH_3$

c. $C_6H_5-CH(OH)-CH_3$

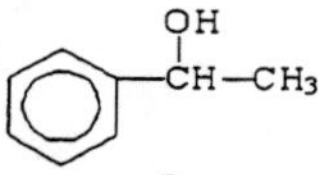

d. 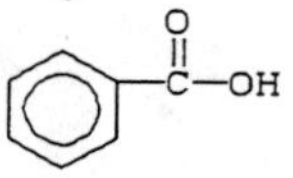 $C_6H_5-C(=O)-OH$

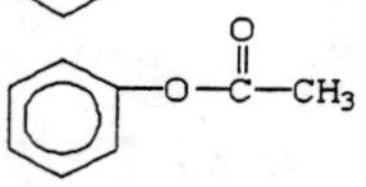

e. 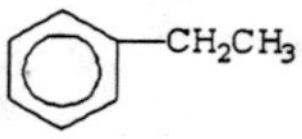 $C_6H_5-CH_2CH_3$

Answer: b Section: 3

14. What is the <u>major</u> organic product of the following reaction?

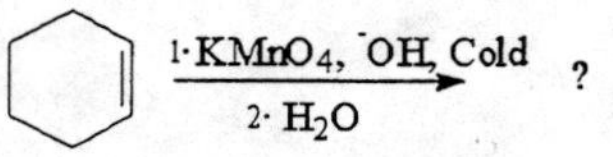

Cyclohexene $\xrightarrow[\text{2- } H_2O]{\text{1-} KMnO_4,\ ^{-}OH,\ \text{Cold}}$?

a. 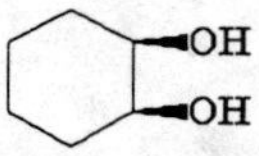 cyclohexane-1,2-diol (both OH wedged)

b. 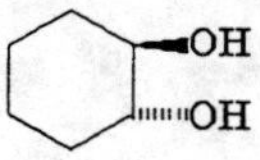 cyclohexane-1,2-diol (one OH wedged, one OH hashed)

c. 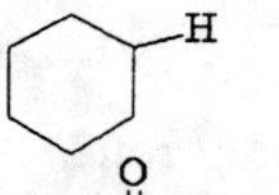 cyclohexane (—H)

d. 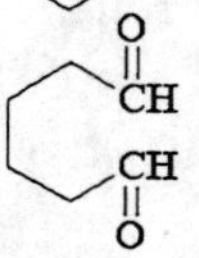 hexanedial (OHC–(CH$_2$)$_4$–CHO)

e. All of the above

Answer: a Section: 4

15. Which of the following reagents effectively cleaves a carbon-carbon double bond?

a. Cl_2/hv
b. $R-\overset{O}{\overset{||}{C}}-OOH$
c. H_2SO_4
d. cold $KMnO_4$
e. hot $KMnO_4$

Answer: e Section: 6

16. An unknown alkene was treated with ozone, and then with H_2O_2 forming the following product:

$$CH_3-\overset{O}{\overset{\|}{C}}-CH_2-CH_2-CH_2-CH_2-\overset{O}{\overset{\|}{C}}-OH$$

What is the structure of the unknown?

a.

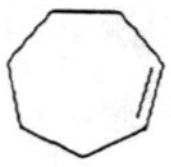

b. 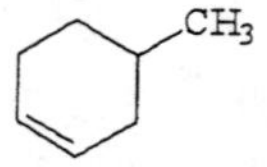 CH_3

c. 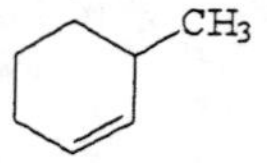 CH_3

d. 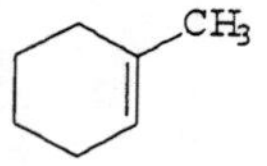 CH_3

e.

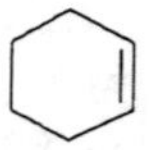

Answer: d Section: 6

SHORT ANSWER

17. Is it easier to reduce a carbon-oxygen double bond, or a carbon-carbon double bond using catalytic hydrogenation?

Answer: A carbon-carbon double bond is easier to reduce. Section: 6

MULTIPLE CHOICE

18. What is the major organic product(s) when the following molecule is treated with ozone, and then with Zn/H_2O?

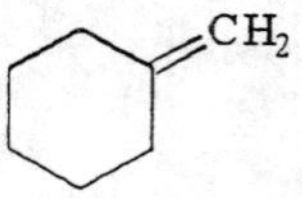

CH_2

a. 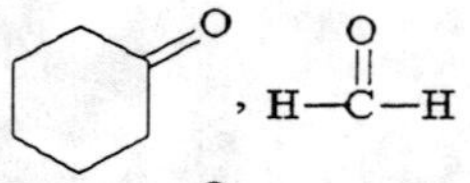 , H—C(=O)—H

b. , CO_2,H_2O

c. COOH , CH_2O

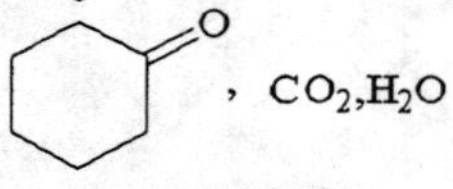

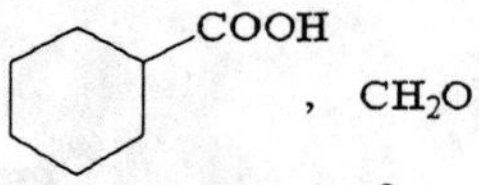

d. 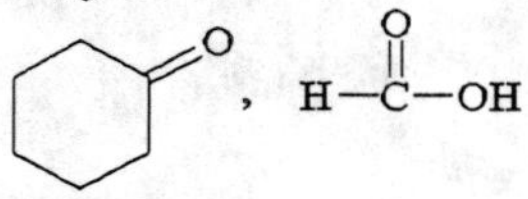 , H—C(=O)—OH

e. 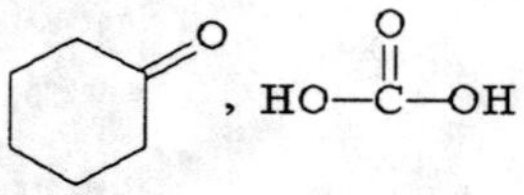 , HO—C(=O)—OH

Answer: a Section: 6

19. Which of the following reagents best facilitates the reaction shown below?

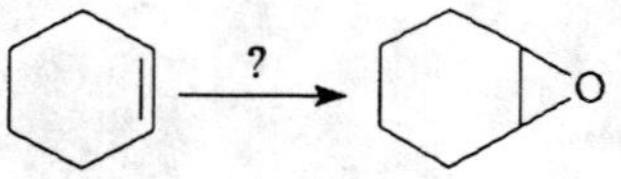

?

a. 1.O_3/ 2.Zn, H_2O

b. 1.O_3/ 2.H_2O_2

c. CH_3—C(=O)—OOH

d. $KMnO_4$ cold

e. $KMnO_4$ hot

Answer: c Section: 8

20. What is the major product when the following compound reacts with MMPP?

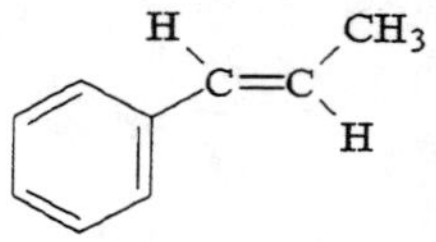

a.

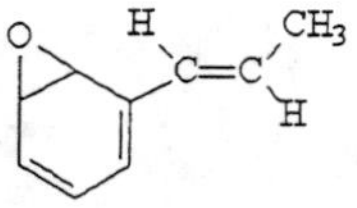

b.

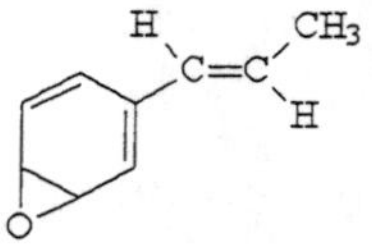

c.

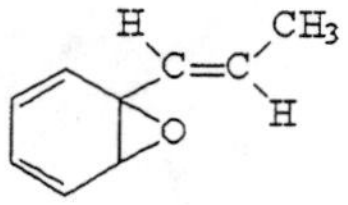

d.

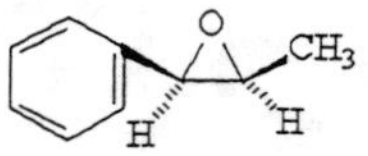

e.

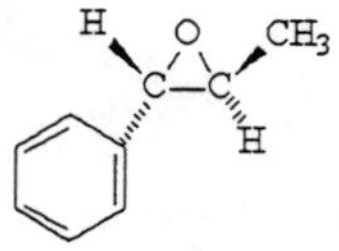

Answer: e Section: 8

21. Which of the following reagents is <u>not</u> an oxidizing agent?

a. $KMnO_4$
b. CrO_3
c. NADH
d. PCC
e. HIO_4

Answer: c Section: 11

22. Which of the following reagents can be used to oxidize 1° alcohols to aldehydes?

a. $KMnO_4$
b. MnO_2
c. $K_2Cr_2O_7$
d. H_2O_2
e. PCC

Answer: e Section: 2

23. Which of the following reagents is best used for the conversion shown below?

a. 1.$NaBH_4$ /2.D_3O^+
b. 1.$NaBD_4$ /2.H_3O^+
c. 1.$NaBH_4$ /2.D_3O^+
d. 1.H_2 /Pt/2.D_2O
e. 1.$LiAlH_4$ /2.H_3O^+

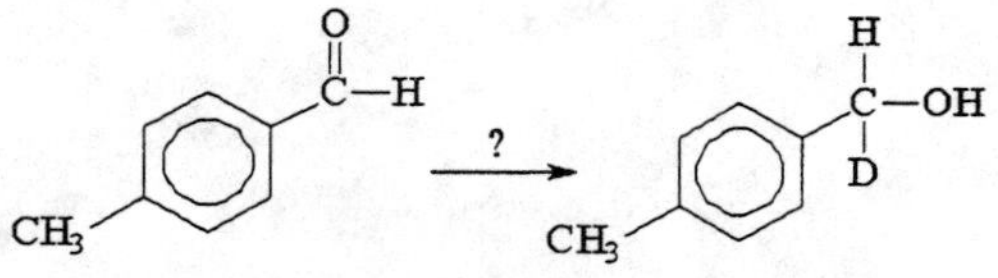

$$CH_3-C_6H_4-CHO \xrightarrow{?} CH_3-C_6H_4-CH(D)(OH)$$

Answer: b Section: 1

SHORT ANSWER

24. What is the product of the following reduction reaction?

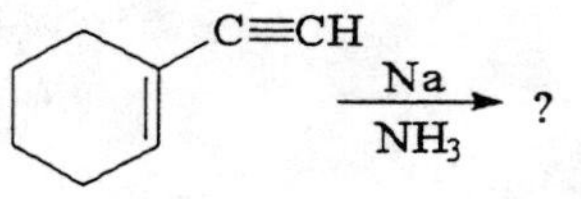

$$\text{1-ethynylcyclohexene } (C{\equiv}CH) \xrightarrow[NH_3]{Na} ?$$

Answer:

The product is

$$\text{1-vinylcyclohexene } (CH{=}CH_2)$$

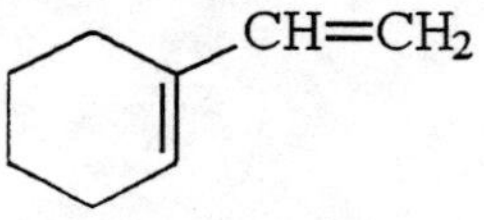

Section: 1

25. When pyruvic acid is reduced using 1.$NaBH_4$ /2.H_3O^+, what is the expected product?

$$CH_3-\overset{O}{\overset{\|}{C}}-\overset{O}{\overset{\|}{C}}-OH \xrightarrow[2.\ H_3O^+]{1.\ NaBH_4} ?$$

Answer:
$NaBH_4$ can reduce only the keto group. The product is lactic acid.

$$CH_3-\overset{OH}{\overset{|}{CH}}-\overset{O}{\overset{\|}{C}}-OH$$

Section: 1

26. Would you expect alkenes and alkynes to be reduced by $NaBH_4$? Why?

Answer:
No, alkenes and alkynes cannot be reduced by $NaBH_4$ because they do not possess a partial positive charge.
Section: 1

27. What is PCC and what is it used for?

Answer:
PCC is pyridinium chlorochromate, a chromic anhydride (CrO_3) and an equivalent of HCl dissolved in pyridine. It is a mild oxidizing agent used to convert alcohols to aldehydes.
Section: 2

28. Propose a mechanism for the following oxidation reaction known as the Baeyer-Villiger Oxidation:

$$C_6H_5-\overset{O}{\overset{\|}{C}}-CH_3 \xrightarrow{R-\overset{O}{\overset{\|}{C}}-OOH} C_6H_5-O-\overset{O}{\overset{\|}{C}}-CH_3 + R-\overset{O}{\overset{\|}{C}}-OH$$

Answer:

$$C_6H_5-\overset{O}{\overset{\|}{C}}-CH_3 + R-\overset{O}{\overset{\|}{C}}-\ddot{O}\ddot{O}H \longrightarrow C_6H_5-C(:\ddot{O}H)(CH_3)-O-O-\overset{O}{\overset{\|}{C}}-R$$

$$\downarrow$$

$$C_6H_5-O-\overset{O}{\overset{\|}{C}}-CH_3 + R-\overset{O}{\overset{\|}{C}}-OH$$

Section: 3

29. Give the structure of the alkyne that gives the following products upon ozonolysis.

$$HO-\overset{O}{\overset{\|}{C}}-CH_2CH_2-\overset{O}{\overset{\|}{C}}-OH,\ 2\ CH_3-\overset{O}{\overset{\|}{C}}-OH$$

Answer: $CH_3C{\equiv}C-CH_2CH_2-C{\equiv}CCH_3$ Section: 7

30. What does MMPP stand for? Give the structure of MMPP.

Answer:
It stands for magnesium monoperoxyphthalate.

It is a stable molecule that is used for epoxidation.

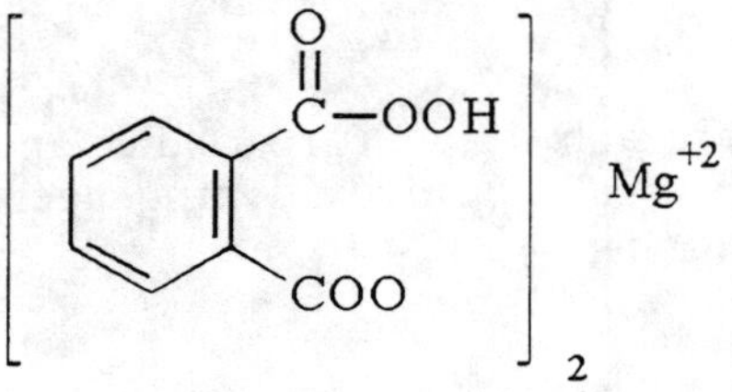

O
C—OOH
Mg^{+2}
COO
2

Section: 8

MULTIPLE CHOICE

31. Classify the reaction below as an oxidation, a reduction, or neither.

$(CH_3)_2CHCH_2OH \longrightarrow (CH_3)_2CHCHO$

a. oxidation
b. reduction
c. neither

Answer: a

32. Classify the reaction below as an oxidation, a reduction, or neither.

$PhCO_2H \longrightarrow PhCH_2OH$

a. oxidation
b. reduction
c. neither

Answer: b

33. Classify the reaction below as an oxidation, a reduction, or neither.

cis-2-pentene $\longrightarrow$ pentane

a. oxidation
b. reduction
c. neither

Answer: b

SHORT ANSWER

34. Provide the structure of the major organic product in the reaction below.

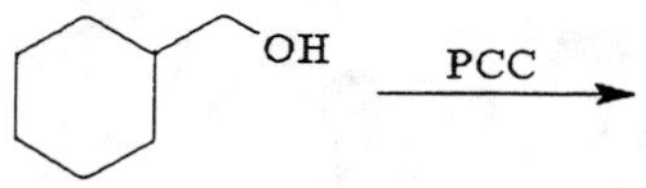

(cyclohexylmethanol) OH $\xrightarrow{PCC}$

Answer:

(cyclohexane) CHO

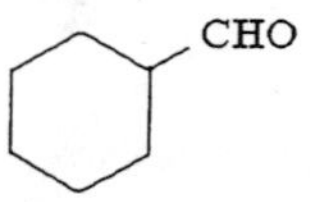

Section: 2

35. Provide the structure of the major organic product in the reaction below.

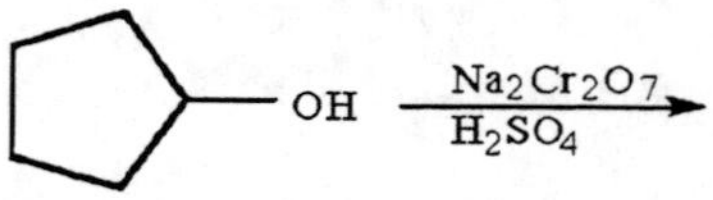

(cyclopentanol) OH $\xrightarrow[H_2SO_4]{Na_2Cr_2O_7}$

Answer:

(cyclopentanone) =O

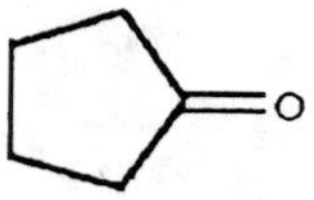

Section: 2

36. Provide the structure of the major organic product in the reaction below.

HO CH_3 $\xrightarrow{PCC}$

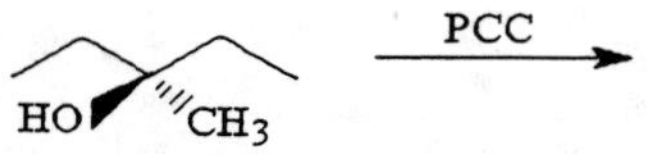

Answer: No reaction. Tertiary alcohols are not oxidized by PCC.

Section: 2

37. Provide the structure of the major organic product in the reaction below.

$(CH_3)_3CCH_2CH_2OH \xrightarrow[H_2SO_4]{Na_2Cr_2O_7}$

Answer: $(CH_3)_3CCH_2CO_2H$ Section: 2

38. Provide the structure of the major organic product in the reaction below.

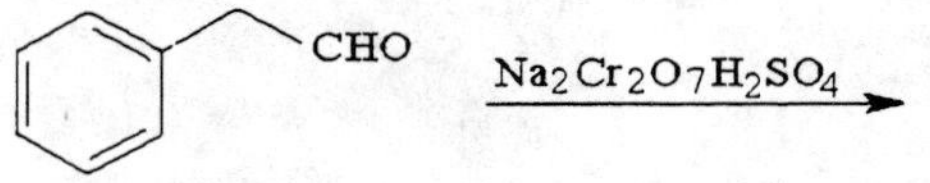

CHO $\xrightarrow{Na_2Cr_2O_7H_2SO_4}$

Answer:

CO_2H

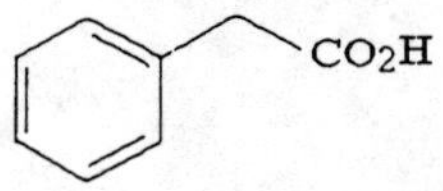

Section: 2

MULTIPLE CHOICE

39. In the chromic acid oxidation of alcohols, the chromium is:

a. reduced from Cr^{+6} to Cr^{+3}.
b. oxidized from Cr^{+6} to Cr^{+3}.
c. reduced from Cr^{+3} to Cr^{+6}.
d. oxidized from Cr^{+3} to Cr^{+6}.
e. none of the above

Answer: a Section: 2

SHORT ANSWER

40. Chromic acid oxidations are believed to proceed through a chromate ester intermediate. In the oxidation of 1-propanol, provide the structure of this ester and give the mechanism for its conversion to the intermediate aldehyde.

Answer:

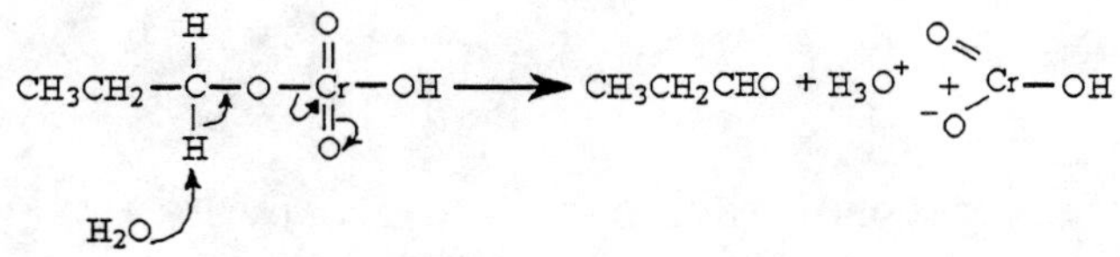

$CH_3CH_2-C(H)(H)-O-Cr(=O)(=O)-OH \longrightarrow CH_3CH_2CHO + H_3O^+ + O=Cr(-O^-)-OH$ (H_2O)

Section: 2

41. The enzyme alcohol dehydrogenase, which is produced by the liver, converts ethanol to ____________, a normal body metabolite.

Answer: acetic acid Section: 11

42. Poisonings by methanol and ethylene glycol are relatively common. How are these poisonings treated clinically?

Answer:
The enzyme alcohol dehydrogenase converts these compounds to toxic substances in the body. If the patient receives intravenous infusions of ethanol, then the ethanol swamps the enzyme and the potentially deadly alcohols pass from the body unoxidized.

Section: 11

43. Provide the structure of the major organic product in the reaction below.

$$CH_3CH_2CH(CH_3)CH_2CO_2CH_3 \xrightarrow[\text{2. } H_3O^+]{\text{1. LiAlH}_4}$$

Answer: $CH_3CH_2CH(CH_3)CH_2CH_2OH$ Section: 1

44. What series of synthetic steps could be used to prepare 1-methylcyclohexanol from cyclohexene?

Answer:
1. H_2O, H^+ or BH_3; H_2O_2, ^-OH or $Hg(OAc)_2$, H_2O; $NaBH_4$
2. $NaCr_2O_7$, H^+ or PCC or $KMnO_4$
3. CH_3MgI
4. H_3O^+

Section: 2

45. What series of synthetic steps could be used to prepare 3-hexanol from 1-propanol?

Answer:
1. PCC
2. $CH_3CH_2CH_2MgBr$
3. H_3O^+

or

1. HBr or PBr_3
2. Mg, Et_2O
3. CH_3CH_2CHO

Section: 2

46. What series of synthetic steps could be used to prepare $(CH_3)_2CHCHO$ from isobutane?

Answer:
1. Br_2, hυ
2. NaOEt, heat
3. BH_3
4. H_2O_2, ^{-}OH
5. PCC

Section: 2

47. What series of synthetic steps could be used to prepare $PhCHOHCH_2CH_3$ from $PhCH_2OH$?

Answer:
1. PCC
2. CH_3CH_2MgBr
3. H_3O^+

Section: 2

48. What series of synthetic steps could be used to carry out the transformation shown below?

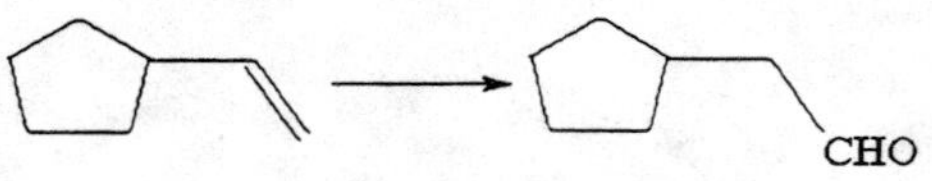

Answer:
1. BH_3
2. H_2O_2, ^{-}OH
3. PCC

Section: 2

49. Draw the major organic product generated in the reaction below. Pay particular attention to regio- and stereochemical detail.

Answer:

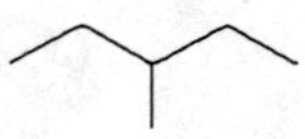

Section: 1

50. Draw the major organic product generated in the reaction below. Pay particular attention to regio- and stereochemical detail.

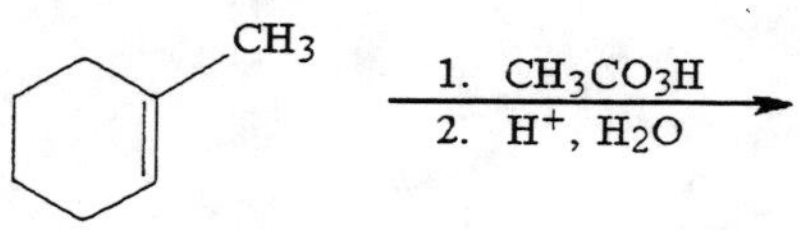

1. CH_3CO_3H
2. H^+, H_2O

Answer:

OH, CH_3, OH + enantiomer

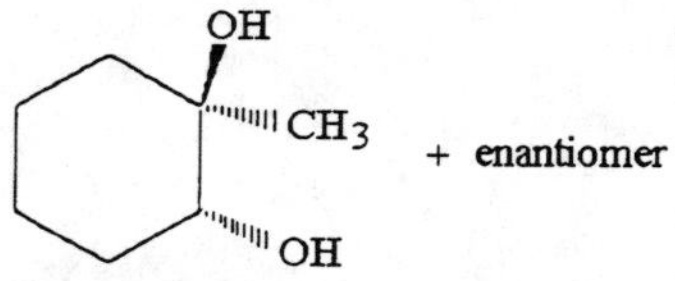

Section: 8

51. Draw the major organic product generated in the reaction below. Pay particular attention to regio- and stereochemical detail.

$PhCO_3H$
CH_2Cl_2

Answer:

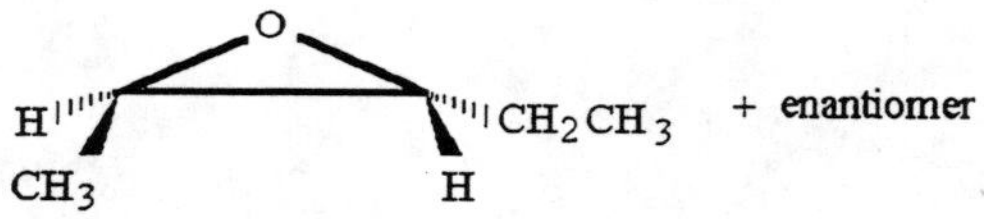

O, H, CH_3, CH_2CH_3, H + enantiomer

Section: 8

52. Draw the major organic product generated in the reaction below. Pay particular attention to regio- and stereochemical detail.

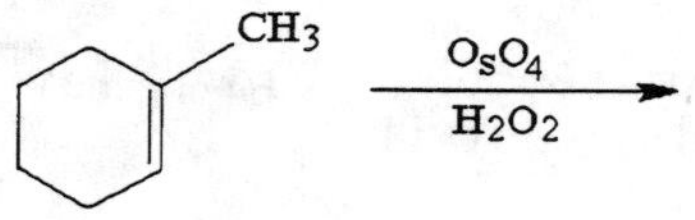

OsO_4
H_2O_2

Answer:

OH, CH_3, OH + enantiomer

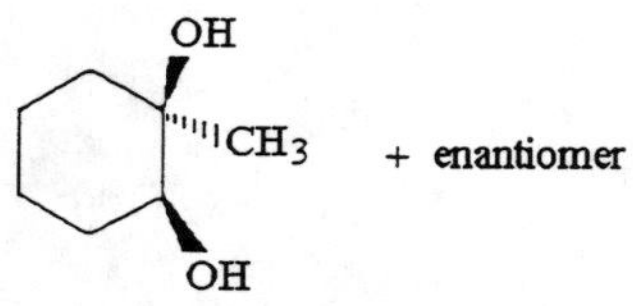

Section: 4

53. Draw the major organic product generated in the reaction below. Pay particular attention to regio- and stereochemical detail.

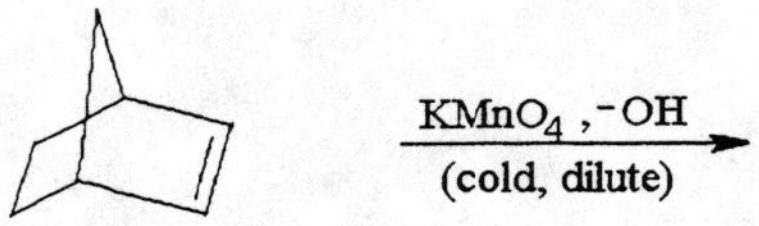

$\xrightarrow[\text{(cold, dilute)}]{KMnO_4,\ ^{-}OH}$

Answer:

OH
OH
H
H

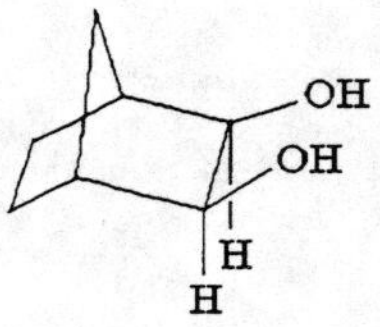

Section: 4

54. Draw the major organic product generated in the reaction below. Pay particular attention to regio- and stereochemical detail.

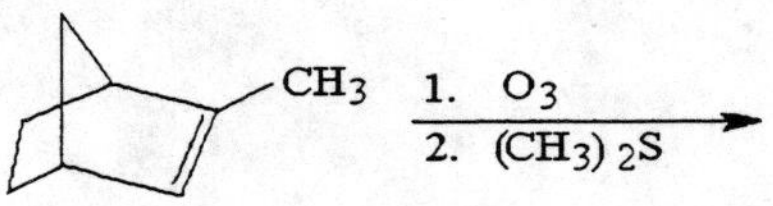

CH_3 $\xrightarrow[\text{2. } (CH_3)_2S]{\text{1. } O_3}$

Answer:

O
CHO
CH_3

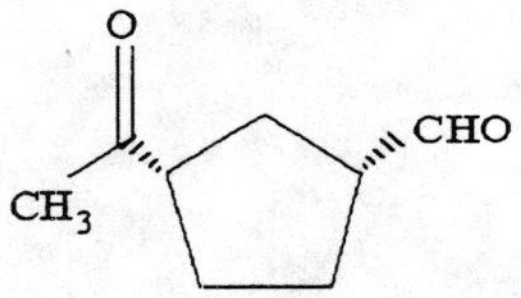

Section: 6

55. Draw the major organic product generated in the reaction below. Pay particular attention to regio- and stereochemical detail.

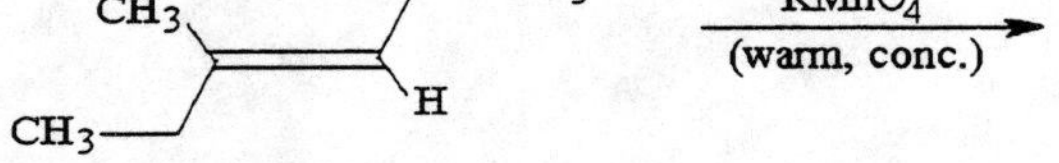

CH_3 CH_3 H CH_3 $\xrightarrow[\text{(warm, conc.)}]{KMnO_4}$

Answer:

$CH_3C(=O)CH_2CH_3$ + $CH_3CH_2C(=O)OH$

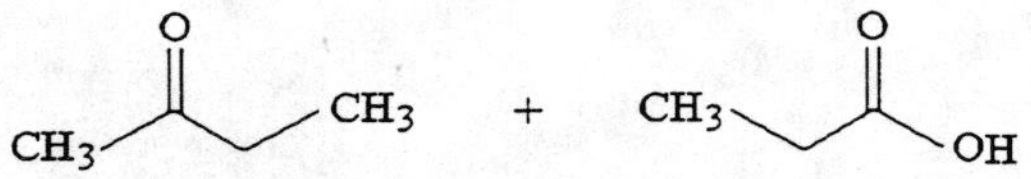

Section: 6

56. Draw the major organic product generated in the reaction below. Pay particular attention to regio- and stereochemical detail.

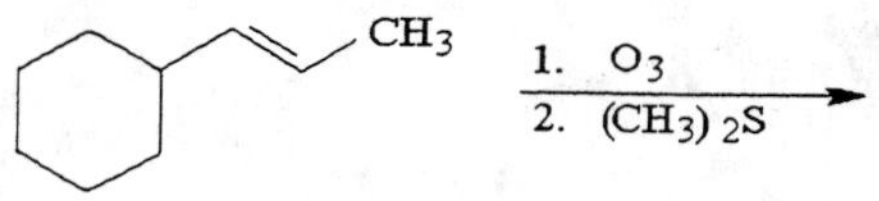

CH_3 1. O_3 2. $(CH_3)_2S$

Answer:

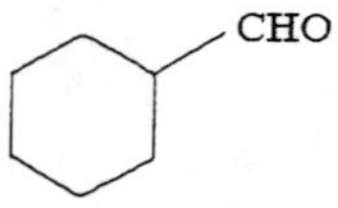

CHO + CH_3CHO

Section: 6

57. Provide the reagents necessary to complete the following transformation.

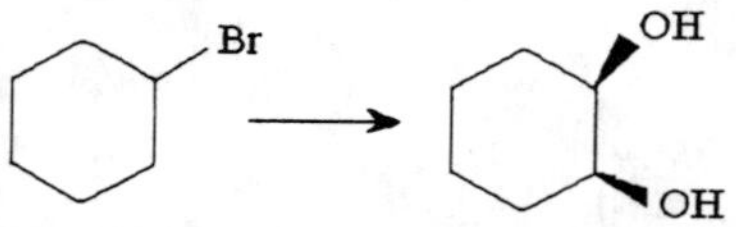

Br → OH OH

Answer:
1. $NaOCH_3$, CH_3OH
2. OsO_4, H_2O_2 or cold, dilute $KMnO_4$, ^-OH

Section: 4

58. Give the structure of the alkene which would yield the following products upon ozonolysis-reduction.

$CH_3CH_2CH_2CH_2CHO$ + CH_2O

Answer: $CH_3CH_2CH_2CH_2CH_2{=}CH_2$ Section: 6

59. Give the structure of the alkene which would yield the following products upon ozonolysis-reduction.

CH_3COCH_3 + CH_3CH_2CHO

Answer: $(CH_3)_2C{=}CHCH_2CH_3$ Section: 6

60. β-Ocimene is a natural product with a pleasant odor. Based on the information below, deduce the structure of β-ocimene.

β-Ocimene ($C_{10}H_{16}$) $\xrightarrow{H_2,\ Pt}$ 2, 6-dimethyloctane

β-Ocimene $\xrightarrow[2.\ (CH_3)_2S]{1.\ O_3}$ $CH_2O + CH_3COCH_3 + CH_3COCHO + OHCCH_2CHO$

Answer:

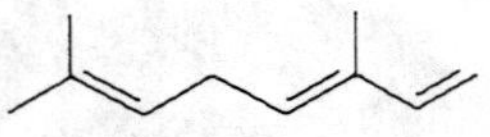

Section: 6

Chapter 18: Carbonyl Compounds III

MULTIPLE CHOICE

1. Which of the following compounds does not have an alpha hydrogen?

   a. 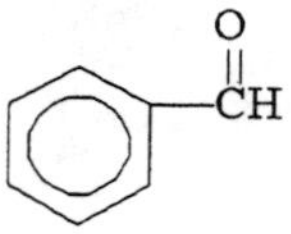 $C_6H_5-\overset{O}{\overset{\|}{C}}H$

   b. $H-\overset{O}{\overset{\|}{C}}-H$

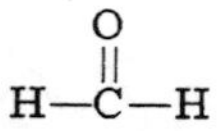

   c. O

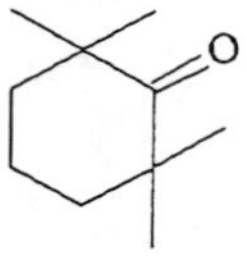

   d. 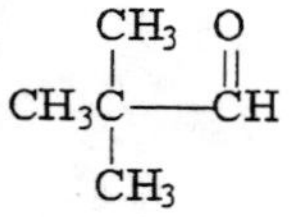 $CH_3\overset{CH_3}{\underset{CH_3}{C}}-\overset{O}{\overset{\|}{C}}H$

   e. All of the above

   Answer: e Section: 1

2. Which of the following underlined alpha hydrogens are the most acidic?

   a. $\underline{CH_3}-\overset{O}{\overset{\|}{C}}H$

   b. $CH_3-\overset{O}{\overset{\|}{C}}-\underline{CH_2}-\overset{O}{\overset{\|}{C}}H$

   c. $CH_3-\overset{O}{\overset{\|}{C}}-\underline{CH_2}-\overset{O}{\overset{\|}{C}}-OCH_3$

   d. $\underline{CH_3}-\overset{O}{\overset{\|}{C}}-CH_3$

   e. $\underline{CH_3}-\overset{O}{\overset{\|}{C}}-O-CH_3$

   Answer: b Section: 1

3. Which of the labeled hydrogen atoms in the following structure is the most acidic?

   a. 1
   b. 2
   c. 3
   d. 4
   e. 5

   H(1)–C_6H_4(H(2))–CH_2CH_2COH (C=O), labels: CH_2 3, CH_2 4, OH 5

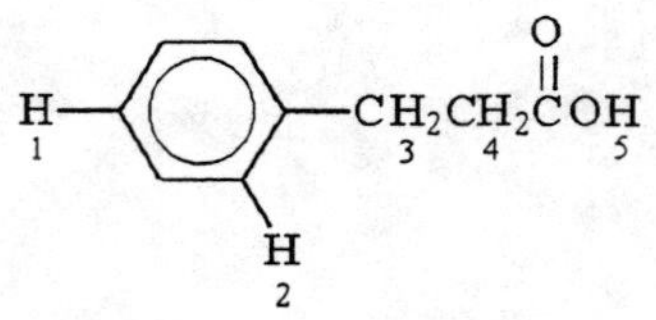

   Answer: e Section: 1

4. Which of the labeled hydrogen atoms in the following structure is the most acidic?

   a. 1
   b. 2
   c. 3
   d. 4
   e. 5

   C_6H_4(H(1), H(2))–C(=O)–$CH_2CH_2CH_3$, labels: 3, 4, 5

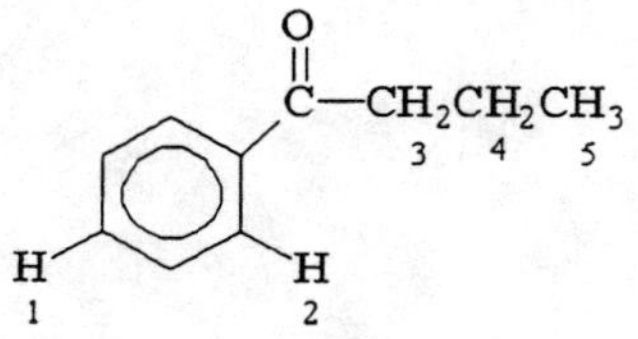

   Answer: c Section: 1

5. Which of the labeled hydrogen atoms in the following structure is the most acidic?

   a. 1
   b. 2
   c. 3
   d. 4
   e. 5

   H(1)–C_6H_4(H(2))–CH_2-C(=O)–CH_2-C(=O)–CH_3, labels: CH_2 3, CH_2 4, CH_3 5

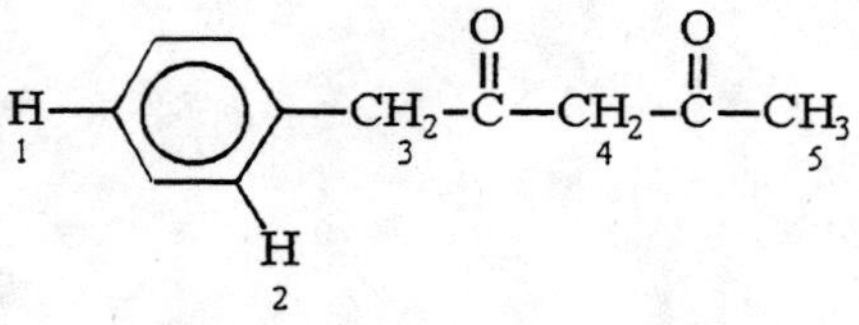

   Answer: d Section: 1

6. Which of the following is a β-diester?

   a. $CH_3O\overset{O}{\overset{\|}{C}}CH_2\overset{O}{\overset{\|}{C}}OCH_3$

   b. $CH_3\overset{O}{\overset{\|}{C}}CH_2\overset{O}{\overset{\|}{C}}CH_3$

   c. $CH_3\overset{O}{\overset{\|}{C}}CH_2\overset{O}{\overset{\|}{C}}OCH_3$

   d. $CH_3O\overset{O}{\overset{\|}{C}}OCH_3$

   e. $CH_3O\overset{O}{\overset{\|}{C}}\overset{O}{\overset{\|}{C}}H$

   Answer: a Section: 1

7. When compared to the keto form, the enol form of which of the following compounds is most stable?

   a.

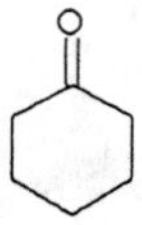

   b. 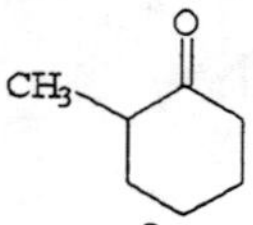 CH_3

   c.

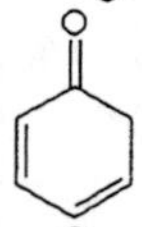

   d.

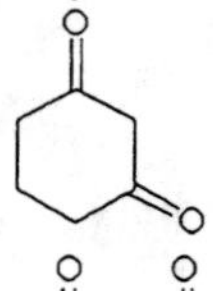

   e. 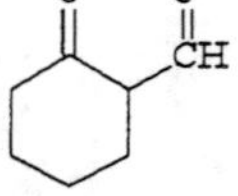 $\overset{O}{\overset{\|}{C}}H$

   Answer: c Section: 2

8. Which of the following is the most stable enol form of 1,3-cyclohexanedione?

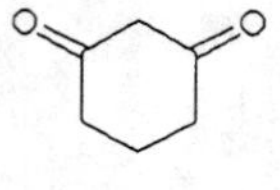

a.

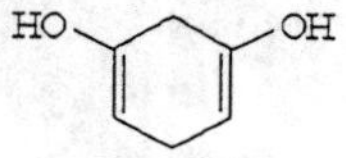

b.

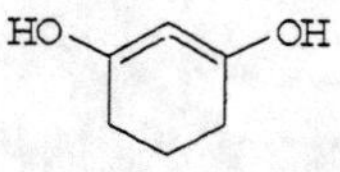

c.

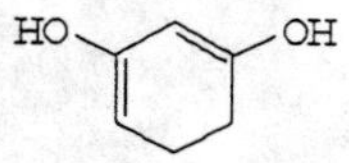

d.

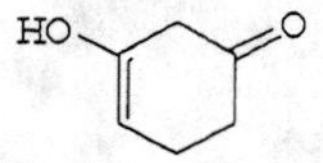

e.

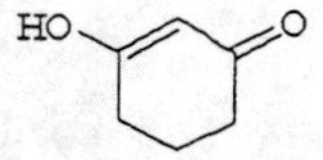

Answer: e Section: 2

9. How would you accomplish the following conversion?

a. H^+/H_2O; HC≡N
b. -OH/NaC≡N
c. Br_2/H^+, H_2O; NaC≡N
d. HBr; NaC≡N
e. $LiAlH_4$; HC≡N; $KMnO_4$

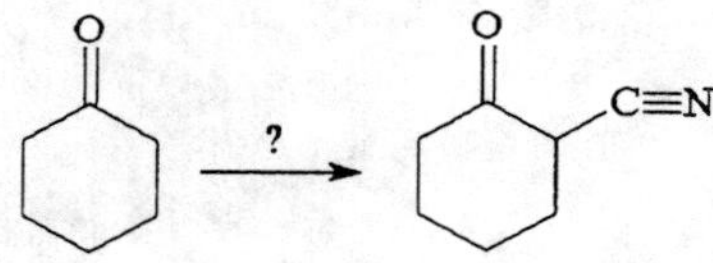

Answer: c Section: 4

10. Which of the following compounds will give a positive iodoform test?

a. propanal
b. 2-pentanone
c. 3-pentanone
d. benzophenone
e. cyclohexanone

Answer: b Section: 4

11. Which is the major organic product from the following reaction?

$$CH_3CH_2\overset{\overset{\large O}{||}}{C}OH \xrightarrow[2.\ H_3O^+]{1.\ Br_2,\ PBr_3}$$

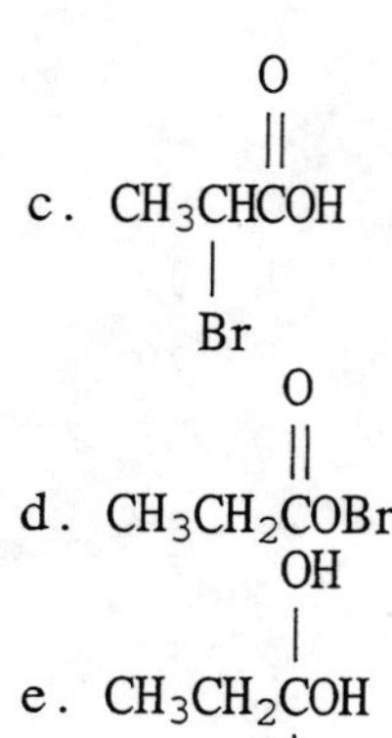

a. $BrCH_2CH_2\overset{\overset{O}{||}}{C}OH$

b. $CH_3CH_2\overset{\overset{O}{||}}{C}Br$

c. $CH_3\underset{\underset{Br}{|}}{C}H\overset{\overset{O}{||}}{C}OH$

d. $CH_3CH_2\overset{\overset{O}{||}}{C}OBr$

e. $CH_3CH_2\underset{\underset{Br}{|}}{\overset{\overset{OH}{|}}{C}}OH$

Answer: c Section: 5

12. Which of the following is the best method for accomplishing the reaction shown below?

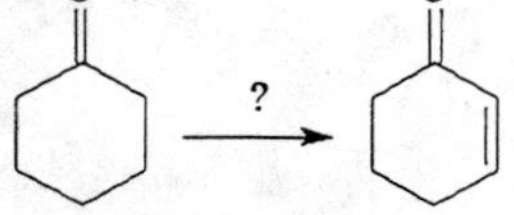

a. HO^-; Br_2; KOH/alcohol

b. LDA/THF; 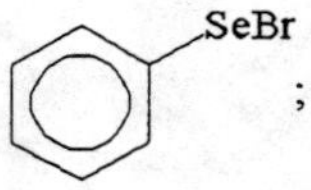 SeBr; H_2O_2

c. HO^-; 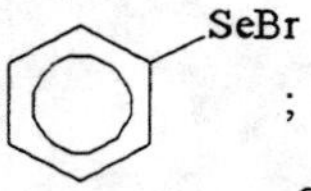 SeBr; H_2O_2

d. LDA/THF; 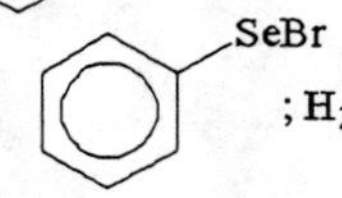 SeBr; H_2O

e. LDA/THF; Br_2; H_2O_2

Answer: b Section: 6

13. What is the <u>major</u> organic product of the following reaction?

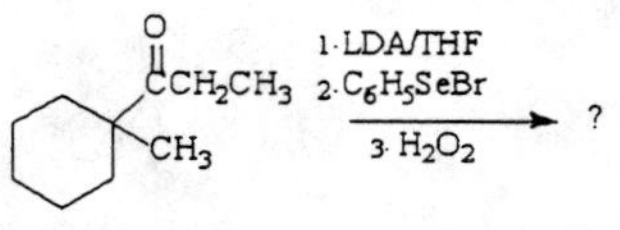

CCH_2CH_3, CH_3; 1. LDA/THF 2. C_6H_5SeBr 3. H_2O_2 → ?

a.

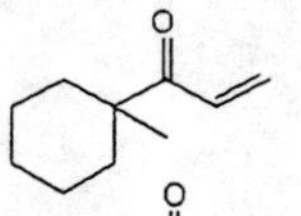

b.

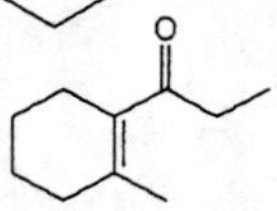

c.

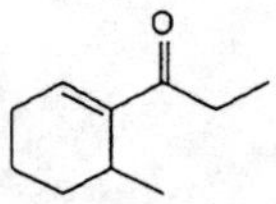

d. 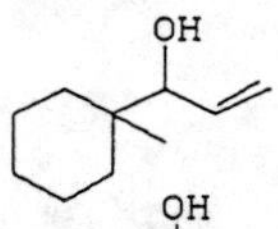 OH

e. 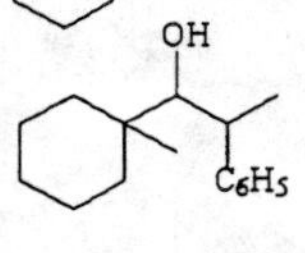 OH, C_6H_5

Answer: a Section: 6

14. What is the major organic product of the following reaction?

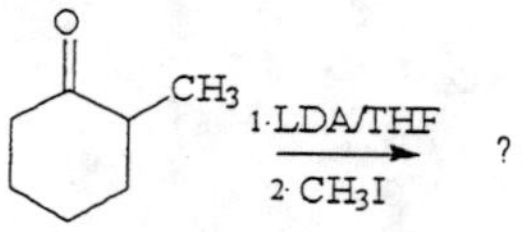

2-methylcyclohexanone $\xrightarrow[2.\ CH_3I]{1.\ LDA/THF}$?

a. 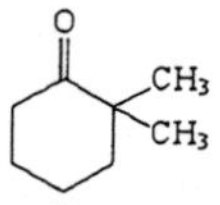 2,2-dimethylcyclohexanone (CH_3, CH_3)

b. 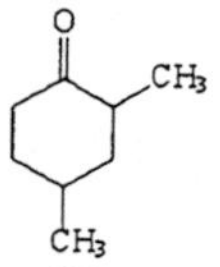 2,4-dimethylcyclohexanone (CH_3, CH_3)

c. 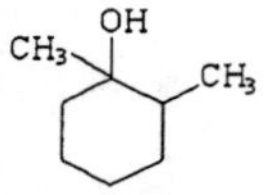 1,2-dimethylcyclohexanol (OH, CH_3, CH_3)

d. 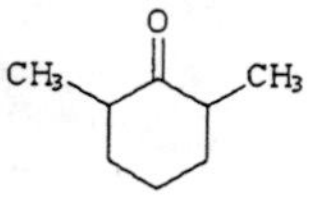 2,6-dimethylcyclohexanone (CH_3, CH_3)

e. 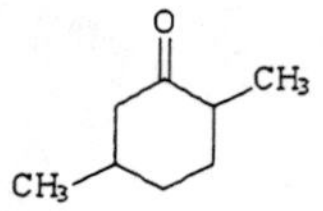 2,5-dimethylcyclohexanone (CH_3, CH_3)

Answer: d Section: 8

15. What is the major organic product of the following reaction?

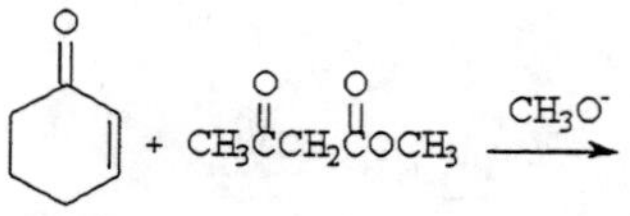

2-cyclohexenone + $CH_3\overset{O}{\overset{\|}{C}}CH_2\overset{O}{\overset{\|}{C}}OCH_3$ $\xrightarrow{CH_3O^-}$

a. cyclohexanone with, at C-3, $-\overset{|}{C}H\overset{O}{\overset{\|}{C}}OCH_3$ bearing $O{=}\overset{|}{C}-CH_3$

b. cyclohexanone with, at C-3, $-\overset{|}{C}H\overset{O}{\overset{\|}{C}}OCH_3$ bearing $HO-\overset{|}{C}H-CH_3$

c. cyclohex-2-en-1-ol with OH and, at C-1, $-\overset{|}{C}H\overset{O}{\overset{\|}{C}}OCH_3$ bearing $O{=}\overset{|}{C}-CH_3$

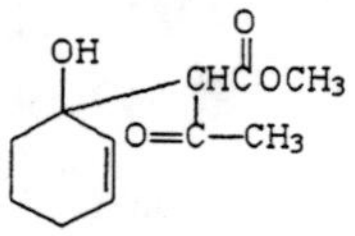

d. 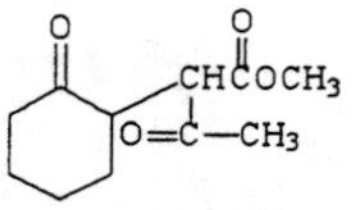 cyclohexanone with, at C-2, $-\overset{|}{C}H\overset{O}{\overset{\|}{C}}OCH_3$ bearing $O{=}\overset{|}{C}-CH_3$

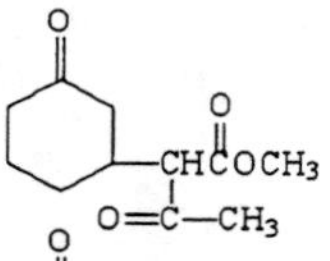

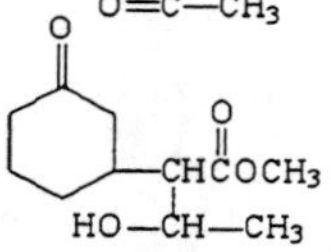

e. 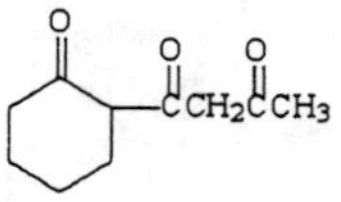 cyclohexanone with, at C-2, $-\overset{O}{\overset{\|}{C}}CH_2\overset{O}{\overset{\|}{C}}CH_3$

Answer: a Section: 9

16. Which of the following methods can be used to prepare the following compound?

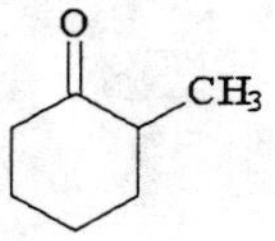

[2-methylcyclohexanone structure]

a. 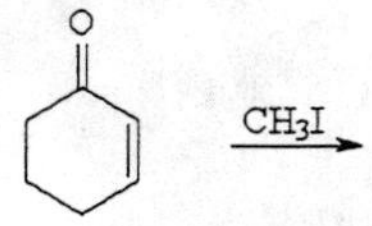 [2-cyclohexenone] $\xrightarrow{CH_3I}$

b. 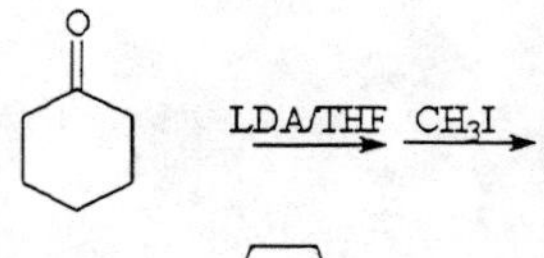 [cyclohexanone] $\xrightarrow{LDA/THF}$ $\xrightarrow{CH_3I}$

c. 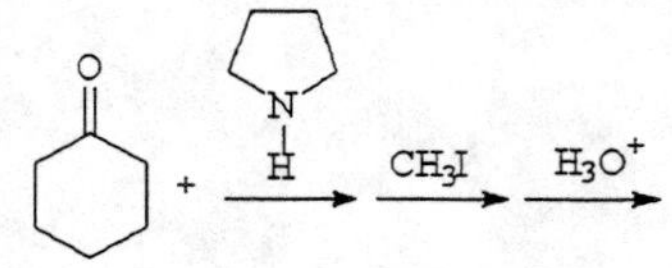 [cyclohexanone] + [pyrrolidine, N–H] $\longrightarrow$ $\xrightarrow{CH_3I}$ $\xrightarrow{H_3O^+}$

d. b and c

e. a, b and c

Answer: d Section: 8

17. What is the <u>major</u> organic product from the following reaction?

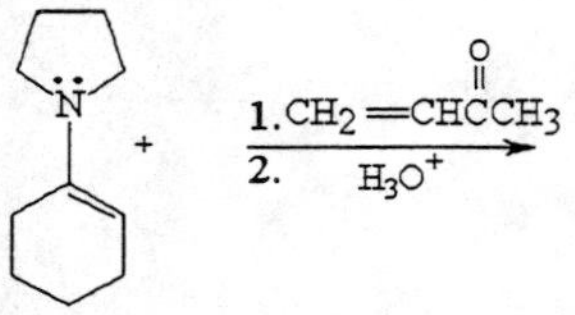

[1-(pyrrolidin-1-yl)cyclohexene] + $\xrightarrow[2.\ H_3O^+]{1.\ CH_2{=}CHCCH_3\ (C{=}O)}$

a. 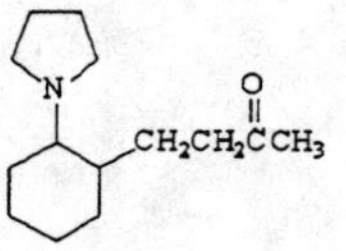 [pyrrolidinyl cyclohexane with $CH_2CH_2CCH_3$ (C=O)]

b. 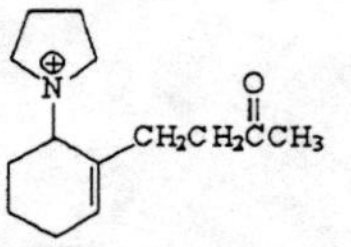 [pyrrolidinium (⊕N) cyclohexene with $CH_2CH_2CCH_3$ (C=O)]

c. [pyrrolidinium (⊕N) cyclohexene with C(OH)(CH_3)CH=CH_2]

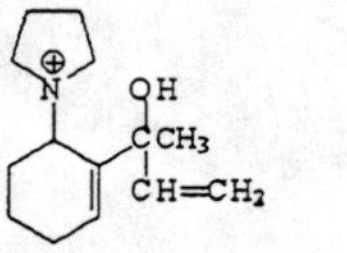

d. 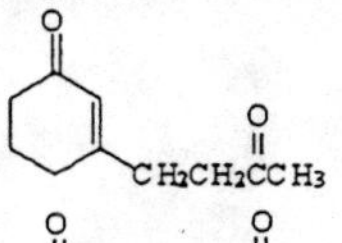 [cyclohexenone with $CH_2CH_2CCH_3$ (C=O)]

e. 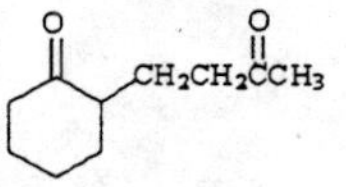 [2-($CH_2CH_2CCH_3$ (C=O))cyclohexanone]

Answer: e Section: 9

18. Which of the following aldehydes does <u>not</u> undergo an aldol addition when mixed with base?

a. $H-\overset{O}{\overset{\|}{C}}-H$

b. $C_6H_5CH_2\overset{O}{\overset{\|}{C}}H$

c. $CH_3\overset{O}{\overset{\|}{C}}H$

d. $CH_3CH_2\overset{O}{\overset{\|}{C}}H$

e. $C_6H_{11}\overset{O}{\overset{\|}{C}}H$

Answer: a Section: 10

19. Which of the following statements describes the first step in the mechanism of the aldol condensation?

a. An alpha hydrogen is abstracted by the base to form an enolate anion.
b. A nucleophilic base attacks the carbonyl carbon atom.
c. The carbonyl oxygen is protonated by the base ion.
d. The alpha hydrogen is abstracted by an acid to the enolate anion.
e. The carbonyl oxygen of one aldehyde attacks the carbonyl carbon of another.

Answer: a Section: 10

20. What is the major organic product of the following aldol addition?

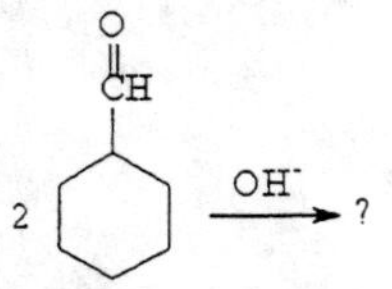

2 (cyclohexanecarbaldehyde) $\xrightarrow{OH^-}$?

a. 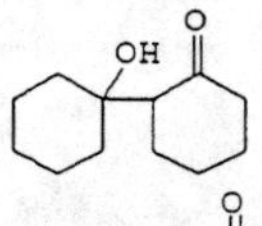 (1-hydroxycyclohexyl)cyclohexanone structure: OH, O

b. 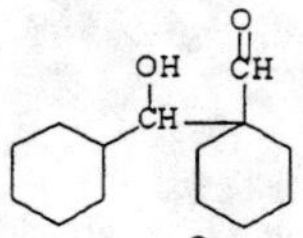 OH, CH, CH=O structure

c. 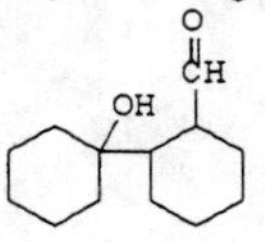 OH, CH=O structure

d. 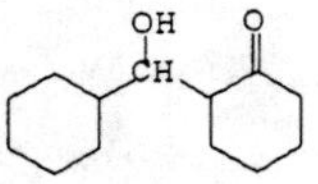 OH, CH, O structure

e. 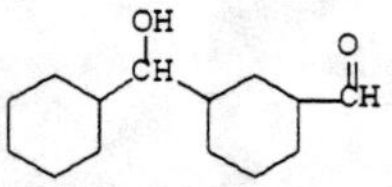 OH, CH, CH=O structure

Answer: b Section: 10

21. What is the major organic product of the following addition?

$$CH_3CH_2\overset{O}{\overset{\|}{C}}H + H-\overset{O}{\overset{\|}{C}}-H \xrightarrow{^-OH} ?$$

(excess)

a. $CH_3CH_2CH(OH)\overset{O}{\overset{\|}{C}}H$

b. $CH_3C(OH)(CH_3)-\overset{O}{\overset{\|}{C}}H$

c. $HOCH_2CH_2CH_2\overset{O}{\overset{\|}{C}}H$

d. $CH_3CH(CH_2OH)\overset{O}{\overset{\|}{C}}H$

e. 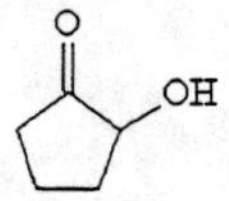 2-hydroxycyclopentanone: O, OH

Answer: d Section: 12

22. What is the major product of the following condensation?

$$2\ CH_3CH_2\overset{O}{\overset{\|}{C}}OCH_2CH_3 \xrightarrow[2\cdot\ H_3O^+]{1\cdot CH_3CH_2O^-} ?$$

a. $CH_3CH_2\overset{OH}{\overset{|}{C}}H\underset{CH_3}{\underset{|}{C}}H\overset{O}{\overset{\|}{C}}OCH_2CH_3$

b. 2-oxocyclohexyl–$\overset{O}{\overset{\|}{C}}OCH_2CH_3$ (cyclohexanone ring with $\overset{O}{\overset{\|}{C}}OCH_2CH_3$ on the carbon adjacent to C=O)

c. $CH_3CH_2\overset{OH}{\overset{|}{C}}HCH_2CH_2\overset{O}{\overset{\|}{C}}OCH_2CH_3$

d. $CH_3CH_2\overset{O}{\overset{\|}{C}}CH_2CH_2\overset{O}{\overset{\|}{C}}OCH_2CH_3$

e. $CH_3CH_2\overset{O}{\overset{\|}{C}}\underset{CH_3}{\underset{|}{C}}H\overset{O}{\overset{\|}{C}}OCH_2CH_3$

Answer: e Section: 13

23. What materials are needed to prepare the following crossed Claisen condensation?

Target: cyclohexanone with $\overset{O}{\overset{\|}{C}}CH_3$ on the carbon adjacent to C=O

a. cyclohex-2-enone $+\ CH_3\overset{O}{\overset{\|}{C}}Cl \xrightarrow{H_3O^+}$

b. cyclohexanone $+\ CH_3\overset{O}{\overset{\|}{C}}OR \xrightarrow[2\cdot\ H_3O^+]{1\cdot CH_3CH_2O^-}$

c. cyclohexanone $+\ CH_3\overset{O}{\overset{\|}{C}}H \xrightarrow[2\cdot\ H_3O^+]{1\cdot CH_3CH_2O^-}$

d. cyclohexanone $+\ CH_3\overset{O}{\overset{\|}{C}}CH_3 \xrightarrow[2\cdot\ H_3O^+]{1\cdot CH_3CH_2O^-}$

e. cyclohex-2-enone $+\ CH_3\overset{O}{\overset{\|}{C}}OR \xrightarrow[2\cdot\ H_3O^+]{1\cdot CH_3CH_2O^-}$

Answer: b Section: 14

24. What is the major organic product of the following reaction?

1. RO^-
2. H_3O^+

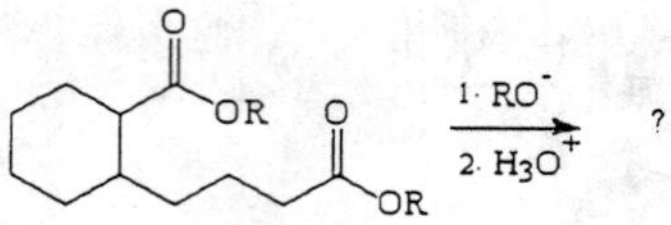

a.

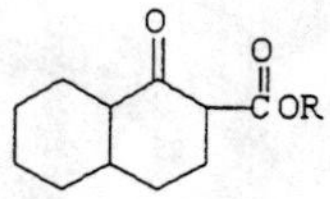

b.

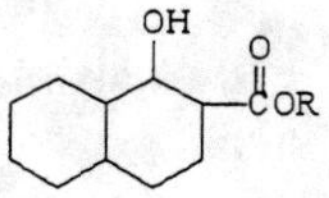

c.

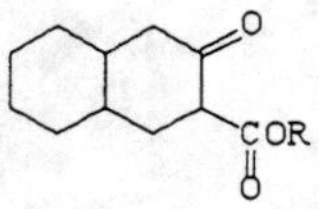

d.

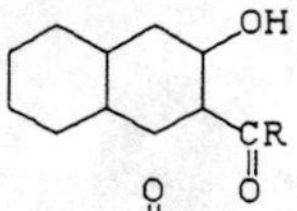

e.

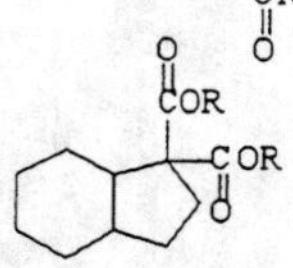

Answer: a Section: 15

25. What is the major organic product of the following reaction?

^-OH, H_2O

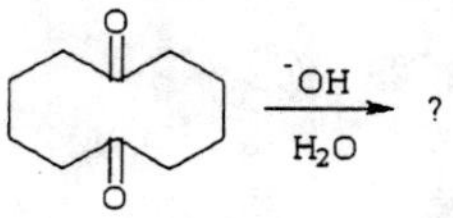

a.

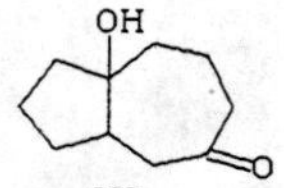

b.

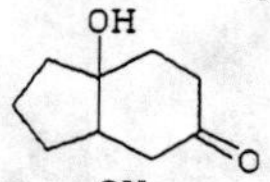

c.

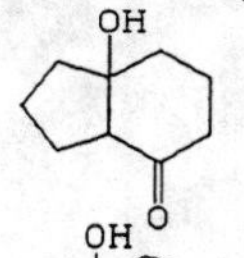

d.

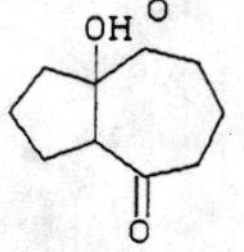

e.

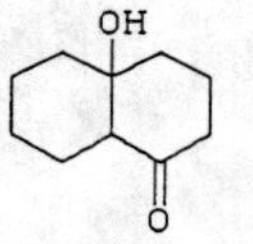

Answer: d Section: 15

26. What materials would you use to prepare the following compound using a <u>Robinson</u> <u>Annulation</u>?

a. 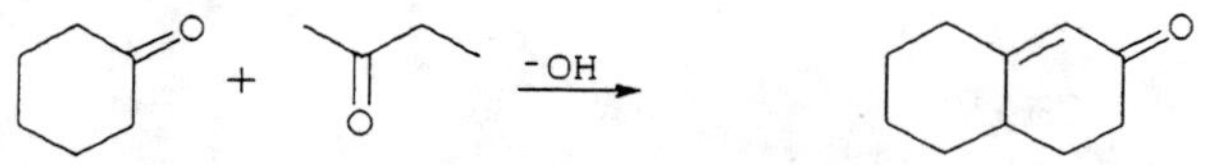 + $\xrightarrow{^{-}OH}$

b. 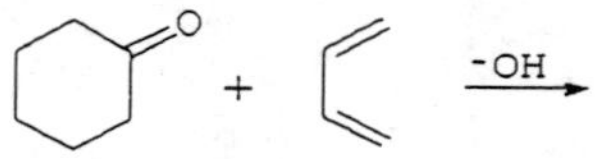 + $\xrightarrow{^{-}OH}$

c. 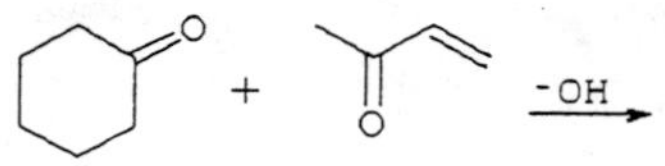 + $\xrightarrow{^{-}OH}$

d. 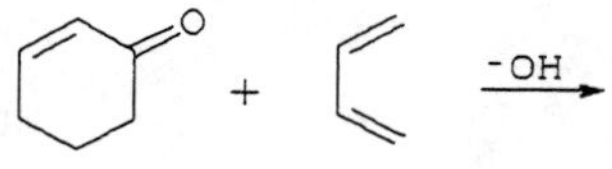 + $\xrightarrow{^{-}OH}$

e. 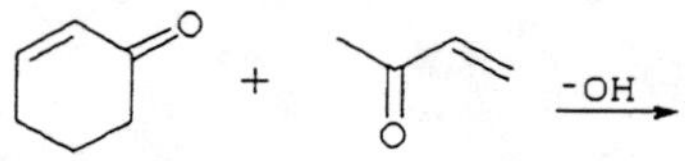 + $\xrightarrow{^{-}OH}$

Answer: c Section: 15

27. What is the <u>major</u> organic product of the following reaction?

$$CH_3CH_2\overset{\overset{\displaystyle O}{\|}}{C}OH \xrightarrow[\text{2. } Br_2\text{, heat}]{\text{1. } Ag^+} ?$$

a. CH_3CH_2Br

b. $CH_3CH_2\overset{\overset{\displaystyle O}{\|}}{C}O^{\ominus}Ag^{\oplus}$

c. $CH_3CH_2\overset{\overset{\displaystyle O}{\|}}{C}Br$

d. $CH_3CH_2\overset{\overset{\displaystyle O}{\|}}{C}OBr$

e. $CH_3\underset{\underset{\displaystyle Br}{|}}{C}H\overset{\overset{\displaystyle O}{\|}}{C}OH$

Answer: a Section: 16

28. What alkyl bromide should be used in the malonic ester synthesis of the following carboxylic acid?

$C_6H_5CH_2CH_2COOH$

a. CH_3Br

b. CH_3CH_2Br

c. $C_6H_5CH_2CH_2Br$

d. C_6H_5Br

e. $C_6H_5CH_2Br$

Answer: e Section: 17

29. What alkyl bromide should be used in the acetoacetic ester synthesis of the following methyl ketone?

$C_6H_5CH_2CH_2CH_2COCH_3$

a. CH_3Br

b. CH_3CH_2Br

c. $C_6H_5CH_2CH_2Br$

d. C_6H_5Br

e. $C_6H_5CH_2Br$

Answer: c Section: 18

30. Which of the following compounds has the most stable enol tautomer?

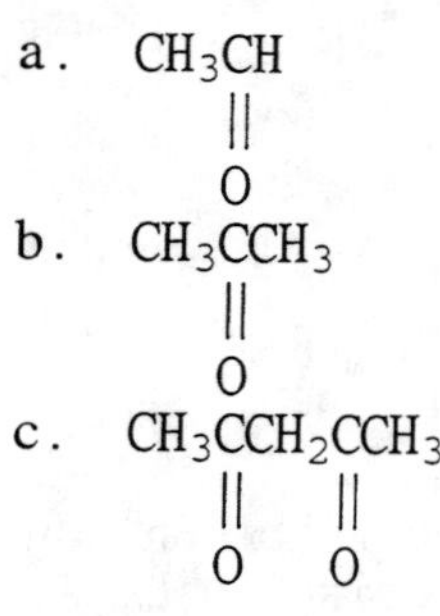

a. $CH_3\underset{\underset{O}{\|}}{C}H$

b. $CH_3\underset{\underset{O}{\|}}{C}CH_3$

c. $CH_3\underset{\underset{O}{\|}}{C}CH_2\underset{\underset{O}{\|}}{C}CH_3$

d. $CH_3\underset{\underset{O}{\|}}{C} - \underset{\underset{O}{\|}}{C}CH_3$

e. $CH_3\underset{\underset{O}{\|}}{C} - \underset{\underset{O}{\|}}{C}OH$

Answer:
c

Section: 2

SHORT ANSWER

31. Which of the following keto-enol tautomers is more stable? Explain.

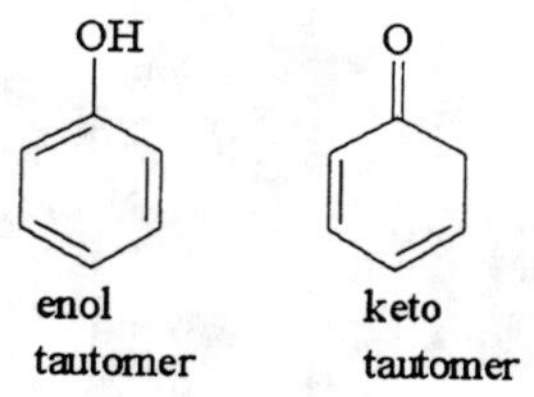

Answer:
The enol form is more stable than the keto form because the enol form is aromatic but the keto form is not.

Section: 2

32. Explain why the following acid cannot be prepared by the malonic ester synthesis:

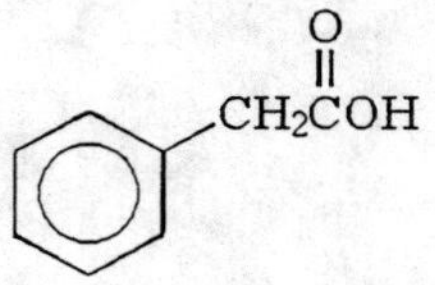

$C_6H_5CH_2COOH$

Answer:
Since bromobenzene is needed to prepare the acid, the reaction would not occur. This is true because it is difficult for a bromine to be a good leaving group if it is directly attached to a benzene ring (see S_N1/S_N2 reactions).

Section: 17

33. (a) The based-catalyzed reaction of 1 mole of acetone with 2 moles of benzaldehyde yields a yellow solid, $C_{17}H_{14}O$. What is the structure of the product?
(b) What will be the result when only 1 mole of benzaldehyde is added?
(c) What will be the result when acetone is not added at all?
(d) What will be the result when benzaldehyde is not added at all?
(e) Why is the self-condensation of acetone unlikely in the presence of benzaldehyde?

Answer:
(a) This is an aldol condensation. (see graph a)
(b) Using only 1 mole of benzaldehyde, the product would be the following (see graph b)
(c) Using no acetone, no reaction will occur since benzaldehyde has no α-hydrogens.
(d) Using no benzaldehyde, the acetone will form an aldol addition product with the following structure (see graph d)
(e) This is because the benzaldehyde carbonyl carbon is more reactive than that of acetone toward nucleophilic attack. This is due to the electron releasing methyl groups on acetone. Also, benzaldehyde is less sterically hindered because it has a hydrogen on one side.

(a)

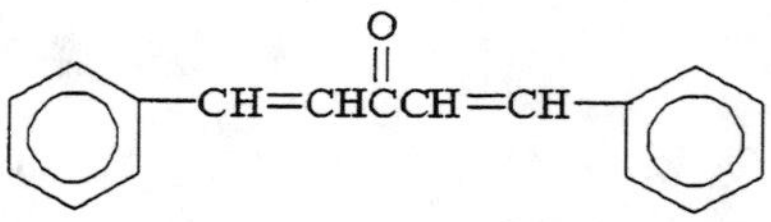

C_6H_5–CH=CHC(=O)CH=CH–C_6H_5

(b)

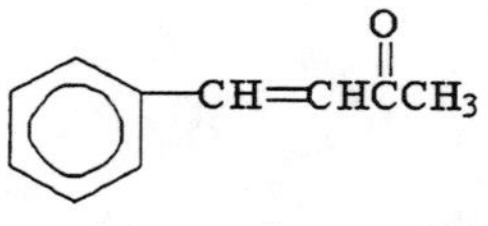

C_6H_5–CH=CHC(=O)CH_3

(d)

$CH_3C(OH)(CH_3)CH_2C(=O)CH_3$

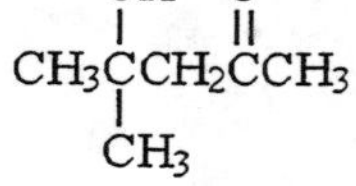

Section: 12

34. Propose a step-wise mechanism that would explain the following conversion:

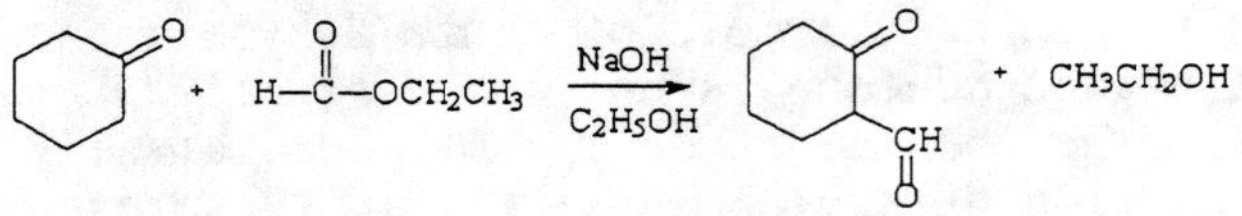

Cyclohexanone + $H-\overset{O}{\overset{\|}{C}}-OCH_2CH_3$ $\xrightarrow[C_2H_5OH]{NaOH}$ 2-formylcyclohexanone ($-\overset{}{C}H=O$) + CH_3CH_2OH

Answer:

Cyclohexanone + $\overset{\ominus}{O}H$ → enolate ↔ carbanion; carbanion + $HCOCH_2CH_3$ (C=O) → tetrahedral intermediate ($CH-OCH_2CH_3$, $O^{\ominus}$) → 2-formylcyclohexanone + CH_3CH_2OH

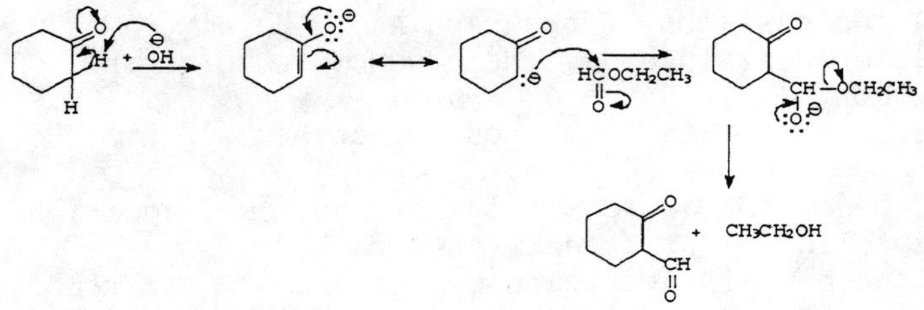

Section: 14

35. Show how an enolate can add to a carbonyl.

Answer:

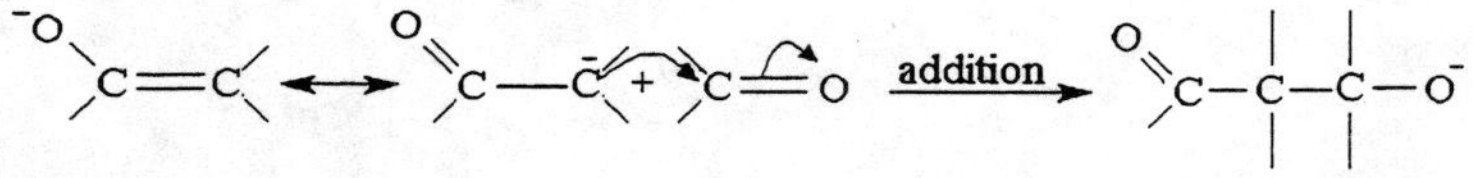

$^{-}O-C=C$ ↔ $O=C-\overset{-}{C}$ + $C=O$ $\xrightarrow{\text{addition}}$ $O=C-C-C-O^{-}$

Section: 10

36. Provide a detailed, stepwise mechanism for the base-catalyzed enolization of acetaldehyde.

Answer:

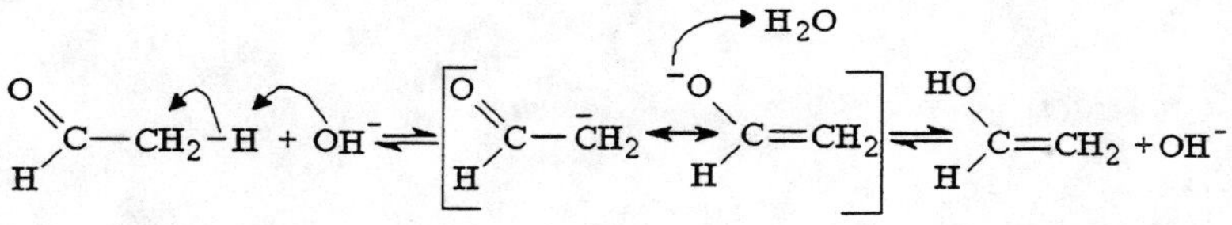

$O=C(H)-CH_2-H + OH^{-} \rightleftharpoons [O=C(H)-\overset{-}{C}H_2 \leftrightarrow {}^{-}O-C(H)=CH_2] \rightleftharpoons HO-C(H)=CH_2 + OH^{-}$ (with H_2O)

Section: 2

37. Provide a detailed, stepwise mechanism for the acid-catalyzed enolization of acetaldehyde.

Answer:

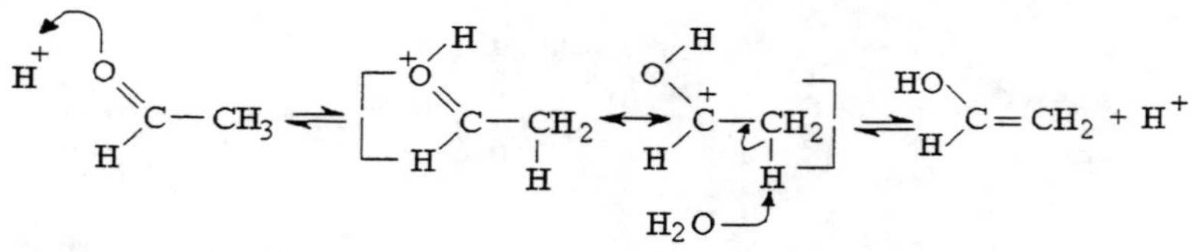

Section: 2

MULTIPLE CHOICE

38. Which of the following reagents will quantitatively convert an enolizable ketone to its enolate salt?

a. lithium hydroxide
b. lithium diisopropylamide
c. methyllithium
d. diethylamine
e. pyridine

Answer: b Section: 7

SHORT ANSWER

39. Provide a detailed, stepwise mechanism for the α-bromination of acetone in base.

Answer:

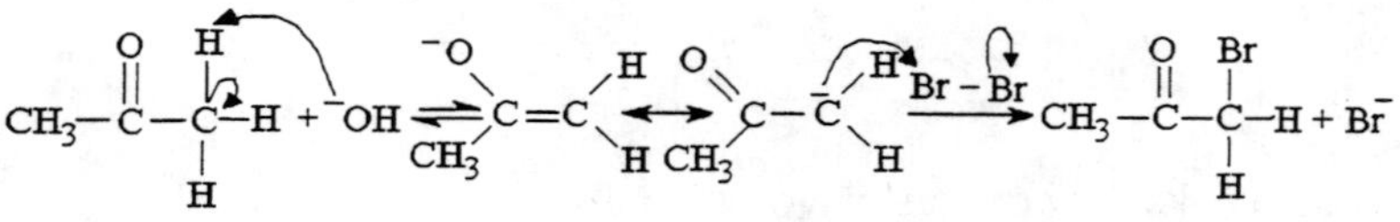

Section: 4

40. Provide a detailed, stepwise mechanism for the formation of acetate and bromodiiodomethane from bromoacetone, hydroxide and iodine.

Answer:

$$CH_3-\overset{O}{\overset{\|}{C}}-\underset{Br}{\overset{H}{C}}-H + {}^-OH \rightleftharpoons \left[\underset{CH_3}{{}^-O}\!\!\!\;C{=}C\underset{Br}{\overset{H}{}} \longleftrightarrow O{=}C-\underset{Br}{\overset{H}{C}}{}^- \right] \xrightarrow{I-I} CH_3-\overset{O}{\overset{\|}{C}}-\underset{Br}{\overset{I}{C}}-H + I^-$$

$$CH_3-\overset{O}{\overset{\|}{C}}-\underset{Br}{\overset{I}{C}}-H + {}^-OH \rightleftharpoons \left[\underset{CH_3}{{}^-O}\!\!\!\;C{=}C\underset{Br}{\overset{I}{}} \longleftrightarrow \underset{CH_3}{O{=}C}-\underset{Br}{\overset{I}{C}}{}^- \right] \xrightarrow{I-I} CH_3-\overset{O}{\overset{\|}{C}}-\underset{Br}{\overset{I}{C}}-I + I^-$$

$$CH_3-\overset{O}{\overset{\|}{C}}-\underset{Br}{\overset{I}{C}}-I + {}^-OH \rightleftharpoons CH_3-\underset{OH}{\overset{{}^-O}{C}}-CBrI_2 \rightleftharpoons CH_3-\overset{O}{\overset{\|}{C}}-OH \quad {}^-\!:\underset{Br}{\overset{I}{C}}-I$$

$$\longrightarrow CH_3-\overset{O}{\overset{\|}{C}}-O^- + CHI_2Br$$

Section: 4

MULTIPLE CHOICE

41. What is the carbon nucleophile which attacks molecular bromine in the acid-catalyzed α-bromination of a ketone?

a. an enolate
b. a Grignard reagent
c. an acetylide
d. a carbocation
e. an enol

Answer: e Section: 4

42. The Hell-Volhard-Zelinsky reaction involves:

a. the α-bromination of carboxylic acids.
b. the α-bromination of ketones.
c. the bromination of alcohols.
d. the oxidation of aldehydes to acids.
e. none of the above.

Answer: a Section: 5

43. Which of the following will alkylate a lithium enolate most rapidly?

a. methyl bromide
b. isopropyl bromide
c. neopentyl bromide
d. bromobenzene
e. 2-methylbromobenzene

Answer: a Section: 7

SHORT ANSWER

44. What iminium salt is produced in the reaction shown below?

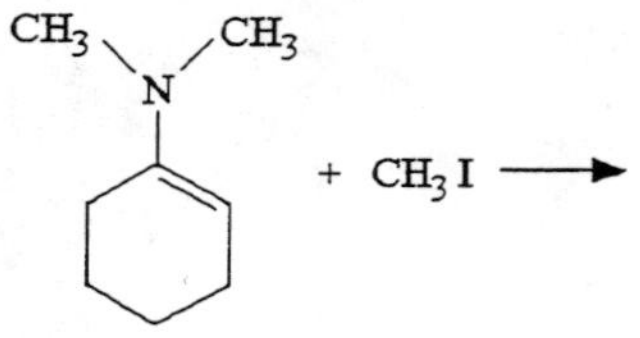

CH_3 CH_3 N + CH_3I →

Answer:

CH_3 + CH_3 N I^- CH_3

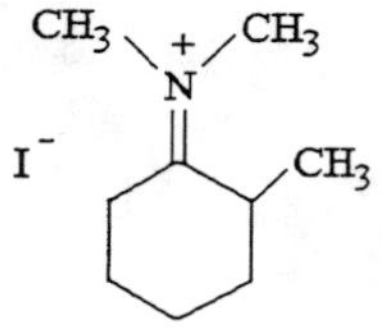

Section: 8

MULTIPLE CHOICE

45. An enolate attacks an aldehyde and the resulting product is subsequently protonated. What type of reaction is this?

a. a Fischer esterification
b. an acid-catalyzed aldol condensation
c. a base-mediated aldol condensation
d. a Hell-Volhard-Zelinsky reaction
e. a Selman-Jones reaction

Answer: c Section: 10

46. Which of the following could result from the dehydration of an aldol?

a. 4-methyl-3-penten-2-one
b. 4-methyl-4-penten-2-one
c. 4-methyl-5-hexen-2-one
d. 4-methyl-4-hexen-2-one
e. 3-methyl-4-penten-2-one

Answer: a Section: 11

47. In theory a poorly planned crossed aldol reaction can produce how many different aldol regioisomers?

a. 1
b. 2
c. 3
d. 4
e. 5

Answer: d Section: 12

SHORT ANSWER

48. Provide the sequence of steps necessary to synthesize the compound shown below from cyclohexene.

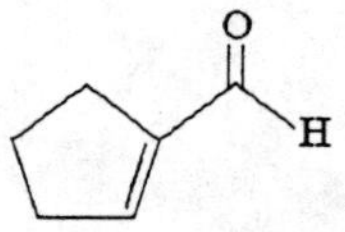

Answer:
1. O_3
2. $S(CH_3)_2$
3. NaOH, Δ

Section: 15

49. What two molecules were condensed in an aldol reaction to produce $(CH_3)_3CCH{=}CHCOCH_3$?

Answer: $(CH_3)_3CCHO$ and CH_3COCH_3 Section: 12

50. What two molecules were condensed in an aldol reaction to produce PhCH=CHCOPh?

Answer: PhCHO and $PhCOCH_3$ Section: 12

51. Provide the structure of the Claisen product in the self condensation of methyl phenylacetate.

Answer:

$C_6H_5CH_2-C(=O)-CH(C_6H_5)-C(=O)OCH_3$

Section: 13

52. Starting with cyclohexene and employing a Dieckmann cyclization show how the compound below can be prepared.

(2-carbomethoxycyclopentanone: $CH_3O-C(=O)-$ on cyclopentanone)

Answer:
1. $KMnO_4$, ^-OH, Δ
2. H^+
3. CH_2N_2 or $SOCl_2$ followed by CH_3OH
4. $NaOCH_3$

Section: 15

53. What two starting materials yield $OHCCH_2CO_2CH_2CH_3$ as the cross Claisen condensation product?

Answer: $HCO_2CH_2CH_3$ and $CH_3CO_2CH_2CH_3$ Section: 14

54. What product is formed in the crossed Claisen condensation between methyl benzoate and cyclohexanone?

Answer:

(2-benzoylcyclohexanone)

Section: 14

55. What product results when malonic ester is treated with the following sequence of reagents:
1. $NaOCH_2CH_3$;
2. $PhCH_2Br$;
3. H_3O+, Δ?

Answer: $PhCH_2CH_2CO_2H$ Section: 17

56. When compound X is heated, $PhCOCH(CH_3)_2$ and CO_2 are produced. Offer a structure for compound X.

Answer: $PhCOC(CH_3)_2CO_2H$ Section: 16

MULTIPLE CHOICE

57. In the Michael reaction, addition to the α,β-unsaturated carbonyl occurs in a:

a. 1,2-fashion.
b. 1,3-fashion.
c. 1,4-fashion.
d. 1,5-fashion.
e. Diels-Alder reaction.

Answer: c Section: 9

SHORT ANSWER

58. Provide a synthesis of the compound shown below from 2-cyclopenten-1-one and acetoacetic ester.

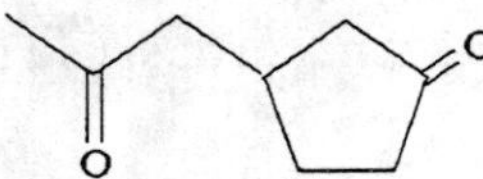

Answer:
From acetoacetic ester:
1. NaOEt, EtOH
2. 2-cyclopenten-1-one
3. H_3O^+, Δ
Section: 18

59. Provide the structure of the major organic product which results when $PhCO_2CH_2CH_3$ and $CH_3CH_2CO_2CH_2CH_3$ are heated in the presence of sodium ethoxide and the compound generated is subsequently treated with cold, dilute acid.

Answer:

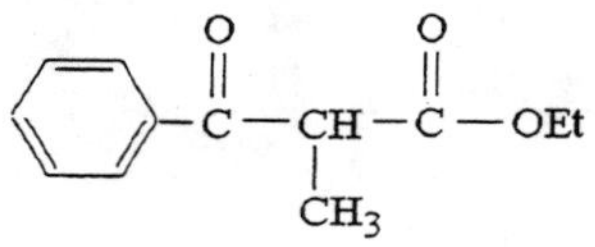

$C_6H_5-C(=O)-CH(CH_3)-C(=O)-OEt$

60. Provide a detailed, stepwise mechanism for the transformation shown below.

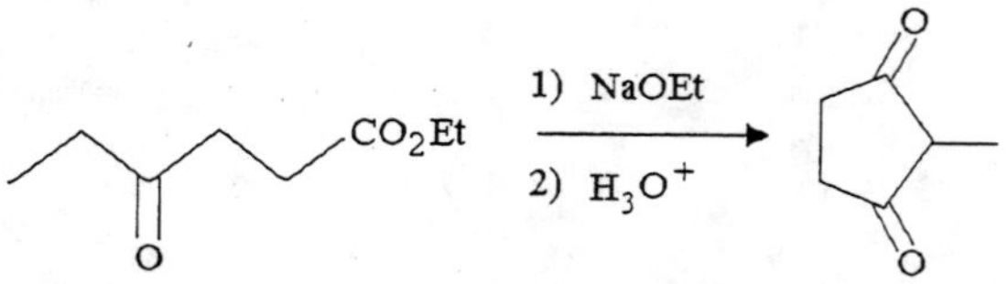

1) NaOEt
2) H_3O^+

Answer:

EtO⁻ H H CO_2Et → H OEt C O → O OEt O →

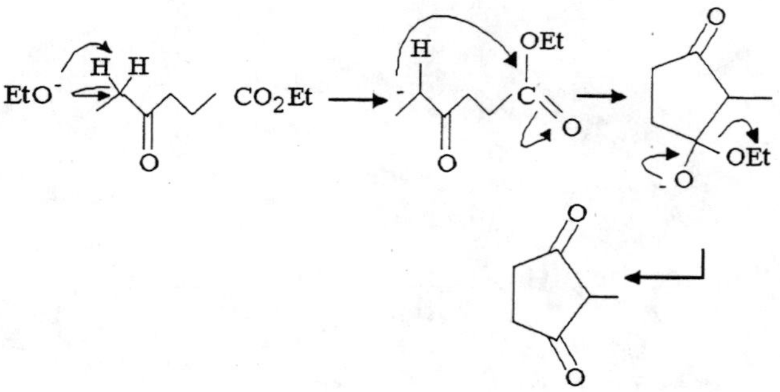

Chapter 19: Carbohydrates

MULTIPLE CHOICE

1. Which of the following represents the general formula of a carbohydrate?

   a. C_nH_{2n+2}
   b. C_nH_{2n}
   c. C_nH_{2n-2}
   d. $C_n(H_2O)_n$
   e. $C_n(HO)_n$

   Answer: d

2. Which of the following best represents the process of photosynthesis?

   a. disproportionation
   b. oxidation/reduction
   c. exothermic
   d. exergonic
   e. elimination

   Answer: b

3. A carbohydrate composed of three to ten sugar molecules is called a(n):

   a. single carbohydrate
   b. disaccharide
   c. oligosaccharide
   d. polysaccharide
   e. monosaccharide

   Answer: c Section: 1

4. Which of the following best describes the sugar D-galactose,

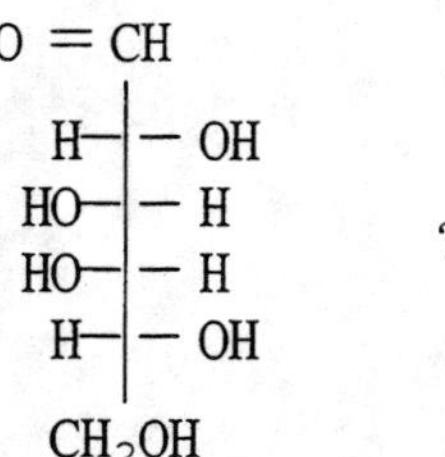

?

   a. D-Aldohexose
   b. D-Ketohexose
   c. L-Aldohexose
   d. L-Ketohexose
   e. D-Aldopentose

   Answer: a Section: 1

5. Which of the following compounds is a D-sugar?

a.
$CH_3C{=}O$
H— — OH
H— — OH
HO— — H
CH_2OH

b.
O = CH
H— — OH
HO— — H
CH_2OH

c.
O = CH
HO— —H
HO— —H
CH_2OH

d.
CH_2OH
HO— —H
O = CH

e.
CH_2OH
H— — OH
O=CH

Answer: d Section: 2

6. What is the relationship between the following compounds?

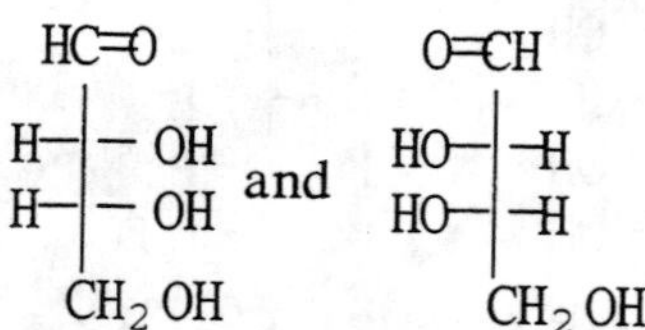

D-erythrose L-erythrose

a. conformational isomers
b. constitutional isomers
c. identical
d. enantiomers
e. diastereomers

Answer: d Section: 3

7. Which of the following corresponds to the definition of an aldopentose?
   I. It is a monosaccharide.
   II. It contains a CHO group.
   III. It is a disaccharide.
   IV. It is an oligosaccharide.

   a. I. and II.
   b. II. and III.
   c. I. and IV.
   d. I. and III.
   e. I., II., and III.

   Answer: a Section: 1

8. At which carbon are the following sugars epimers of each other?

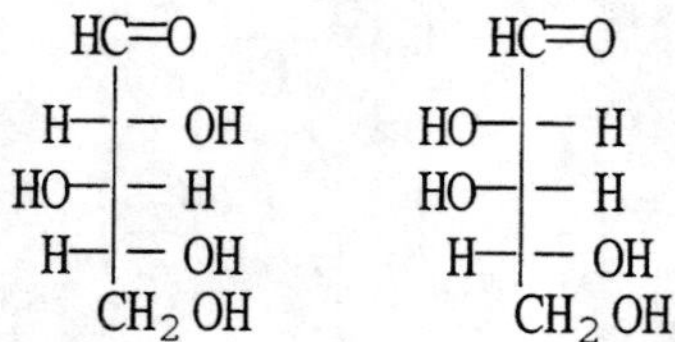

   a. C-1
   b. C-2
   c. C-3
   d. C-4
   e. C-5

   Answer: b Section: 3

9. At which carbon are the following sugars epimers of each other?

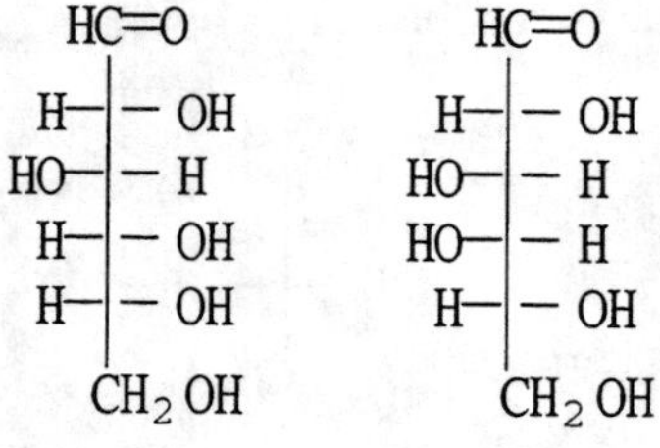

D-Glucose D-Galactose

   a. C-1
   b. C-2
   c. C-3
   d. C-4
   e. C-5

   Answer: d Section: 3

10. At which carbon are the following sugars epimers of each other?

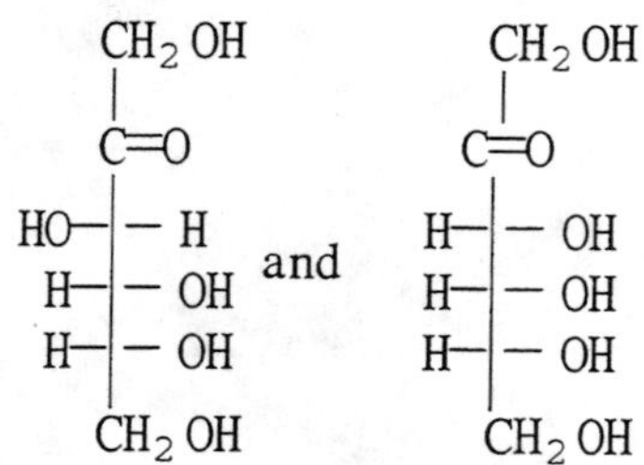

D-Fructose D-Psicose

a. C-1
b. C-2
c. C-3
d. C-4
e. C-5

Answer: c Section: 4

11. How many chirality centers are there in an aldohexose?

a. 2
b. 3
c. 4
d. 5
e. 6

Answer: c Section: 3

12. How many chirality centers are there in a 2-ketohexose?

a. 2
b. 3
c. 4
d. 5
e. 6

Answer: b Section: 3

13. How many stereoisomers are possible for a ketohexose?

a. 2
b. 4
c. 8
d. 16
e. 32

Answer: c Section: 4

14. Which of the following is/are the product(s) obtained from the reduction of D-fructose?

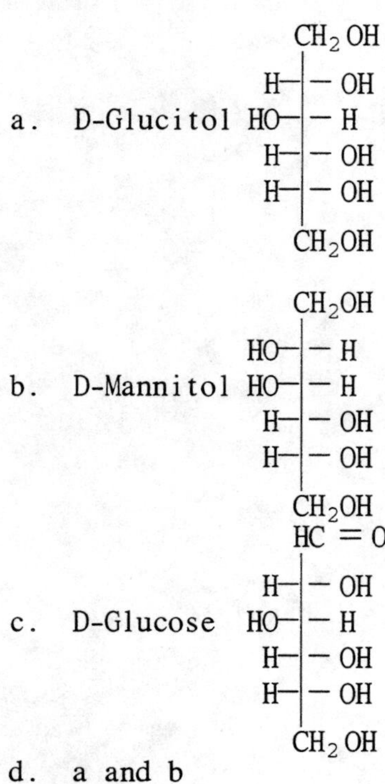

d. a and b
e. a and c

Answer: d Section: 5

15. Which of the following monosaccharides will form D-glucaric acid upon oxidation with nitric acid?

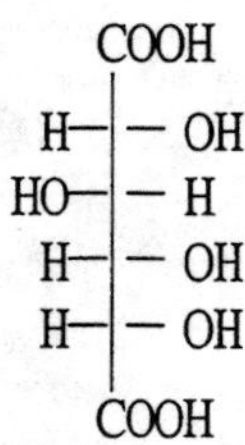

a. D-Glucose

HC = O
H— —OH
HO— —H
H— —OH
H— —OH
CH_2OH

b. L-Glucose

$$\begin{array}{c} CH_2OH \\ H-|-OH \\ H-|-OH \\ HO-|-H \\ H-|-OH \\ HC=O \end{array}$$

c. D-Glucitol

$$\begin{array}{c} CH_2OH \\ H-|-OH \\ HO-|-H \\ H-|-OH \\ H-|-OH \\ CH_2OH \end{array}$$

d. a and b
e. a and c

Answer: e Section: 5

16. Which of the following pairs of monosaccharides will form the <u>same</u> osazone when reacted with phenylhydrazine?

a. D-Glucose and L-Glucose
b. D-Glucose and D-Gulose
c. D-Glucose and D-Fructose
d. D-Glucose and D-Allose
e. D-Glucose and D-Galactose

Answer: c Section: 6

17. Which monosaccharides are formed in the Kiliani-Fischer synthesis starting with D-xylose?

$$\begin{array}{c} HC=O \\ H-|-OH \\ HO-|-H \\ H-|-OH \\ CH_2OH \end{array}$$

a. D-Glucose and D-Mannose
b. D-Gulose and D-Idose
c. D-Galactose and D-Talose
d. D-Allose and D-Altrose
e. D-Ribose and D-Arabinose

Answer: b Section: 7

18. Which two monosaccharides can be degraded by the Ruff Degradation to D-arabinose?

```
     HC=O
      |
  HO—|— H
   H—|— OH
   H—|— OH
     CH2OH
```

a. D-Allose and D-Altrose
b. D-Gulose and D-Idose
c. D-Galactose and D-Talose
d. D-Erythrose and D-Threose
e. D-Glucose and D-Mannose

Answer: e Section: 8

19. An aqueous solution of glucose behaves as an aldehyde because:

a. it is hydrolyzed by water to the free aldehyde.
b. it is a ketone, but is in equilibrium with the aldehyde form.
c. glucose is actually a cyclic aldehyde.
d. its cyclic hemiacetal, the predominant form, is in equilibrium with the free aldehyde form.
e. it can be oxidized with periodic acid.

Answer: d Section: 10

20. C-2 epimeric aldohexoses give the same:

a. osazone.
b. aldaric acid when oxidized with HNO_3.
c. products with the Kiliani-Fischer synthesis.
d. a and b
e. b and c

Answer: a Section: 6

21. Which of the following statements best describes the meaning of <u>mutorotation</u>?

a. A rapid exchange between the α and β forms
b. A rapid exchange between the D and L forms
c. A slow exchange between hydrogen and deuterated hydrogen
d. A slow change in optical rotation to reach an equilibrium value
e. A slow change in absolute configuration to reach an equilibrium value

Answer: d Section: 10

22. Which of the following is the cyclic hemiacetal formed from 4-hydroxyheptanal?

a. CH_3–(ring O)–OH

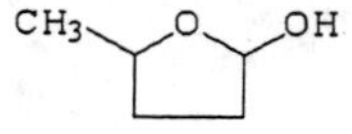

b. CH_3CH_2–(ring O)–OH

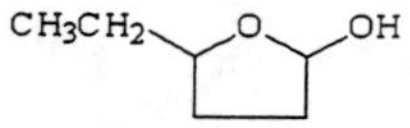

c. $CH_3CH_2CH_2$–(ring O)–OH

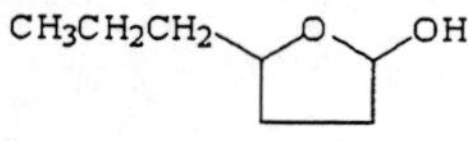

d. OH, $CH_3CH_2CH_2$–

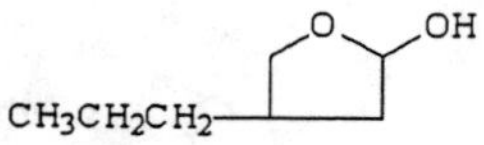

e. $CH_3CH_2CH_2$–(ring O), OH

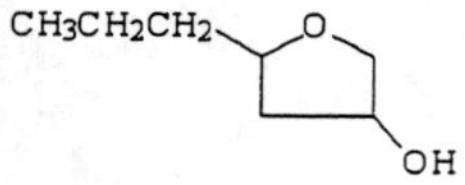

Answer: c Section: 10

23. Which of the following is the Haworth projection of β-D-galactose?

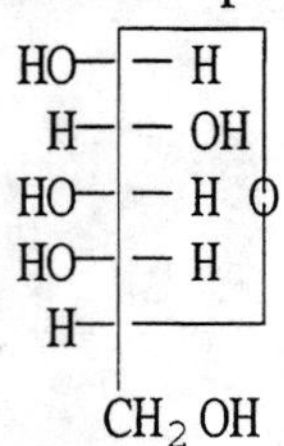

a.

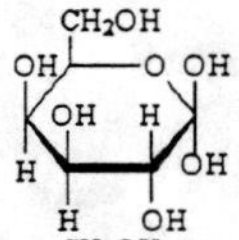

b.

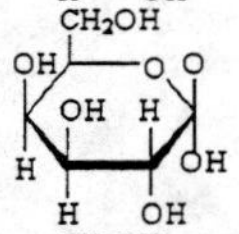

c.

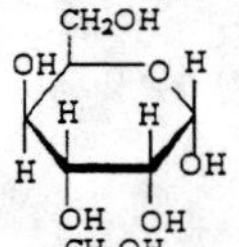

d.

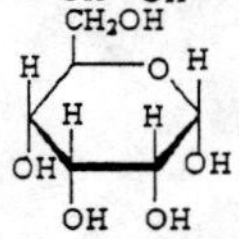

e.

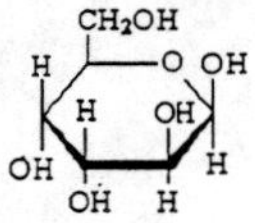

Answer: a Section: 10

24. Which of the following statements best describes the meaning of a <u>glycoside</u>?

a. It is the mirror image of a sugar.
b. It is the hemiacetal of a sugar.
c. It is the acetal of a sugar.
d. It is the enantiomer of a sugar.
e. It is the enol-keto form of a sugar.

Answer: c Section: 12

25. Which of the following is methyl-α-D-glucoside?

a.

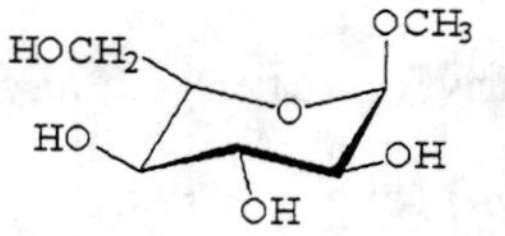

b.

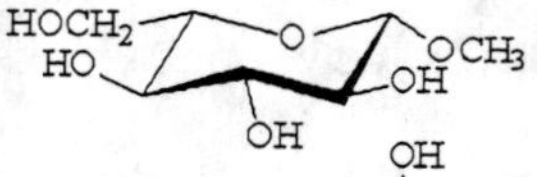

c.

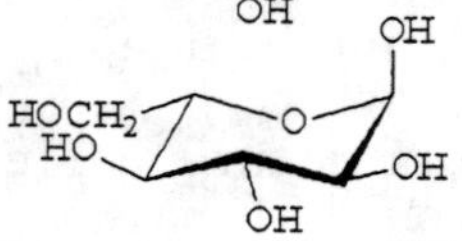

d.

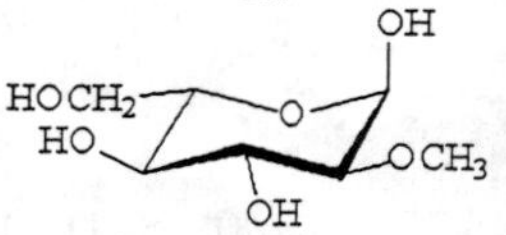

e.

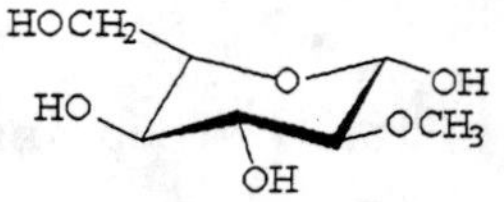

Answer: a Section: 12

26. Which of the following compounds is a non-reducing sugar?

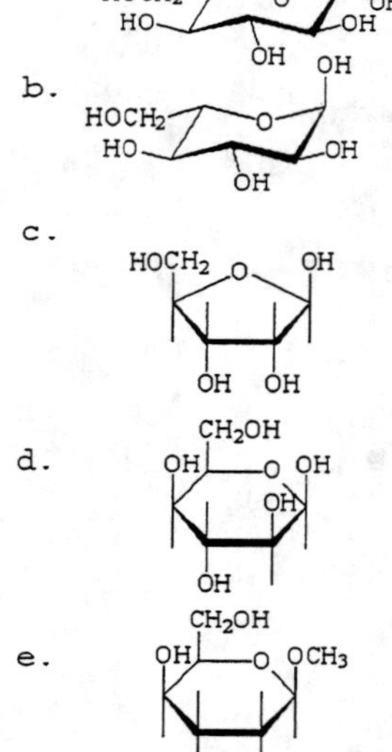

Answer: e Section: 12

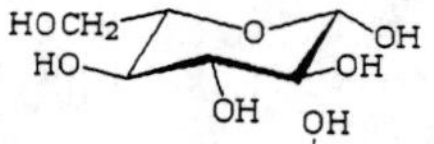

27. Which of the following statements best describes the difference between amylose and amylopectin?

    a. Amylose is a branched polysaccharide while amylopectin is a chain polysaccharide.
    b. Amylose is a straight-chain polysaccharide while amylopectin is a branched polysaccharide.
    c. Amylose contains α-1,6-glycosidic linkage which amylopectin does not contain.
    d. Amylose is composed of thousands of D-glucose units while amylopectin is composed of thousands of D-galactose units.
    e. Amylose is one of the largest molecules found in nature while amylopectin is one of the smallest molecules found in nature.

    Answer: b Section: 4

28. Which of the following statemwnts best describes the main structural difference between cellulose and chitin?

    a. The C-2 carbon in chitin contains an N-acetyl amino group instead of an OH group.
    b. The C-2 carbon in cellulose contains an N-acetyl amino group instead of an OH group.
    c. Chitin is a β-1,4-glycosidic linkage while cellulose is a α-1,4-glycosidic linkage.
    d. Chitin is a α-1,4-glycosidic linkage while cellulose is a α-1,4-glycosidic linkage.
    e. Chitin is a branched polysaccharide while cellulose is a chain polysaccharide.

    Answer: a Section: 14

29. Which of the following is the Haworth projection of a α-D-tagatose?

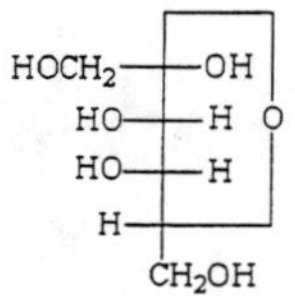

a.

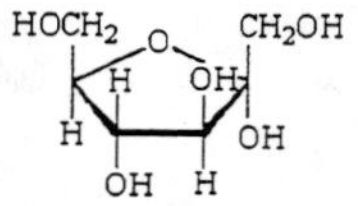

b.

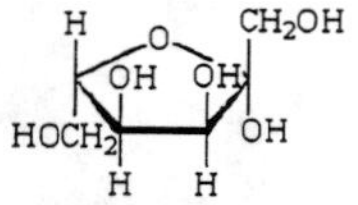

c.

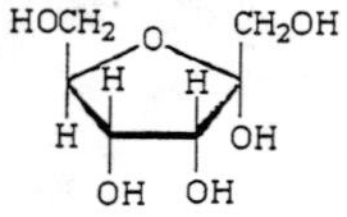

d.

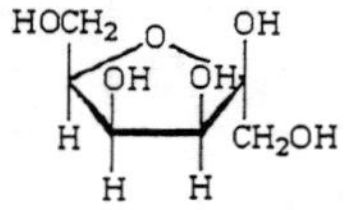

e.

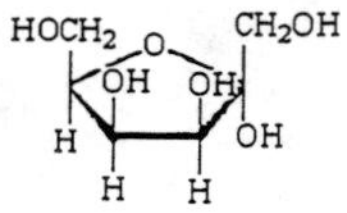

Answer: e

30. Which of the following blood types is known as the universal donor?

a. A
b. B
c. AB
d. O
e. ABO

Answer: d Section: 16

31. Which of the following blood types is known as the universal acceptor?

a. A
b. B
c. AB
d. O
e. ABO

Answer: c Section: 16

32. Which of the following is true about sucrose?

    a. It hydrolyzes to fructose and glucose.
    b. It is a reducing sugar.
    c. It is a monosaccharide.
    d. It undergoes mutorotation in water.

    Answer: a Section: 13

SHORT ANSWER

33. Is the following structure of glucose a D or L configuration? Explain.

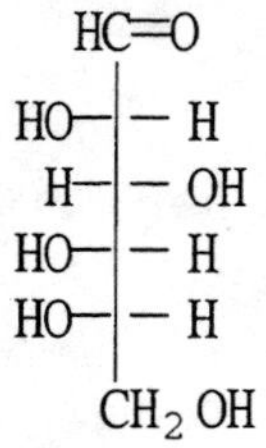

```
     HC=O
      |
HO—   | — H
 H—   | — OH
HO—   | — H
HO—   | — H
      |
     CH2 OH
```

Answer:
This is L-glucose because the OH group attached to the bottom-most stereogenic center is on the left.

Section: 2

34. When D-tagatose is added to a basic aqueous solution, an equilibrium mixture of three monosaccharides is obtained. Identify the monosaccharides.

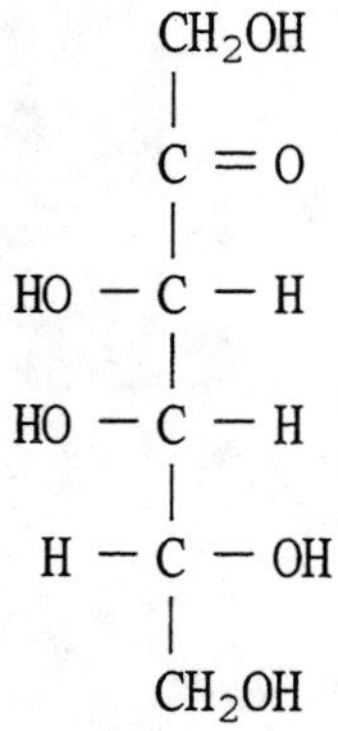

```
       CH2OH
        |
        C = O
        |
  HO  — C — H
        |
  HO  — C — H
        |
   H  — C — OH
        |
       CH2OH
```

D-tagatose

Answer:

```
   CH2OH              CHOH              H-C=O            H-C=O
     |                 ||                 |                |
   C = O             C - OH         HO - C - H        H - C - OH
     |                 |                  |                |
HO - C - H        HO - C - H        HO - C - H       HO - C - H
     |      ⊖OH        |      ⊖OH         |       +        |
HO - C - H  ⇌    HO - C - H   ⇌    HO - C - H       HO - C - H
     |      H2O        |      H2O         |                |
 H - C - OH        H - C - OH         H - | - OH       H - | - OH
     |                 |                  |                |
   CH2OH             CH2OH              CH2OH            CH2OH

D-tagatose       enol form of        D-talose       D-galactose
                  D-tagatose
```

Therefore the three monosaccharides are: D-tagatose, D-galactose, and D-talose.

Section: 5

35. Why is β-D-glucopyranose more stable in nature than α-D-glucopyranose?

Answer:
The five hydroxy groups in β-D-glucopyranose are located on the equatorial position of the chair conformation. This arrangement is more stable than the chair conformation in which one of the hydroxy groups of -D-glucopyranose is in the axial position.

Section: 11

36. What is the main structural difference between amylose and cellulose?

Answer:
The main difference is that amylose is composed of glucose units joined by α-1,4-glycosidic linkages while cellulose is composed of glucose units joined by β-1,4-glycosidic linkages.

Section: 14

Chapter 19: Carbohydrates

MULTIPLE CHOICE

37. All chiral D-sugars rotate plane-polarized light:

    a. clockwise.
    b. counterclockwise.
    c. +20.0° .
    d. in a direction that cannot be predicted but must be determined experimentally.
    e. since they are optically inactive.

    Answer: d Section: 2

SHORT ANSWER

38. Draw the Fischer projection for the open-chain form of D-glucose.

    Answer:

```
      CHO
       |
   H --|-- OH
  HO --|-- H
   H --|-- OH
   H --|-- OH
       |
     CH2OH
```

    Section: 3

39. Draw the Fischer projection for the open-chain form of D-fructose.

    Answer:

```
     CH2OH
       |
     C = O
       |
  HO --|-- H
   H --|-- OH
   H --|-- OH
       |
     CH2OH
```

    Section: 4

40. Draw the Fischer projection for the open-chain form of D-erythrose.

    Answer:

```
      CHO
       |
   H --|-- OH
   H --|-- OH
       |
     CH2OH
```

    Section: 5

MULTIPLE CHOICE

41. In solution, glucose exists as:

    a. the open-chain form only.
    b. the cyclic hemiacetal form only.
    c. the cyclic acetal form only.
    d. an equilibrium mixture of the open-chain form and cyclic acetal forms.
    e. an equilibrium mixture of the open-chain form and cyclic hemiacetal forms.

    Answer: e Section: 10

42. When a monosaccharide reacts to give the pyranose form from its open-chain form, how many distinct pyranose forms are possible?

    a. 1
    b. 2
    c. 2^n, where n is the number of carbons present
    d. 4n + 2, where n is the number of carbons present
    e. 4

    Answer: b Section: 10

43. Six-membered cyclic hemiacetals and five-membered cyclic hemiacetals are called respectively:

    a. mannoses and xyloses.
    b. maltoses and arabinoses.
    c. pyranoses and fructoses.
    d. glyoses and fructoses.
    e. none of the above

    Answer: e Section: 10

44. Anomers of D-glucopyranose differ in their stereochemistry at:

    a. C1.
    b. C2.
    c. C3.
    d. C4.
    e. C5.

    Answer: a Section: 10

45. A pyranose with the hydroxyl group on the anomeric carbon pointing up in the Haworth structure is designated:

    a. a'.
    b. b'.
    c. α.
    d. β.
    e. γ.

    Answer: d Section: 10

SHORT ANSWER

46. When pure α-D-glucopyranose is dissolved in water, the optical rotation of the resulting solution changes over a period of time. What is the name of this phenomenon and why does it occur?

    Answer:
    The phenomenon is called mutarotation. In solution, α-D-glucopyranose is converted to the equilbrium mixture of α and β anomers via the open-chain form.
    Section: 10

47. Draw the Haworth structure of α-D-ribofuranose

    Answer:

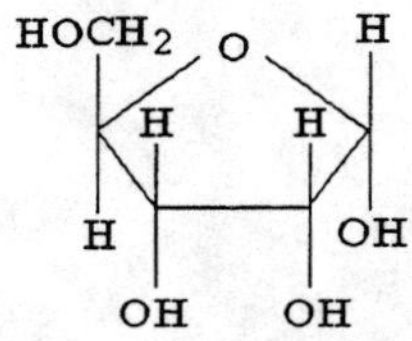

    Section: 10

48. Draw the Haworth structure of β-D-glucopyranose

    Answer:

CH$_2$OH
H
O
OH
H
H
OH
HO
H
H
OH

    Section: 10

49. Draw the more stable chair conformer of α-D-glucopyranose.

Answer:

H CH$_2$OH O
HO H H
HO H OH H
H OH

Section: 10

MULTIPLE CHOICE

50. Reduction of a 2-ketohexose with $NaBH_4$ yields:

a. a single aldohexose.
b. a mixture of acetals.
c. a mixture of alditols.
d. a mixture of cyclic hemiacetals.
e. a single pyranose.

Answer: c Section: 5

SHORT ANSWER

51. Provide the Fischer projection of the open-chain form of the aldonic acid which results when L-glucose is treated with bromine water.

Answer:

COOH
HO— —H
H— —OH
HO— —H
HO— —H
CH_2OH

Section: 5

MULTIPLE CHOICE

52. When D-threose is treated with $NaBH_4$,:

a. a 70:30 mixture of enantiomeric alditols results.
b. a 50:50 mixture of enantiomeric alditols results.
c. a meso alditol is produced.
d. the product mixture contains two diastereomeric alditols.
e. an optically active alditol is produced.

Answer: e Section: 5

Chapter 19: Carbohydrates

SHORT ANSWER

53. An optically active D-aldopentose produced an optically inactive aldaric acid upon treatment with HNO_3. When this aldopentose was subjected to a Ruff degradation, a D-aldotetrose was generated. This aldotetrose gave an optically inactive alditol upon reduction with $NaBH_4$. Use these data to provide the structure of the starting D-aldopentose.

Answer:

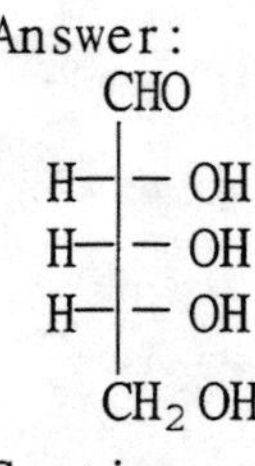

Section: 5

54. An optically active D-aldopentose produced an optically active alditol upon treatment with $NaBH_4$. When this aldopentose was subjected to a Ruff degradation, a D-aldotetrose was generated. This aldotetrose gave an optically active aldaric acid upon oxidation with HNO_3. Use these data to provide the structure of the starting D-aldopentose.

Answer:

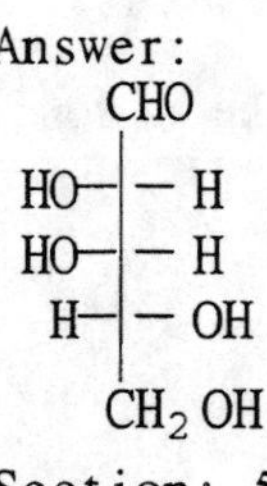

Section: 5

55. An optically active D-aldopentose produced an optically inactive alditol upon treatment with $NaBH_4$. When this aldopentose was subjected to a Ruff degradation, a D-aldotetrose was generated. This aldotetrose gave an optically active aldaric acid upon oxidation with HNO_3. Use these data to provide the structure of the starting D-aldopentose.

Answer:

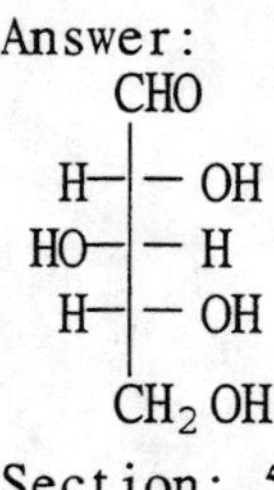

Section: 5

56. Provide the structure of the product which results when D-ribose is treated with bromine water.

Answer:

```
      COOH
   H—|— OH
   H—|— OH
   H—|— OH
      CH2OH
```

Section: 5

MULTIPLE CHOICE

57. Which of the following would give a positive Tollen's test?

a. α-D-glucopyranose
b. methyl β-D-glucopyranoside
c. sucrose
d. methyl α-D-ribofuranoside
e. none of the above

Answer: a Section: 12

SHORT ANSWER

58. Under what conditions is the methyl glycoside of galactose prepared?

Answer: CH_3OH, H^+ (nonaqueous) Section: 12

MULTIPLE CHOICE

59. Two D-aldopentoses give the same D-aldotetrose upon Ruff degradation. The two aldopentoses are:

a. enantiomers.
b. meso compounds.
c. C2 epimers.
d. C4 epimers.
e. D-glucose and D-galactose.

Answer: c Section: 8

SHORT ANSWER

60. What nucleophilic carbon species is used in the Kiliani-Fischer synthesis?

Answer: cyanide Section: 7

Chapter 20: Amino Acids, Peptides, and Proteins

MULTIPLE CHOICE

1. The monomeric units that make up peptides and protein polymers are:

   a. Nucleic acids
   b. Amino acids
   c. Oligosaccharides
   d. Amylopectins
   e. Celluloses

   Answer: b

2. Which labeled bond in the following molecule is known as the peptide bond?

```
                          O                 O
                          ||                ||
               1      2   ||  3    4        ||
     — NH —— CH —— C —— N —— CH —— C —
              5|           |      |
               R           H      R
```

   a. 1
   b. 2
   c. 3
   d. 4
   e. 5

   Answer: c Section: 7

3. What are enzymes?

   a. Saccharides that catalyze chemical reactions
   b. Nucleic acids that catalyze chemical reactions
   c. Unsaturated fats that catalyze chemical reactions
   d. DNA molecules that catalyze chemical reactions
   e. Proteins that catalyze chemical reactions

   Answer: e

4. Which of the following molecules is the skeletal structure of an amino acid?

a. 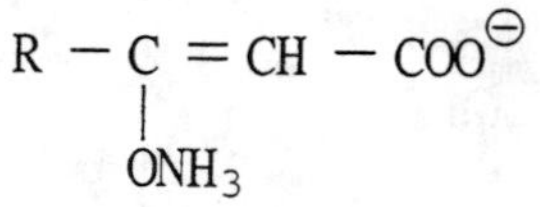 $R - C = CH - COO^{\ominus}$
   |
   ONH_3

b. $R - CH - CH_2 - COO^{\ominus}$
   |
   NH_3
   $\oplus$

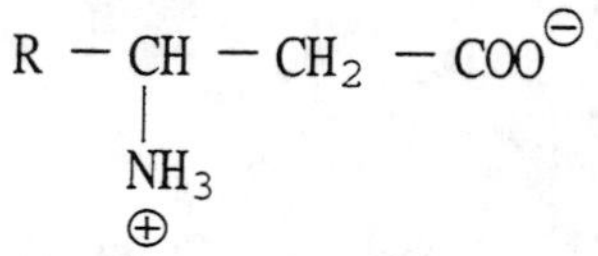

c. $R - CH - COO^{\ominus}$
   |
   NH_3
   $\oplus$

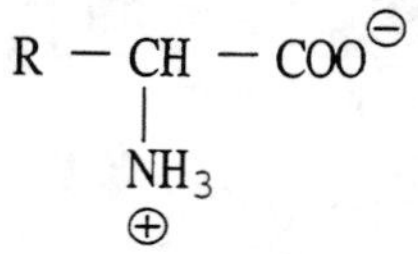

d. 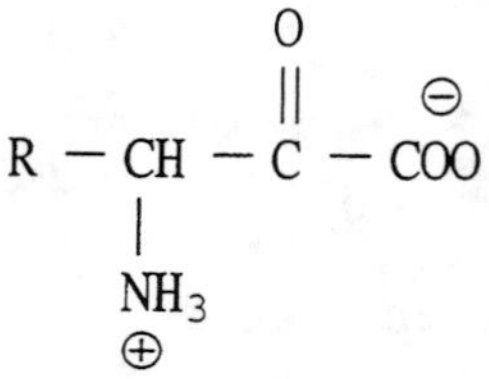 O
   || $\ominus$
   $R - CH - C - COO$
   |
   NH_3
   $\oplus$

e. 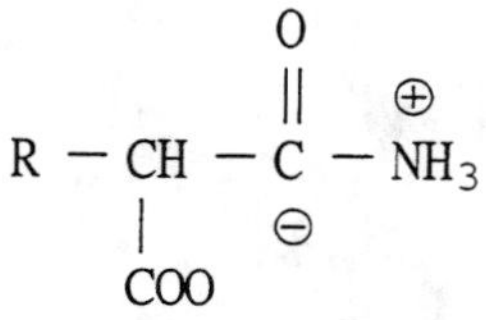 O
   || $\oplus$
   $R - CH - C - NH_3$
   | $\ominus$
   COO

Answer: c Section: 1

5. Which of the following amino acids has an aliphatic R group?

a. Serine
b. Cysteine
c. Asparagine
d. Tyrosine
e. Leucine

Answer: e Section: 1

6. Which of the following amino acids has an aromatic R group?

a. Serine
b. Cysteine
c. Asparagine
d. Tyrosine
e. Leucine

Answer: d Section: 1

7. Which of the following amino acids has a sulfur in the R group?

   a. Serine
   b. Cysteine
   c. Asparagine
   d. Tyrosine
   e. Leucine

   Answer: b Section: 1

8. Which of the following amino acids has a heterocyclic R group?

   a. Glycine
   b. Threonine
   c. Proline
   d. Aspartic acid
   e. Arginine

   Answer: c Section: 1

9. The α-carbon of all the amino acids is a stereogenic center except for:

   a. Glycine
   b. Threonine
   c. Proline
   d. Aspartic acid
   e. Arginine

   Answer: a Section: 1

10. Which of the following is an L-amino acid?

   a. Fischer projection: top $COO^{\ominus}$; left $\overset{\oplus}{H_3N}$; right H; bottom H

   $$\begin{array}{ccc} & COO^{\ominus} & \\ \overset{\oplus}{H_3N} & - | - & H \\ & H & \end{array}$$

   b. Fischer projection: top $COO^{\ominus}$; left $\overset{\oplus}{H_3N}$; right H; bottom R

   $$\begin{array}{ccc} & COO^{\ominus} & \\ \overset{\oplus}{H_3N} & - | - & H \\ & R & \end{array}$$

   c. Fischer projection: top $COO^{\ominus}$; left H; right $\overset{\oplus}{NH_3}$; bottom R

   $$\begin{array}{ccc} & COO^{\ominus} & \\ H & - | - & \overset{\oplus}{NH_3} \\ & R & \end{array}$$

d. 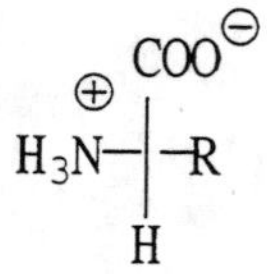 $H_3\overset{\oplus}{N}-C(COO^{\ominus})(H)-R$

e. 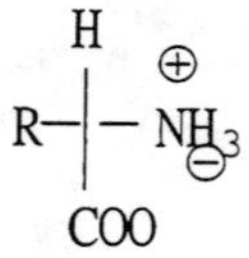 $R-C(H)(COO^{\ominus})-\overset{\oplus}{N}H_3$

Answer: b Section: 2

11. Which of the following best represents the structure of an amino acid in basic solution (pH=11)?

a. 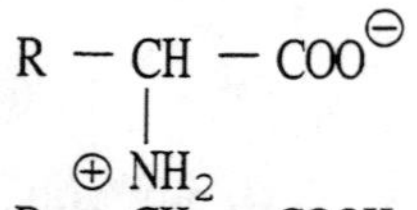 $R-CH(\overset{\oplus}{N}H_2)-COO^{\ominus}$

b. 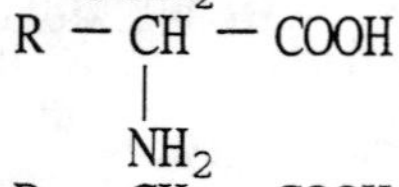 $R-CH(NH_2)-COOH$

c. 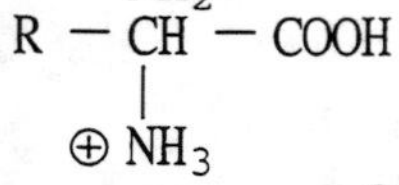 $R-CH(\overset{\oplus}{N}H_3)-COOH$

d. 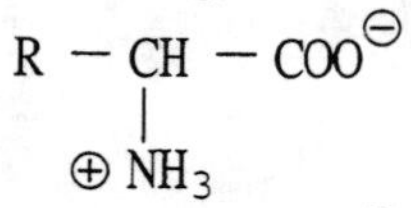 $R-CH(\overset{\oplus}{N}H_3)-COO^{\ominus}$

e. 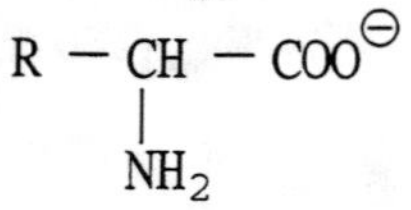 $R-CH(NH_2)-COO^{\ominus}$

Answer: e Section: 3

12. Which of the following is a <u>zwitterion</u>?

a. $R-CH(\overset{\oplus}{N}H_2)-COO^{\ominus}$

b. $R-CH(NH_2)-COOH$

c. $R-CH(\overset{\oplus}{N}H_3)-COOH$

d. $R-CH(\overset{\oplus}{N}H_3)-COO^{\ominus}$

e. $R-CH(NH_2)-COO^{\ominus}$

Answer: d Section: 3

13. Which of the following amino acids has a side chain with an ionizable proton and can exist in four different forms, depending on the pH of the solution?

a. Proline (pyrrolidine ring, $N^{\oplus}H_2$, ring carbon bearing $COO^{\ominus}$)

b. Serine $HO-CH_2-CH(\overset{\oplus}{N}H_3)-COO^{\ominus}$

c. Histidine (imidazole ring: HN, N–H) $-CH_2-CH(\overset{\oplus}{N}H_3)-COO^{\ominus}$

d. Alanine $CH_3-CH(\overset{\oplus}{N}H_3)-COO^{\ominus}$

e. Glycine $H-CH(\overset{\oplus}{N}H_3)-COO^{\ominus}$

Answer: c Section: 3

14. What is the pI of glycine? The structure and pKa values are shown below.

a. 7.26
b. 5.97
c. 3.63
d. 7.50
e. 11.94

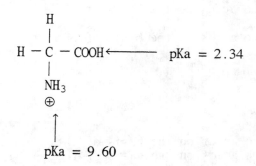

H
|
H — C — COOH ← pKa = 2.34
|
$NH_3^{\oplus}$
↑
pKa = 9.60

Answer: b Section: 4

15. What is the pI of arginine? The structure and pKa values are shown below.

a. 10.76
b. 7.90
c. 5.61
d. 7.33
e. 9.67

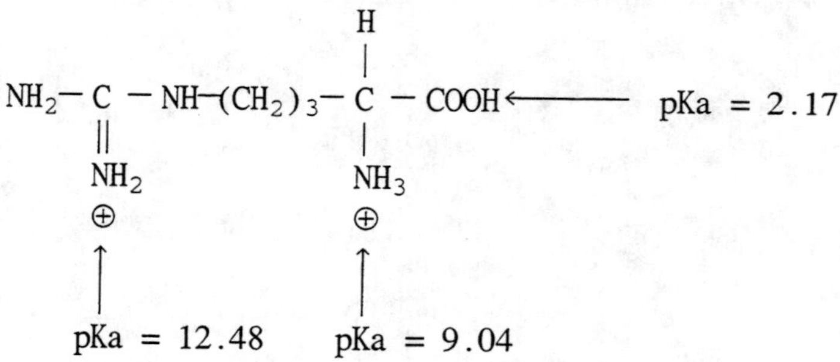

Answer: a Section: 4

16. What is <u>electrophoresis</u>?

    a. A technique that separates amino acids on the basis of their polarity
    b. A technique that separates amino acids on the basis of their solubility in water
    c. A technique that separates amino acids on the basis of their pI values
    d. A technique that separates amino acids on the basis of pKa of α-COOH values
    e. A technique that separates amino acids on the basis of pKa of α-$^{+}NH_3$ values

    Answer: c Section: 5

17. What is <u>thin-layer chromatography</u>?

    a. A technique that separates amino acids on the basis of their polarity
    b. A technique that separates amino acids on the basis of their solubility in water
    c. A technique that separates amino acids on the basis of their pI values
    d. A technique that separates amino acids on the basis of pKa of α-COOH values
    e. A technique that separates amino acids on the basis of pKa of α-$^{+}NH_3$ values

    Answer: a Section: 5

18. Which of the following amino acids will be closest to the origin when separated by thin-layer chromatography?

    a. Serine $HO-CH_2-CH(^{\oplus}NH_3)-COO^{\ominus}$

    b. Alanine $CH_3-CH(^{\oplus}NH_3)-COO^{\ominus}$

    c. Cysteine $HS-CH_2-CH(^{\oplus}NH_3)-COO^{\ominus}$

    d. Aspartate $^{\ominus}OOCCH_2-CH(^{\oplus}NH_3)COO^{\ominus}$

    e. Phenylalanine $C_6H_5-CH_2-CH(^{\oplus}NH_3)-COOH$

    Answer: d Section: 5

19. Which of the following amino acids will be retained longest in the column when separated by cation-exchange chromatography?

a. Serine $HO-CH_2-CH(NH_3^{\oplus})-COO^{\ominus}$

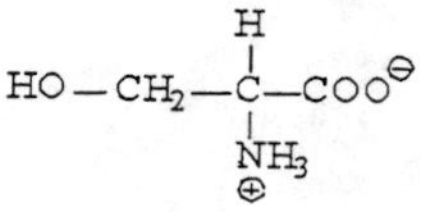

b. Alanine $CH_3-CH(NH_3^{\oplus})-COO^{\ominus}$

c. Cysteine $HS-CH_2-CH(NH_3^{\oplus})-COO^{\ominus}$

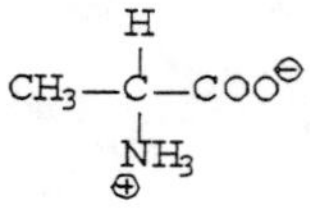

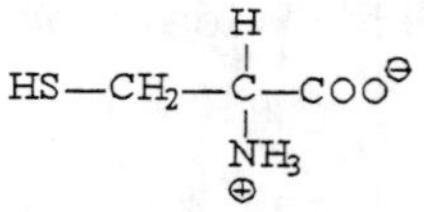

d. Lysine $H_3\overset{\oplus}{N}-CH_2CH_2CH_2CH_2CH(NH_3^{\oplus})-COOH$

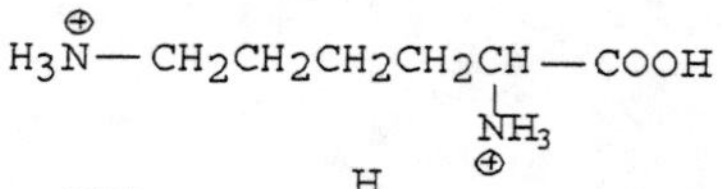

e. Phenylalanine $C_6H_5-CH_2-CH(NH_3^{\oplus})-COOH$

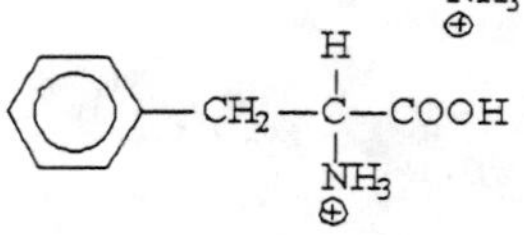

Answer: d Section: 5

20. What is the major organic product of the following reaction?

$$CH_3-\overset{O}{\overset{\|}{C}}-OH \xrightarrow[2.\ H_3O^+]{1.\ Br_2/P} \xrightarrow[NH_3]{excess} ?$$

a. $CH_3\overset{O}{\overset{\|}{C}}NH_2$

b. $CH_2(NH_3^{\oplus})COO^{\ominus}$

c. $CH_2(Br)\overset{O}{\overset{\|}{C}}NH_2$

d. $CH_2(NH_3^{\oplus})\overset{O}{\overset{\|}{C}}NH_2$

e. $CH_3C(Br)=NH$

Answer: b Section: 6

21. Which of the following is an acceptable name for the peptide sequence shown below?

Glu-Glu-His-Val-Cys

a. 1,2-Diglutamylhistidylvalylcysteine
b. Glutamylglutamylhistidylvalylcysteine
c. 1-Glutamyl-2-glutamylhistidylvalylcysteine
d. Diglutamylhistidylvalylcysteine
e. Biglutamylhistidylvalylcysteine

Answer: b Section: 7

22. What is the product obtained from the mild oxidation of the thiol shown below?

$$\underset{\text{Thiol}}{2\ \text{R–SH}} \xrightarrow[\text{oxidation}]{\text{mild}}$$

a. R-S-R
b. R-OH
c. R-S-S-R
d. R-OO-R
e. R-S-OH

Answer: c Section: 7

23. Which of the following structures can polypeptides have?

a. Primary structure
b. Secondary structure
c. Tertiary structure
d. Quaternary structure
e. All of the above

Answer: e Section: 11

24. Which of the following is the first step in the determination of the primary structure of proteins?

a. Determining the number and kind of amino acids in the peptide
b. Reducing the disulfide bridges in the protein
c. Protecting the N-terminal of the peptide
d. Protecting the C-terminal of the peptide
e. Hydrolyzing the protein with dilute acid

Answer: b Section: 12

25. Phenyl isothiocyanate (PITC) is commonly known as Edman's reagent. Edman's reagent is widely used to identify which of the following?

    a. The number and kind of amino acids in the protein
    b. The disulfide bridges in the protein
    c. C-terminal amino acid of the protein
    d. N-terminal amino acid of the protein
    e. The number of peptide bonds in the protein

    Answer: d Section: 12

26. Carboxypeptidase is an enzyme used in the determination of the primary structure of a protein. Specifically, which of the following does it identify?

    a. The number and kind of amino acids in the protein
    b. The disulfide bridges in the protein
    c. C-terminal amino acid of the protein
    d. N-terminal amino acid of the protein
    e. The number of peptide bonds in the protein

    Answer: c Section: 12

27. A Decapeptide undergoes partial hydrolysis to give peptides whose amino acids are shown below. Reaction of the decapeptide with Edman's reagent releases PTH-leu. What is the sequence of the decapeptide?

    (1) Pro, Ser
    (2) Gly, Gly
    (3) Met, Ala, Leu
    (4) Gly, Ala
    (5) Glu, Ser, Val, Pro
    (6) Glu, Pro, Gly
    (7) Met, Leu
    (8) His, Val

    a. Leu-Met-Ala-Gly-Gly-Glu-Pro-Ser-His-Val
    b. Leu-Met-Ala-Gly-Gly-Glu-Ser-Pro-Val-His
    c. Leu-Ala-Met-Gly-Gly-Glu-Pro-Ser-Val-His
    d. Leu-Met-Gly-Ala-Gly-Glu-Pro-Ser-Val-His
    e. Leu-Met-Ala-Gly-Gly-Glu-Pro-Ser-Val-His

    Answer: e Section: 12

28. Which of the following may characterize the <u>secondary structure</u> of proteins?

    a. Conformation of the protein backbone
    b. α-Helix
    c. Parallel β-pleated sheet
    d. Anti-parallel β-pleated sheet
    e. All of the above

    Answer: e Section: 13

29. Which of the following is the <u>quaternary structure</u> of proteins concerned with?

   a. Sequence of amino acids in the peptide chain
   b. Description of the way the peptide chains are arranged with respect to each other
   c. Location of the disulfide bridges in the peptide chain
   d. Conformation of the protein backbone
   e. Three-dimensional arrangement of all atoms in the protein

   Answer: b Section: 15

30. Which of the following protein structures does <u>denaturation</u> destroy?

   a. Primary and secondary structures
   b. Secondary and tertiary structures
   c. Tertiary and quaternary structures
   d. Secondary, tertiary, and quaternary structures

   Answer: d Section: 16

31. Which of the following are capable of denaturing proteins?

   a. Organic solvents
   b. Detergents
   c. Extreme pH
   d. Heat
   e. All of the above

   Answer: e Section: 16

SHORT ANSWER

32. Explain what essential amino acids are and list all of them.

   Answer:
   An essential amino acid is one that humans must obtain from their diet because it cannot be synthesized at all or in sufficient quantity in the body. They are valine, leucine, isoleucine, threonine, methionine, lysine, arginine, phenylalanine, histidine, and tryptophan.

   Section: 1

33. In nature, which is the most commonly found isomer, L-amino acids or D-amino acids?

   Answer:
   Unlike monosaccharides where the D-isomers are more common, most amino acids found in nature have the L-configuration.

   Section: 2

34. Draw the form in which glutamate exists at pH = 0.

Answer:
At pH = 0, groups will be in their acidic form:

$$HO-\overset{\overset{O}{\|}}{C}-CH_2CH_2-\underset{\underset{\underset{\oplus}{NH_3}}{|}}{CH}-\overset{\overset{O}{\|}}{C}-OH$$

Section: 3

35. Define what is meant by isoelectric point (pI) and give an example.

Answer:
Isoelectric point is the pH of an amino acid at which it has no net electric charge. For amino acids that have no ionizable side chain, the pI value is the average of its two pKa's. If the amino acid has an ionizable side chain, the pI value is the average of the pKa's of similarly ionizable groups.

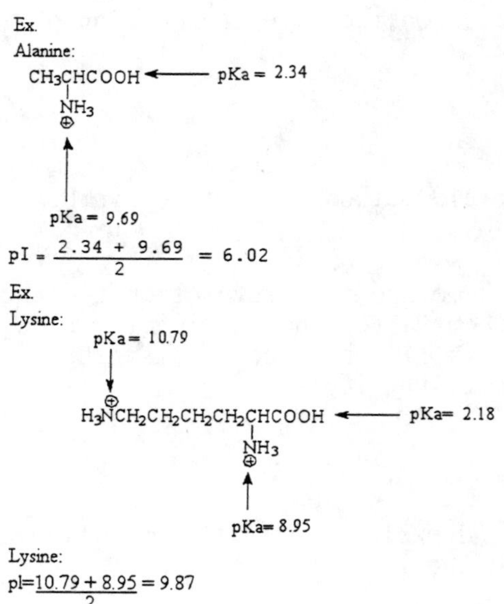

Section: 4

36. Rate the following amino acids in decreasing order of migration towards the cathode (negative electrode) when separated by electrophoresis in a solution of pH= 7.3.

I. Lysine: pI= 9.87
II. Alanine: pI= 6.02
III. Aspartate: pI= 5.95

Answer: I > II > III Section: 5

37. What is an amino acid analyzer?

Answer:
An instrument that automates ion-exchange chromatography. After the amino acids are separated, ninhydrin is added to each fraction. The concentration of the amino acid can be determined by measuring the absorbance at 570 nm. In this way, the identity and the relative amount of each amino acid in a protein can be determined.

Section: 5

38. Explain what is meant by kinetic resolution and give an example.

Answer:
A separation technique in which an enzyme (such as pig kidney aminoacylase) is used to separate a racemic mixture of amino acids into its enantiomers. The separation depends on the rate of reaction of the enzyme with the two enantiomers.

Section: 6

39. Explain the importance of the Merrifield method in peptide synthesis.

Answer:
In 1969, Bruce Merrifield developed a method that revolutionized the synthesis of peptides by producing them faster and in higher yields. This method became known as automated solid-phase peptide synthesis. The method won him the Nobel Prize in chemistry in 1984.

Section: 10

40. Describe what a β-pleated sheet is and explain the difference between parallel and anti-parallel β-pleated sheets.

Answer:
A β-pleated sheet is a type of secondary structure of a protein in which the polypeptide backbone is extended in a zigzag structure. In parallel sheets, adjacent chains run in the same direction, whereas adjacent chains run in opposite directions in anti-parallel sheets.

Section: 13

MULTIPLE CHOICE

41. Which of the following terms best describes the side chain of valine?

a. acidic
b. basic
c. charged, polar
d. uncharged, polar
e. nonpolar

Answer: e Section: 1

SHORT ANSWER

42. Provide the Fischer projection of L-aspartic acid.

Answer:

COOH
H_2N—|—H
CH_2-COOH

Section: 1

43. Draw the form of L-lysine which is present at biological pH.

Answer:

COO^-
H_3N^+—|—H
$(CH_2)_4$-NH_3^+

Section: 1

44. Draw the form of L-tryptophan which is present at biological pH.

Answer:

COO^-
$H_3\overset{+}{N}$—|—H
CH_2
N
H

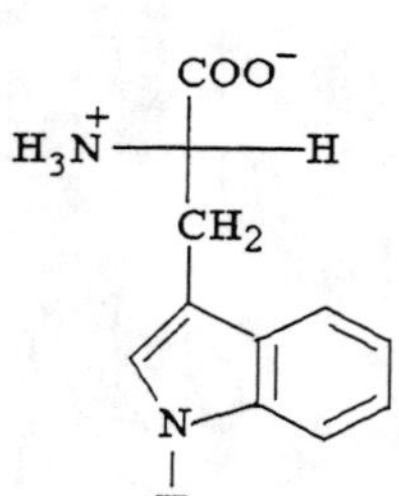

Section: 1

45. What is the net charge of arginine in a solution of Ph 1.0?

    Answer: +2 Section: 1

46. Draw structures for the forms of glycine present in basic, neutral, and acidic solutions.

    Answer:

$H_2N-CH_2-COO^-$	$H_3\overset{+}{N}-CH_2-COO^-$	$H_3\overset{+}{N}-CH_2-COOH$
in basic solution	in neutral solution	in acidic solution

    Section: 3

MULTIPLE CHOICE

47. Which of the following amino acids has its isoelectric point at the lowest pH?

    a. glutamic acid
    b. lysine
    c. valine
    d. glycine
    e. methionine

    Answer: a Section: 5

SHORT ANSWER

48. Describe how electrophoresis works.

    Answer:
    A sample of an amino acid mixture is placed in the center of an acrylamide gel. Two electrodes are placed in contact with the edges of the gel and a high electrical potential is applied across the electrodes. Based on their charge, species then migrate through the gel toward the appropriate electrode and are thereby separated.
    Section: 5

49. Draw the dipeptide Val-Tyr.

    Answer:

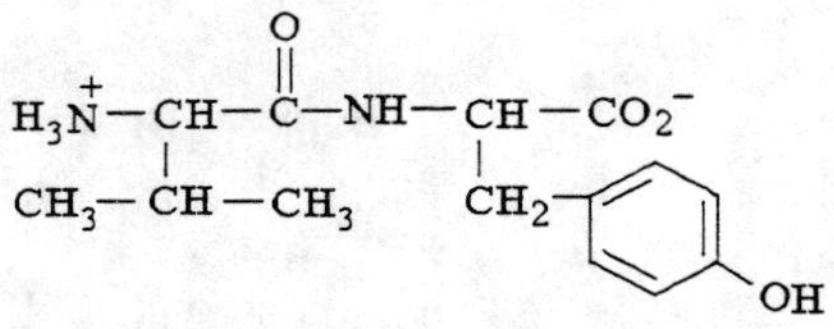

    $$H_3\overset{+}{N}-\underset{\displaystyle CH_3-CH-CH_3}{CH}-\overset{\displaystyle O}{\overset{\|}{C}}-NH-\underset{\displaystyle CH_2-C_6H_4-OH}{CH}-CO_2^-$$

    Section: 7

MULTIPLE CHOICE

50. When a disulfide linkage is formed, the compound containing this new linkage has been __________.

    a. hydrolyzed
    b. dehydrated
    c. electrolyzed
    d. oxidized
    e. reduced

    Answer: d Section: 7

SHORT ANSWER

51. Draw the structure of the tetrapeptide Ser-Leu-Phe-Pro.

    Answer:

$$H_3\overset{+}{N}-\underset{\underset{CH_2OH}{|}}{CH}-\overset{\overset{O}{\|}}{C}-NH-\underset{\underset{\underset{\underset{CH_3}{|}}{CH_2CHCH_3}}{|}}{CH}-\overset{\overset{O}{\|}}{C}-NH-\underset{\underset{CH_2\text{-}Ph}{|}}{CH}-\overset{\overset{O}{\|}}{C}-N-CH-CO_2^-$$

    Section: 7

MULTIPLE CHOICE

52. What peptide remains after two sequential Edman degradations of the pentapeptide Ser-Pro-Gly-Phe-Arg?

    a. Pro-Gly-Phe
    b. Ser-Pro-Gly
    c. Gly-Phe-Arg
    d. Pro-Gly
    e. Phe-Arg

    Answer: c Section: 12

SHORT ANSWER

53. How is the enzyme chymotrypsin used in peptide structure determination?

    Answer:
    Chymotrypsin cleaves the peptide chain at the carboxyl groups of phenylalanine, tyrosine, and tryptophan. Each of these shortened chains is then sequenced and the original peptide is sequenced by determining fragment overlap.

    Section: 12

54. Which type of protein, globular or fibrous, tends to function primarily as structural parts of an organism?

    Answer: fibrous Section: 11

55. Give a detailed, stepwise mechanism for the formation of the activated acyl derivative from the reaction of an amino acid with N,N'-dicyclohexylcarbodiimide.

    Answer:

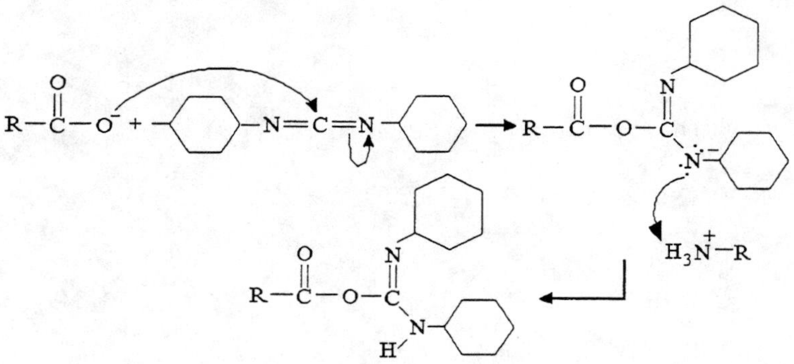

    Section: 9

56. Provide the structure of Thr-Gln-Met.

    Answer:

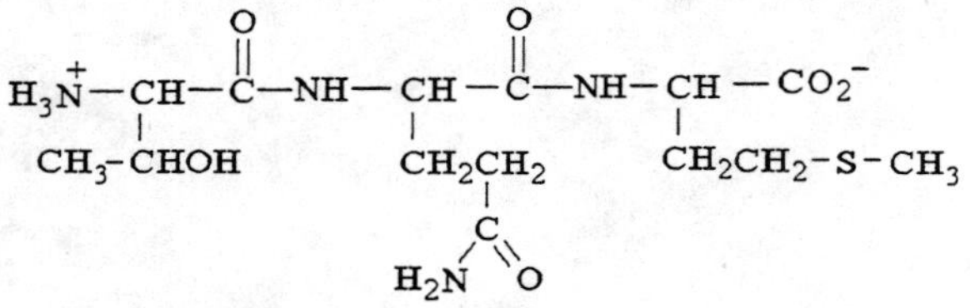

    Section: 7

57. Complete hydrolysis of a hexapeptide gives 2 Gly, Leu, Phe, Pro, and Tyr. Reaction of the peptide with phenylisothiocyanate gave the phenylthiohydantoin of Pro. Partial hydrolysis of the peptide gave the following fragments: Phe-Gly-Tyr, Gly-Phe-Gly, Pro-Leu-Gly, Leu-Gly-Phe. What is the structure of the peptide?

    Answer: Pro-Leu-Gly-Phe-Gly-Tyr Section: 12

58. What is the major force responsible for the formation of an α-helix in protein secondary structure?

    Answer: hydrogen bonding Section: 13

59. What is the name of the amino acid produced when propanoic acid is subjected to the following sequence of reagents: 1. PBr_3, Br_2; 2. H_2O; 3. NH_3, Δ?

Answer:
a. alanine
b. aspartic acid
c. glutamic acid
d. valine
e. asparagine

Section: 6

60. In a globular protein, would the side chain of aspartic acid most probably be oriented toward the interior of the protein or outward toward the aqueous surroundings? Explain.

Answer:
The aspartic acid side chain is negatively charged at biological pH. This negatively charged group is highly hydrophilic and will most likely be oriented so that interaction with water is possible.

Section: 14

Chapter 21: Catalysis

MULTIPLE CHOICE

1. Which of the following is the best definition of a catalyst?

   a. It increases the rate of a chemical reaction.
   b. It decreases the rate of a reaction.
   c. It decreases activation energy.
   d. It increases activation energy.
   e. a and c

   Answer: e

2. Which of the following statements is not true about catalysts?

   a. They can increase the rate of a chemical reaction.
   b. They can change the mechanism of a chemical reaction.
   c. They are changed in a chemical reaction.
   d. They participate in a chemical reaction.
   e. They can decrease the activation energy of a chemical reaction.

   Answer: c

3. Which of the following statements is/are true about catalysts?

   a. They can be used only once.
   b. They participate in the chemical reaction, but are not changed or consumed.
   c. They change the stability of reactants and products.
   d. They change the amount of product formed in a chemical reaction.
   e. All of the above

   Answer: b

4. Which of the following is/are true about base-catalysts?

   a. They increase the rate of a reaction by removing a proton from a reactant.
   b. They increase the rate of a reaction by providing a pathway with a more stable transition state.
   c. They facilitate the formation of a negatively charged tetrahedral intermediate.
   d. a and b only
   e. a, b and c

   Answer: e Section: 4

5. Which of the following is a step in the mechanism of the reaction shown below?

$$CH_3CH_2{-}Cl + \overset{\ominus}{O}H \xrightarrow{I^{\ominus}} CH_3CH_2OH + Cl^{\ominus}$$

a. 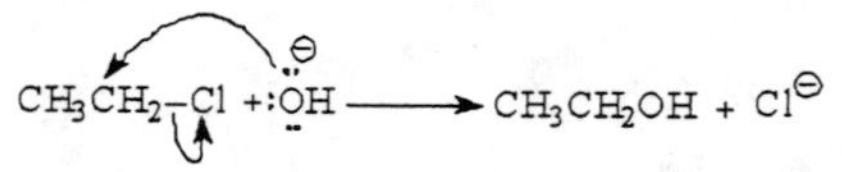 $CH_3CH_2{-}Cl + :\ddot{O}H^{\ominus} \longrightarrow CH_3CH_2OH + Cl^{\ominus}$

b. 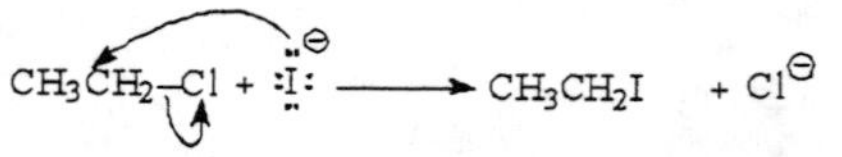 $CH_3CH_2{-}Cl + :\ddot{I}:^{\ominus} \longrightarrow CH_3CH_2I + Cl^{\ominus}$

c. 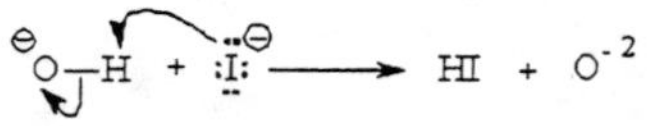 $^{\ominus}O{-}H + :\ddot{I}:^{\ominus} \longrightarrow HI + O^{-2}$

d. 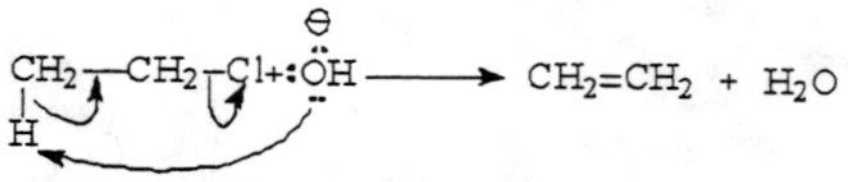 $CH_2(H){-}CH_2{-}Cl + :\ddot{O}H^{\ominus} \longrightarrow CH_2{=}CH_2 + H_2O$

e.  $CH_2{=}CH_2 + H{-}I \longrightarrow CH_3\overset{\oplus}{C}H_2 + I^{\ominus}$

Answer: b Section: 2

6. Which of the following is an intermediate in the mechanism of hydrolysis of the ester shown below?

$$CH_3{-}\overset{O}{\overset{\|}{C}}{-}OCH_3 + H_2O \overset{H^+}{\rightleftharpoons} CH_3{-}\overset{O}{\overset{\|}{C}}{-}OH + CH_3OH$$

a. $CH_3{-}\overset{\oplus OH}{\overset{\|}{C}}{-}OH$

b. $CH_3{-}C(OH)(\overset{\oplus}{O}H_2){-}OCH_3$

c. $CH_3{-}C(OH)(OH){-}OCH_3$

d. $CH_3{-}\overset{\oplus OH}{\overset{\|}{C}}{-}OCH_3$

e. All of the above

Answer: e Section: 3

7. Which of the following represents the first step(s) in the mechanism involving general-base catalysis for the reaction shown below?

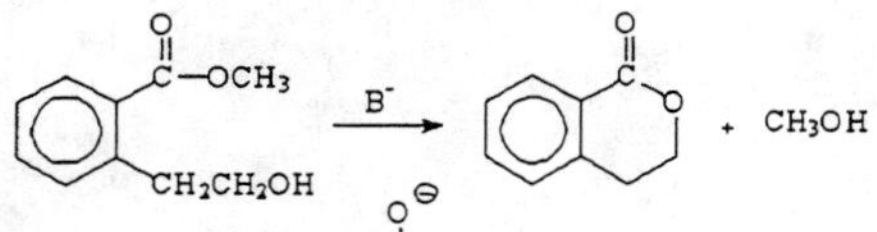

a.

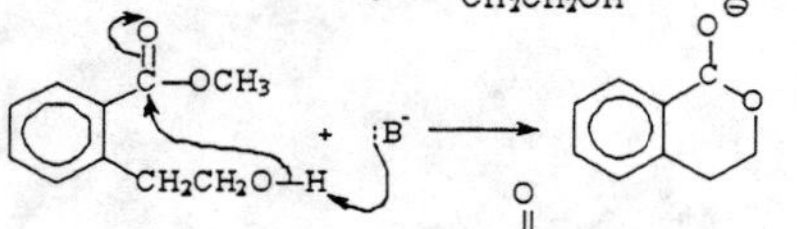

b.

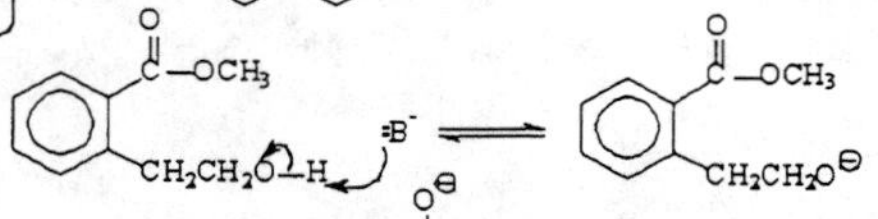

c.

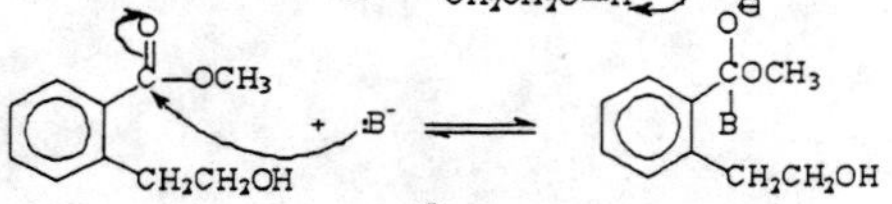

d.

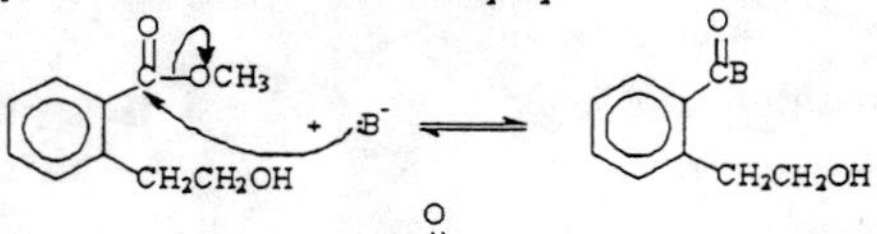

e.

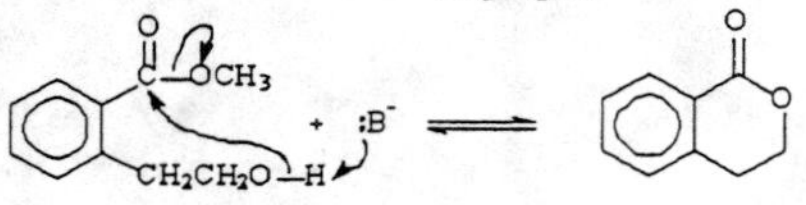

Answer: a Section: 4

8. Which of the following is not true about metal-ion catalysts?

a. They coordinate with atoms with non-bonded electrons.
b. They can make a reaction center more susceptible to receiving electrons and stabilize a developing charge on the transition state.
c. They can make a leaving group a weaker base and therefore a better leaving group.
d. They act as Lewis bases.
e. They can increase the electrophilicity of a reaction center.

Answer: d Section: 5

9. Which of the following is an intermediate in the metal-ion catalyzed reaction shown below?

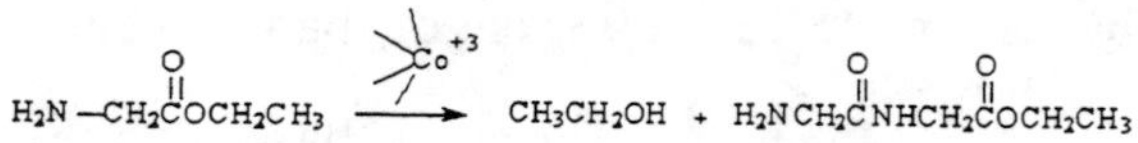

$$H_2N-CH_2\overset{O}{\overset{\|}{C}}OCH_2CH_3 \xrightarrow{Co^{+3}} CH_3CH_2OH + H_2NCH_2\overset{O}{\overset{\|}{C}}NHCH_2\overset{O}{\overset{\|}{C}}OCH_2CH_3$$

a. Co^{+3} coordinated to carbonyl O ($\oplus$) of $C(-CH_2NH_2)(OCH_2CH_3)$

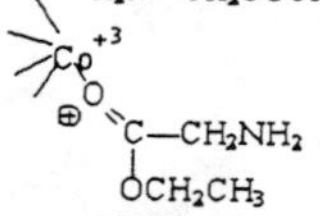

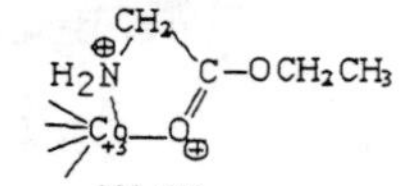

b. Chelate ring: $H_2\overset{\oplus}{N}-CH_2-C(-OCH_2CH_3)=\overset{\oplus}{O}-Co^{+3}$ (Co bonded to N)

c. Co^{+3} ring with CH_2CH_2, O, and $\overset{\oplus}{O}=C(CH_2NH_2)$

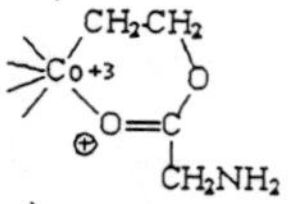

d. 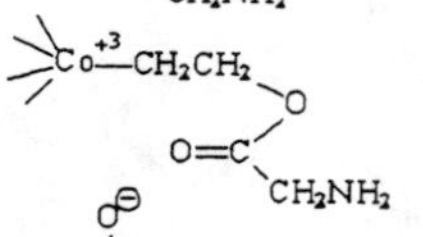 $Co^{+3}-CH_2CH_2-O-C(=O)-CH_2NH_2$

e. 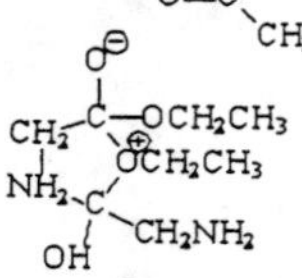 $O^{\ominus}-C(CH_2)(OCH_2CH_3)(\overset{\oplus}{O}CH_2CH_3)$; $NH_2-C(OH)(CH_2NH_2)$

Answer: b Section: 5

10. Which of the following are affected when an intramolecular catalyst is used?

a. Number of collisions between the two per unit time
b. Fraction of collisions with sufficient energy
c. Fraction of collisions with proper orientation
d. Rate of the reaction
e. All of the above

Answer: e Section: 6

11. What is the main purpose for using metal-ions as catalysts in the following hydrolysis reaction?

    a. Increases the rate by making water a stronger base, thereby increasing its electrophilicity
    b. Increases the rate by making water a stronger base, thereby increasing its nucleophilicity
    c. Increases the rate by making water a stronger acid, thereby increasing its nucleophilicity
    d. Increases the rate by making water a stronger acid, thereby increasing its electrophilicity
    e. Increases the rate by making water a stronger acid, thereby decreasing its nucleophilicity

    C—OCH_3 HN N OH $+ H_2O$ $\xrightarrow[Co^{+2}]{Ni^{+2}\ \text{or}}$ C—OH HN N OH $+ CH_3OH$

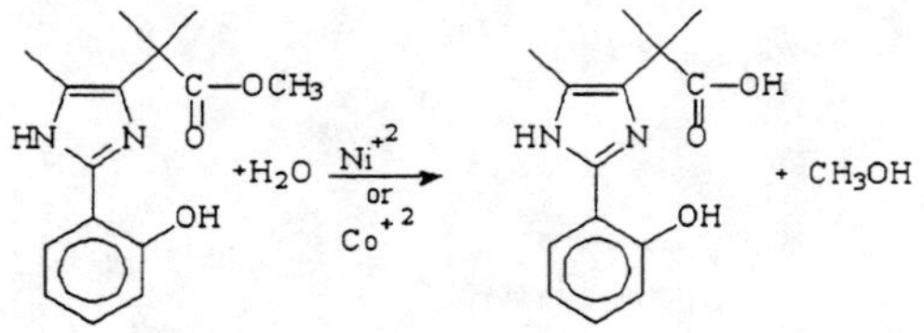

    Answer: c Section: 5

12. Which of the following is true of globular proteins?

    a. They may be proteins with predominantly quaternary structure.
    b. They may be circular strands of short peptides.
    c. They may be three-dimensional chains of short peptides.
    d. They may be enzymes that catalyze biological reactions.
    e. They may be large strands of proteins and amylopectins.

    Answer: d Section: 8

13. How do enzymes differ from non-biological catalysts?

    a. They have specificity for the substrate.
    b. They have lower molecular weights than non-biological catalysts.
    c. They require higher pressure to be functional.
    d. They require lower pH to be functional.
    e. They are more polar than non-biological catalysts.

    Answer: a Section: 8

14. What does the model termed "lock-and-key" stand for?

    a. The hydrogen bonds in the α-helix fit each other like a key fits a lock.
    b. The amino acids in a peptide chain fit each other like a key fits a lock.
    c. The substrate fits the active site of the enzyme like a key fits a lock.
    d. The non-polar sites of proteins fit each other like a key fits a lock.
    e. The N-terminal and the C-terminal ends of a protein fit each other like a key fits a lock.

    Answer: c Section: 8

15. Which of the following statements account for the remarkable catalytic ability of enzymes?

    a. Proper positioning and specificity of active site and substrate
    b. Proper positioning of amino acid side chains that serve as catalytic groups
    c. Ability of the groups on the enzyme to stabilize transition states
    d. a and b
    e. All of the above

    Answer: e Section: 8

16. Which of the following is not true about the enzyme carboxypeptidase A?

    a. Arg 145 and Tyr 248 are involved in the catalytic process of the enzyme.
    b. The metal ion in carboxypeptidase A is partially liganded to the oxygen of the carbonyl group that will be hydrolyzed.
    c. Carboxypeptidase A catalyzes the hydrolysis of the C-terminal peptide bond in proteins.
    d. Carboxypeptidase A is a metalloenzyme.
    e. The metal ion in carboxypeptidase is a Cu^{+2}.

    Answer: e Section: 8

17. Which of the labeled bonds in the following peptide chain will break when carboxypeptidase A is used to hydrolyze the chain?

```
                 O                 O                 O
       5         || 4      3       || 2          1   ||    ⊖
    — NH — CH — C — NH — CH — C — NH — CH — C — O
            |                  |                  |
            R                  R                  R
```

   a. 1
   b. 2
   c. 3
   d. 4
   e. 5

   Answer: b Section: 8

18. Which of the following is not true about serine proteases?

   a. They are called serine proteases because they are able to break the peptide chain at the serine linkage.
   b. Endopeptidases and serine proteases are synonyms with each other.
   c. Trypsin, chymotrypsin, and elastase are all serine proteases.
   d. Serine proteases catalyze the hydrolysis of protein peptide chains.
   e. They are called serine proteases because they all have a serine residue that participates in the catalytic process.

   Answer: a Section: 8

19. Which of the following sets of amino acids is found in all serine proteases?

   a. Arginine, histidine, serine
   b. Arginine, aspartate, serine
   c. Proline, histidine, serine
   d. Aspartate, histidine, serine
   e. Aspartate, proline, serine

   Answer: d Section: 8

20. Which of the following is the main difference among the active sites of all serine proteases?

   a. The presence of serine residue at the active site
   b. The size and charge of the pocket at the active site
   c. The presence of histidine residue at the active site
   d. The presence of aspartate at the active site
   e. The hydrolysis of the peptide bond

   Answer: b Section: 8

21. Which of the following intermediates is produced in the second step of the mechanism for bovine chymotrypsin hydrolysis of a peptide bond?

    a. Epoxy-enzyme intermediate
    b. Carboxy-enzyme intermediate
    c. Acyl-enzyme intermediate
    d. Phenoxy-enzyme intermediate
    e. Oxy-enzyme intermediate

    Answer: c Section: 8

22. Which of the following statements best describes what lysozyme is?

    a. An enzyme that destroys the cell walls of bacteria
    b. An enzyme that repairs the cell walls of bacteria
    c. An enzyme that hydrolyzes the NAM-NAG bond
    d. a and c
    e. b and c

    Answer: d Section: 8

23. Which of the following residues are found at the active site of lysozyme?

    a. His 57, Asp 102
    b. His 57, Asp 52
    c. Glu 35, His 57
    d. Asp 102, Glu 35
    e. Glu 35, Asp 52

    Answer: e Section: 8

24. After lysozyme binds its substrate, the enzyme undergoes a conformational change that distorts residue D in the substrate. Which of the following best describes this distortion?

    a. From a chair to a half-chair conformation
    b. From a half-chair to a chair conformation
    c. From a chair to a boat conformation
    d. From a boat to a chair conformation
    e. From a boat to a half-chair conformation

    Answer: a Section: 8

25. Which of the following is not true about the enzyme aldolase?

    a. It catalyzes the cleavage of D-glucose into two molecules each containing 3-carbons.
    b. It catalyzes the cleavage of D-galactose into two molecules each containing 3-carbons.
    c. It catalyzes the cleavage of a six carbon compound into two three carbon compounds.
    d. It catalyzes the cleavage of D-fructose 1,6-diphosphate into glyceraldehyde-3-phosphate and dihydroxyacetone phosphate.
    e. The enzyme is called aldolase because the reverse reaction is an aldol condensation.

    Answer: b Section: 8

26. Which of the following residues acts as a general-base catalyst in the cleavage of the C_3-C_4 bond in the mechanism of the aldolase-catalyzed reaction?

    a. Methionine
    b. Histidine
    c. Cysteine
    d. Asparagine
    e. Lysine

    Answer: c Section: 8

27. Which of the following residues acts as a general-acid catalyst in aldolase but as a general-base catalyst in chymotrypsin?

    a. Methionine
    b. Histidine
    c. Cysteine
    d. Asparagine
    e. Lysine

    Answer: b Section: 8

28. What are transition-state analogs?

    a. Artificial enzymes that are structurally similar to the transition state of the enzyme-catalyzed reaction
    b. Artificial enzymes that are used as antigens to stimulate the synthesis of complementary antibodies
    c. Artificial enzymes that are used as antigens to inhibit the synthesis of complementary antibodies
    d. a and b
    e. a and c

    Answer: d Section: 9

SHORT ANSWER

29. List four common types of catalysts used in organic reactions and give one example of each.

Answer:

1. Nucleophilic catalysts: Ex $I^{\ominus}$, imidazole (:N ... :NH)

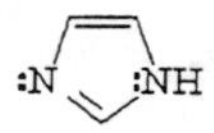

2. Acid catalysts: Ex $H^{\oplus}$, $AlCl_3$

3. Base catalysts: Ex $^{\ominus}OH$, NH_3

4. Metal-ion catalysts: Ex Cu^{+2}, Al^{+3}, Co^{+3}, Cu^{+2} etc.

Section: 1

30. Propose a step-by-step mechanism for the general-acid catalyzed reaction shown below:

$$CH_3-\overset{O}{\overset{\|}{C}}-OCH_3 + H_2O \xrightleftharpoons{HA} CH_3-\overset{O}{\overset{\|}{C}}-OH + CH_3OH$$

Answer:

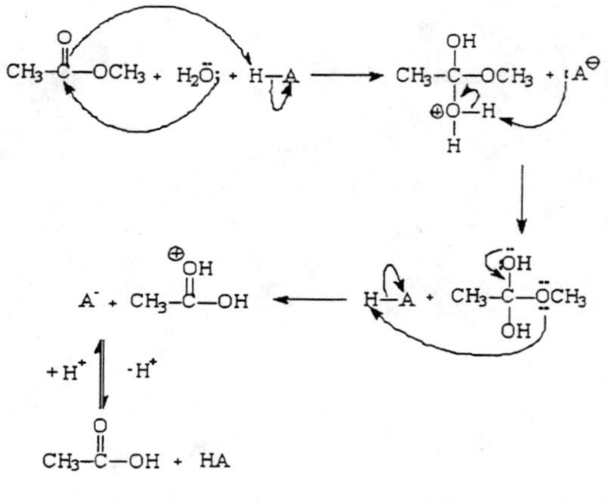

$CH_3-\overset{O}{\overset{\|}{C}}-OCH_3 + H_2\ddot{O}: + H-A \longrightarrow CH_3-\overset{OH}{\overset{|}{C}}(-OCH_3)(-\overset{\oplus}{O}H-H) + :A^{\ominus}$

$\downarrow$

$H-A + CH_3-\overset{\ddot{O}H}{\overset{|}{C}}(-\ddot{O}CH_3)(-OH) \longrightarrow A^{-} + CH_3-\overset{\oplus OH}{\overset{\|}{C}}-OH$

$+H^{+} \rightleftharpoons -H^{+}$

$CH_3-\overset{O}{\overset{\|}{C}}-OH + HA$

Section: 3

31. Show why typical metal-bound water has a considerably lower pKa than that of water.

Answer:
The metal ion binds to water making it easier for water to lose a proton, forming metal-bound hydroxide ion which is a stronger nucleophile than water. Typical pKa values for metal-bound water are between 5.7 - 12.7, much lower than that of 15.6 for water.

metal—OH_2 ⇌ metal--$\overset{\ominus}{O}H$ + $H^{\oplus}$

Section: 5

32. What factor(s) would affect the rate of an intramolecular reaction?

Answer:
There are three main factors:
1. Number of collisions between the two molecules per unit time
2. Fraction of collisions with sufficient energy
3. Fraction of collisions with proper orientation

Section: 6

33. Explain why an intramolecular reaction resulting in the formation of five- or six-membered rings occurs more readily than the corresponding intermolecular reaction.

Answer:
An intramolecular reaction has the advantage of having the reacting groups tied together in the same molecule. Consequently, the reacting groups have a greater chance of colliding.

Section: 6

34. Explain what is meant by anchimeric assistance and give an example.

Answer:
In Greek, anchimeric means "adjacent parts." This is the same thing as saying intramolecular catalyst.

Ex.

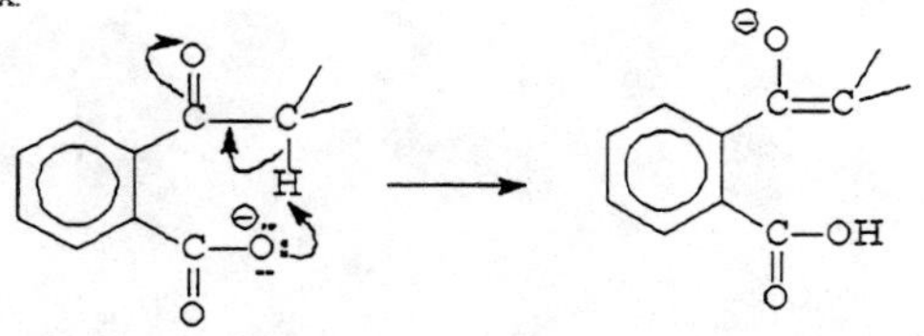

Section: 7

35. Explain the large difference in the relative rates of the following hydrolysis reactions:

I. Phenyl acetate + H_2O → phenol + $CH_3-C(=O)-O^{\ominus}$

Rel. Rate= 1

II. Aspirin + H_2O → salicylate + $CH_3-C(=O)-O^{\ominus}$

Aspirin

Rel. Rate= 150

Answer:
This large difference in the relative rates is due to the fact that the ortho-carboxy substituent (in aspirin) is an intramolecular general-base catalyst. It increases the nucleophilicity of H_2O, thereby increasing the rate of formation of the tetrahedral intermediate.

Section: 7

36. Explain the phenomenon known as molecular recognition. Give an example.

Answer:
Molecular recognition is the recognition of one molecule by another molecule.

Ex.
The specificity of the active site of the enzyme for its substrate. This results from the conformation of the protein and the particular amino acid side chains that make up the active site. There are many other kinds of molecular recognition other than enzyme-substrate; hormone-receptors and inhibitor-enzyme.

Section: 8

37. Explain what is meant by the induced fit model. Give an example.

Answer:
Binding energy from the substrate-enzyme interaction is used to induce a conformational change in the enzyme to better fit the substrate.

Ex.
The enzyme carboxypeptidase A fits this model.

Section: 8

38. Explain what is meant by site-specific mutagenesis.

Answer:
It is a technique for determining the relationship between protein function and structure. This is done by substituting one amino acid of a protein for another and seeing what the effect is.

Section: 8

39. What are NAM and NAG and what do they do?

Answer:
NAM is N-acetylmuramic acid and NAG is N-acetylglucosamine. They are molecules that make up the cell walls of bacteria.

Section: 8

40. Why is histidine a versatile catalytic group?

Answer:
Because the imidazole ring in histidine has a pKa of 6.8 which is close to the physiological pH of 7.2; thus it can exist in both its acidic and basic forms.

Section: 8

MULTIPLE CHOICE

41. "Being in the right place... can be the most powerful factor of all in determining how fast a reaction goes." In terms of reaction rate theory, this statement refers to the _________ for a reaction.

a. probability factor
b. energy factor
c. resonance effect
d. inductive effect

Answer: a Section: 6

42. The compound $CH_3OCH_2CH_2CH_2CH_2Br$ undergoes a solvolysis reaction is aqueous ethanol solution. Which of the following is evidence that a neighboring group effect is occuring?

a. $CH_3CH_2CH_2CH_2Br$ reacts more slowly than $CH_3OCH_2CH_2CH_2CH_2Br$ does.
b. The reaction is faster when the concentration of water is higher.
c. The reaction is first order.
d. HBr is a product of the reaction.

Answer: a Section: 6

43. A pure sample of the <u>trans</u>-2-bromocyclopentanol enantiomer shown below reacts with HCl. The reaction proceeds by way of a bromonium ion. What is/are the product(s) of the reaction?

a. I only
b. II only
c. I and III
d. II and IV

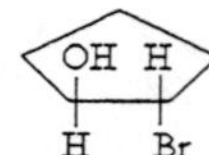

I.
Cl H
H Br

II.
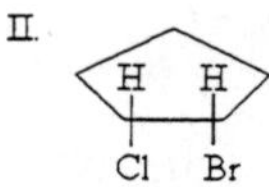

III.
H Cl
Br H

IV.
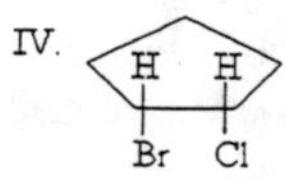

Answer: c Section: 7

Chapter 22: The Organic Mechanisms of the Coenzymes

MULTIPLE CHOICE

1. Which of the following statements best describes what cofactors are?

   a. They are metal ions that assist enzymes in carrying out their catalytic reactions.
   b. They are organic molecules that assist enzymes in carrying out their catalytic reactions.
   c. They are organic molecules called coenzymes.
   d. They are organic molecules derived from compounds commonly known as vitamins.
   e. All of the above

   Answer: e

2. Which of the following statements is <u>not</u> true of coenzymes?

   a. Coenzymes catalyze chemical reactions similar to enzymes.
   b. Coenzymes may function as oxidizing and reducing agents.
   c. Coenzymes play a variety of chemical roles exactly like those played by the amino acid side chains of enzymes.
   d. Coenzymes may function as nucleophiles or strong bases needed in a reaction.
   e. Coenzymes may function as activators needed in a reaction.

   Answer: c

3. Which of the following is/are coenzyme(s) formed from the vitamin niacin?

   a. NAD^+
   b. $NADP^+$
   c. NADH
   d. NADPH
   e. All of the above

   Answer: e Section: 2

4. Which of the following is/are the coenzyme(s) formed from the vitamin thiamine (B_1)?

   a. TPP
   b. CoASH
   c. PLP
   d. THF
   e. FAD

   Answer: a Section: 4

5. Which of the following is/are water-soluble vitamins?

   a. Vitamin B_{12}
   b. Riboflavin (vitamin B_2)
   c. Thiamine (vitamin B_1)
   d. Pyridoxine (vitamin B_6)
   e. All of the above

   Answer: e

6. Which of the following is <u>not</u> a lipid-soluble vitamin?

   a. Vitamin A
   b. Vitamin D
   c. Vitamin E
   d. Vitamin C
   e. Vitamin K

   Answer: d

7. Which of the following vitamins is the only lipid-soluble vitamin that is known to function as a coenzyme?

   a. Vitamin A
   b. Vitamin D
   c. Vitamin B_1
   d. Vitamin C
   e. Vitamin K

   Answer: e

8. Which of the following vitamins is the only water-soluble vitamin that <u>does not</u> function as a coenzyme?

   a. Vitamin B_{12}
   b. Riboflavin (B_2)
   c. Vitamin C
   d. Thiamine (B_1)
   e. Pyridoxine (B_6)

   Answer: c

9. Which of the following vitamins are radical inhibitors or antioxidants?

   a. Niacin
   b. Vitamins B_1 and B_{12}
   c. Vitamins B_2 and B_6
   d. Vitamins A and D
   e. Vitamins C and E

   Answer: e

10. Which of the following vitamins can be synthesized by all mammals except humans and guinea pigs?

    a. Vitamin B_{12}
    b. Riboflavin (B_2)
    c. Vitamin C
    d. Thiamine (B_1)
    e. Pyridoxine (B_6)

    Answer: c

11. In which of the following classes of reactions is a coenzyme required?

    a. Oxidation reactions
    b. Two carbon transfer
    c. Acyl transfer
    d. Reduction reactions
    e. a, b, and d

    Answer: e Section: 1

12. Which of the following statements is/are true about metabolism?

    a. Reactions that living organisms carry out to obtain energy and to synthesize required compounds
    b. Reactions that organic chemists carry out to obtain energy and to synthesize required compounds
    c. Reactions that consist of catabolic and anabolic reactions
    d. a and c
    e. b and c

    Answer: d Section: 1

13. Which of the following is <u>not true</u> about catabolic reactions?

    a. Catabolic reactions require NAD^+ for a coenzyme.
    b. Catabolic reactions are most often oxidation reactions.
    c. Catabolic reactions are most often reduction reactions.
    d. Catabolic reactions require an oxidizing coenzyme.
    e. Catabolic reactions break down complex molecules into simpler ones.

    Answer: c Section: 1

14. At what position of the pyridine ring in the following coenzyme will oxidation-reduction reactions take place?

a. 1
b. 2
c. 3
d. 4
e. 5

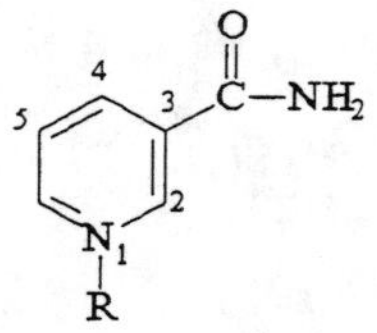

(Structure: pyridine ring with positions 1–5 labelled; N at position 1 bearing R; $-\overset{O}{\overset{\|}{C}}-NH_2$ at position 3)

Answer: d Section: 2

15. What is the major organic product of this oxidation-reduction reaction?

$$CH_3-\underset{H_S}{\overset{H_R}{C}}-OH + NAD^+ \xrightarrow[\text{dehydrogenase}]{\text{alcohol}} ?$$

a. $R-\underset{H}{C}=O$

b. $CH_3-\underset{O}{\underset{\|}{C}}-OH_S$

c. $CH_3-\underset{O}{\underset{\|}{C}}-OH_R$

d. $CH_3-\underset{H_R}{C}=O$

e. $CH_3-\underset{H_S}{C}=O$

Answer: e Section: 2

16. Generally speaking, which of the following oxidation-reduction reactions utilize NAD^+ and $NADP^+$ as oxidation coenzymes?

 a. Alcohols oxidized to ketones
 b. Alcohols oxidized to aldehydes
 c. Alcohols oxidized to carboxylic acids
 d. a and b
 e. a, b, and c

 Answer: e Section: 2

17. What are the products obtained from the reaction of NADH with FMN catalyzed by NADH dehydrogenase as shown below?

 a. $NAD^+ + FMNH_2$
 b. $NAD^+ + FMNH$
 c. NAF + DHN
 d. $NAF + DNH_2$
 e. $N_2A_2D_2 + FMNH_2$

$$\text{NADH} + \text{FMN} \xrightarrow[\text{dehydrogenase}]{\text{NADH}} ? + ?$$

 Answer: a Section: 3

18. At what position on the following coenzyme (FAD/FMN) will oxidation-reduction reactions take place?

 a. 6
 b. 2
 c. 3
 d. 4
 e. 5

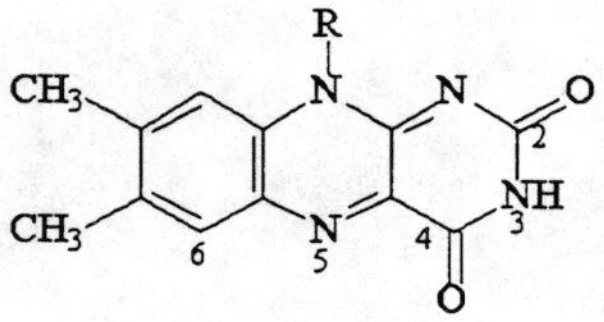

R
CH3 N N O
2
NH
CH3 6 N 4 3
5
O

 Answer: e Section: 3

19. Which of the following substrates requires the coenzyme thiamine pyrophosphate (TPP) for catalyzing its decarboxylation reaction?

a. β-keto acid, $CH_3-C(=O)-CH_2-COO^{\ominus}$

b. α-keto acid, $CH_3-C(=O)-COO^{\ominus}$

c. γ-keto acid, $CH_3-C(=O)-CH_2CH_2-COO^{\ominus}$

d. α-hydroxy acid, $CH_3-CH(OH)-COO^{\ominus}$

e. β-hydroxy acid, $CH_3-CH(OH)-CH_2-COO^{\ominus}$

Answer: b Section: 4

20. Which of the following statements best describes what an "electron sink" is?

a. A free radical initiator for enzyme catalysis
b. A site for electron localization
c. A site for electron delocalization
d. A site that donates electrons for enzyme catalysis
e. A site that accepts electrons for enzyme catalysis

Answer: c Section: 4

21. In addition to TPP, which of the following coenzymes is/are required to facilitate the following reaction?

$$\underset{\text{Pyruvate}}{CH_3-C(=O)-COO^-} + CoASH \xrightarrow[\text{system}]{\text{Pyruvate dehydrogenase}} \underset{\text{Acetyl CoA}}{CH_3-C(=O)-S-CoA} + CO_2$$

a. Lipoate
b. Coenzyme A
c. FAD
d. NAD^+
e. All of the above

Answer: e Section: 4

22. Which of the following coenzymes is required for the enzyme catalyzed reaction shown below?

$$\text{Xylose-5-P} + \text{Ribose-5-P} \xrightarrow{\text{transketolase}} \text{Glyceraldehyde-3-P} + \text{Sedoheptulose-7-P}$$

a. TPP
b. PLC
c. THF
d. NADH
e. NADPH

Answer: a Section: 4

23. Which of the following reactions does the coenzyme biotin catalyze?

a. Decarboxylation of an α-keto acid
b. Decarboxylation of an α-carboxy acid
c. Carboxylation of an α-carboxy acid
d. Decarboxylation of a carbon adjacent to an ester or keto group
e. Carboxylation of a carbon adjacent to an ester or keto group

Answer: e Section: 5

24. Which of the following are required species for biotin-requiring enzymes?

a. ATP
b. Mg^{+2}
c. HCO_3^-
d. a and b
e. a, b and c

Answer: e Section: 5

25. Which of the following is the sequence of steps followed by biotin-requiring enzymes, as listed below?
   I. Transfer of the carboxy group from carboxybiotin to the substrate
   II. Activation of bicarbonate by ATP
   III. Formation of carboxybiotin

a. I, II, III
b. I, III, II
c. II, I, III
d. II, III, I
e. III, II, I

Answer: d Section: 5

26. What does the term "Pyridoxal" stand for?

    a. Pyridine aldehyde
    b. Pyrrol oxyaldehyde
    c. Pyrrolidine oxyaldehyde
    d. Peroxy aldehyde
    e. Pyrimidoxy aldehyde

    Answer: a Section: 6

27. Which of the following reactions of amino acids is <u>not</u> catalyzed by PLP-requiring enzymes?

    a. decarboxylation
    b. transamination
    c. α,β-elimination
    d. α-keto esterification
    e. $C\alpha$-$C\beta$ bond cleavage

    Answer: d Section: 6

28. Which of the following reactions is called an α,β-elimination of an amino acid?

    a. $$CH_3-\underset{\underset{\oplus}{NH_3}}{\underset{|}{CH_2}}-COO^{\ominus} \xrightarrow[PLP]{enzyme} CH_3CH_2\overset{\oplus}{NH_3} + CO_2$$

    b. $$\underset{X}{\underset{|}{CH_2}}-\underset{\underset{\oplus}{NH_3}}{\underset{|}{CH}}-COO^{\ominus} \xrightarrow[PLP]{enzyme} CH_3\overset{\overset{O}{\|}}{C}COO^{\ominus} + X^{\ominus} + \overset{\oplus}{NH_4}$$

    c. $$\underset{\underset{\oplus}{NH_3}}{\underset{|}{CH_2}}-\overset{\ominus}{COO} + \overset{\ominus}{OOC}-CH_2\overset{\overset{O}{\|}}{C}-\overset{\ominus}{COO} \xrightarrow[PLP]{enzyme} CH_3-\overset{\overset{O}{\|}}{C}-\overset{\ominus}{COO} + \overset{\ominus}{OOC}-CH_2-\overset{\overset{\oplus}{NH_3}}{\overset{|}{CH}}-\overset{\ominus}{COO}$$

    d. $$R-\overset{\overset{H}{|}}{\underset{\underset{\oplus}{NH_3}}{\underset{|}{C}}}-\overset{\overset{O}{\|}}{C}-O^{\ominus} \xrightarrow[PLP]{enzyme} R-\overset{\overset{H}{|}}{\underset{\underset{\oplus}{NH_3}}{\underset{|}{C}}}-COO^{\ominus} + {}^{\ominus}OOC-\overset{\overset{H}{|}}{\underset{\underset{\oplus}{NH_3}}{\underset{|}{C}}}-R$$

e. 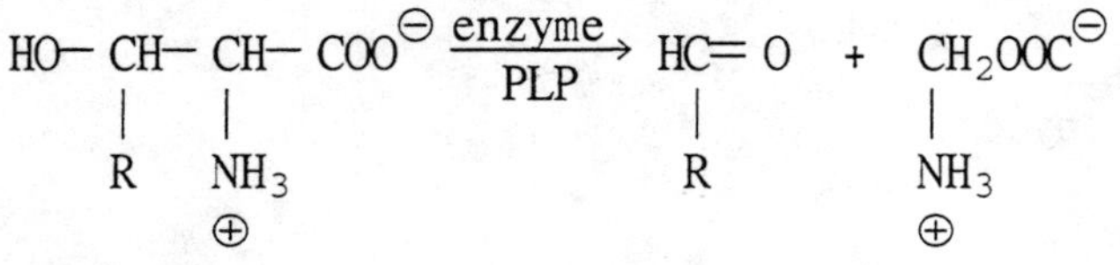 $HO-CH(R)-CH(NH_3^{\oplus})-COO^{\ominus} \xrightarrow[PLP]{enzyme} HC(R){=}O + CH_2(NH_3^{\oplus})OOC^{\ominus}$

Answer: b Section: 6

29. Which of the following statements is <u>not</u> true about the structure of vitamin B_{12}?

a. The structure was determined by Dorothy Hodgkin in 1955 using X-ray crystallography.
b. The structure has an adenosyl group.
c. The structure has a cyano group, HO^-, or H_2O.
d. The cyano group is coordinated to cobalt.
e. The structure cannot be synthesized by plants and animals.

Answer: b Section: 7

30. Which of the following best represents the chemical reaction catalyzed by B_{12}-dependent enzymes?

a. 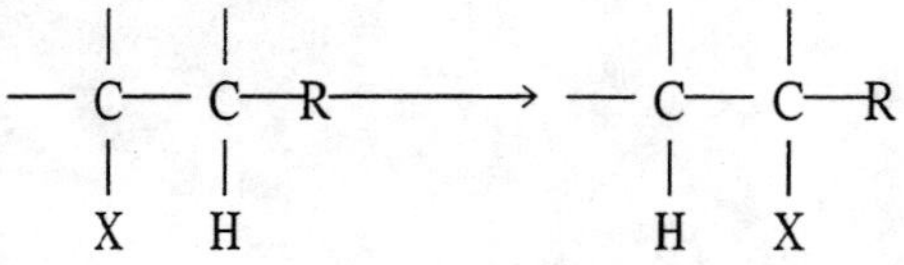 $-C(X)-C(H)-R \longrightarrow -C(H)-C(X)-R$

b. $-C(X)-C(H)-R \longrightarrow -C{=}C-R$

c. $-C(X)-C(H)-R \longrightarrow -C(H)-C(H)-R$

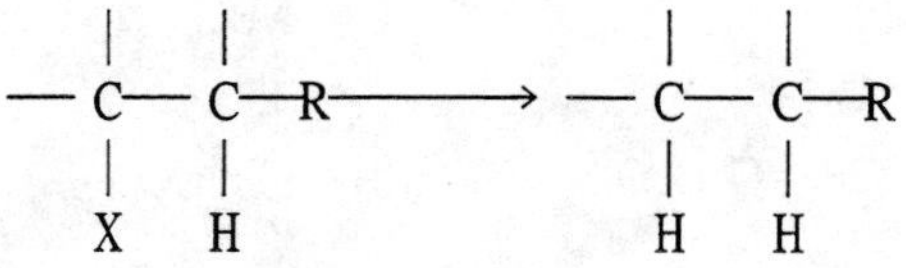

d. 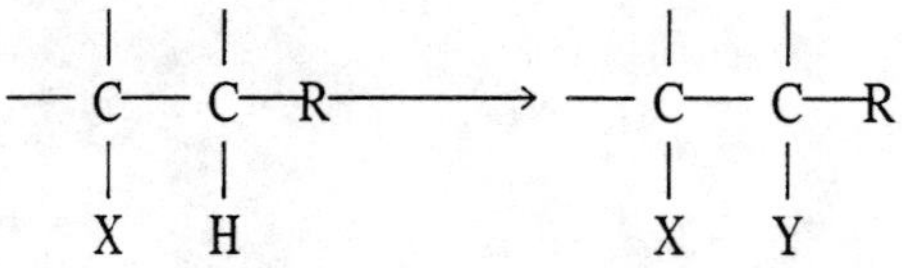 $-C(X)-C(H)-R \longrightarrow -C(X)-C(Y)-R$

e. 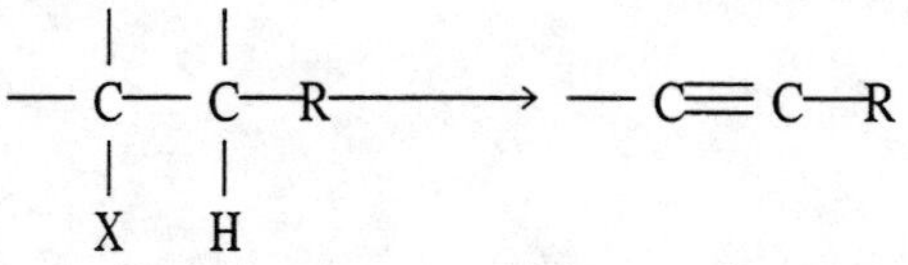 $-C(X)-C(H)-R \longrightarrow -C{\equiv}C-R$

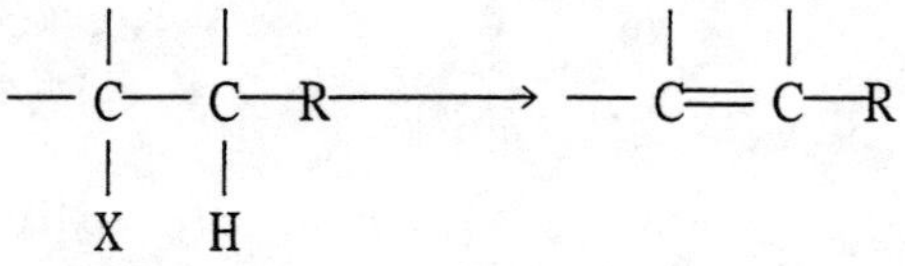

Answer: a Section: 7

31. Which of the following statements is <u>not</u> true about the coenzyme tetrahydrofolate (THF)?

   a. There are six different THF-coenzymes.
   b. The coenzyme is used by enzyme-catalyzing reactions that donate a group containing a single carbon to their substrate.
   c. THF results from the reduction of two double bonds of the precursor vitamin (folate).
   d. Mammals synthesize the vitamin (folate) but bacteria cannot.
   e. Homocysteine methyl transferase and glycinamide ribonucleotide are examples of enzymes that require THF-coenzyme.

   Answer: d Section: 8

32. The thymine molecule used for the biosynthesis of DNA is synthesized from uracil as shown below.

   The structure of dTMP is a methylated dUMP. What is the structure of dTMP?

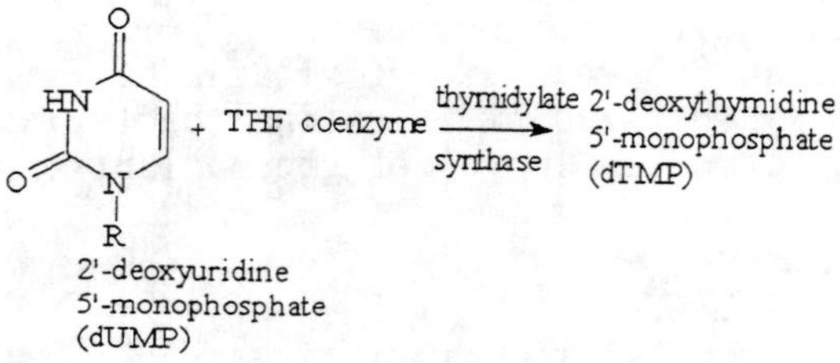

   2'-deoxyuridine 5'-monophosphate (dUMP) + THF coenzyme —(thymidylate synthase)→ 2'-deoxythymidine 5'-monophosphate (dTMP)

   a. 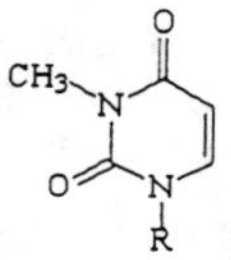
   b. 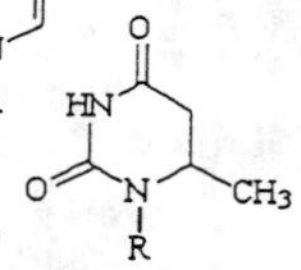
   c. 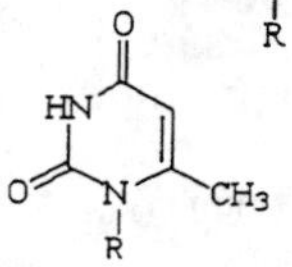
   d. 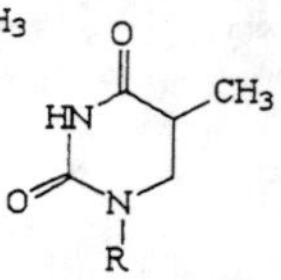
   e.

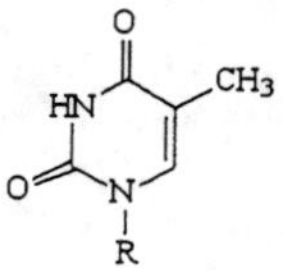

   Answer: e Section: 8

33. Which of the following statements is/are true about 5-fluorouracil,

a. It is an anticancer drug.
b. It inhibits the enzyme thymidylate synthase.
c. It is an example of a mechanism-based inhibitor.
d. It is an example of a suicide inhibitor.
e. All of the above

Answer: e Section: 8

34. Which of the following statements is <u>not</u> true about vitamin K?

a. The letter K comes from koagulation.
b. Vitamin K is required for proper Ca^{+2} binding in blood clotting.
c. Kitase is the active coenzyme of vitamin K.
d. Vitamin K is synthesized by intestinal bacteria.
e. Vitamin K is formed in leaves of green plants.

Answer: c Section: 9

35. Which of the following statements is/are true about the vitamin KH_2?

a. It is the active coenzyme form of vitamin K.
b. Vitamin K is a quinone while vitamin KH_2 is a hydroquinone.
c. Vitamin KH_2 is the coenzyme for the enzyme that catalyzes the carboxylation of glutamate side chains in proteins to form γ-carboxyglutamates.
d. a and c
e. a, b, and c

Answer: e Section: 9

SHORT ANSWER

36. What are holoenzymes and apoenzymes?

Answer:
Holoenzymes are enzymes with their cofactor while apoenzymes are enzymes without their cofactor.

37. Which type of vitamins, water-soluble or lipid-soluble, can we overdose on?

Answer:
One can overdose on lipid-soluble vitamins because they are not easily eliminated by the body and can accumulate in cell membranes and other non-polar components of the body.

38. Explain what is meant by a nucleotide and give an example.

Answer:
A nucleotide is composed of a heterocyclic molecule attached to the β -position of a phosphorylated ribose.
Ex.
NAD^+ (nicotinamide adenine dinucleotide) is composed of two nucleotides linked together through their phosphate group.

Section: 2

39. What are catabolic and anabolic reactions?

Answer:
Catabolic reactions are reactions where complex molecules break down into simple molecules and energy.

Anabolic reactions are reactions where simple molecules react with energy to produce complex biomolecules.

Section: 1

40. Can an NAD^+/NADH dependent enzyme tell the difference between the pro-R hydrogen and pro-S hydrogen of a substrate? Explain.

Answer:
The enzyme can tell the difference between R and S hydrogens because the enzyme has a specific binding site for its coenzyme which blocks one of the coenzyme's sides, leaving the other side open for the substrate to bind to. For example, alcohol dehydrogenase removes only the pro-R hydrogen of ethanol.

Section: 2

41. Explain what is meant by a prosthetic group. Give an example.

Answer:
A tightly bound coenzyme is known as a prosthetic group.
Ex.
Flavin adenine dinucleotide (FAD) or
Flavin mononucleotide (FMN)

Section: 3

42. Propose a mechanism for the following reaction.

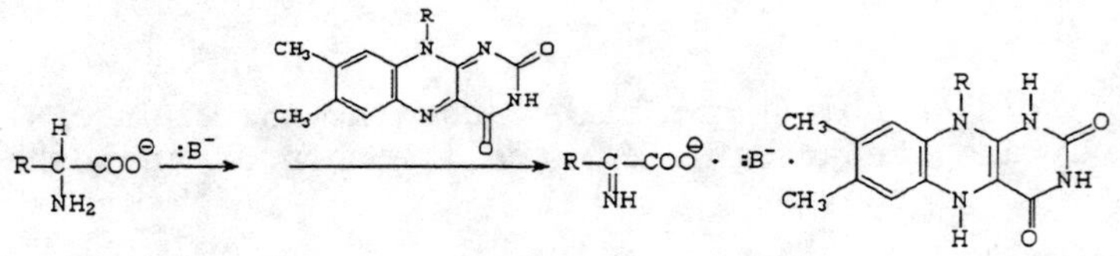

R–CH(NH$_2$)–COO$^{\ominus}$:B$^-$ → (flavin) → R–C(=NH)–COO$^{\ominus}$ · :B$^-$ · (reduced flavin, N–H)

Answer:

R–CH(NH$_2$)–COO$^{\ominus}$ + :B$^-$ → R–C$^{\ominus}$(NH$_2$)–COO$^{\ominus}$ + flavin (H–B) → flavin N-adduct R–C(NH$_2$)(COO$^{\ominus}$)–N + :B$^-$ + H–B → R–C(=NH)–COO$^{\ominus}$ + :B$^-$ + reduced flavin

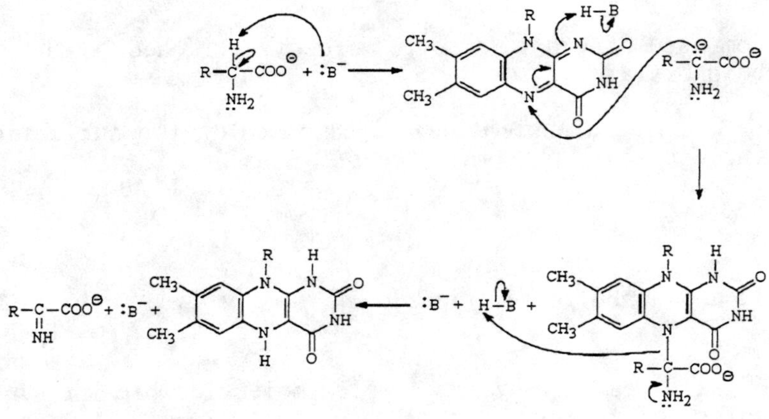

Section: 3

43. Propose a mechanism for the following reaction.

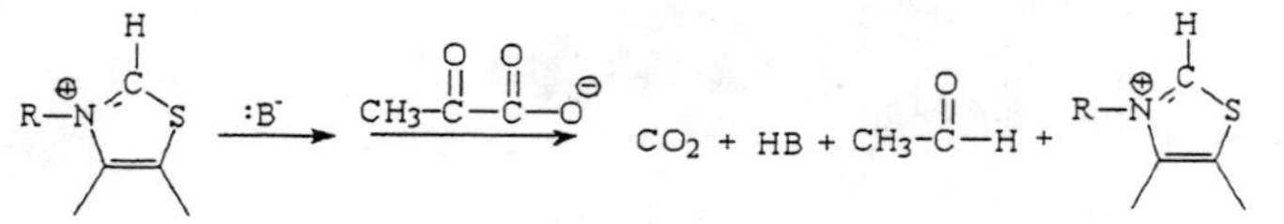

Answer:

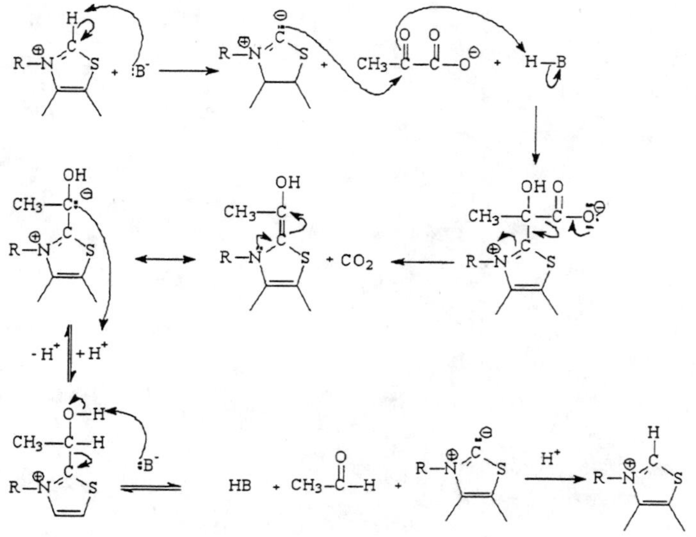

Section: 4

44. Show the transimination (trans-Schiffization) reaction that happens in all PLP-requiring enzymes in which the amino acid substrate forms a Schiff base with PLP, thereby freeing the lysine side chain.

Answer:

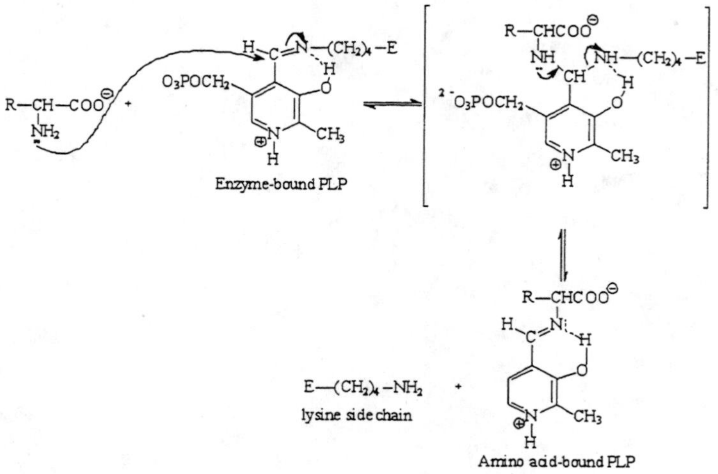

Enzyme-bound PLP

lysine side chain

Amino acid-bound PLP

Section: 6

45. Propose a mechanism for the following coenzyme B_{12}-dependent enzyme-catalyzed reaction.

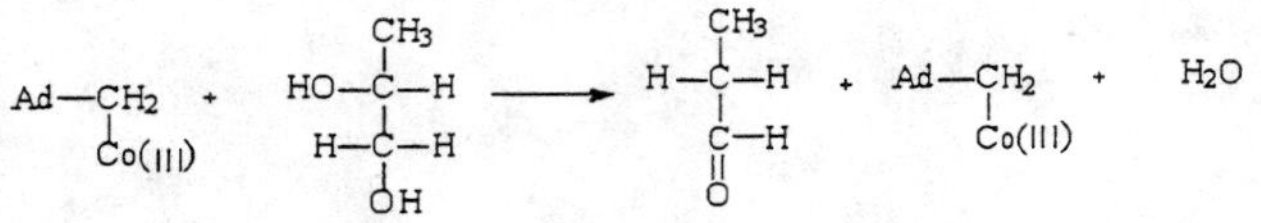

Answer:

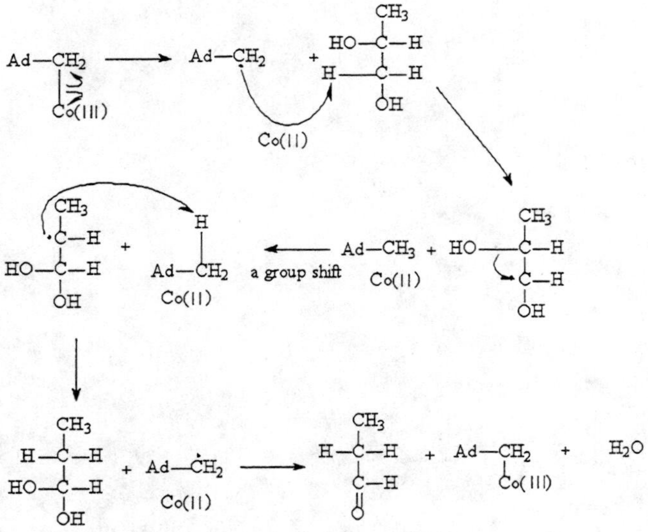

Section: 7

46. Propose a mechanism for the vitamin KH_2-catalyzed carboxylation of glutamate shown below:

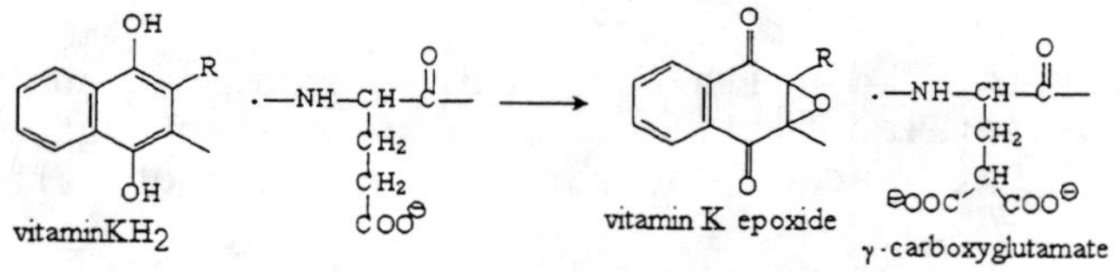

vitaminKH$_2$ → vitamin K epoxide, γ-carboxyglutamate

Answer:

vitamin K epoxide

γ-carboxyglutamate

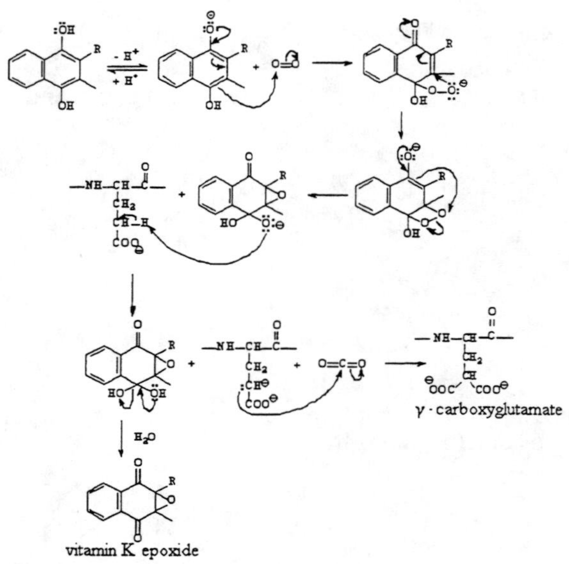

Section: 9

Chapter 23: Lipids

MULTIPLE CHOICE

1. Which of the following statements is <u>not</u> true about fatty acids?

   a. Fatty acids are carboxylic acids with long hydrocarbon side chains.
   b. The double bonds in unsaturated fatty acids are always conjugated.
   c. Most naturally occuring fatty acids contain even numbers of carbons and are unbranched.
   d. Fatty acids can be saturated or unsaturated.
   e. Physical properties of fatty acids depend on the length of the hydrocarbon chain and the degree of unsaturation.

   Answer: b Section: 1

2. Which of the following statements is <u>true</u> for both cholesterol and vitamin A?

   a. Both are vitamins.
   b. Both are proteins.
   c. Both are lipids.
   d. Both are steroids.
   e. Both are fatty acids.

   Answer: c Section: 1

3. Which of the following best explains why the melting points of saturated fats increase with increasing molecular weight?

   a. Decreased polarity
   b. Decreased hydrogen bonding
   c. Increased hydrogen bonding
   d. Decreased intermolecular van der Waal's interactions
   e. Increased intermolecular van der Waal's interactions

   Answer: e Section: 1

4. Which of the following statements best describes the structure of waxes?

   a. Long-chain unsaturated carboxylic acids.
   b. Long-chain saturated carboxylic acids.
   c. Long-chain esters.
   d. Short-chain esters.
   e. Long-chain anhydrides.

   Answer: c Section: 2

5. Which of the following molecules is known as a triglyceride?

a. 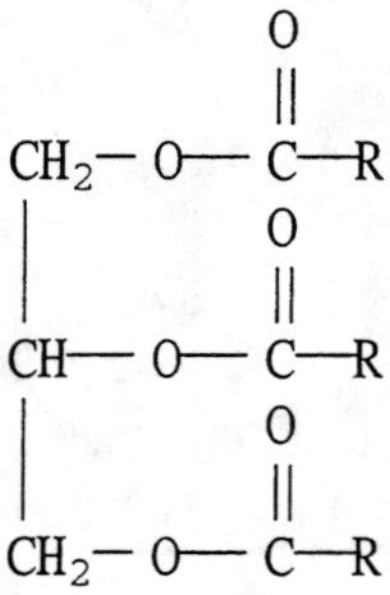
$$\begin{array}{l} CH_2-O-\overset{\overset{\displaystyle O}{\|}}{C}-R \\ |\\ CH-O-\overset{\overset{\displaystyle O}{\|}}{C}-R \\ | \\ CH_2-O-\overset{\overset{\displaystyle O}{\|}}{C}-R \end{array}$$

b.
$$\begin{array}{l} CH_2-\overset{\overset{\displaystyle O}{\|}}{C}-OR \\ |\\ CH-\overset{\overset{\displaystyle O}{\|}}{C}-OR \\ | \\ CH_2-\overset{\overset{\displaystyle O}{\|}}{C}-OR \end{array}$$

c.
$$\begin{array}{l} CH_2-\overset{\overset{\displaystyle O}{\|}}{C}-R \\ |\\ CH-\overset{\overset{\displaystyle O}{\|}}{C}-R \\ | \\ CH_2-\overset{\overset{\displaystyle O}{\|}}{C}-R \end{array}$$

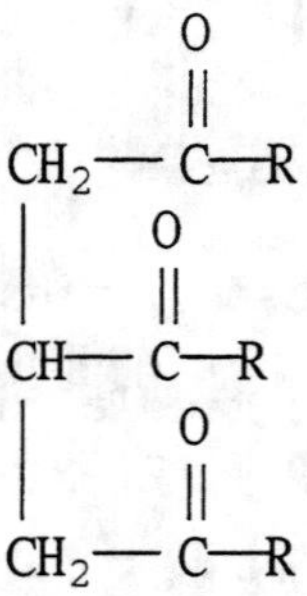

d. 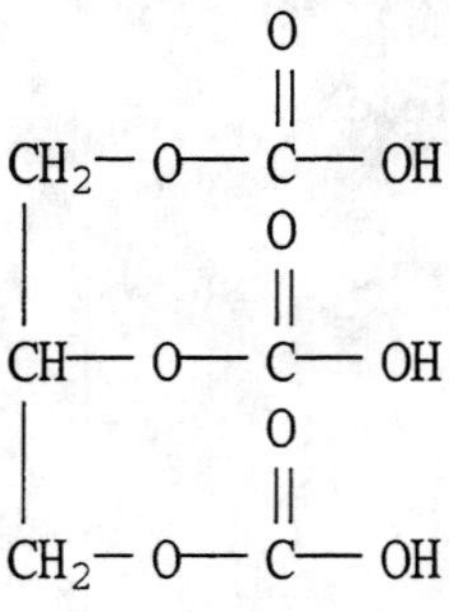
$$\begin{array}{l} CH_2-O-\overset{\overset{\displaystyle O}{\|}}{C}-OH \\ |\\ CH-O-\overset{\overset{\displaystyle O}{\|}}{C}-OH \\ | \\ CH_2-O-\overset{\overset{\displaystyle O}{\|}}{C}-OH \end{array}$$

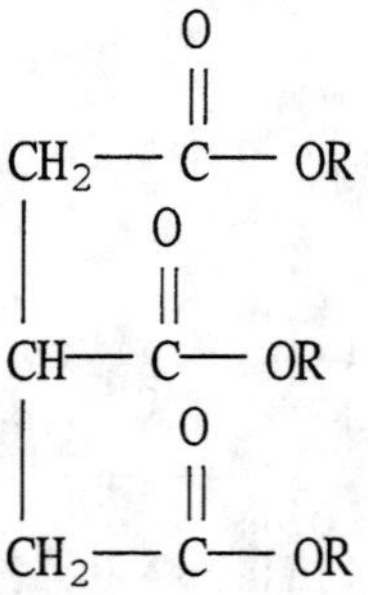

e. 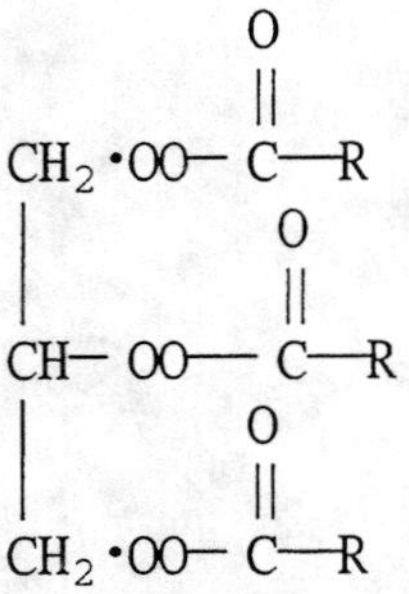 Structure:

$$\begin{array}{l} \qquad\qquad\quad O \\ \qquad\qquad\quad \| \\ CH_2{\cdot}OO-C-R \\ | \qquad\qquad\quad O \\ | \qquad\qquad\quad \| \\ CH-OO-C-R \\ | \qquad\qquad\quad O \\ | \qquad\qquad\quad \| \\ CH_2{\cdot}OO-C-R \end{array}$$

Answer: a Section: 3

6. Which of the following statements is <u>not</u> true about triacylglycerols?

   a. When solids and semisolids at room temperature, they are called fats.
   b. When liquids at room temperature, they are called oils.
   c. When hydrolyzed, they produce glycerol and carboxylate salts.
   d. Triacylglycerols with low melting points are composed of saturated fatty acids, causing them to be liquids at room temperature.
   e. Commonly known soap is the salt of a long-chain fatty acid which can be obtained from triacylglycerols.

Answer: d Section: 3

7. Which of the following statements is <u>not</u> true about phospholipids?

   a. They are similar to triacylglycerols except that the middle OH group of glycerol reacts with a phosphate rather than with a fatty acid.
   b. They constitute a major component of cell membranes.
   c. Phosphatidic acid is a phospholipid.
   d. The C-2 carbon of glycerol in phosphoacylglycerols has the R configuration.
   e. Cephalin and lecithin are examples of phospholipids.

Answer: a Section: 4

8. Which of the following molecules is known as phosphatidic acid?

a.
```
                 O
                 ||
CH2 — O — C — R1
 |              O
 |              ||
CH — O — P — OH
 |               |
 |               O⊖
 |                 O
 |                 ||
CH2 — O — C — R2
```

b.
```
                 O
                 ||
CH2 — O — C — R1
 |               O
 |               ||
CH — O — P — OH
 |               |
 |               O⊖
 |                 O
 |                 ||
CH2 — O — C — R2
```

c.
```
                 O
                 ||
CH2 — O — C — R1
 |              O
 |              ||
CH — O — C — R2
 |              O
 |              ||
CH2 — O — P — OH
                |
                O⊖
```

d.
```
                 O
                 ||
CH2 — O — P — OH
 |               |
 |               O⊖
 |                 O
 |                 ||
CH — O — C — R1
 |                O
 |                ||
CH2 — O — C — R2
```

```
              O
              ||
e.  CH2— O—P— OR
    |         |
    |         O⊖
    |         O
    |         ||
    CH— O— C—R1
    |         O
    |         ||
    CH2— O— C—R2
```

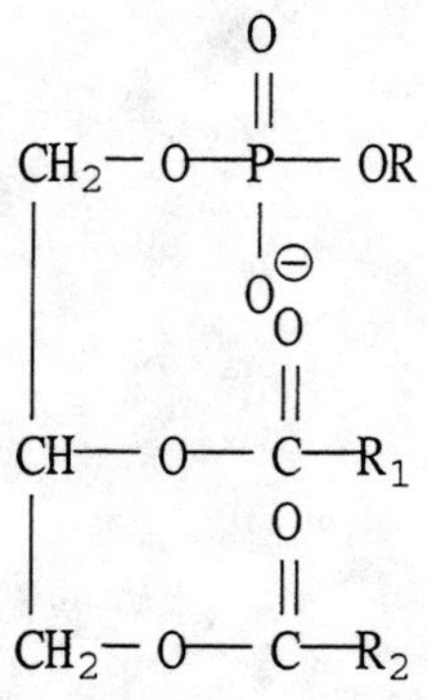

Answer: c Section: 4

9. Which of the following is the skeleton for prostaglandins?

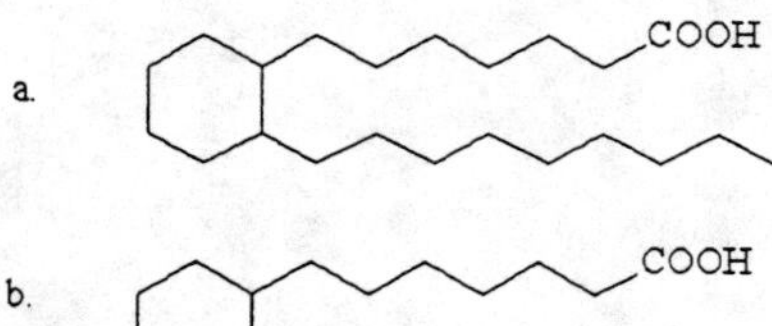

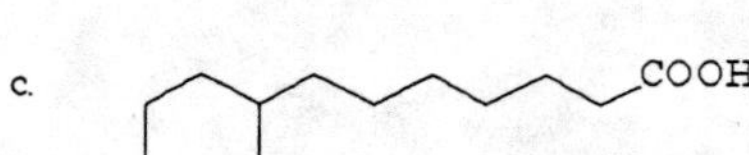

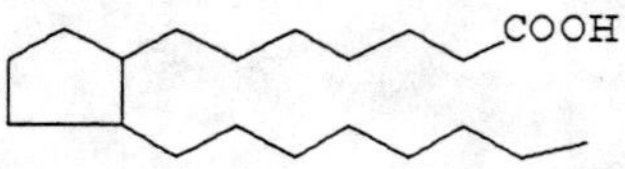

d.

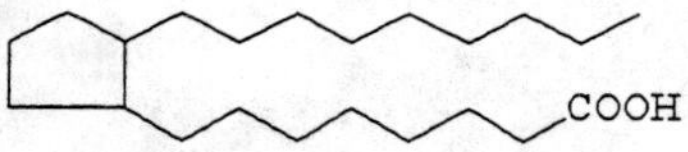

COOH

e.

$CONH_2$

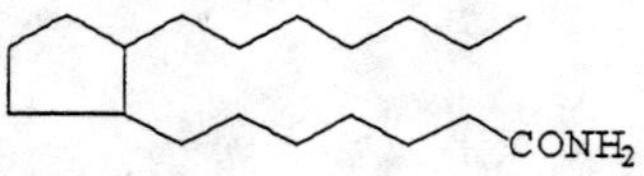

Answer: c Section: 5

10. Which of the following fatty acids cannot be synthesized by mammals but must be included in the diet because it is needed to synthesize arachidonic acid which in turn will synthesize prostaglandins?

b. Palmitic acid COOH

c. Arachidic acid COOH

d. Linoleic acid COOH

e. Oleic acid COOH

Answer: d Section: 1

11. Why is aspirin used as an anti-inflammatory drug?

a. Aspirin enhances the production of prostaglandins.
b. Aspirin inhibits the production of prostaglandins.
c. Aspirin promotes the activity of cyclooxygenase.
d. Aspirin transfers a hydroxy group to a serine hydroxyl group of the enzyme.
e. Aspirin causes an increase in the rate of blood clotting.

Answer: b Section: 5

12. Which of the following statements is true about terpenes?

a. They are a class of lipids that can be isolated from plants.
b. They contain carbon atoms in multiples of five.
c. They are composed of isoprene units joined together in a head-to-tail fashion.
d. Tetraterpenes have 40 carbon atoms.
e. All of the above

Answer: e Section: 6

13. How many isoprene units are in a sesquiterpene?

a. 1
b. 2
c. 3
d. 4
e. 5

Answer: c Section: 6

14. How many isoprene units are in β-carotene?

    a. 4
    b. 5
    c. 6
    d. 7
    e. 8

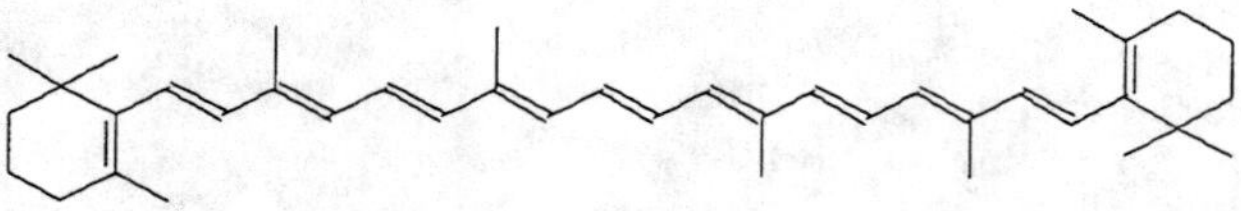

Answer: e Section: 6

15. The five-carbon compound used for the biosynthesis of terpenes is called isopentenyl pyrophosphate. Which of the following is the structure of isopentenyl pyrophosphate?

    a. 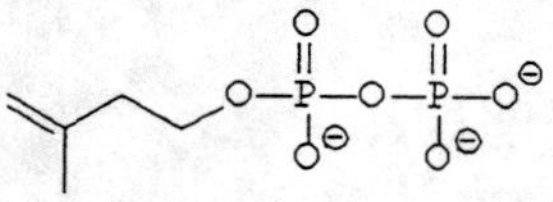
    b. 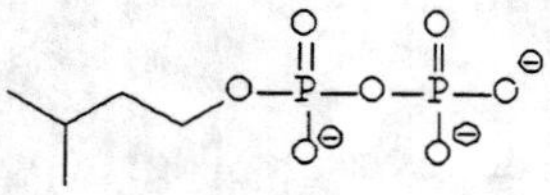
    c. 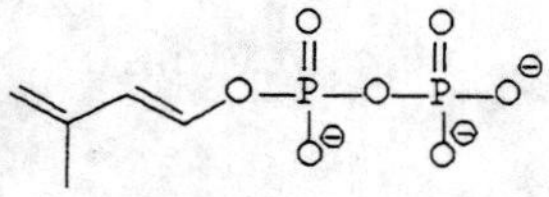
    d. 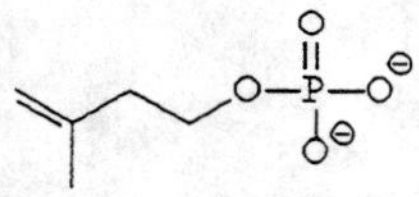
    e.

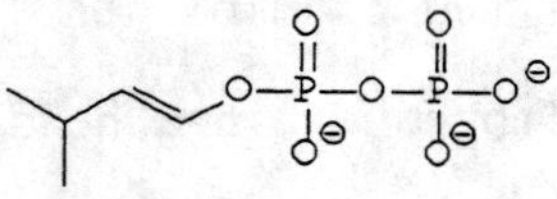

Answer: a Section: 8

16. Squalene is an important precursor of cholesterol which is the precursor of all other steroids. Squaline is prepared from two molecules of farnesyl pyrophosphate that are joined head-to-head.

    What is the structure of farnesyl pyrophosphate?

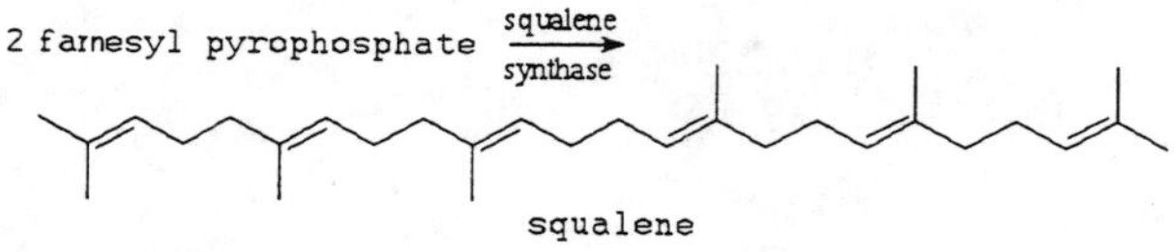

2 farnesyl pyrophosphate $\xrightarrow[\text{synthase}]{\text{squalene}}$

squalene

a. 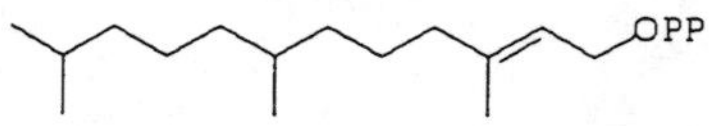 OPP

b. 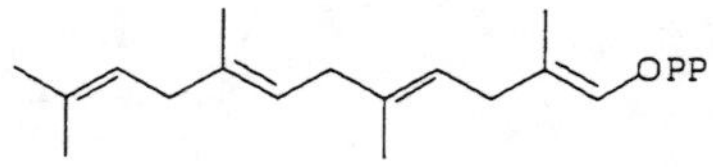 OPP

c. 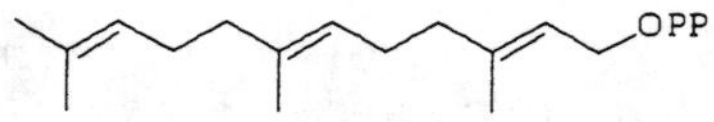 OPP

d. 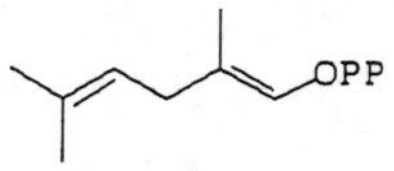 OPP

e. 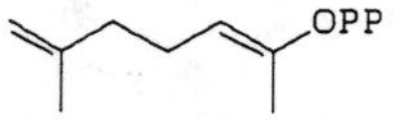 OPP

Answer: c Section: 10

17. Which of the following statements is not true about steroids?

    a. They are lipids.
    b. They are hormones.
    c. They are synthesized in the glands and delivered by the bloodstream to tissues in order to stimulate or inhibit certain processes.
    d. All steroids contain a tetracyclic ring system.
    e. In steroids, the B, C, and D rings are all cis fused.

    Answer: e Section: 9

18. Which of the following is the precursor for all steroids?

    a. Cholesterol
    b. Cortisone
    c. Progesterone
    d. Estrone
    e. Androsterone

    Answer: a Section: 9

19. How many chirality centers does cholesterol have?

    a. 4
    b. 5
    c. 6
    d. 7
    e. 8

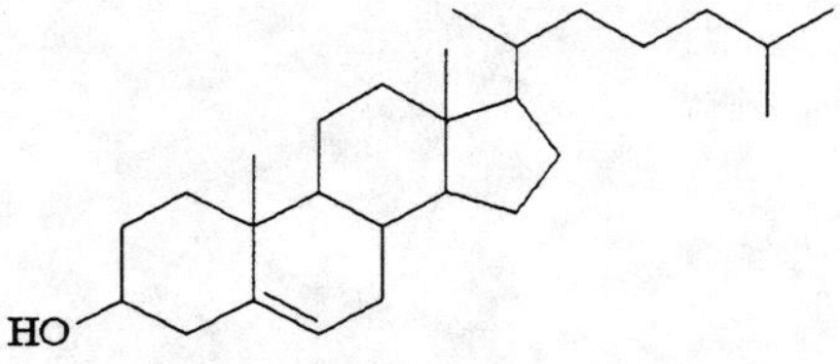

Cholesterol

Answer: e Section: 9

20. Which of the following classes of hormones are involved in preparing the lining of the uterus for implantation of an ovum and are essential for the maintenance of pregnancy?

    a. Estrogen
    b. Androgen
    c. Mineralocorticoids
    d. Glucocorticoids
    e. Progestins

Answer: e Section: 9

21. To which class of hormones does the steroid cortisone belong?

    a. Estrogen
    b. Androgen
    c. Mineralocorticoids
    d. Glucocorticoids
    e. Progestins

Answer: d Section: 9

22. Which of the following statements best describes the structural difference between testosterone, a male sex hormone, and progesterone, a female hormone?

    a. They only differ in the substituent at C-3.
    b. They only differ in the substituent at C-13 and C-17.
    c. They only differ in the substituent at C-13.
    d. They only differ in the substituent at C-17.
    e. They only differ in the substituent at C-3 and C-13.

Answer: b Section: 9

23. Which of the following statements best describes what anabolic steroids are?

    a. They are synthetic steroids.
    b. Steroids that aid in the development of muscles.
    c. When taken in high doses, they regulate heartbeat, kidney and liver functions, and increase muscle mass.
    d. a and b
    e. a, b, and c

    Answer: d Section: 9

SHORT ANSWER

24. Palmitic acid (hexadecanoic acid) has a melting point of 63° C. Palmitoleic acid (9-hexadecenoic acid) has a melting point of 32° C. Explain this difference in melting points.

    Answer:
    Palmitoleic acid, an unsaturated fat, has a much lower melting point than the saturated palmitic acid, although they both have comparable molecular weights. This is true because the double bonds in an unsaturated fatty acid have the cis configuration. This produces a bend in the chains preventing them from packing together as tightly as saturated fats. This results in a decrease in intermolecular interactions which lowers the melting point.
    Section: 1

25. Phosphoacylglycerols form membranes by arranging themselves in a lipid bilayer. What does lipid bilayer mean?

    Answer:
    It means that the polar heads of the phosphoacylglycerols are on the outside of the bilayer, while the interior of the bilayer consists of the fatty acid chains and cholesterol.
    Section: 4

26. Give the chemical structure for sphingosine, a molecule that is found in sphingolipids.

    Answer:
    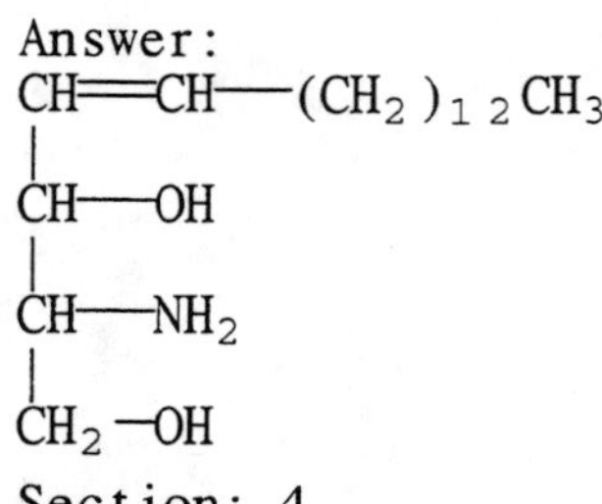

    Section: 4

27. What is the difference between a sphingomyelin and a cerebroside? Give an example.

Answer:
A sphingomyelin is a sphingolipid that is also a phospholipid. Cerebroside is a sphingolipid that is not a phospholipid. The bottom OH in a cerebroside is bonded to a sugar residue in a β-linkage.

Sphingomyelin (A phospholipid)	Cerebroside (Not a Phospholipid)
$CH = CH - (CH_2)_{12}CH_3$	$CH = CH - (CH_2)_{12}CH_3$
|	|
$CH - OH$	$CH - OH$
|	|
$CH - NH_2$	$CH - NH_2$
|	|
$CH_2 - O$ — phosphate	$CH_2 - O$ — Sugar

Sphingomyelin
(A phospholipid)

Cerebroside
(Not a Phospholipid)

Section: 4

28. In the prostaglandin called $PGF_{2\alpha}$, what does the 2α stand for? Give an example.

Answer:
The two stands for the number of double bonds in the side chain, and the indicates that the two OH groups are on the same side of the ring.

Ex

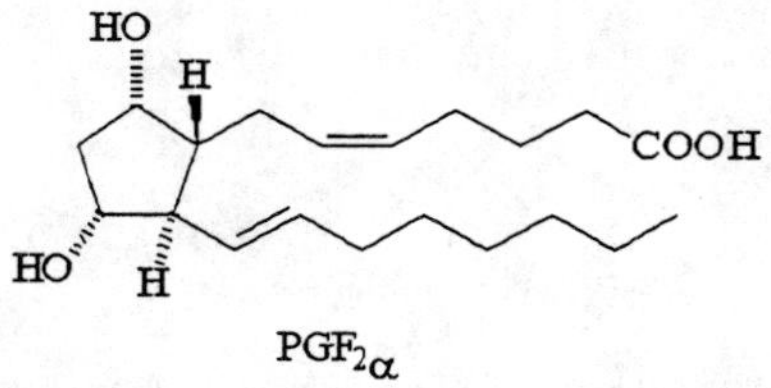

HO
H
COOH
HO
H

$PGF_{2\alpha}$

Section: 5

29. Show the steps involved in the formation of prostaglandin PGE_2 from arachidonic acid:

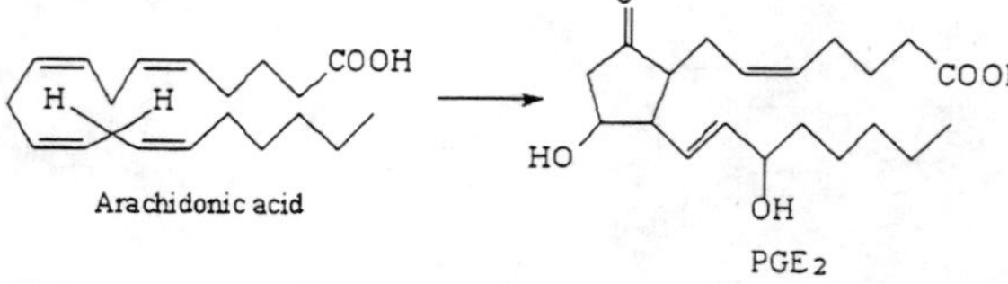

Arachidonic acid → PGE_2

Answer:

Arachidonic acid —cyclooxygenase→ ... hydroperoxidase → PGH_2 → PGE_2

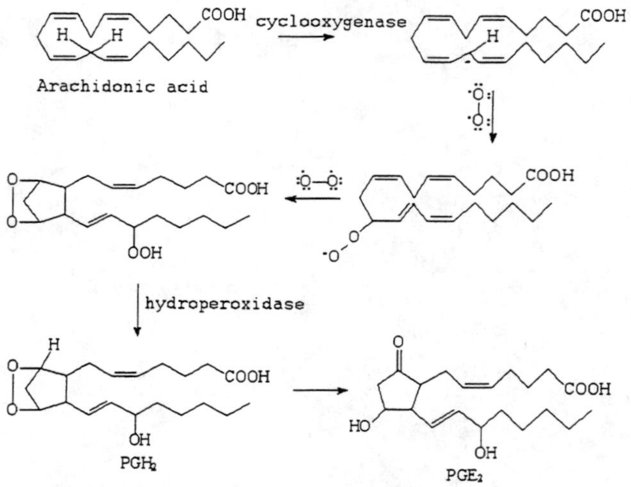

Section: 5

30. In the rod cells of the eye, retinol (vitamin A) can be oxidized and then isomerized into an aldehyde and a cis double bond at C-11 to form a molecule called 11Z-retinal. Propose a structure for 11Z-retinal.

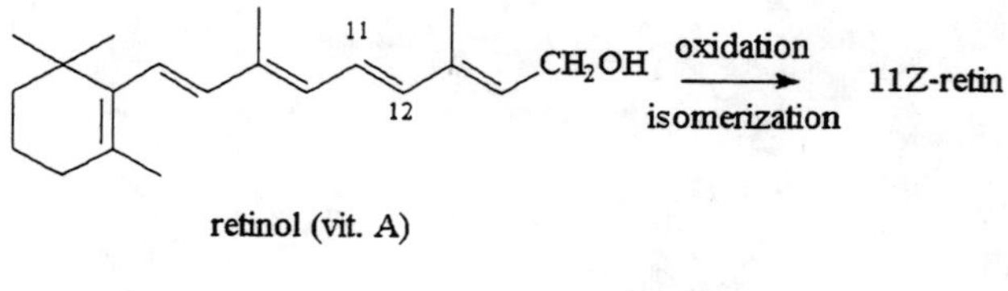

retinol (vit. A) —oxidation, isomerization→ 11Z-retin

Answer:

11Z-retinal

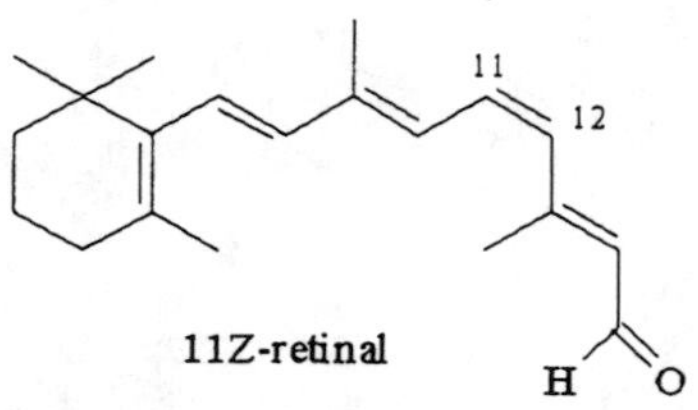

Section: 7

31. Propose a mechanism for the reaction, of dimethylallyl pyrophosphate with isopentenyl pyrophosphate to form geranyl pyrophosphate.

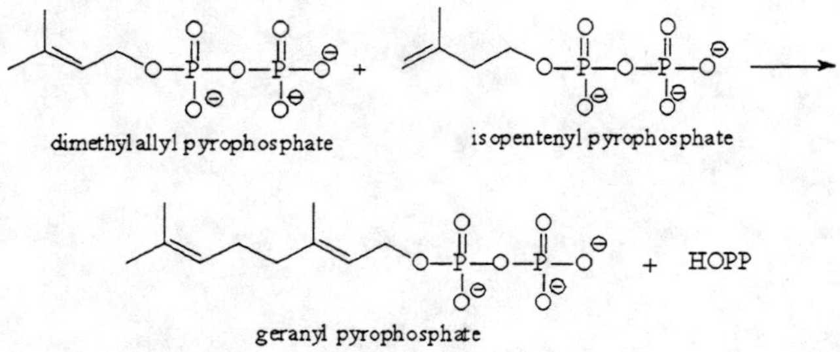

dimethylallyl pyrophosphate + isopentenyl pyrophosphate ⟶

geranyl pyrophosphate + HOPP

Answer:

$^{\ominus}$:ÖPP +

geranyl pyrophosphate + HOPP

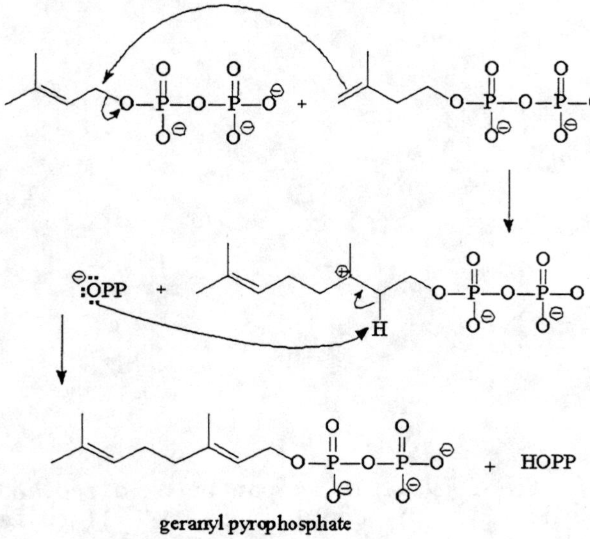

Section: 8

32. Draw the steroid ring system with the six-membered rings in the chair conformation.

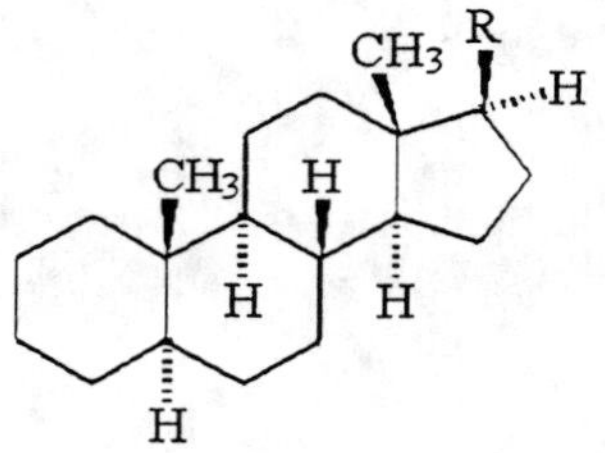

Answer:

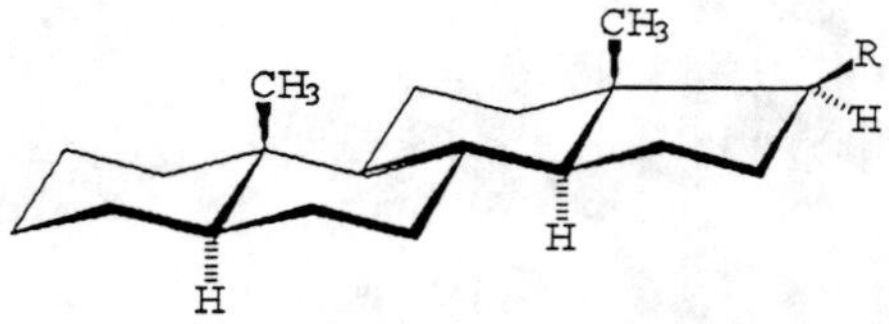

Section: 9

33. Show the numbering system of the steroid ring.

Answer:

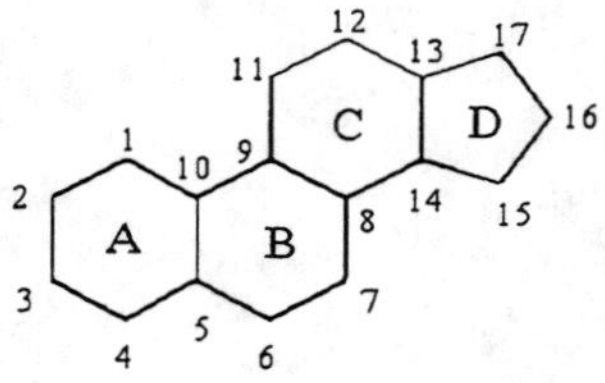

Section: 9

34. Show the biosynthetic pathway in the synthesis of lanosterol (a precursor to cholesterol) from squalene.

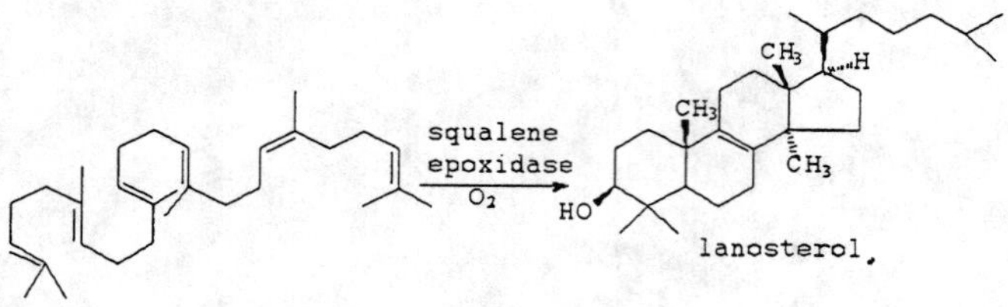

squalene epoxidase O_2

CH_3 H CH_3 CH_3 HO

lanosterol.

Answer:

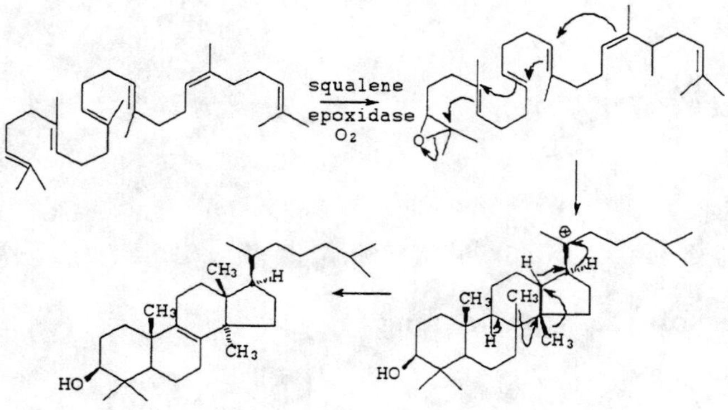

squalene epoxidase O_2

CH_3 H CH_3 CH_3 HO

lanosterol

H H CH_3 CH_3 H CH_3 HO

Section: 10

35. Define <u>lipid</u>

Answer:
Lipids are substances that can be extracted from cells and tissues by nonpolar organic solvents.

MULTIPLE CHOICE

36. Which of the following terms best describes the compound below?
$CH_3(CH_2)_{26}CO_2CH_2(CH_2)_{32}CH_3$

a. a fat
b. a wax
c. a terpene
d. an unsaturated triglyceride
e. an oil

Answer: b Section: 2

37. Which of the following terms best describes the compound below?

```
              O
              ||
 CH2— O— C—(CH2)18CH3
             O
 |           ||
 CH— O— C—(CH2)16CH3
             O
 |           ||
 CH2— O— C—(CH2)18CH3
```

a. a wax
b. a saturated triglyceride
c. a terpene
d. a prostaglandin
e. a lecithin

Answer: b Section: 3

38. Which of the following terms best describes the compound below?

a. a steroid
b. an unsaturated triglycaride
c. a soap
d. a complex lipid
e. a terpene

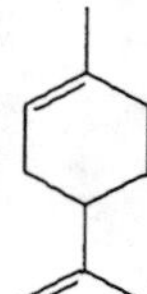

Answer: e Section: 6

39. Which of the following terms best describes the compound below?

a. a prostaglandin
b. a fatty acid
c. a steriod
d. a cephalin
e. an essential oil

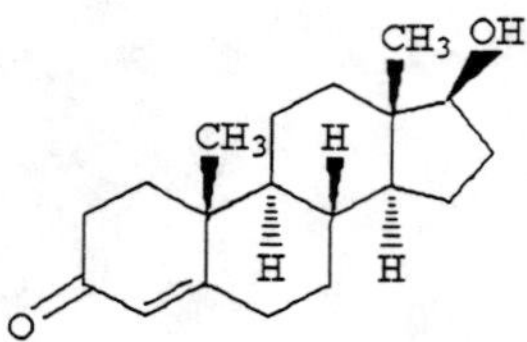

Answer: c Section: 9

40. Which of the following terms best describes the compound below?

$CH_3(CH_2)_{12}CO_2H$

a. a fatty acid
b. an oil
c. a wax
d. a soap
e. a phosphatidic acid

Answer: a Section: 1

41. Which of the following terms best describes the compound below?

$CH_3(CH_2)_7CH{=}CH(CH_2)_7CO_2H$

a. an unsaturated fatty acid
b. a triglyceride
c. a synthetic detergent
d. a micelle
e. isoprene

Answer: a Section: 1

SHORT ANSWER

42. Was the compound shown below more likely isolated from an animal or a plant? Explain.

$$\begin{array}{l} CH_2{-}O{-}\overset{\overset{\displaystyle O}{\|}}{C}{-}(CH_2)_7CH{=}CH(CH_2)_7CH_3 \\ | \\ CH{-}O{-}\overset{\overset{\displaystyle O}{\|}}{C}{-}(CH_2)_7CH{=}CHCH_2CH{=}CH(CH_2)_4CH_3 \\ | \\ CH_2{-}O{-}\overset{\overset{\displaystyle O}{\|}}{C}{-}(CH_2)_7CH{=}CH(CH_2)_7CH_3 \end{array}$$

Answer:
The compound shown is a polyunsaturated triglyceride. Such triglycerides are liquid at room temperature (ie, are oils) and are more likely to have been isolated from a plant.

Section: 3

MULTIPLE CHOICE

43. Which of the following terms best describes the compound below?

```
                O
                ||
CH2— O— C—(CH2)14CH3
 |              O
 |              ||
CH— O— C—(CH2)14CH3
 |              O
 |              ||
CH2— O—P— OCH2CH2NH3+
                |
                O-
```

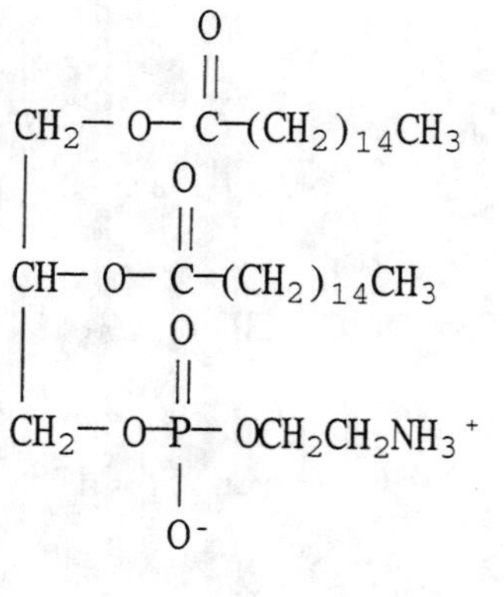

a. a phospholipid
b. a phosphogylceride
c. a cephalin
d. a molecule which contains a polar head group
e. all of the above

Answer: e Section: 4

44. Which of the following terms best describes the compound below?

a. a lecithin
b. a diterpene
c. a prostaglandin
d. a glyceride
e. a steroid

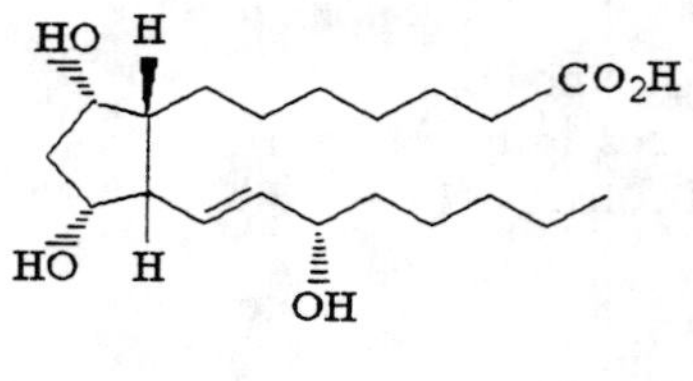

HO H
CO_2H
HO H
OH

Answer: c Section: 5

SHORT ANSWER

45. When spermaceti is heated in aqueous potassium hydroxide, $CH_3(CH_2)_{14}CH_2OH$ and $CH_3(CH_2)_{14}CO_2^- K^+$ are formed. Provide the structure of spermaceti.

Answer: $CH_3(CH_2)_{14}CO_2CH_2(CH_2)_{14}CH_3$ Section: 2

46. Lauric acid has the formula $CH_3(CH_2)_{10}CO_2H$. Provide the structure of the triglyceride which was formed from three equivalents of lauric acid.

Answer:

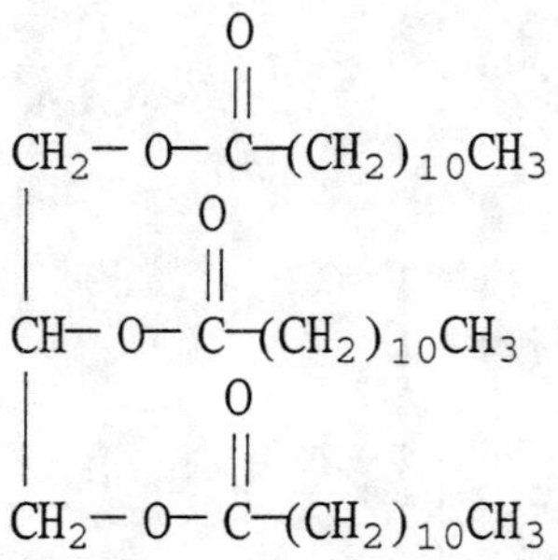

$$\begin{array}{l} CH_2-O-\overset{\overset{\displaystyle O}{\|}}{C}-(CH_2)_{10}CH_3 \\ | \\ CH-O-\overset{\overset{\displaystyle O}{\|}}{C}-(CH_2)_{10}CH_3 \\ | \\ CH_2-O-\overset{\overset{\displaystyle O}{\|}}{C}-(CH_2)_{10}CH_3 \end{array}$$

Section: 3

47. How is a fat distinguished from an oil, (a) based on a physical property difference and (b) based on a difference in molecular structure?

Answer:
Fats have higher melting points than oils and tend to have fewer carbon-carbon double bonds (sites of unsaturation) present.

Section: 3

48. Why and in what fashion does the presence of a double bond in a fatty acid affect its melting point relative to the system where the double bond is absent?

Answer:
The presence of a double bond lowers the melting point. The double bond introduces a "kink" in the carbon chain which retards efficient molecular packing and thus hinders solidification.

Section: 3

49. What does the term <u>polyunsaturated</u> mean when use to describe a triglyceride?

Answer:
This term indicates the presence of several carbon-carbon double bonds in the fatty acid residues of the triglyceride.

Section: 3

50. When a vegetable oil is partially hydrogenated, how is its structure altered and how is its melting point affected?

Answer:
H_2 is added across some of the carbon-carbon double bonds present. This decreases the number of sites of unsaturation and consequently increases the degree of saturation. The melting point increases.

Section: 3

51. Draw the structure of the product which results when the compound shown below is completely hydrogenated using excess H_2.

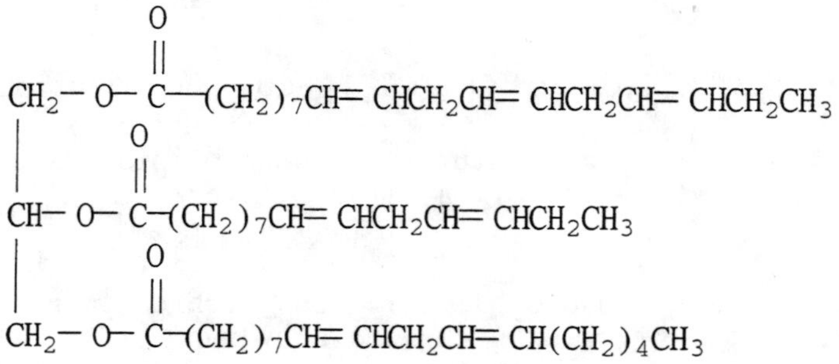

$$\begin{array}{l} CH_2-O-\overset{O}{\overset{\|}{C}}-(CH_2)_7CH=CHCH_2CH=CHCH_2CH=CHCH_2CH_3 \\ | \\ CH-O-\overset{O}{\overset{\|}{C}}-(CH_2)_7CH=CHCH_2CH=CHCH_2CH_3 \\ | \\ CH_2-O-\overset{O}{\overset{\|}{C}}-(CH_2)_7CH=CHCH_2CH=CH(CH_2)_4CH_3 \end{array}$$

Answer:

$$\begin{array}{l} CH_2-O-\overset{O}{\overset{\|}{C}}-(CH_2)_{16}CH_3 \\ | \\ CH-O-\overset{O}{\overset{\|}{C}}-(CH_2)_{16}CH_3 \\ | \\ CH_2-O-\overset{O}{\overset{\|}{C}}-(CH_2)_{16}CH_3 \end{array}$$

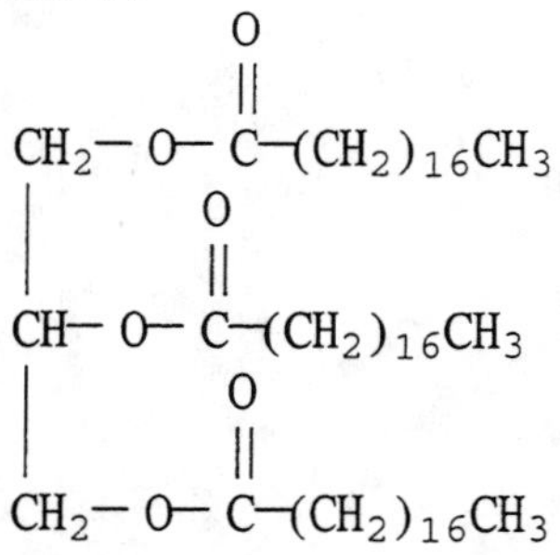

Section: 3

52. Provide the structures of the major organic products which result when tristearin (shown below) is heated in an aqueous solution of sodium hydroxide.

$$
\begin{array}{l}
\qquad\qquad\quad O \\
\qquad\qquad\quad \| \\
CH_2-O-C-(CH_2)_{16}CH_3 \\
| \qquad\qquad\; O \\
| \qquad\qquad\; \| \\
CH-O-C-(CH_2)_{16}CH_3 \\
| \qquad\qquad\; O \\
| \qquad\qquad\; \| \\
CH_2-O-C-(CH_2)_{16}CH_3
\end{array}
$$

Answer: $HOCH_2CHOHCH_2OH$ and $CH_3(CH_2)_{16}CO_2{}^-Na^+$ Section: 3

53. Explain which structural features of phospholipids make them suitable components from which to form a lipid bilayer.

Answer:
Phospholipids have a hydrophilic region which contains an ionized phosphatidic acid or ester group and a hydrophobic portion which contains two fatty acid esters.

Section: 4

54. Provide a sketch of a lipid bilayer. Label the polar and nonpolar regions.

Answer:

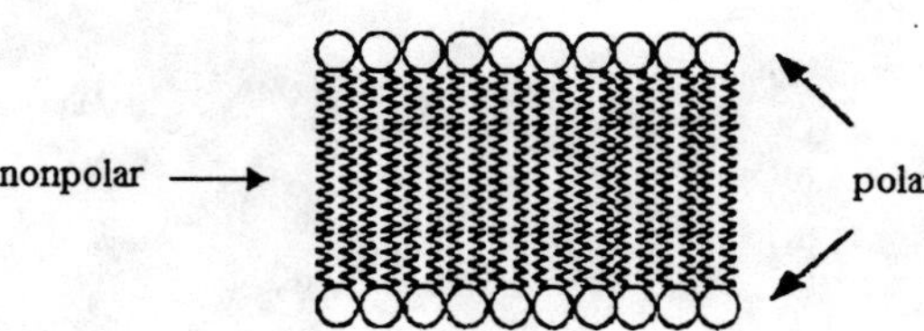

Section: 4

MULTIPLE CHOICE

55. Most of the principal sex hormones in humans are:

    a. triglycerides.
    b. prostaglandins.
    c. steroids.
    d. phospholipids.
    e. terpenes.

    Answer: c Section: 9

SHORT ANSWER

56. Draw a neat conformational structure of the compound below.

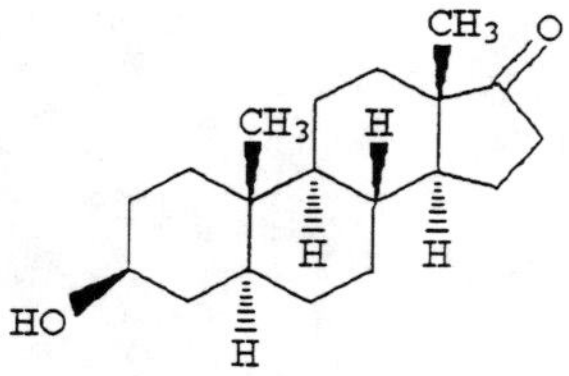

    Answer:

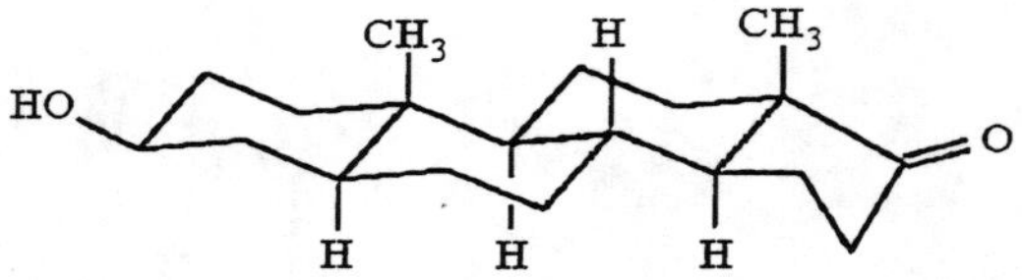

57. How would you prepare $CH_3(CH_2)_{14}CH_2NH_2$ from palmitic acid, $CH_3(CH_2)_{14}CO_2H$?

    Answer:
    1. $SOCl_2$
    2. NH_3
    3. $LiAlH_4$

58. A terpene is believed to have one of the structures below. Based on the isoprene rule, which structure is more likely. Explain your reasoning.

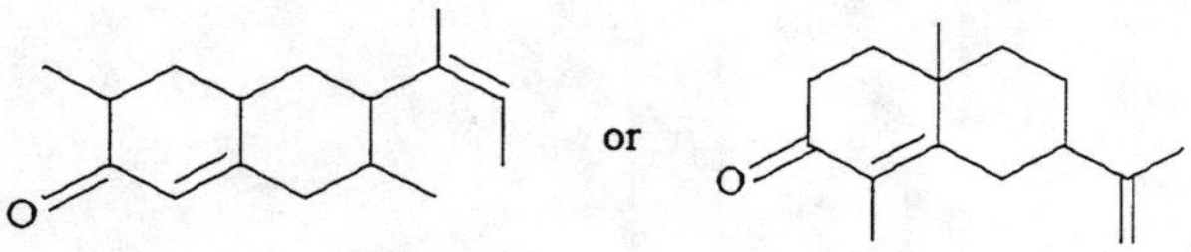

Answer:
This compound contains an integral number of isoprene units.

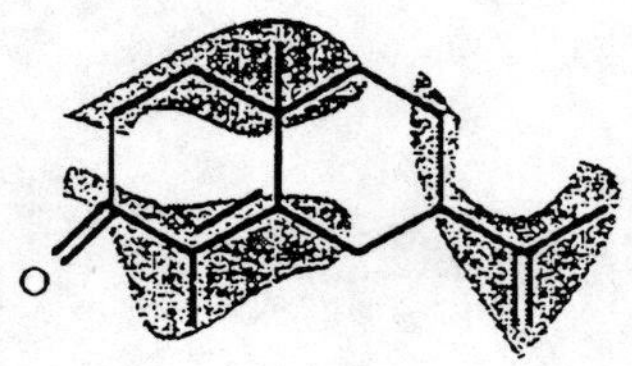

Section: 6

59. Circle the isoprene units present in menthol below.

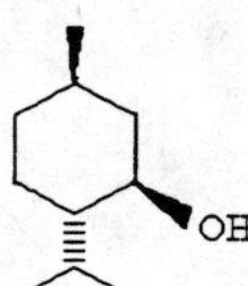

Answer:

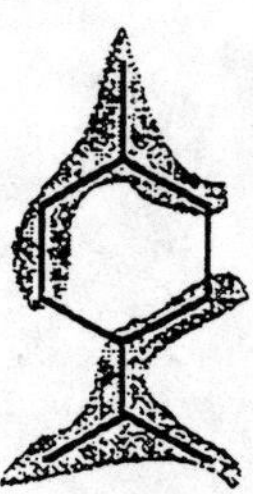

Section: 6

60. Would the addition of HCl to the double bond of oleic acid, $CH_3(CH_2)_7CH{=}CH(CH_2)_7CO_2H$, occur with high regioselectivity? Explain.

Answer:
No. The two possible carbocation intermediates are very similar in structure and stability and would thus be formed at comparable rates.

Chapter 24: Nucleosides, Nucleotides, and Nucleic Acids

MULTIPLE CHOICE

1. Which of the following statements best explains what nucleic acids are?

   a. They are acidic in nature.
   b. They are the carriers of genetic information.
   c. DNA and RNA are collectively known as nucleic acids.
   d. Nucleic acids are chains of five-membered ring sugars, each bonded to a heterocyclic amine, linked by phosphate groups.
   e. All of the above

   Answer: e

2. Which of the following phosphate groups is found in nucleic acids?

   a. An orthophosporic acid $HO-P(=O)(OH)-OH$

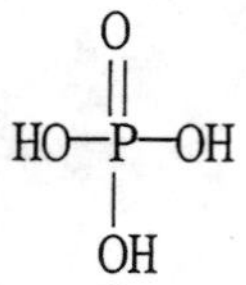

   b. A phosphomonoester $HO-P(=O)(OH)-OR$

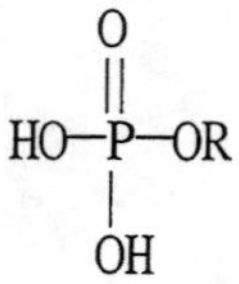

   c. A phosphodiester $RO-P(=O)(OH)-OR$

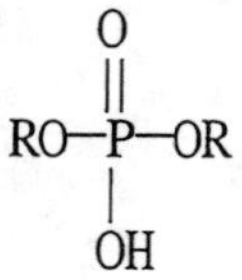

   d. A phosphotiester $RO-P(=O)(OR)-OR$

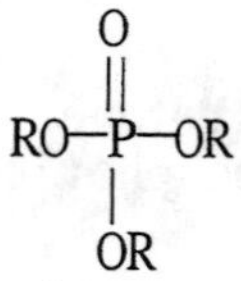

   e. A phosphate ion $^{\ominus}O-P(=O)(O^{\ominus})-O^{\ominus}$

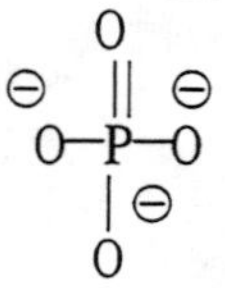

   Answer: c Section: 7

3. Which of the following are known as pyrimidines?

   a. Cytosine and thymine
   b. Adenine and cytosine
   c. Adenine and thymine
   d. Guanine and thymine
   e. Adenine and guanine

   Answer: a Section: 1

4. Which of the following bases distinguish DNA from RNA?

   a. Adenine and cytosine
   b. Guanine and uracil
   c. Thymine and uracil
   d. Uracil and guanine
   e. Adenine and thymine

   Answer: c Section: 1

5. Which of the following is the name of the nucleoside made from adenine and found in DNA?

   a. adenosine
   b. 2'-deoxyadenosine
   c. 2-deoxyadenosine
   d. 2'-oxyadenosine
   e. 2'-deoxyadenine

   Answer: b Section: 1

6. What does the name ATP stand for?

   a. Adenine triphosphate
   b. Adenosine tetraphosphate
   c. Adenine tetraphosphate
   d. Adenosine triphosphate
   e. Adenine triphosphoric acid

   Answer: d Section: 2

7. When phosphoric acid is heated it loses water to form an anhydride known as pyrophosphate or pyrophosphoric acid. What is the structure of pyrophosphoric acid?

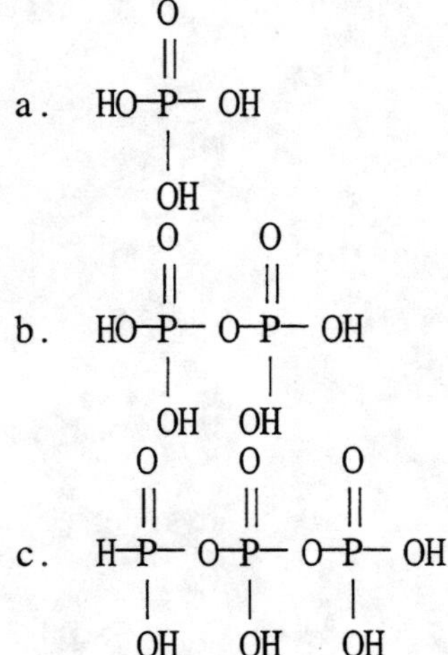

d.

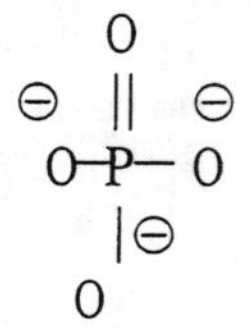

$$\ominus O-\overset{\overset{O}{||}}{\underset{\underset{O}{|\ominus}}{P}}-O^{\ominus}$$

e.

$$HO-\overset{\overset{O}{||}}{\underset{\underset{OH}{|}}{P}}-\overset{\overset{O}{||}}{\underset{\underset{OH}{|}}{P}}-OH$$

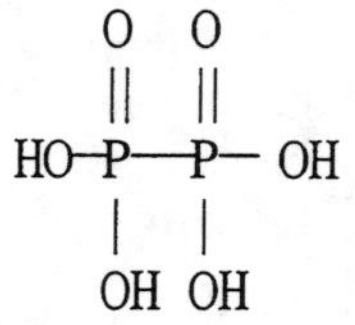

Answer: b   Section: 1

8. Which of the following statements best explains the source of energy behind the ATP molecule?

   a. The energy released when a phosphoanhydride bond of ATP is broken
   b. The energy released when adenosine binds to the phosphate
   c. The energy released when ribose binds to the phosphate
   d. The energy released when adenine binds to ribose
   e. The energy released when ATP is reduced

   Answer: a   Section: 2

9. Which of the following is the first step in the mechanism of the reaction shown below?

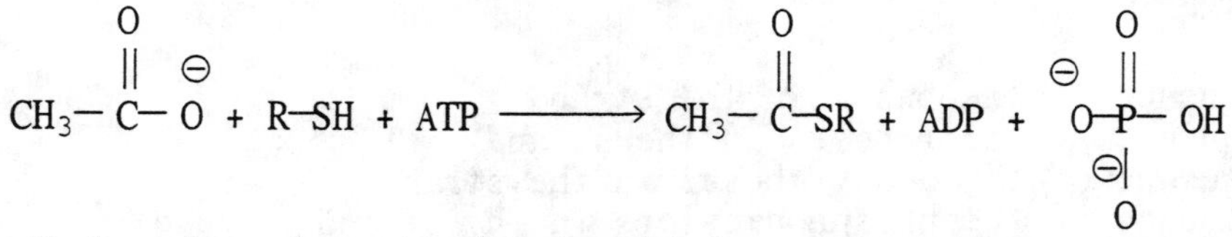

$$CH_3-\overset{\overset{O}{||}}{C}-O^{\ominus} + R-SH + ATP \longrightarrow CH_3-\overset{\overset{O}{||}}{C}-SR + ADP + {}^{\ominus}O-\overset{\overset{O}{||}}{\underset{\underset{O}{\ominus|}}{P}}-OH$$

   a. Carboxylate ion attacks the thiol.
   b. Carboxylate ion attacks the α-phosphorous of ATP.
   c. Carboxylate ion attacks the γ-phosphorous of ATP.
   d. Thiol molecule attacks the β-phosphorous of ATP.
   e. Thiol molecule attacks the acyl group.

   Answer: c   Section: 3

10. Which of the following reactions converts nucleotide subunits into nucleic acids?

   a. Nucleophilic attack on the α-phosphorous of ATP
   b. Nucleophilic attack on the β-phosphorous of ATP
   c. Nucleophilic attack on the γ-phosphorous of ATP
   d. a and b
   e. a, b and c

   Answer: a   Section: 3

11. Which of the following metal ions complexes with cellular ATP to neutralize some of its negative charges at the active site of an enzyme in order that nucleophilic substitution reactions proceed rapidly?

    a. Manganese
    b. Iron
    c. Cobalt
    d. Magnesium
    e. Calcium

    Answer: d Section: 3

12. Which of the following reaction sequences best describes the formation of the nucleic acid in the biosynthesis of DNA?

    a. Nucleophilic attack of a 5'-OH group of one nucleotide on the β-phosphorous of another
    b. Nucleophilic attack of a 5'-OH group of one nucleotide on the α-phosphorous of another
    c. Nucleophilic attack of a 3'-OH group of a nucleotide on the γ-phosphorous of another
    d. Nucleophilic attack of a 3'-OH group of one nucleotide on the β-phosphorous of another
    e. Nucleophilic attack of a 3'-OH group of one nucleotide on the α-phosphorous of another

    Answer: e Section: 7

13. Which of the following constitutes the primary structure of a nucleic acid?

    a. The sequence of the bases in the strand
    b. The sequence of the riboses in the strand
    c. The sequence of the deoxyriboses in the strand
    d. The sequence of the phosphodiesters in the strand
    e. The sequence of the 5'-OH groups in the strand

    Answer: a Section: 7

14. Which of the following statements is not true about the structure of DNA as proposed by Watson and Crick?

    a. The number of adenines in DNA is equal to the number of thymines.
    b. The number of cytosines is equal to the number of guanines.
    c. DNA consists of two strands of nucleic acids with the sugar-phosphate backbone on the inside and the bases on the outside.
    d. The chains in DNA are held together by hydrogen bonding.
    e. Adenine always pairs with thymine and guanine always pairs with cytosine.

    Answer: c Section: 7

15. How many hydrogen bonds exist between the cytosine/guanine base pairs?

    a. 1
    b. 2
    c. 3
    d. 4
    e. 5

    Answer: c Section: 7

16. Which of the following is the conventional method of writing the base sequence of a portion of a DNA chain?

    a. 5'-end--bases--5'-end
    b. 5'-end--bases--3'-end
    c. 3'-end--bases--5'-end
    d. 3'-end--bases--3'-end
    e. 5'-end--bases--2'-end

    Answer: b Section: 7

17. In which of the following helical forms does mainly all DNA in living organisms exist?

    a. Z-helix
    b. A-helix
    c. B-helix
    d. b and c
    e. a, b and c

    Answer: c Section: 8

18. How many billion base pairs are there in the total DNA of a human cell?

    a. 3
    b. 5
    c. 8
    d. 11
    e. 13

    Answer: a Section: 9

19. Which of the following statements best describes the meaning of transcription?

    a. The synthesis of rRNA from a DNA blueprint
    b. Transcription is synonymous with replication
    c. The synthesis of identical copies from a DNA blueprint
    d. The synthesis of mRNA from a DNA blueprint
    e. The synthesis of tRNA from a DNA blueprint

    Answer: d Section: 10

20. In an entire DNA strand, what percentage consists of genetic information and what percentage consists of introns?

    a. 90% genetic information, 10% introns
    b. 10% genetic information, 90% introns
    c. 50% genetic information, 50% introns
    d. 100% genetic information, 0% introns
    e. 90% genetic information, 10% exon

    Answer: b Section: 10

21. Which of the following statements is <u>not</u> true about RNA?

    a. It is single stranded.
    b. It is smaller than DNA.
    c. It has fewer base pairs than DNA.
    d. It exists exclusively in the nucleus of the cell.
    e. There are three kinds: rRNA, mRNA, and tRNA.

    Answer: d Section: 10

22. What is the composition of a ribosome?

    a. 40% protein, 60% rRNA
    b. 60% protein, 40% tRNA
    c. 40% protein, 60% tRNA
    d. 40% protein, 60% mRNA
    e. 100% protein, 0% rRNA

    Answer: a Section: 11

23. Which of the following is true about tRNA?

    a. All tRNA's have a CCG sequence at the 3'-end.
    b. All tRNA's have a CCG sequence at the 5'-end.
    c. All tRNA's have a CCA sequence at the 5'-end.
    d. All tRNA's have a CCA sequence at the 3'-end.
    e. All tRNA's have a CAC sequence at the 3'-end.

    Answer: d Section: 11

24. Which of the following base sequences would most likely be recognized by a restriction endonuclease?

    a. ACGCGT
    b. ACGGGT
    c. ACACGT
    d. ACATCGT
    e. CCAACC

    Answer: a Section: 14

25. Which of the following is a phosphoamidite?

a. 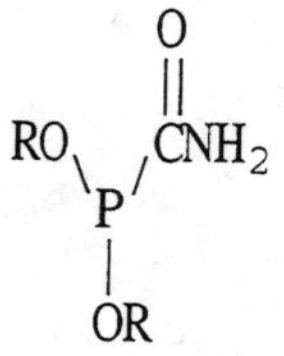 $(RO)_2P{-}C(=O)NH_2$ (P bonded to RO, OR and CNH_2 with C=O)

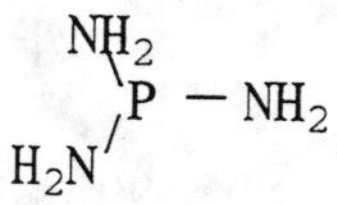

b. $P(NH_2)_3$

c. $(RO)_2P{-}NR_2$

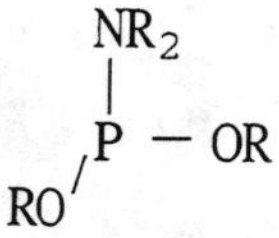

d. 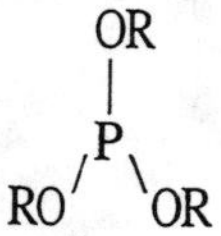 $P(OR)_3$

e. 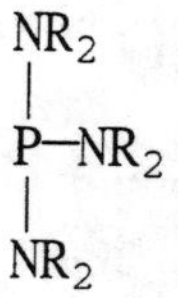 $P(NR_2)_3$

Answer: c Section: 15

26. A new method, called the H-phosphonate method, is replacing the phosphoamidite method for preparing oligonucleotides. Which of the following is the first step in the H-phosphonate method?

a. Oxidation with iodine
b. The 5'-OH group of the nucleoside reacts with the anhydroside.
c. The protecting group is removed by ammonium hydroxide.
d. The DMTr group is removed with mild acid.
e. The monomer, H-phosphonate, is activated by reacting it with acyl chloride to form an anhydride.

Answer: e Section: 15

27. Which of the following is <u>not</u> true about retroviruses?

    a. The genetic information is contained in RNA.
    b. The genetic information flows from DNA to RNA.
    c. Drugs have been designed to interfere with the synthesis of DNA by retroviruses.
    d. AZT is a design drug that was synthesized to combat the HIV retrovirus.
    e. The retrovirus uses the sequence of bases in RNA as a template to synthesize DNA which enters and infects the host cell.

    Answer: b Section: 16

SHORT ANSWER

28. What is a nucleoside? Give an example.

    Answer: A nucleoside is a base plus a sugar.

    Ex. NH_2 N N N HO O OH OH

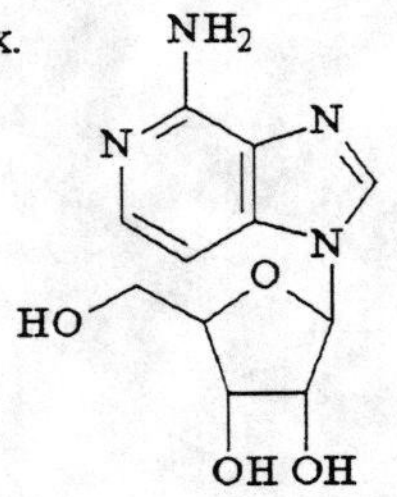

    Section: 1

29. What is a nucleotide? Give an example.

    Answer:
    A nucleotide is a nucleoside with one of its OH groups bonded to phosphoric acid in an ester linkage. Nucleotides of RNA are called ribonucleotides while the nucleotides of DNA are called deoxyribonucleotides.

    Ex. NH_3 N O N O P O O O O OH OH

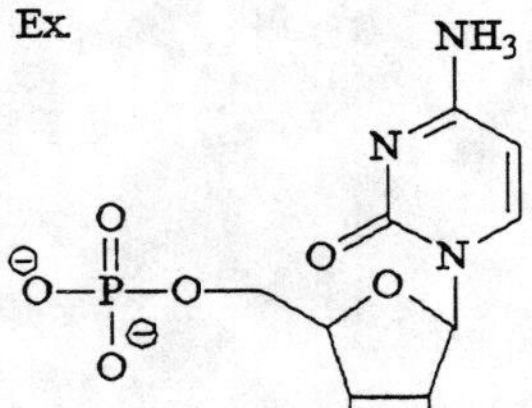

    cytidine 5'-monophosphate

    Section: 1

30. Draw the chemical structure of adenosine triphosphate, (ATP).

Answer:

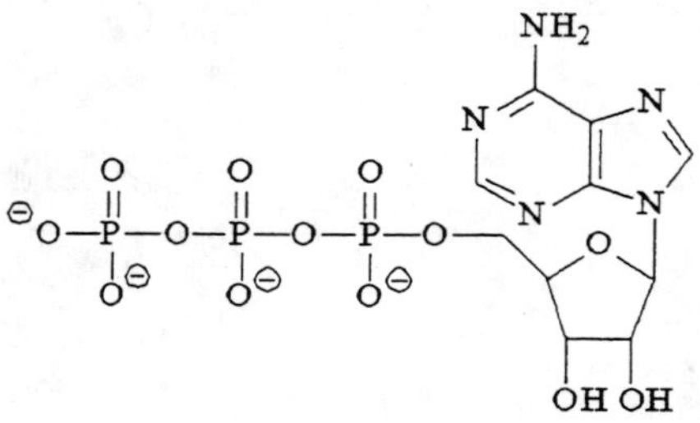

Section: 2

31. Propose a mechanism for the conversion of D-glucose to D-glucose-6-phosphate using ATP as a source of the phosphate.

Answer:

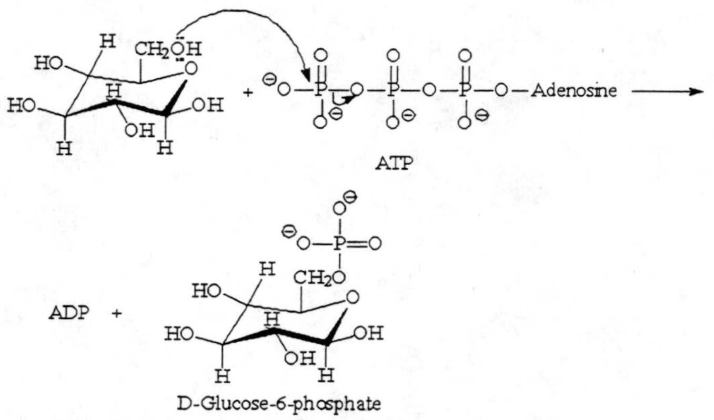

ATP

ADP +

D-Glucose-6-phosphate

Section: 2

32. Propose a mechanism for the following reaction:

$$CH_3-\overset{O}{\overset{\|}{C}}-O^{\ominus} + RSH + ATP \longrightarrow CH_3-\overset{O}{\overset{\|}{C}}-SR + ADP + {}^{\ominus}O-\overset{O}{\overset{\|}{P}}(-O^{\ominus})-OH$$

Answer:

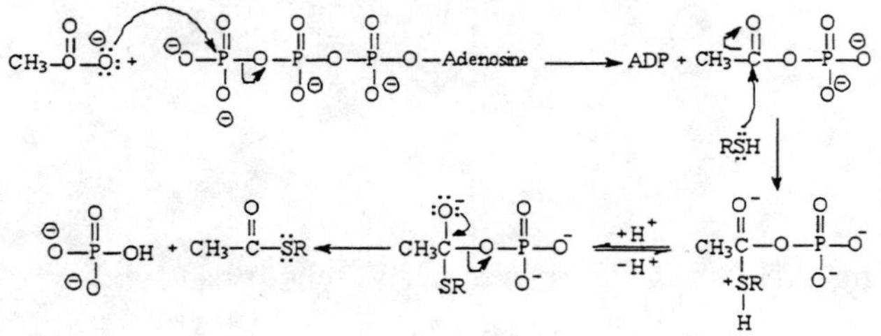

Section: 3

33. A commonly known nucleotide is adenosine 3',5'-phosphate (cAMP). What is the structure of cAMP?

Answer:

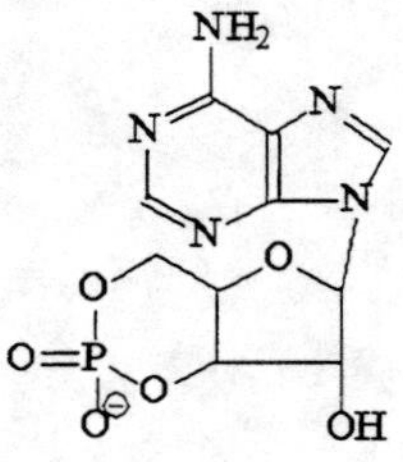

cAMP

Section: 6

34. Propose a mechanism for the formation of cyclic AMP from ATP.

Answer:

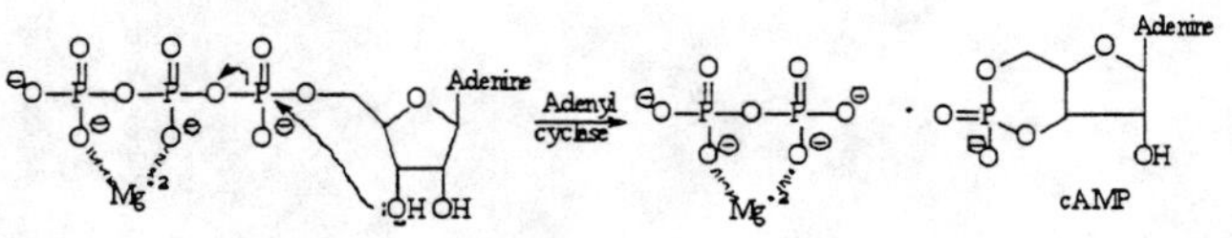

Section: 6

35. What are stacking interactions in DNA and what purpose do they serve?

Answer:
Stacking interactions are weak attractive forces (van der Waal's forces) that contribute significantly to the stability of the double helix. Stacking interactions are strongest between two purines and weakest between two pyrimidines.

Section: 7

36. What is meant by semiconservative DNA replication?

Answer:
When DNA strands are replicated, two daughter molecules of DNA are produced. Each daughter molecule contains one of two original strands plus a newly synthesized strand. This is called semiconservative DNA replication.

Section: 7

37. Assuming that the human genome with its 3 billion base pairs is entirely in a B-helix, how long is the DNA in a human cell?

Answer:
Recall that B-DNA has 10 base pairs and a rise of 3.3 angstrom.
Therefore, the length of the DNA would be:

$$\frac{3 \times 10^9}{10} \frac{\text{base pairs}}{\text{base pairs}} \times 3.3 \text{ angstrom} \times 10^{-8} \frac{\text{cm}}{\text{angstrom}} = 9.9 \text{ cm}$$

Section: 7

38. What is a codon? Give an example.

Answer:
A codon is a sequence of three bases that specifies a particular amino acid that is to be incorporated into a protein.
Ex. CUU is the code for leucine.

Section: 12

39. Methionine is the first amino acid incorporated into the heptapeptide specified by the following mRNA:
5'=AUGGACCCCGUUAUUAAACAC-3'. What is the sequence of the amino acid in the heptapeptide?

Answer:
Using the genetic code table, the sequence is
Met-Asp-Pro-Val-Ile-Lys-His.

Section: 12

40. What are Maxam-Gilbert and Sanger sequencing?

 Answer: They are methods used to sequence restriction fragments of DNA.

 Section: 14

41. What are antigene and antisense agents?

 Answer:
 They are synthetic polymers that bind to specific sequences of a nucleic acid DNA sequence. When the polymer binds to DNA, it is called an antigene agent, and when the polymer binds to RNA, it is called an antisense agent.

 Section: 16

42. What are the four common ribonucleosides?

 Answer: cytidine, uridine, adenosine, and guanosine Section: 1

43. What is the major structural difference between a ribonucleoside and a ribonucleotide?

 Answer:
 A ribonycleotide is a ribonucleoside which has been phosphorylated at its 5' carbon

 Section: 1

44. Besides a possible difference in base structure, what is the major structural difference between ribo- and deoxyribnucleosides?

 Answer:
 Deoxyribonycleosides have a hydrogen instead of a hydroxyl group at the C2 position of the sugar.

 Section: 1

45. Name the two pyrimidine bases which serve as components in deoxyribonucleotides.

 Answer: cytosine and thymine Section: 1

46. Name the two purine bases which serve as components in ribonucleosides.

 Answer: adenine and guanine Section: 1

47. Show the hydrogen bonding which occurs when guanine and cytosine form a base pair.

Answer:

guanine
cytosine

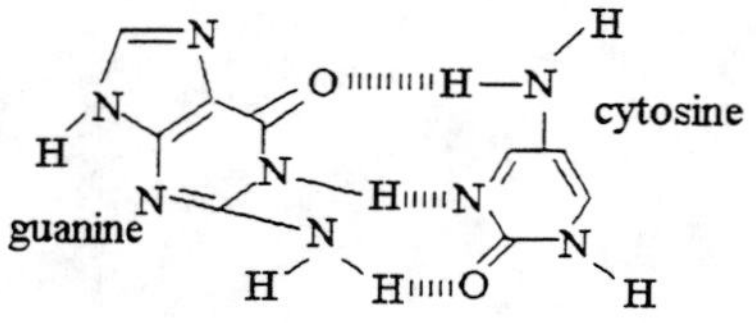

Section: 7

48. Show the hydrogen bonding which occurs when adenine and thymine form a base pair.

Answer:

adenine
thymine

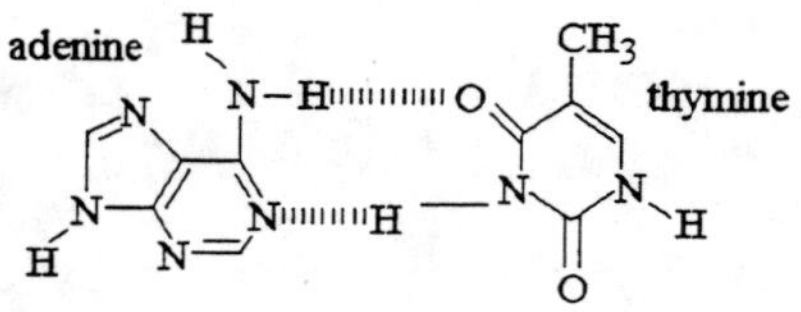

Section: 7

Chapter 25: Synthetic Polymers

MULTIPLE CHOICE

1. The joining together of many monomers to form a very large molecule is called?

   a. Polymerization
   b. Monomerization
   c. Dimerization
   d. Hydroxylation
   e. Homogenization

   Answer: a

2. The first polymer, invented by A. Parke in 1856, was

   a. Polyethlene
   b. Polyester
   c. Polystyrene
   d. Celluloid
   e. Teflon

   Answer: d

3. What is the structure of the monomer from which the following polymer is made?

$$-\underset{\displaystyle CH_3}{\underset{|}{CH}}-\underset{\displaystyle CH_3}{\underset{|}{CH}}-\underset{\displaystyle CH_3}{\underset{|}{CH}}-\underset{\displaystyle CH_3}{\underset{|}{CH}}-\underset{\displaystyle CH_3}{\underset{|}{CH}}-\underset{\displaystyle CH_3}{\underset{|}{CH}}-\underset{\displaystyle CH_3}{\underset{|}{CH}}-\underset{\displaystyle CH_3}{\underset{|}{CH}}-$$

   a. CH_3CH_3
   b. $CH_2{=}CH_2$
   c. $CH_3CH{=}CH_2$
   d. $CH_2{=}CHCH{=}CH_2$
   e. $CH_3CH{=}CHCH_3$

   Answer: e Section: 2

4. What is the structure of the monomer from which the following polymer is made?

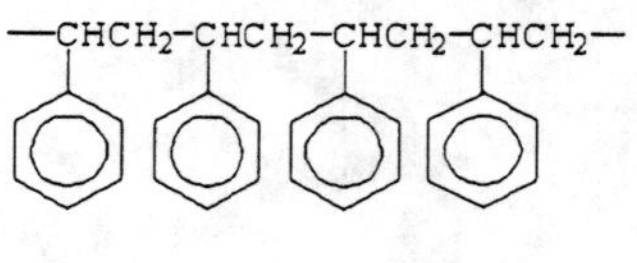

—CHCH$_2$-CHCH$_2$-CHCH$_2$-CHCH$_2$— (each CH bearing a benzene ring)

a. benzene ring–CH_2CH_3

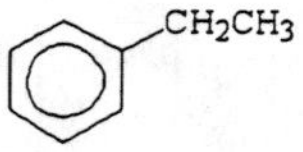

b. 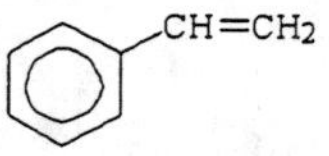 benzene ring–$CH{=}CH_2$

c. 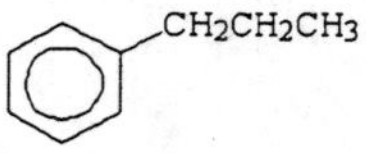 benzene ring–$CH_2CH_2CH_3$

d. 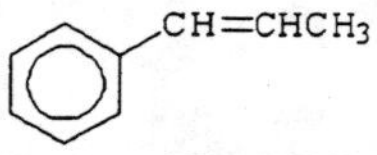 benzene ring–$CH{=}CHCH_3$

e. 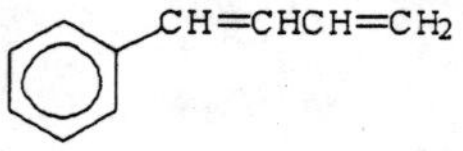 benzene ring–$CH{=}CHCH{=}CH_2$

Answer: b Section: 2

5. Which of the following addition polymers results from the reaction below?

$$nCF_2{=}CF_2 \xrightarrow{\text{catalyst}} ?$$

a. [=CF-CF=]n
b. [CF_3 -CF_3]n
c. [-CF_2 -CH=CHCF_2 -]n
d. [-CF_2 -CF_2 -]n
e. [-CF_2=CF_2 -]n

Answer: d Section: 2

6. Which dimer (step-growth dimerization) will form from the following monomer?

$$2\ \underset{\mathrm{NH_2}}{\underset{|}{\mathrm{CH_2}}}-\overset{\mathrm{O}}{\overset{\|}{\mathrm{C}}}-\mathrm{OH} \longrightarrow$$

   a. $\underset{\mathrm{NH_2}}{\underset{|}{\mathrm{CH_2}}}\overset{\mathrm{O}}{\overset{\|}{\mathrm{C}}}\mathrm{O}\overset{\mathrm{O}}{\overset{\|}{\mathrm{C}}}\underset{\mathrm{NH_2}}{\underset{|}{\mathrm{CH_2}}}$

   b. $\underset{\mathrm{NH_2}}{\underset{|}{\mathrm{CH_2}}}\overset{\mathrm{O}}{\overset{\|}{\mathrm{C}}}\mathrm{ONH}\overset{\mathrm{O}}{\overset{\|}{\mathrm{C}}}\mathrm{OH}$

   c. $\underset{\mathrm{NH_2}}{\underset{|}{\mathrm{CH_2}}}\overset{\mathrm{O}}{\overset{\|}{\mathrm{C}}}\mathrm{NHCH_2}\overset{\mathrm{O}}{\overset{\|}{\mathrm{C}}}\mathrm{OH}$

   d. $\mathrm{HO}\overset{\mathrm{O}}{\overset{\|}{\mathrm{C}}}\mathrm{CH_2NHNHCH_2}\overset{\mathrm{O}}{\overset{\|}{\mathrm{C}}}\mathrm{OH}$

   e. $\mathrm{HO}\overset{\mathrm{O}}{\overset{\|}{\mathrm{C}}}\mathrm{CH_2N}=\mathrm{NCH_2}\overset{\mathrm{O}}{\overset{\|}{\mathrm{C}}}\mathrm{OH}$

Answer: c Section: 6

7. Which of the following statements best describes what alpha olefins are?

   a. Polysubstituted ethylenes
   b. Monosubstituted ethylenes
   c. Disubstituted ethylenes
   d. Ethylenes
   e. Isoprenes

Answer: b Section: 2

8. The choice of mechanism in chain-growth polymerization depends on which of the following?

   a. Structure of the monomer and initiator used
   b. Structure of the polymer and initiator used
   c. Temperature and pressure of the reaction
   d. a and b
   e. a, b, and c

   Answer: a Section: 2

9. Chain-growth polymerization proceeds by which of the following mechanisms?

   a. Radical polymerization
   b. Cationic polymerization
   c. Anionic polymerization
   d. a and b
   e. a, b, and c

   Answer: e Section: 2

10. Which of the following species can best serve as a radical initiator for radical polymerization?

    a. ROH
    b. ROR
    c. ROOR
    d. RCOOR
    e. RCOOH

    Answer: c Section: 2

11. Which of the following is the first chain propagating step in the radical polymerization of ethylene?

    a. $ROOR + CH_2{=}CH_2 \longrightarrow ROCH_2CH_2OR$
    b. $RO\cdot + CH_2{=}CH_2 \longrightarrow ROCH_2\dot{C}H_2$
    c. $RO\cdot + CH_2{=}CH_2 \longrightarrow ROCH{=}CH_2 + H\cdot$
    d. $ROOR + CH_2{=}CH_2 \longrightarrow 2\ ROH + \dot{C}H{=}\dot{C}H$
    e. $2RO\cdot + CH_2{=}CH_2 \longrightarrow 2\ ROH + \dot{C}H{=}\dot{C}H$

    Answer: b Section: 2

12. Which of the following is a possible step to terminate a chain in the radical polymerization of ethylene?

a. $ROOR + CH_2{=}CH_2 \longrightarrow ROCH_2CH_2OR$
b. $CH_2{=}CH_2 + CH_2{=}CH_2 \longrightarrow CH_3CH{=}CHCH_3$
c. $RO\cdot + CH_2{=}CH_2 \longrightarrow ROCH_2{=}\dot{C}H_2$
d. $ROCH_2\dot{C}H_2 + CH_2{=}CH_2 \longrightarrow ROCH_2CH_2CH_2\dot{C}H_2$
e. $Ro{-}[CH_2CH_{2n}]{-}CH_2\dot{C}H_2 \longrightarrow$

$$RO{-}[CH_2CH_2]{-}CH_2CH_2CH_2CH_2{-}[CH_2CH_2]{-}OR$$

Answer: e Section: 2

13. Which of the following statements best explains the process known as <u>chain transfer</u>?

a. A process that can control the rate of production of the polymer
b. A process that can control the molecular weight of the polymer
c. A process that can control the reactivity of the monomer
d. A process that can slow down the polymerization reaction
e. A process that can inhibit the polymerization reaction

Answer: b Section: 2

14. Chain-growth polymerization exhibits a marked prefrence for:

a. tail-to-tail addition
b. head-to-head addition
c. head-to-tail addition
d. head-to-middle addition
e. middle-to-tail addition

Answer: c Section: 2

15. Which of the following are characteristic of <u>linear</u> polyethylene?

a. They are branched polymers.
b. They are known as high-density polyethylene.
c. They are soft and flexible polymers.
d. a and b
e. a and c

Answer: b Section: 2

16. Which of the following species can best serve as an initiator for cationic polymerization?

 a. ROOR
 b. ROH
 c. ROR
 d. $AlCl_3$
 e. RCOOR

 Answer: d Section: 2

17. Which of the following is the chain-initiating step in the cationic polymerization of propene?

 a. $BF_3 + CH_2{=}CH(CH_3) \longrightarrow F_2BCH{=}CH(CH_3) + HF$

 b. $BF_3 + CH_2{=}CH(CH_3) \longrightarrow CH_2{=}CBF_2(CH_3) + HF$

 c. $BF_3 + CH_2{=}CH(CH_3) \longrightarrow CH_2{=}CH(CH_2BF_2) + HF$

 d. $BF_3 + CH_2{=}CH(CH_3) \longrightarrow \overset{\oplus}{C}H_2CH(CH_3)\overset{\ominus}{B}F_3$

 e. $BF_3 + CH_2{=}CH(CH_3) \longrightarrow F_3\overset{\ominus}{B}CH_2\overset{\oplus}{C}H(CH_3)$

 Answer: e Section: 2

18. Which of the following is more likely to undergo both radical <u>and</u> cationic polymerization?

 a. $CH_2{=}CH_2$
 b. $C_6H_5CH{=}CH_2$ (benzene ring bearing CH=CH$_2$)
 c. $CH_2{=}CHCl$
 d. $CH_2{=}CHC{\equiv}N$
 e. $CH_2{=}C(CH_3)COOR$

 Answer: b Section: 2

19. Which of the following monomers has the greatest ability to undergo cationic polymerization?

a. 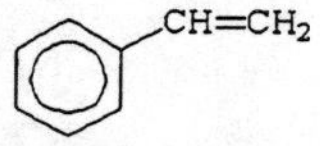 $CH{=}CH_2$

b. 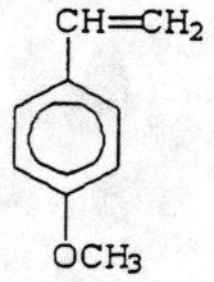 $CH{=}CH_2$
OCH_3

c. 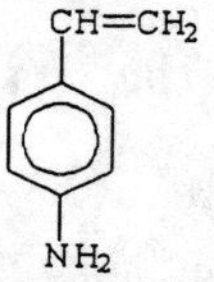 $CH{=}CH_2$
NH_2

d. 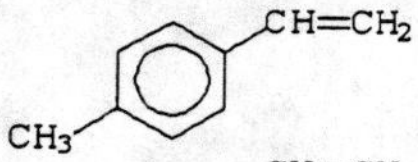 $CH{=}CH_2$
CH_3

e. 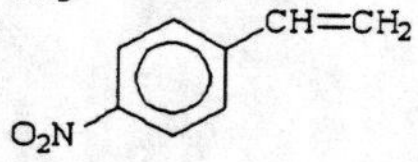 $CH{=}CH_2$
O_2N

Answer: c Section: 2

20. Which of the following monomers has the greatest ability to undergo anionic polymerization?

a. 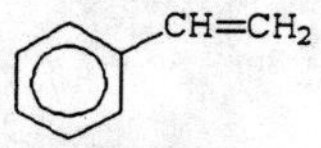 $CH{=}CH_2$

b. 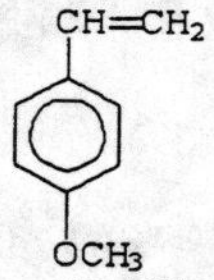 $CH{=}CH_2$
OCH_3

c. 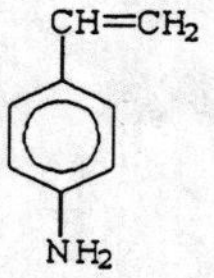 $CH{=}CH_2$
NH_2

d. 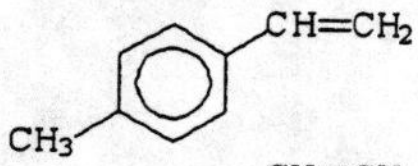 $CH{=}CH_2$
CH_3

e. 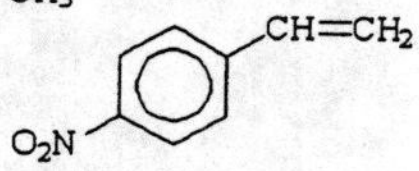 $CH{=}CH_2$
O_2N

Answer: e Section: 2

21. Which of the following species can best serve as an initiator for anionic polymerization?

a. $CH_3CH_2CH_2CH_2Li$
b. BF_3
c. HOOH
d. $CH_3CH_2CH_2CH_2OH$
e. $CH_3CH_2CH_2CH_2OCH_2CH_2CH_2CH_3$

Answer: a Section: 2

22. Which of the following is more likely to form non-terminated polymeric chains called <u>living polymers</u>?

a. Cationic polymerization
b. Free radical polymerization
c. Anionic polymerization
d. a and b
e. a and c

Answer: c Section: 2

23. A substituted ethylene polymer in which the substituents are randomly oriented is called:

a. A Ditactic polymer
b. An Eutatic polymer
c. An Isotactic polymer
d. A Syndiotatic polymer
e. An Atactic polymer

Answer: e Section: 3

24. Which of the following polymers is arranged in an isotactic configuration?

a.

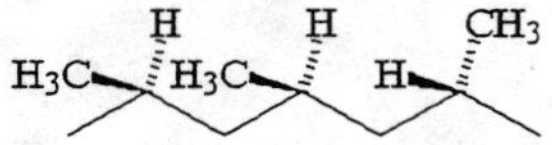

b.

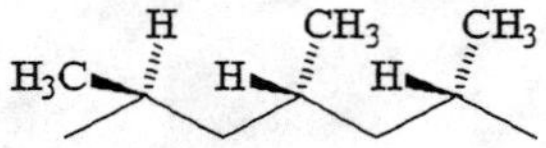

c.

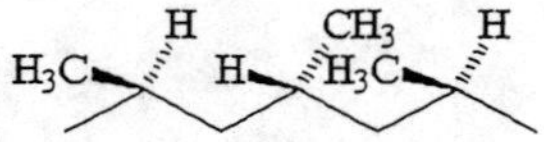

d.

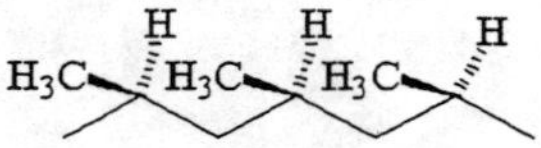

e.

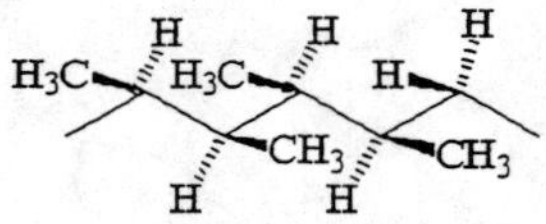

Answer: d Section: 3

25. Which of the following initiators are known as Ziegler-Natta catalysts, which were discovered in 1953 in order to control the stereochemistry of polymers?

a. Al/Ti initiators
b. Ni/Cd initiators
c. Zn/Hg initiators
d. Pb/C initiators
e. Pb/PbO_2 initiators

Answer: a Section: 3

26. What is the difference between natural rubber and Gutta-percha rubber?

a. Natural rubber is hard and brittle while gutta-percha is soft and rubbery.
b. Natural rubber is natural while gutta-percha is synthetic rubber.
c. Natural rubber has a cis configuration while gutta-percha has a trans configuration.
d. Natural rubber and gutta-percha rubber are enantiomers.
e. The monomer for natural rubber is isoprene while the monomer for gutta-percha is 1,3-butadiene.

Answer: c Section: 4

27. Which of the following statements best describes what vulcanization means?

    a. A process by which rigid and non-sticky rubber can be made soft and sticky
    b. A chemical process discovered by Charles Schultz in 1944
    c. The heating of rubber with sulfur
    d. A chemical process that destroys cross-linking in rubber
    e. All of the above

    Answer: c Section: 4

28. Which of the following are copolymers?

    a. An Alternating copolymer
    b. A Block copolymer
    c. A Random copolymer
    d. A Graft copolymer
    e. All of the above

    Answer: e Section: 5

29. Which of the following are step-growth polymers?

    a. Polyethylene
    b. Polyester
    c. Polypropylene
    d. Polystyrene
    e. Plexiglass

    Answer: b Section: 6

30. What are aramides?

    a. Aromatic polyester polymers
    b. Aromatic amine polymers
    c. Alkyl halide polymers
    d. Aromatic polyamide polymers
    e. Aliphatic polyamide polymers

    Answer: d Section: 6

31. What monomers formed the following polymer?

$$-\overset{O}{\overset{\|}{C}}-C_6H_4-\overset{O}{\overset{\|}{C}}-NH-C_6H_4-NH\overset{O}{\overset{\|}{C}}-C_6H_4-NH-$$

a. $H_2N-C_6H_4-NH_2 + HO\overset{O}{\overset{\|}{C}}-C_6H_4-\overset{O}{\overset{\|}{C}}OH$

b. $2\ HO\overset{O}{\overset{\|}{C}}-C_6H_4-NH_2$

c. $H_2N\overset{O}{\overset{\|}{C}}-C_6H_4-NH_2 + HO-C_6H_4-\overset{O}{\overset{\|}{C}}OH$

d. $2\ H_2N\overset{O}{\overset{\|}{C}}-C_6H_4-\overset{O}{\overset{\|}{C}}NH_2$

e. $HO-C_6H_4-OH + H_2N\overset{O}{\overset{\|}{C}}-C_6H_4-\overset{O}{\overset{\|}{C}}NH_2$

Answer: a Section: 6

32. Which of the following are characteristics of thermosetting polymers?

a. Strong and rigid
b. High degree of cross-linking
c. Can be remelted by heating
d. a and b
e. a, b, and c

Answer: d Section: 7

33. What are <u>elastomers</u>?

a. Elastic organic molecules that are added to polymers to make them harder
b. Plastics that stretch and then revert to their original shape
c. Polymers that are soft and sticky
d. Plastics that are hard and brittle
e. Elastic molecules that are added to polymers to make them stretch more

Answer: b Section: 7

34. Which of the following is characteristic of oriented polymers?

a. Weaker than steel
b. Chains are arranged in parallel fashion
c. Chains are arranged in perpendicular fashion
d. Polymer chains are made from sulfur atoms
e. All of the above

Answer: b Section: 7

SHORT ANSWER

35. List the two main groups of polymers and give one example of each.

Answer:
Polymers are divided into two main groups:
a. Biopolymers- ex. DNA
b. Synthetic polymers- ex. Polyester

36. List the two major classes of synthetic polymers and give one example of each.

Answer:
a. Addition polymers, also known as chain-growth polymers- ex. Polystyrene
b. Condensation polymers, also known as step-growth polymers- ex. Nylon

Section: 1

37. List the three phases in the mechanism of chain-growth polymerization.

Answer:
1. Initiation
2. Propagation
3. Termination

Section: 2

38. Show the three methods by which the following radical can be terminated.

$$RO—CH_2—\dot{C}H_2$$

Answer:

a. $2\ RO—CH_2\dot{C}H_2 \longrightarrow RO—CH_2CH_2CH_2CH_2—OR$

b. $2\ RO—CH_2\dot{C}H_2 \longrightarrow RO—CH{=}CH_2 + ROCH_2CH_3$

c. $RO—CH_2\dot{C}H_2$ + impurity $\longrightarrow RO—CH_2CH_2—$ impurity

Section: 2

39. Explain why some monomers undergo chain-growth polymerization by a radical mechanism.

Answer:
A monomer with a substituent (X) that is able to stabilize the growing radical species by resonance will undergo chain-growth polymerization by a free radical mechanism.

Section: 2

40. What are common factors that enter into the choice of radical initiators for chain-growth polymerization?

Answer:
1. A relatively weak oxygen-oxygen bond
2. Desired solubility of the initiator
3. Temperature of the polymerization reaction

Section: 2

41. Show the mechanism for the formation of a segment of polystyrene containing two molecules of styrene and initiated by hydrogen peroxide.

Answer:

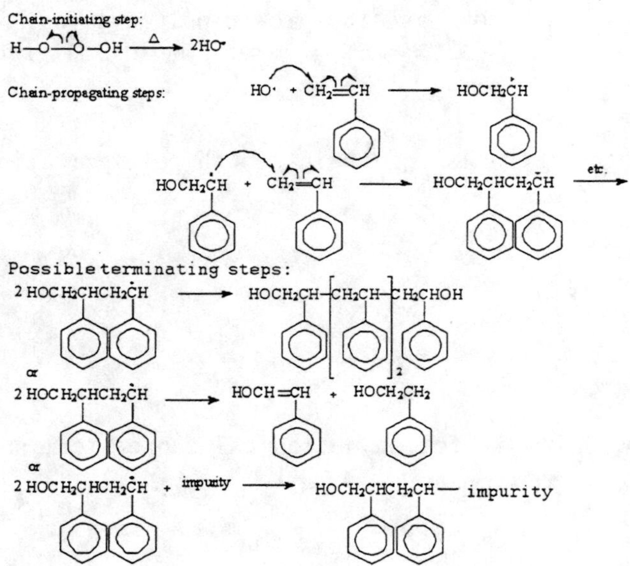

Section: 2

42. When 3-methyl-1-butene

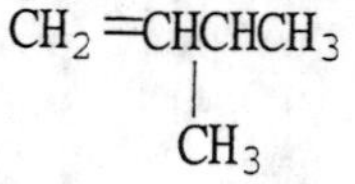

$$CH_2{=}CH\underset{\displaystyle CH_3}{\underset{|}{C}}HCH_3$$

undergoes cationic polymerization, the following expected carbocation is formed.

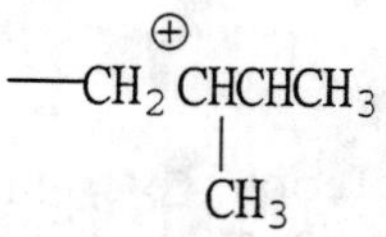

$$—CH_2\overset{\oplus}{C}H\underset{\displaystyle CH_3}{\underset{|}{C}}HCH_3$$

However, the following carbocation is also found in the reaction.

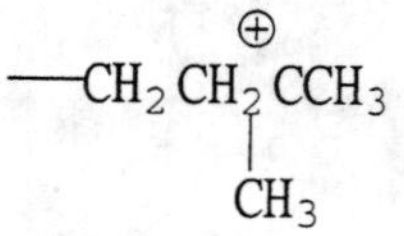

$$—CH_2CH_2\overset{\oplus}{\underset{\displaystyle CH_3}{\underset{|}{C}}}CH_3$$

Explain.

Answer:
The expected carbocation is a 2° carbocation which may rearrange to a 3° carbocation, a more stable carbocation. This is a classic 1,2-hydride shift.

Section: 2

43. Propose a mechanism for the following polymerization reaction:

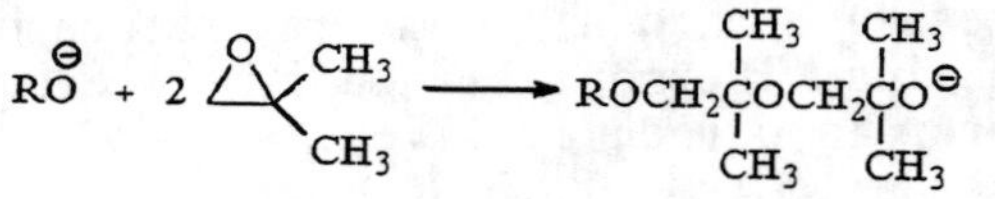

$$R\overset{\ominus}{O} + 2\ \text{(2,2-dimethyloxirane)} \longrightarrow ROCH_2\underset{CH_3}{\overset{CH_3}{C}}OCH_2\underset{CH_3}{\overset{CH_3}{C}}O^{\ominus}$$

Answer:

$$R\ddot{\underset{..}{O}}{:}^{\ominus} + 2\ \text{(2,2-dimethyloxirane)} \longrightarrow ROCH_2\underset{CH_3}{\overset{CH_3}{C}}\ddot{\underset{..}{O}}{:}^{\ominus} + \text{(2,2-dimethyloxirane)} \longrightarrow ROCH_2\underset{CH_3}{\overset{CH_3}{C}}OCH_2\underset{CH_3}{\overset{CH_3}{C}}O^{\ominus}$$

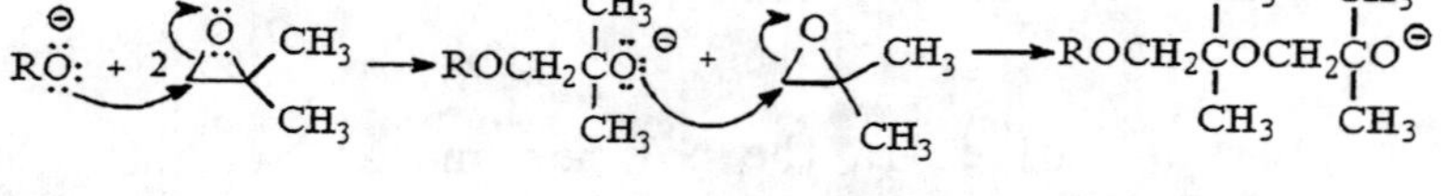

Section: 6

44. When 2,2-dimethyloxirane (see figure below) undergoes anionic polymerization, the nucleophile attacks the least substituted carbon of the epoxide, but when it undergoes cationic polymerization, the nucleophile attacks the most substituted carbon of the epoxide. Explain why this is so.

Answer:
In anionic polymerization, the nucleophile will seek the least hindered carbon similar to an S_N2 reaction. However, in cationic polymerization, the reaction involves a partial carbocation and the attack is on the most hindered (most stable) carbocation.

Section: 6

45. What are crystallites?

Answer:
The regions of polymers in which the chains are highly ordered with respect to one another.

Section: 7

46. Describe what thermoplastic polymers are.

Answer:
Polymers that have both ordered crystalline regions and amorphous, non-crystalline regions. The polymers are hard at room temperature but can be remelted and remolded by heating. Thermoplastic polymers are the plastics that we most often encounter in our daily lives.

Section: 7

47. Describe what plasticizers are and give an example.

Answer:
An organic compound that dissolves in the polymer and allows the polymer chains to slide by each other.

Ex. Dibutyl phthalate,

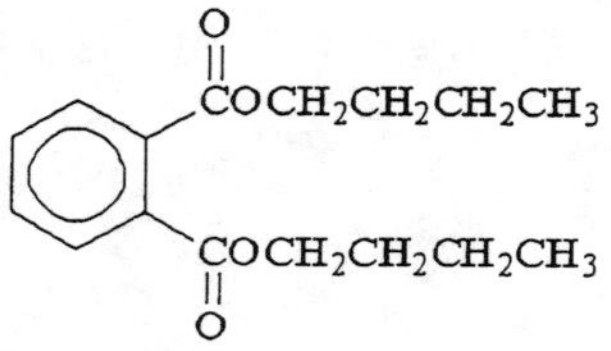

$$\text{C}_6\text{H}_4(\text{COOCH}_2\text{CH}_2\text{CH}_2\text{CH}_3)_2$$

Section: 7

48. What are biodegradable polymers? Give an example.

Answer:
Polymers that can be broken into smaller segments by enzyme-catalyzed reaction using enzymes produced by microorganisms.
Ex. Inserting hydrolyzable esters into polymers can cause weak links in the polymer which are hydrolyzable by enzyme-catalyzed reactions.
Section: 7

49. Show the reaction which occurs when benzoyl peroxide is heated.

Answer:

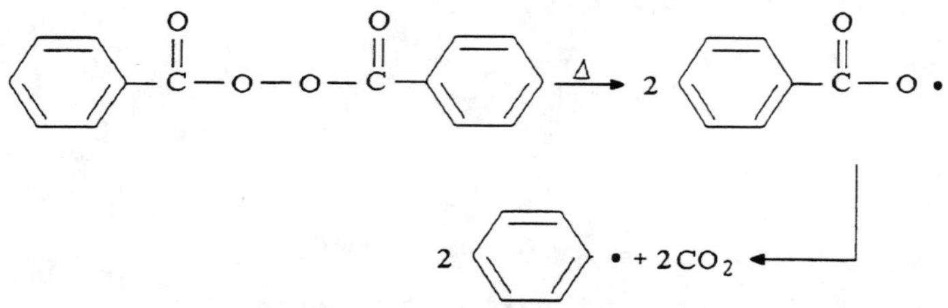

$$\text{C}_6\text{H}_5\text{C(=O)}\text{-O-O-}\text{C(=O)}\text{C}_6\text{H}_5 \xrightarrow{\Delta} 2\ \text{C}_6\text{H}_5\text{C(=O)-O}\bullet \longrightarrow 2\ \text{C}_6\text{H}_5\bullet + 2CO_2$$

Section: 2

50. Show how branching could occur durring the free-radical polymerization of styrene.

Answer:

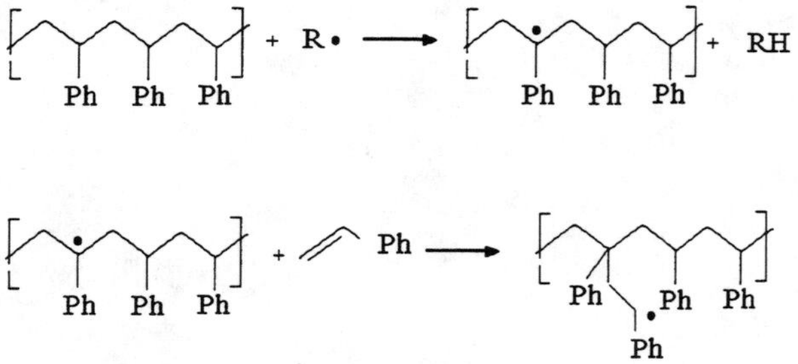

[chain with Ph Ph Ph] + R• → [chain radical with Ph Ph Ph] + RH

[chain radical with Ph Ph Ph] + styrene (=/Ph) → [branched chain with Ph, •Ph, Ph, Ph]

Section: 2

51. Provide a mechanism to show how $H_2C{=}C(CH_3)_2$ is polymerized using BF_3 as the initiatior.

Answer:

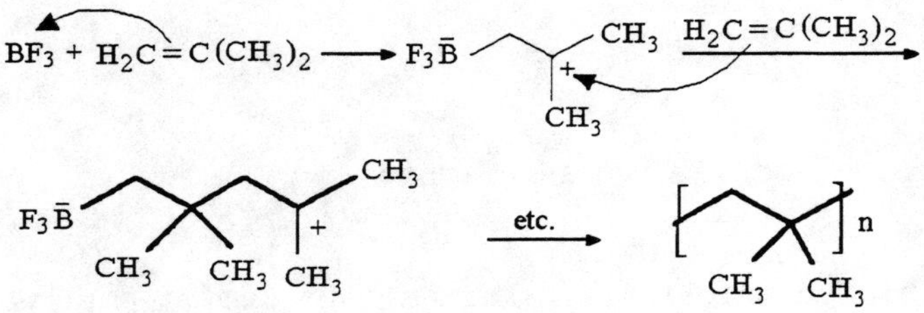

$BF_3 + H_2C{=}C(CH_3)_2 \longrightarrow F_3\bar{B}$–$CH_2$–$\overset{+}{C}(CH_3)_2$ $\xrightarrow{H_2C=C(CH_3)_2}$

$F_3\bar{B}$–$CH_2C(CH_3)_2CH_2\overset{+}{C}(CH_3)_2$ $\xrightarrow{etc.}$ $[CH_2C(CH_3)_2]_n$

Section: 2

52. Provide the structure of Saran, an alternating copolymer of $H_2C{=}CHCl$ and $H_2C{=}CCl_2$.

Answer:

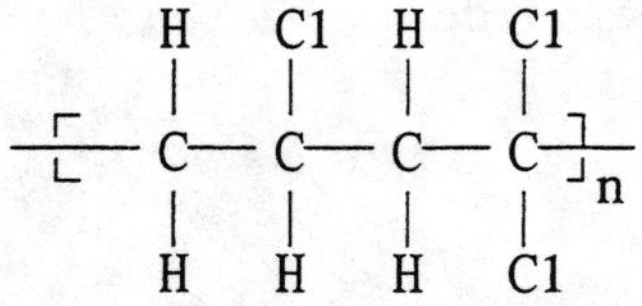

$$-\!\!\left[\mathrm{CH_2{-}CHCl{-}CH_2{-}CCl_2}\right]_n\!\!-$$

Section: 5

MULTIPLE CHOICE

53. What kind of polymer is produced in the following reaction?

a. a polycarbonate
b. a polyurethane
c. a polyester
d. a synthetic rubber
e. a poly (acrylonitrile)

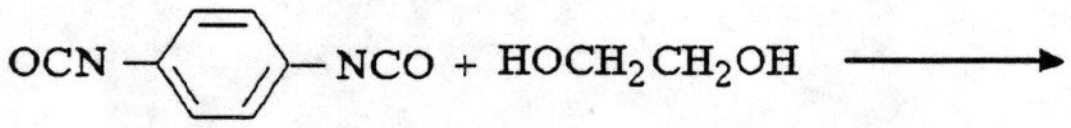

OCN–C_6H_4–NCO + $HOCH_2CH_2OH \longrightarrow$

Answer: b Section: 6

SHORT ANSWER

54. Draw the structure of the polymer produced in the following reaction.

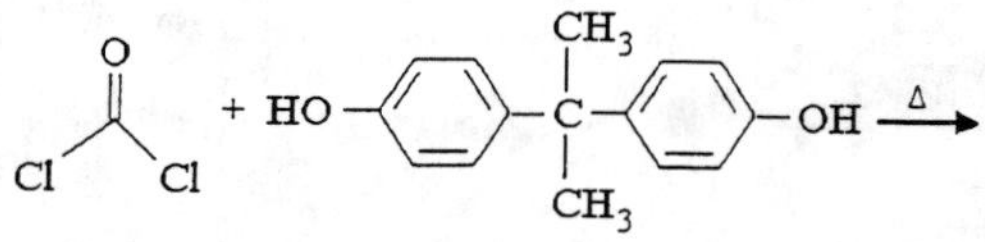

O Cl Cl + HO CH_3 C CH_3 OH $\xrightarrow{\Delta}$

Answer:

O C–O CH_3 C CH_3 O n

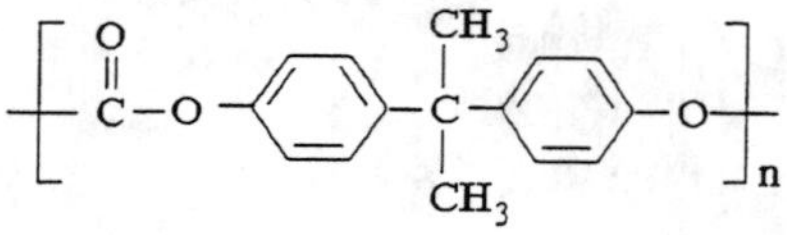

Section: 6

MULTIPLE CHOICE

55. Natural rubber is a:

a. polyamide.
b. polyester.
c. polycarbonate.
d. polyurethane.
e. none of the above

Answer: e Section: 4

SHORT ANSWER

56. Draw the products which would result if poly(ethylene terephthalate) were hydrolyzed in hot aqueous NaOH.

Answer:

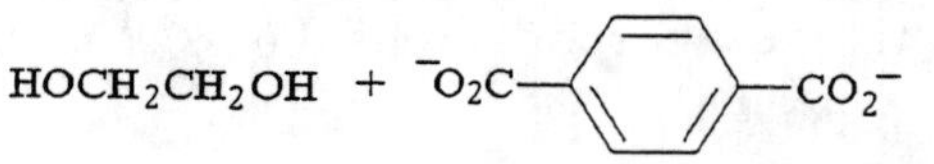

$HOCH_2CH_2OH$ + ^-O_2C– –CO_2^-

Section: 6

57. Propose a synthesis of poly(vinyl alcohol) from $H_2C{=}CHO_2CCH_3$.

Answer:
1. $(PhCO_2)_2$
2. NaOH, H_2O

Section: 2

58. Draw the structure of the polymer produced in the following reaction.

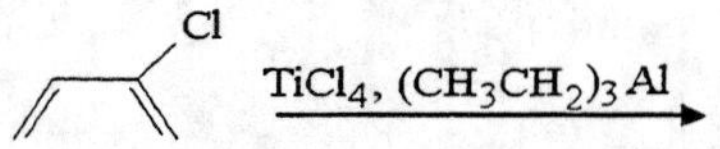

$\xrightarrow{TiCl_4,\ (CH_3CH_2)_3Al}$

Answer:

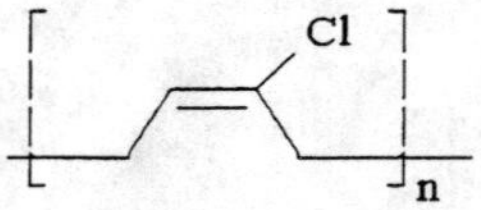

Section: 4

59. Provide a mechanism to show how $H_2C{=}CHCO_2CH_3$ is polymerized using butyllithium as the initiator.

Answer:

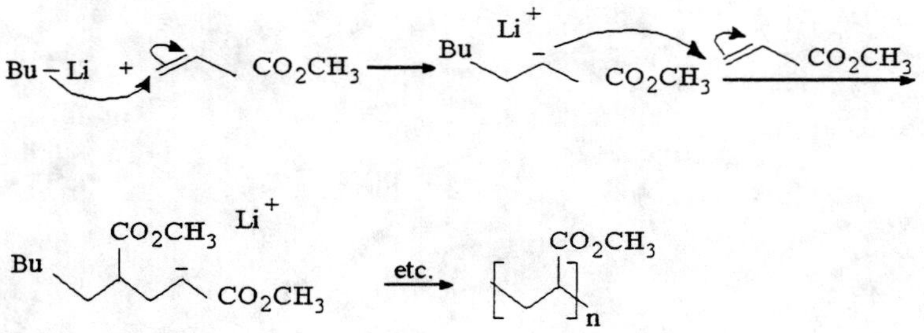

Bu–Li + CO_2CH_3 → Bu CO_2CH_3 Li^+ + CO_2CH_3 →

Bu CO_2CH_3 Li^+ CO_2CH_3 $\xrightarrow{etc.}$ CO_2CH_3 n

Section: 2

60. Why is chain branching less common in anionic polymerization than in cationic polymerization?

Answer:
The intermediate which leads to chain branching is a more substituted carbon ion. Greater substitution stabilizes carbocations but destabilizes carbanions.

Section: 2

Chapter 26: Heterocylic Compounds

MULTIPLE CHOICE

1. Which of the following statements best describes the meaning of the term alkaloid?

   a. Heterocyclic natural products with a nitrogen as a hetero atom
   b. Natural, cyclic hydrocarbons
   c. Heterocyclic natural products with a sulfur as a hetero atom
   d. Heterocyclic natural products with an oxygen as a hetero atom
   e. Natural, cyclic hydrocarbons with alkyl groups

   Answer: a

2. What is the systematic name for the following compound?

   a. aziridine
   b. oxetane
   c. oxirane
   d. azetidine
   e. thiirane

   Answer: c Section: 1

3. What is the systematic name for the following compound?

   a. aziridine
   b. oxetane
   c. oxirane
   d. azetidine
   e. thiirane

   Answer: e Section: 1

4. Which of the following is the structure for azetidine?

a. 

b. 

c. 

d. 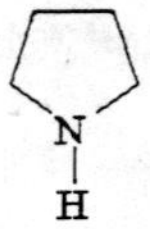

e. 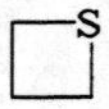

Answer: a Section: 1

5. Which of the following is the structure for N-methylpiperidine?

a. 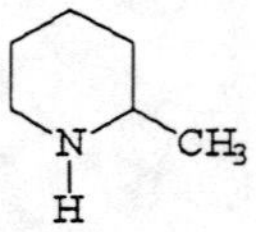

b. 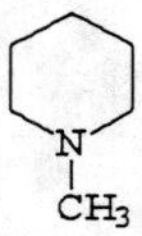

c. 

d. 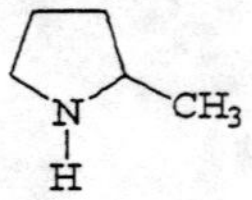

e. 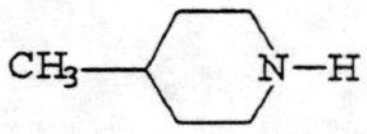

Answer: b Section: 1

6. Which of the following is the structure of 1,4-dioxane?

a. 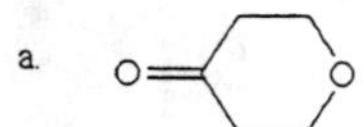

b.

c.

d.

e.

Answer: d Section: 1

7. What is the systematic name of the following compound?

a. 1-Azabicyclo [2.2.2]heptane
b. 2-Azatricyclo [2.2.2]heptane
c. 2-Azabicyclo[2.2.2]octane
d. 2-Azabicyclo[2.2.2]heptane
e. 2-Azabicyclo[2.2.0]octane

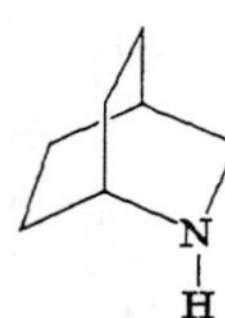

Answer: c Section: 1

8. What is the major product of the following reaction?

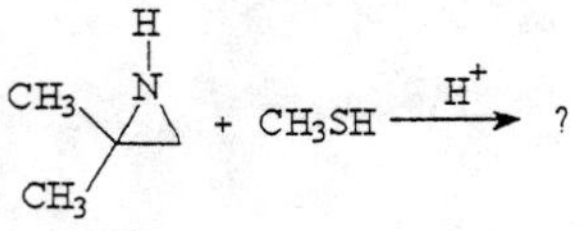

2,2-dimethylaziridine $+ CH_3SH \xrightarrow{H^+}$?

a. 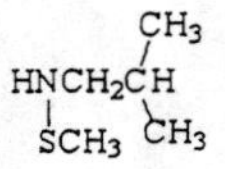 $HNCH_2CH(CH_3)_2$ with SCH_3 on N

b. 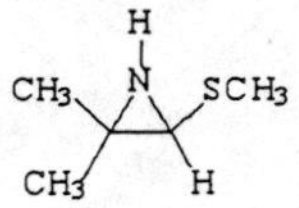 2,2-dimethyl-3-(methylthio)aziridine

c. 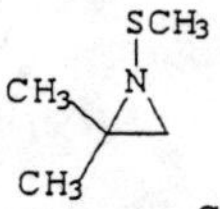 1-(methylthio)-2,2-dimethylaziridine

d. 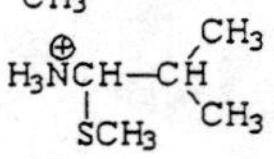 $\overset{\oplus}{H_3N}CH(SCH_3)-CH(CH_3)_2$

e. 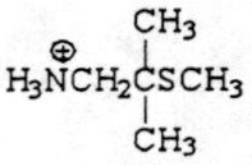 $\overset{\oplus}{H_3N}CH_2C(CH_3)_2SCH_3$

Answer: e Section: 1

9. What is the major product of the following reaction?

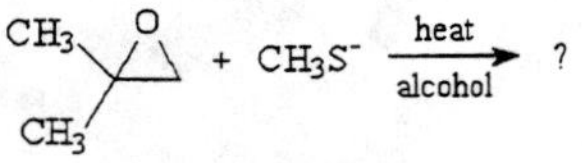

2,2-dimethyloxirane $+ CH_3S^- \xrightarrow[\text{alcohol}]{\text{heat}}$?

a. 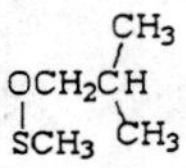 $OCH_2CH(CH_3)_2$ with SCH_3 on O

b. 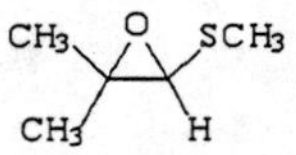 2,2-dimethyl-3-(methylthio)oxirane

c. 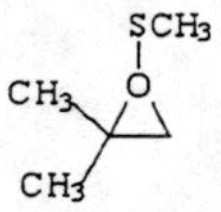 2,2-dimethyloxirane with SCH_3 on O

d. 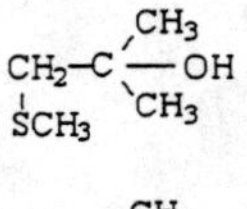 $CH_2(SCH_3)-C(CH_3)_2-OH$

e. 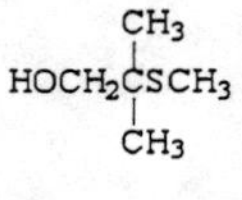 $HOCH_2C(CH_3)_2SCH_3$

Answer: d Section: 1

10. What is the <u>major</u> product of the following reaction?

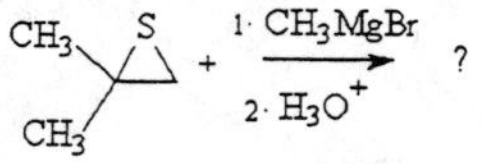

$(CH_3)_2C$–thiirane (2,2-dimethylthiirane) + 1. CH_3MgBr, 2. H_3O^+ → ?

a. 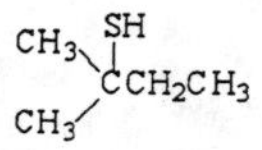 $(CH_3)_2C(SH)CH_2CH_3$

b. 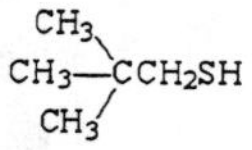 $(CH_3)_3CCH_2SH$

c. 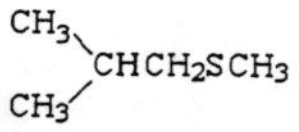 $(CH_3)_2CHCH_2SCH_3$

d. 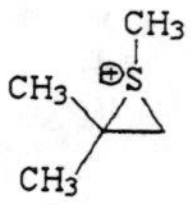 2,2-dimethylthiiranium with CH_3 on $S^{\oplus}$

e. 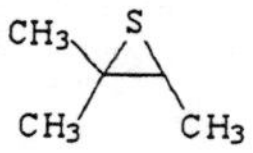 CH_3, CH_3, CH_3-substituted thiirane (S ring with CH_3 and CH_3 on one carbon, CH_3 on the other)

Answer: a Section: 1

11. What is the <u>major</u> product of the following reaction?

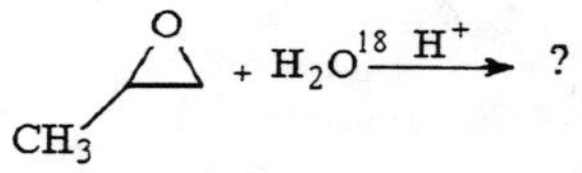

CH_3-epoxide (propylene oxide) + H_2O^{18} $\xrightarrow{H^+}$?

a. 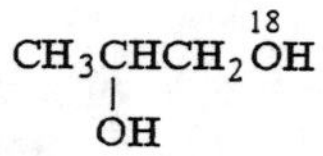 $CH_3CH(OH)CH_2{}^{18}OH$

b. 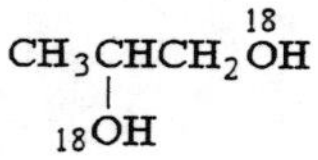 $CH_3CH({}^{18}OH)CH_2{}^{18}OH$

c. 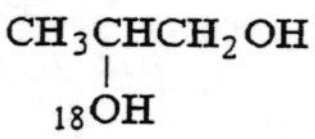 $CH_3CH({}^{18}OH)CH_2OH$

d. 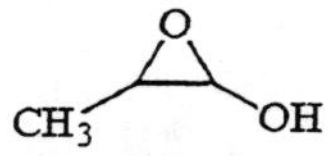 epoxide with ring ${}^{18}O$, CH_3 and OH substituents

e. 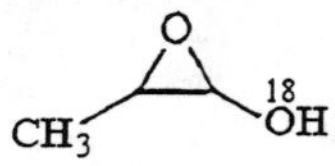 epoxide with ring O, CH_3 and ${}^{18}OH$ substituents

Answer: c Section: 1

12. What is the product of the following reaction?

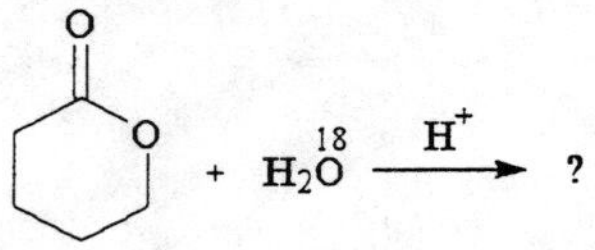

O

O + $H_2\overset{18}{O}$ $\xrightarrow{H^+}$?

a. 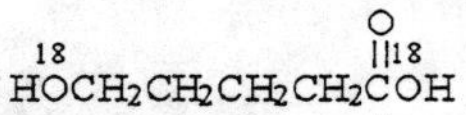 18 O ||18
$HOCH_2CH_2CH_2CH_2COH$

b. 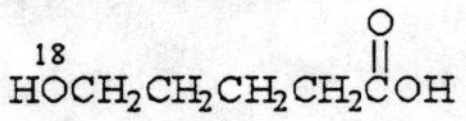 18 O ||
$HOCH_2CH_2CH_2CH_2COH$

c. 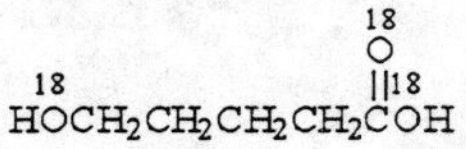 18 18 O ||18
$HOCH_2CH_2CH_2CH_2COH$

d. 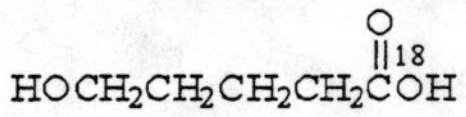 O ||18
$HOCH_2CH_2CH_2CH_2COH$

e. 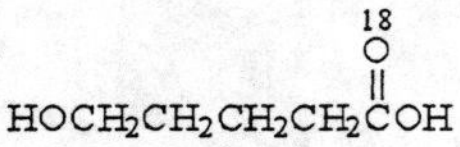 18 O ||
$HOCH_2CH_2CH_2CH_2COH$

Answer: d Section: 1

13. What is the <u>major</u> product of the following reaction?

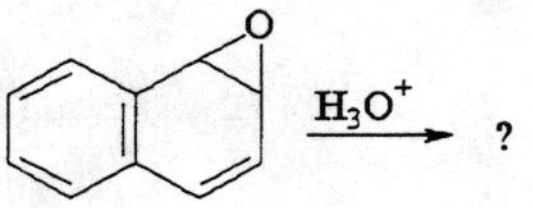

O

$\xrightarrow{H_3O^+}$?

a. 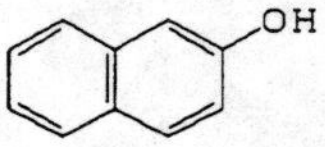 OH

b. 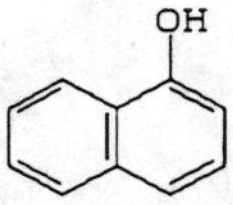 OH

c. 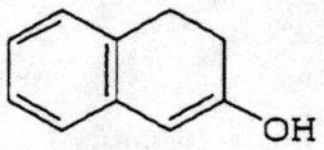 OH

d. 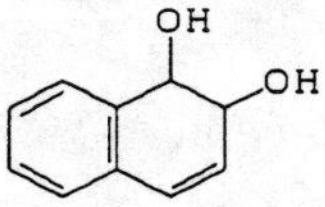 OH OH

e. 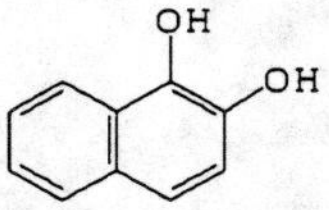 OH OH

Answer: b Section: 1

14. Which of the following aromatic compounds is the most stable?

a.  (thiophene)

b.  (furan)

c. 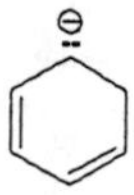 (cyclohexadienyl anion)

d.  (benzene)

e. 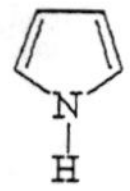 (pyrrole, N–H)

Answer: d Section: 2

15. What is the major product of the following reaction?

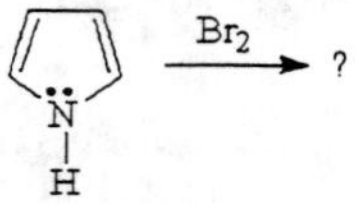

(pyrrole) $\xrightarrow{Br_2}$?

a. 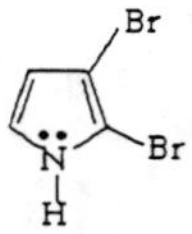 (2,3-dibromopyrrole)

b.  (N-bromopyrrole)

c. 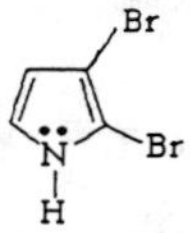 (2,3-dibromopyrrole)

d.  (3-bromopyrrole)

e. 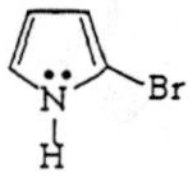 (2-bromopyrrole)

Answer: e Section: 2

16. Which of the following compounds is the <u>most</u> reactive toward electrophilic aromatic substitution?

a.

b.

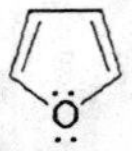

c.

d.

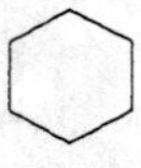

e.

Answer: c Section: 2

17. What is the <u>major</u> product of the following reaction?

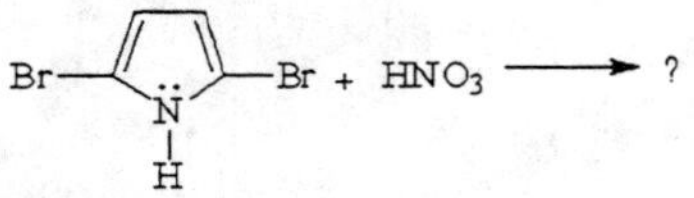

Br— —Br + HNO_3 ⟶ ?

a.

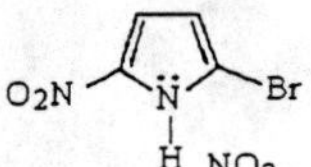

b.

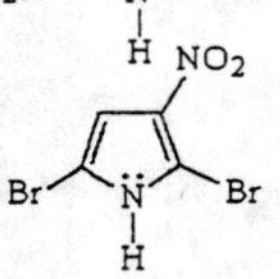

c.

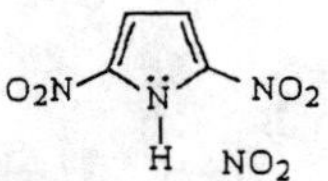

d.

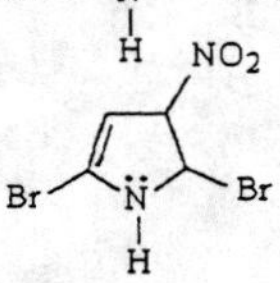

e.

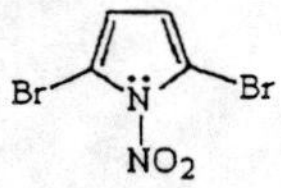

Answer: b Section: 2

18. Which of the following compounds is indole?

a.

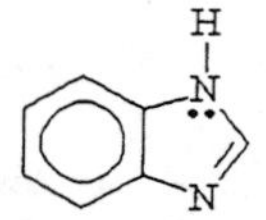

b.

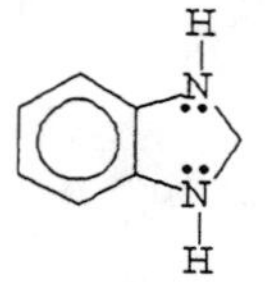

c.

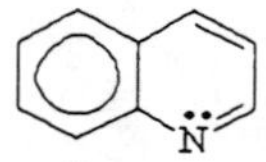

d.

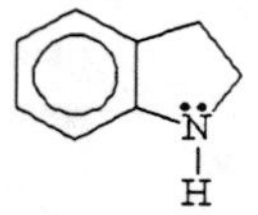

e.

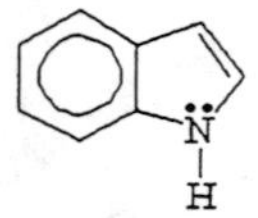

Answer: e Section: 2

19. What is the major product of the following reaction?

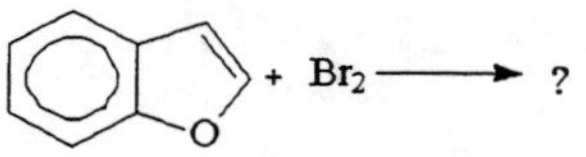

+ Br_2 → ?

Benzofuran

a. —Br

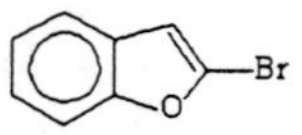

b. Br

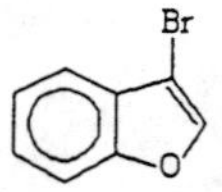

c. Br

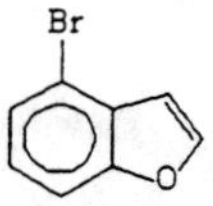

d. Br

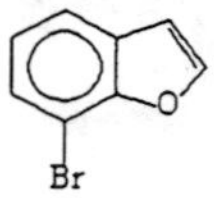

e. Br, Br

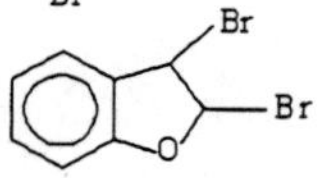

Answer: a Section: 2

20. What is the <u>major</u> product of the following reaction?

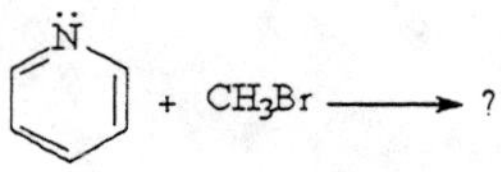

pyridine + CH_3Br ⟶ ?

a. 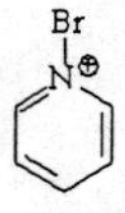 N-bromopyridinium (Br on $N^{\oplus}$)

b. 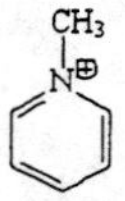 N-methylpyridinium (CH_3 on $N^{\oplus}$)

c. 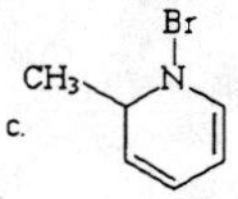 1-bromo-2-methyl-1,2-dihydropyridine (Br on N, CH_3 on C-2)

d. 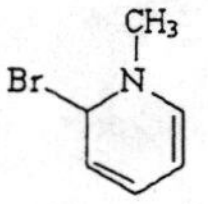 2-bromo-1-methyl-1,2-dihydropyridine (CH_3 on N, Br on C-2)

e. 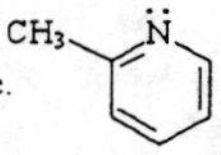 2-methylpyridine (CH_3 on C-2)

Answer: b Section: 3

21. What is the <u>major</u> product of the following reaction?

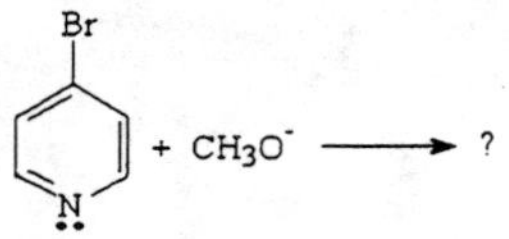

4-bromopyridine + CH_3O^- ⟶ ?

a. 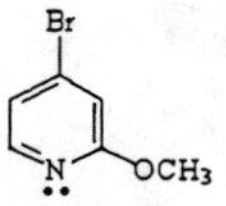 4-bromo-2-methoxypyridine (Br, OCH_3)

b. 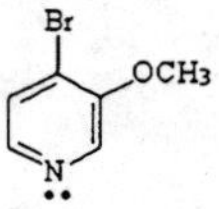 4-bromo-3-methoxypyridine (Br, OCH_3)

c. 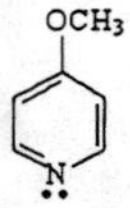 4-methoxypyridine (OCH_3)

d. 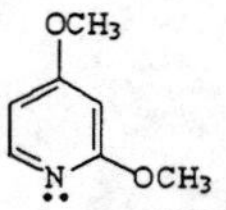 2,4-dimethoxypyridine (OCH_3, OCH_3)

e. 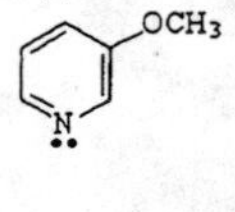 3-methoxypyridine (OCH_3)

Answer: c Section: 3

22. Which of the following is the best method for preparing the following compound?

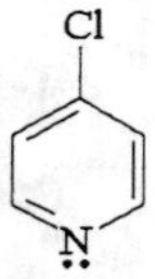

Cl

N

a. + $SOCl_2$ → (Cl, N)

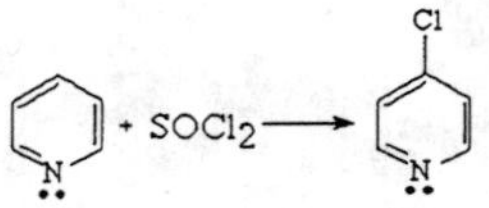

b. (Br, N) + Cl^- → (Cl, N)

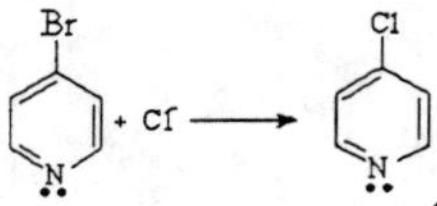

c. (N) + HCl → (Cl, N)

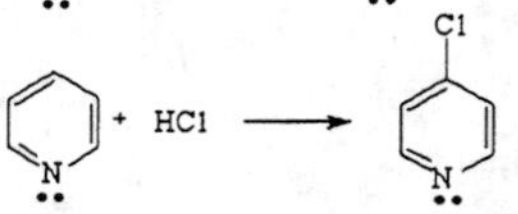

d. (⊕N–O⊖) + Cl_2 / $FeCl_3$ / heat → (Cl, ⊕N–O⊖) PCl_3 → (Cl, N)

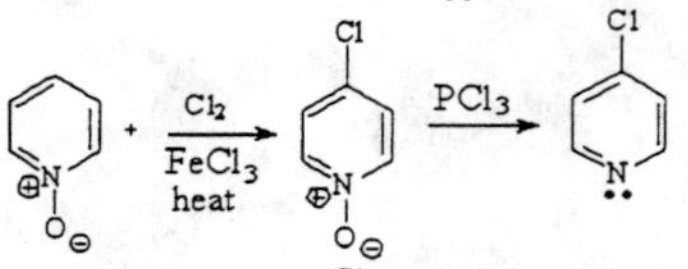

e. (N) + Cl_2 / $FeCl_3$ / heat → (Cl, N)

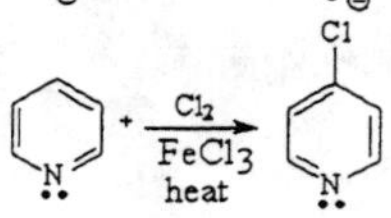

Answer: d Section: 3

23. What are the products of the following reaction?

a. I and II
b. I and III
c. II and IV
d. II and III
e. III and IV

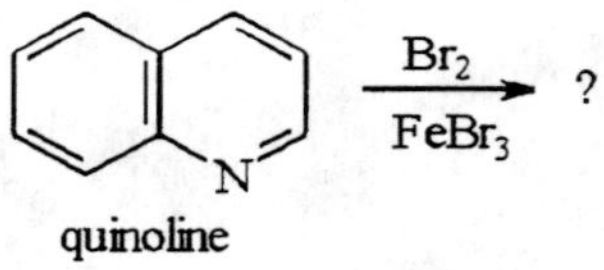

quinoline $\xrightarrow[FeBr_3]{Br_2}$?

I. (N, Br)

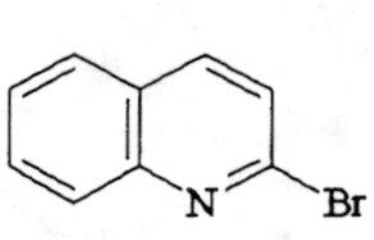

III. (Br, N)

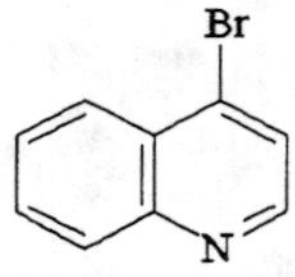

II. (Br, N)

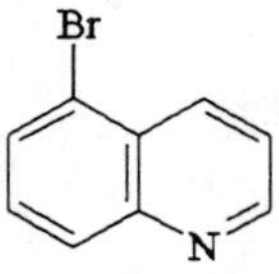

IV. (N, Br)

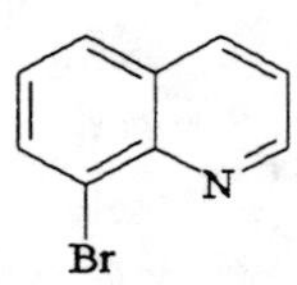

Answer: c Section: 3

24. What is/are the product(s) of the following reaction?

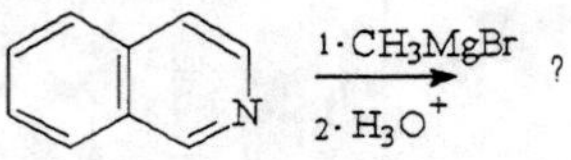

1· CH_3MgBr
2· H_3O^+
?

a. 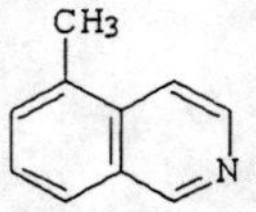 CH_3 N

b. 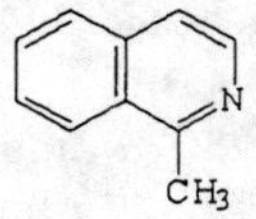 N CH_3

c. 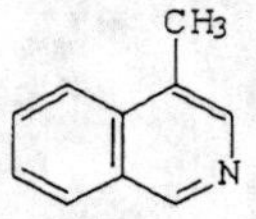 CH_3 N

d. a and b

e. b and c

Answer: b Section: 3

25. What is the <u>major</u> product of the following reaction?

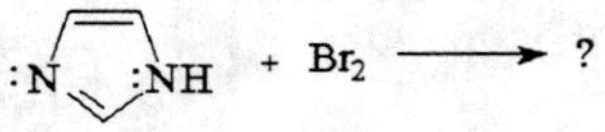

:N :NH + Br_2 → ?

a. 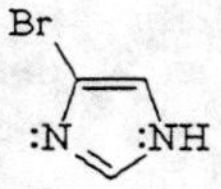 Br :N :NH

b. 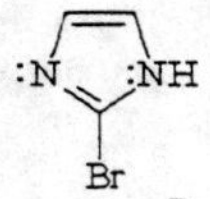 :N :NH Br

c. 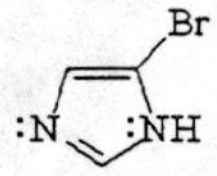 Br :N :NH

d. 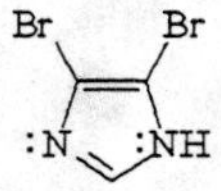 Br Br :N :NH

e. 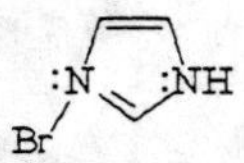 :N :NH Br

Answer: a Section: 4

26. Which of the following is the structure of purine?

a.

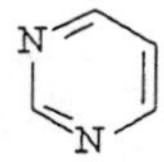

b.

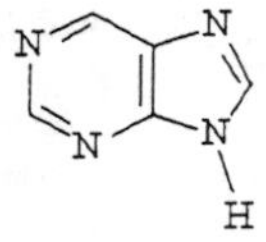

c.

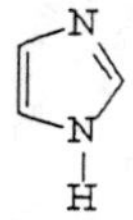

d.

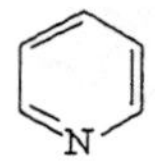

e.

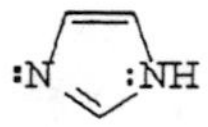

Answer: b Section: 4

27. Which of the following nucleic acids contain the pyrimidine ring structure?

a. adenine, cytosine, and guanine
b. guanine, adenine, and uracil
c. adenine, uracil, and thymine
d. guanine, uracil, and thymine
e. cytosine, uracil, and thymine

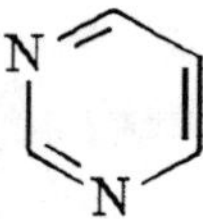

Answer: e Section: 4

28. What is the major product of the following reaction?

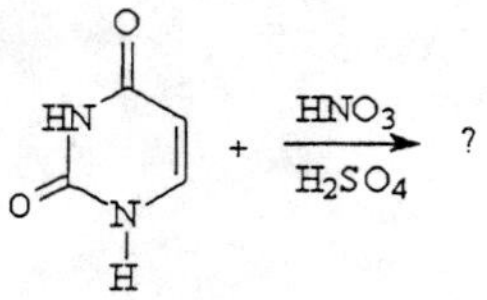

a.

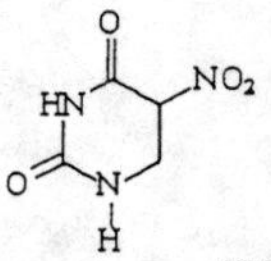

b.

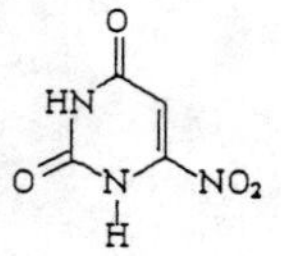

c.

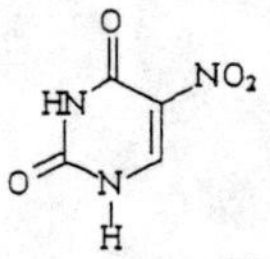

d.

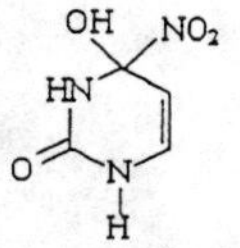

e.

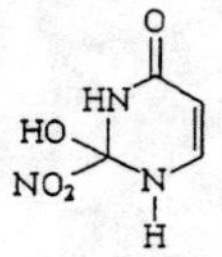

Answer: c Section: 4

29. Which of the following heterocyclics has the lowest pKa value?

a.

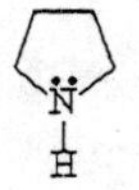

b.

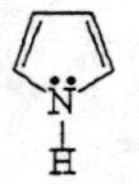

c.

d.

e.

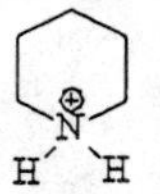

Answer: d Section: 4

SHORT ANSWER

30. Propose a plausible mechanism for the following reaction.

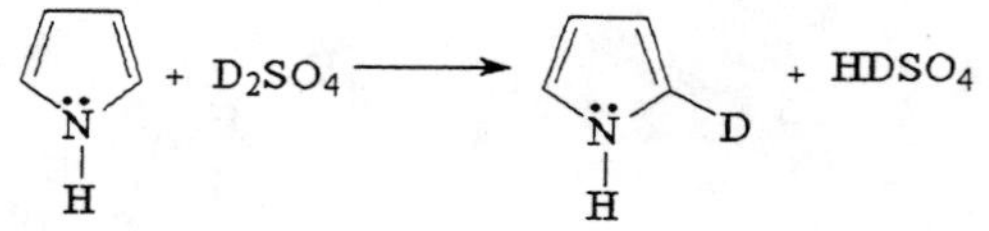

Pyrrole + D_2SO_4 ⟶ 2-deuteriopyrrole + $HDSO_4$

Answer:

Pyrrole + D–O–SO_2–O–D ⟶ (cation, H, D) + DSO_4^- ⟶ 2-deuteriopyrrole + $HDSO_4$

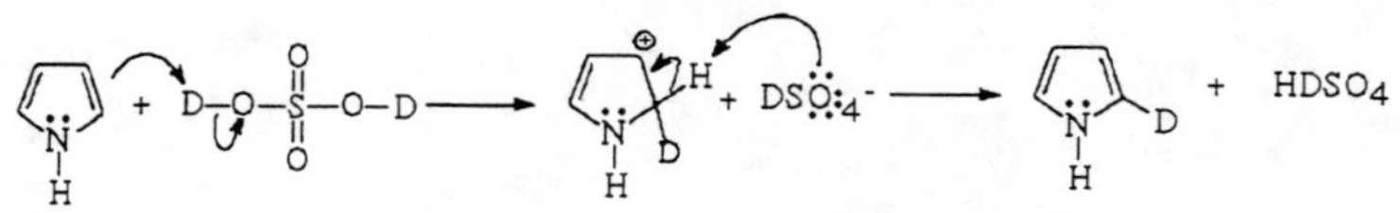

Section: 2

31. Do you expect pyridine to undergo electrophilic substitution reactions? Give an example.

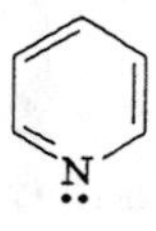

Answer:
Pyridine is very unreactive towards electrophilic substitution reactions. It is even less reactive than nitrobenzene. However, under rigorous conditions, substitution on the C-3 position will occur with low yield.

Ex. Pyridine + Br_2 $\xrightarrow{FeBr_3,\ 300\,°C}$ 3-bromopyridine (30%) + HBr

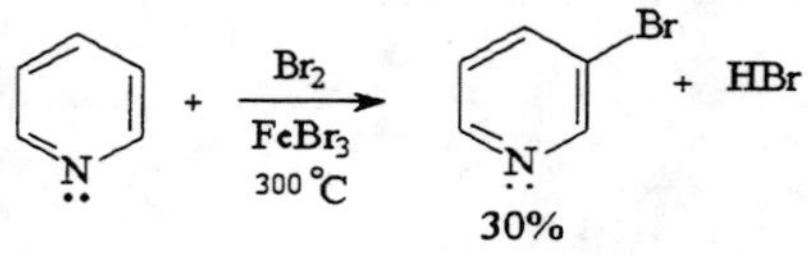

Section: 3

32. Do you expect pyridine to undergo nucleophilic substitution reactions? If yes, on what position would the substitution occur? Give an example.

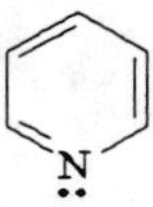

Answer:
Pyridine is more reactive than benzene towards nucleophilic substitution reactions. The substitution takes place on the C-2 and C-4 positions.

Major Minor

Section: 3

33. Imidazole boils at 257° C, while the boiling point of N-methylimidazole is 199° C. Explain the difference in their boiling points.

Imidazole N-methylimidazole

Answer:
Imidazole exhibits hydrogen bonding that N-methylimidazole does not. Therefore, imidazole will boil at a higher temperature than N-methylimidazole.

Section: 4

34. Which of the following compounds is easier to decarboxylate?

I II

Answer:
Compound II would be easier to decarboxylate because the electrons left behind when CO_2 is removed can be delocalized.

Section: 3

35. Arrange the following in decreasing order of reactivity towards electrophilic aromatic substitution.

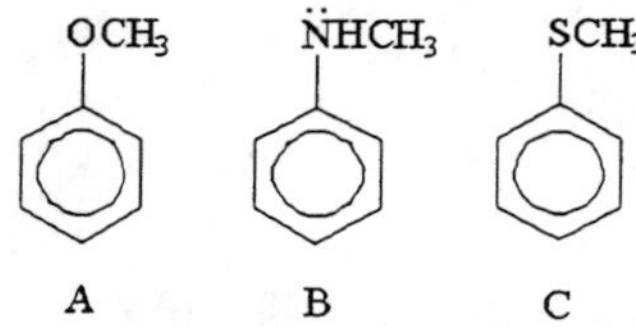

Answer: B>C>A

36. Draw the structure of piperidine.

Answer:

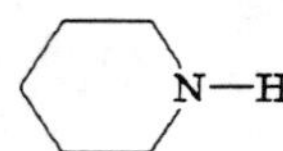

Section: 1

37. Provide the systematic name for the compound below.

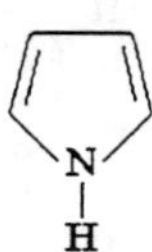

Answer: pyrrole Section: 2

38. Circle the stronger base in the pair below, and briefly explain your choice.

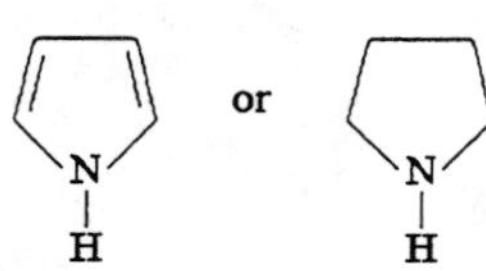

Answer:
Pyrrolidine is the stronger base. Protonation of pyrrole results in loss of aromaticity.

Section: 2

39. Circle the stronger base in the pair below, and briefly explain your choice.

or $(CH_3CH_2)_2NH$

Answer:
Diethylamine is the stronger base. The lone pair electrons in pyridine are in an sp^2 orbital which is lower in energy than the sp^3 orbital of the lone pair in duethylamine.

Section: 3

40. Circle the stronger base in the pair below, and briefly explain your choice.

NH_2 or NH_2

Answer:
Cyclohexylamine is the stronger base. The lone pair electrons in aniline are delocalized into the adjacent aromatic π system.

41. Provide the structure of the major organic product in the reaction below.

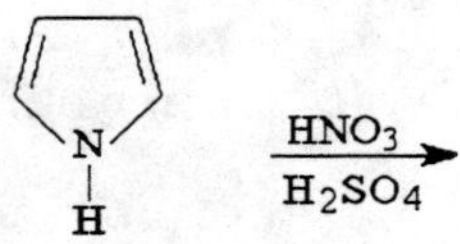

Answer:

NO_2

Section: 2

42. Pyridine does not undergo Friedel-Crafts reactions. Offer an explanation.

    Answer:
    The presence of the electronegative nitrogen atom in the ring deactivates it toward reaction with electrophiles. In fact, the nitrogen lone pair tends to react with electrophiles to produce an even further deactivated ring. It should also be noted that the nitrogen lone pair is in an orbital perpendicular to the ring's π system and could not stabilize the intermediate carbocation in any event.

    Section: 3

MULTIPLE CHOICE

43. When pyrrole undergoes electrophilic aromatic substitution, at which position does substitution occur?

    a. 2-position
    b. 3-position
    c. 4-position

    Answer: a Section: 2

44. Which of the following chloropyridines readily undergo nucleophilic substitution upon treatment with NaCN?

    a. 2-chloropyridine
    b. 3-chloropyridine
    c. 4-chloropyridine
    d. both a and b
    e. both a and c
    f. both b and c

    Answer: e Section: 3

SHORT ANSWER

45. Provide a structural representation of oxetane.

    Answer:

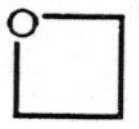

    O

    Section: 1

46. Provide a structural representation of cis-3-ethyl-1,2-epoxycyclopentane.

Answer:

O

CH_2CH_3

Section: 1

47. Provide the major organic product in the reaction below.

1. Cl_2, H_2O
2. NaOH

Answer:

O

48. Provide the major organic product in the reaction below.

$HOCH_2(CH_2)_2CH_2Br \xrightarrow{NaOH}$

Answer:

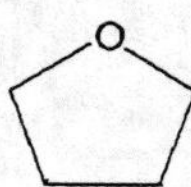

49. Provide the major organic product in the reaction below.

CH_3 CH_3 H CH_3 O $\xrightarrow{CH_3OH,\ H^+}$

Answer:

OH

CH_3 OCH_3

CH_3

CH_3

Section: 1

MULTIPLE CHOICE

50. What results when cis-2-butene is subjected to the following reaction sequence: (1) Cl_2, H_2O, (2) NaOH, (3) H_3O^+?

    a. a meso epoxide
    b. a 1:1 mixture of enantiomeric epoxides
    c. a meso diol
    d. a 1:1 mixture of enantiomeric diols
    e. 2-butanol

    Answer: d Section: 1

SHORT ANSWER

51. What is the stereochemistry of the product of the acid hydrolysis of trans-2,3-epoxybutane?

    Answer: meso diol Section: 1

52. Provide an acceptable name for the compound shown below.

    Answer: 1,4-dioxane Section: 1

MULTIPLE CHOICE

53. Which of the following compounds is <u>not</u> aromatic?

(a) 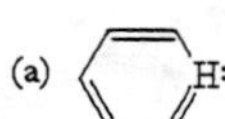

(b) 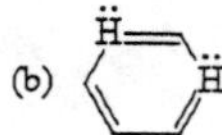

(c) 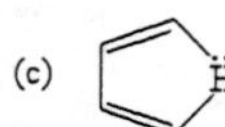

(d) 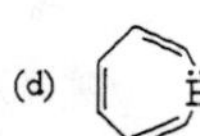

Answer: d Section: 2

SHORT ANSWER

54. Indicate whether the following molecule is aromatic, antiaromatic, or nonaromatic.

Answer: aromatic Section: 2

55. Indicate whether the following molecule is aromatic, antiaromatic, or nonaromatic.

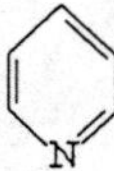

Answer: aromatic Section: 3

56. Indicate whether the following molecule is aromatic, antiaromatic, or nonaromatic.

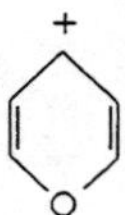

Answer: aromatic

MULTIPLE CHOICE

57. An unknown compound has the formula $C_8H_{10}OS$, and is known to contain a thiophene ring. Thr proton NMR spectrum of this compound is:
δ0.98, triplet, 3H
δ1.74, sextet, 2H
δ2.80, triplet, 2H
δ7.04, multiplet, 1H
δ7.55, multiplet, 2H

What is the structure of this unknown?

a. I
b. II
c. III
d. IV

I. $CH_3CH_2CH_2$–(2,5-thiophenediyl)–C(=O)–H

II. CH_3–(2,5-thiophenediyl)–CH_2CH_2C(=O)–H

III. (2-thienyl)–CH_2CH_2C(=O)–CH_3

IV. (2-thienyl)–C(=O)–$CH_2CH_2CH_3$

Answer: d

58. What is the major product of the reaction sequence shown?

$CH_3CH(Cl)CH_3$ —Mg, ether→ —H_2C---CH_2 (O bridging)→ —H_3O^+→ ?

a. $(CH_3)_2CHCH_2CH_2OH$
b. $(CH_3)_2CHCHOHCH_3$
c. $CH_3CH_2CH_2CH_2CH_2OH$
d. $CH_3CH_2CH_2CHOHCH_3$

Answer: a

59. In the following question choose the best answer.
The tertiary amine

(a) 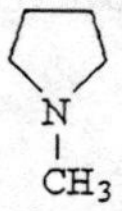 N–CH_3

(b)  $NHCH_3$

(c) $(CH_3)_3NHCH_2$

(d) $(CH_3)_4NBr$

Answer: a Section: 1

60. In the following question choose the best answer.
The most basic amine.

(a) $CH_3CH_2NH_2$

(b) $(CH_3)_3N$

(c)  N

(d) 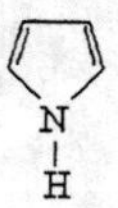 N–H

Answer: a Section: 2

Chapter 27: Pericyclic Reactions

MULTIPLE CHOICE

1. Which of the following statements best describes what is meant by pericyclic reactions?

   a. Reactions in which a nucleophile reacts with an electrophile
   b. Reactions in which an electrophile reacts with a nucleophile
   c. Reactions in which a free radical reacts with another free radical
   d. Oxidation-reduction reactions in which electrons are lost or gained
   e. Reactions that occur as a result of a cyclic reorganization of electrons

   Answer: e

2. Which of the following reactions best describes the Diels-Alder reaction?

   a. Electrocyclic reaction
   b. Cycloaddition reaction
   c. Sigmatropic reaction
   d. Radical reaction
   e. Nucleophilic substitution reaction

   Answer: b Section: 4

3. Which of the following statements best describes the theory of <u>Conservation of Orbital Symmetry</u>?

   a. Molecular orbital of the transition state must be similar to that of the reactant.
   b. Molecular orbital of the transition state must be similar to that of the product.
   c. Only S orbitals from reactants and products are utilized.
   d. Molecular orbitals of reactant and product must have similar symmetry.
   e. Molecular orbitals of reactant and product must have different symmetry.

   Answer: d Section: 1

4. Which of the following pericyclic reactions best describes the following reaction?

   a. Nucleophilic
   b. Electrocyclic
   c. Radical
   d. Sigmatropic
   e. Cycloaddition

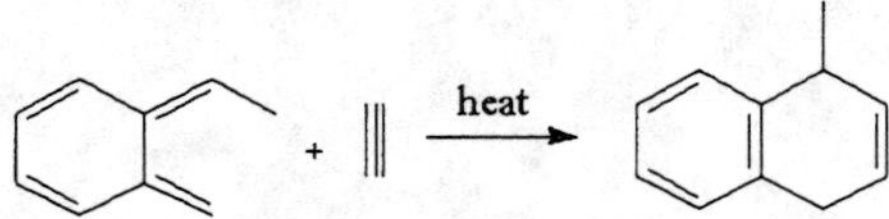

Answer: e Section: 4

5. Which of the following molecular orbitals is produced when in-phase atomic orbitals overlap?

   a. Bonding molecular orbital
   b. Antibonding molecular orbital
   c. Nonbonding molecular orbital
   d. Out-of-phase molecular orbital
   e. Excited energy state of molecular orbital

Answer: a Section: 2

6. Which of the following molecular orbitals are formed from <u>subtractive</u> overlap of atomic orbitals?

   a. Antibonding orbitals
   b. Out-of-phase orbitals
   c. Bonding orbitals
   d. a and b
   e. b and c

Answer: d Section: 2

7. How many molecular orbitals are produced from the overlap of four p atomic orbitals?

   a. 2
   b. 8
   c. 4
   d. 16
   e. 1

Answer: c Section: 2

8. Which of the following sets of atomic orbitals form an asymmetric molecular orbital?

a. 

b.

c.

d.

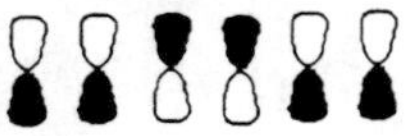

e.

Answer: e Section: 2

9. How many π molecular orbitals does 1,3,5,7-octatetraene have?

a. 2
b. 4
c. 6
d. 8
e. 10

Answer: d Section: 2

10. How many nodes does the highest energy molecular orbital of 1,3,5,7-octatetraene have?

a. 4
b. 5
c. 6
d. 7
e. 8

Answer: e Section: 2

11. What is the <u>major</u> product of the following reaction?

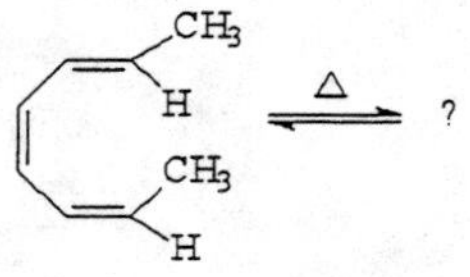

$\xrightleftharpoons{\Delta}$?

a.

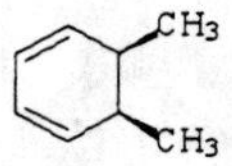

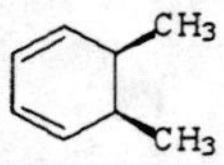

b.

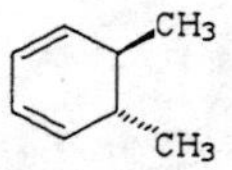

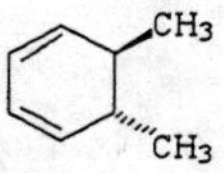

c.

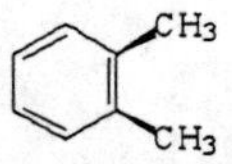

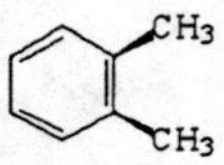

d.

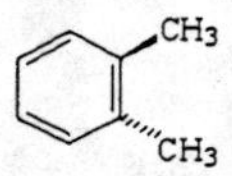

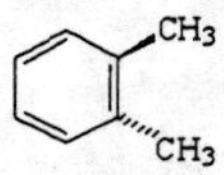

e.

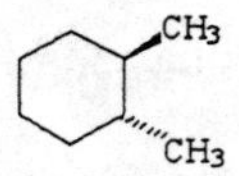

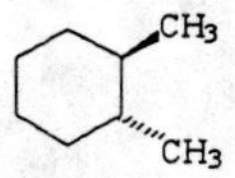

Answer: b Section: 3

12. What is the major product of the following reaction?

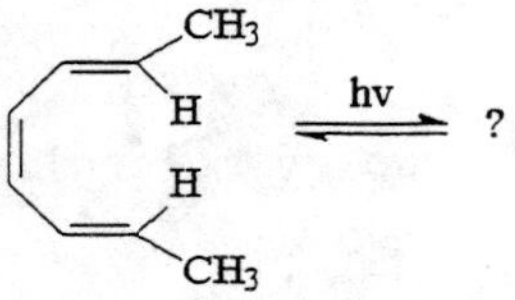

$\xrightleftharpoons{hv}$?

a.

b.

c.

d.

e.

Answer: b Section: 3

13. Which of the following statements best describes the term suprafacial?

    a. Formation of two sigma bonds from the same side of the pi system
    b. Formation of two pi bonds from the same side of the sigma system
    c. Formation of two sigma bonds from opposite sides of the pi system
    d. Formation of two pi bonds from opposite sides of the sigma system
    e. An anti cycloaddition

    Answer: a Section: 4

14. Which of the following ring formations must involve a suprafacial bond formation?

    a. four-membered ring
    b. five-membered ring
    c. six-membered ring
    d. a and c
    e. a, b and c

    Answer: e Section: 4

15. What is the major product of the following reaction?

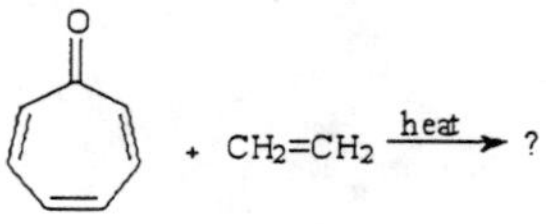

    + $CH_2{=}CH_2$ $\xrightarrow{\text{heat}}$?

    a.

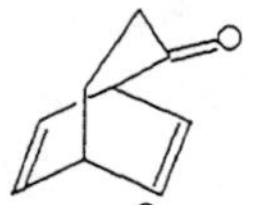

    b.

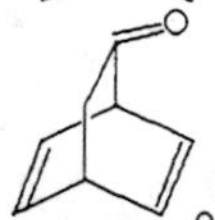

    c.

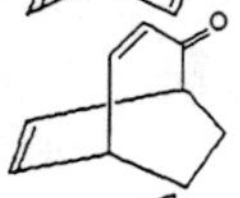

    d.

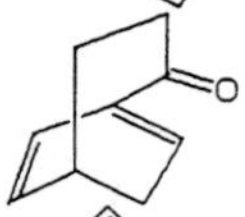

    e.

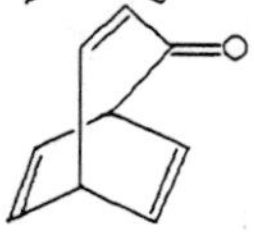

    Answer: c Section: 4

16. Which of the following best describes the sigmatropic rearrangement that occurs in the reaction shown below?

a. [1,3] Sigmatropic rearrangement
b. [2,3] Sigmatropic rearrangement
c. [3,3] Sigmatropic rearrangement
d. [1,5] Sigmatropic rearrangement
e. [1,4] Sigmatropic rearrangement

CH_2 CH_3 △ → CH_3 CH_2

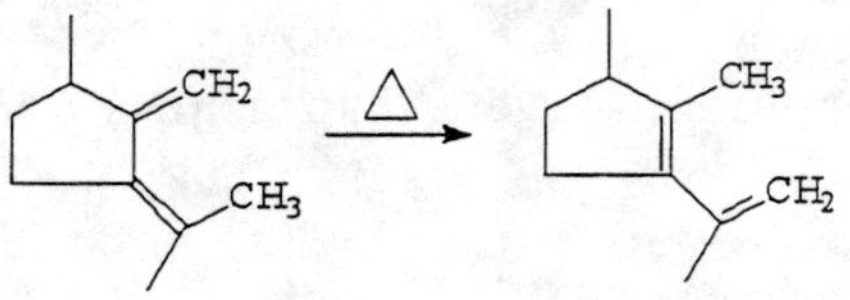

Answer: d Section: 5

17. What is the major product of the following [3,3] sigmatropic rearrangement?

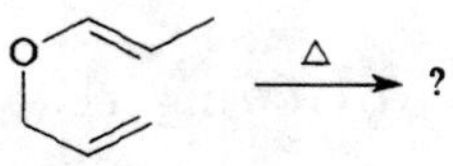

O △ → ?

a. 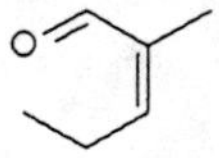 O

b. 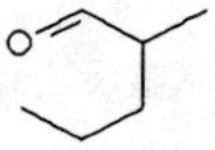 O

c. 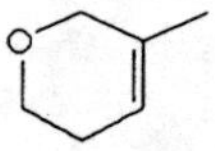 O

d. 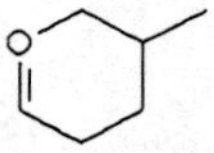 O

e O

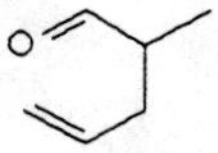

Answer: e Section: 5

18. What is the major product from the following rearrangement?

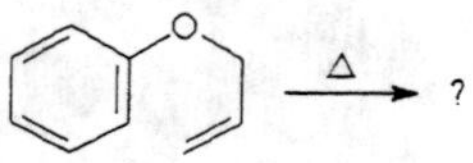

a.

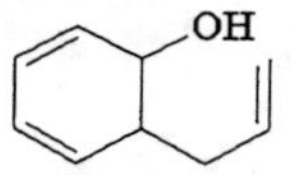

b.

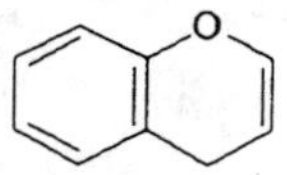

c.

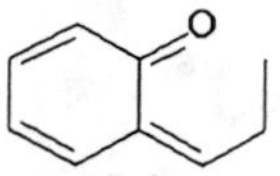

d.

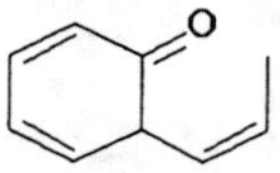

e.

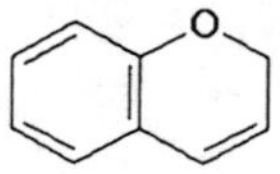

Answer: a Section: 5

19. Under what conditions would a 1,3-hydrogen shift occur in a sigmatropic rearrangement?

a. Thermal conditions
b. Photochemical conditions
c. Suprafacial pathway
d. a and c
e. b and c

Answer: e Section: 5

20. What is/are the product(s) of the following 1,3-hydrogen shift?

a. I
b. II
c. III
d. I and II
e. I and III

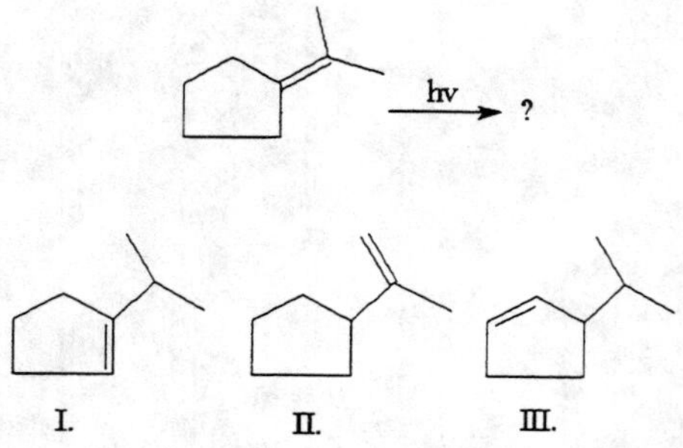

$\xrightarrow{h\nu}$?

I. II. III.

Answer: d Section: 5

21. Under what conditions would a 1,5-hydrogen shift occur in a sigmatropic rearrangement?

a. Thermal conditions
b. Photochemical conditions
c. Suprafacial pathway
d. a and c
e. b and c

Answer: d Section: 5

22. What is the product of the following 1,5-hydrogen shift?

$$CH_3CH{=}CHCH{=}CD_2 \xrightarrow{\Delta} ?$$

a. $CH_2{=}CHCH_2CH{=}CHD$
b. $CH_2{=}CHCH{=}CHCH_2D$
c. $CHD{=}CHCH{=}CHCH_3$
d. $CH_2{=}CHCH_2CH_2CH_2D$
e. $CH_2{=}CHCH{=}CHCD_3$

Answer: b Section: 5

23. What is the product of the following 1,7-hydrogen shift?

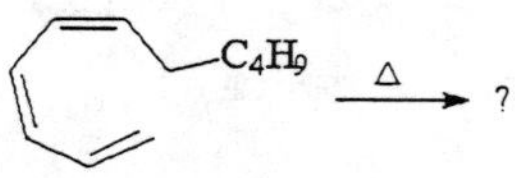

C_4H_9 $\xrightarrow{\Delta}$?

a. 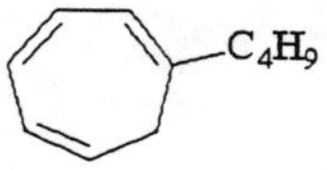 C_4H_9

b. C_4H_9

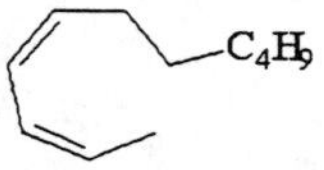

c. C_4H_9

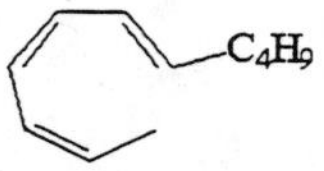

d. 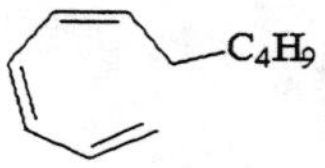 C_4H_9

e. 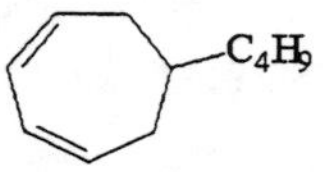 C_4H_9

Answer: c Section: 5

24. What is the product from the following 1,3-carbon shift reaction?

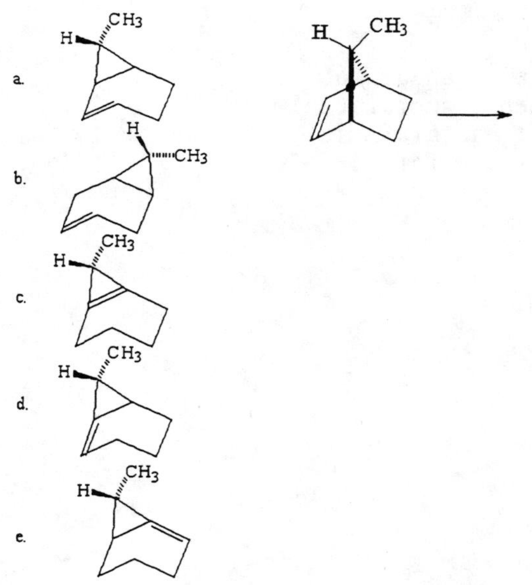

Answer: a Section: 5

25. What is the product obtained from the following [1,5] sigmatropic rearrangement?

$OCH_2CH{=}CHCH{=}CH_2$ (on benzene ring) ⟶ ?

a. $OCH{=}CHCH_2CH{=}CH_2$ (on benzene ring)

b. O

c. O

d. $OCH{=}CHCH{=}CHCH_3$ (on benzene ring)

e. $OCH_2CH{=}CHCH_2CH_3$ (on benzene ring)

Answer: d Section: 5

26. Which of the following best explains the rearrangement when provitamin D_3 rearranges to vitamin D_3?

a. [1,7] cycloaddition rearrangement, antarafacial
b. [1,7] cycloaddition rearrangement, suprafacial
c. [1,7] sigmatropic rearrangement, antarafacial
d. [1,7] sigmatropic rearrangement, suprafacial
e. [1,7] electrocyclic reaction, suprafacial

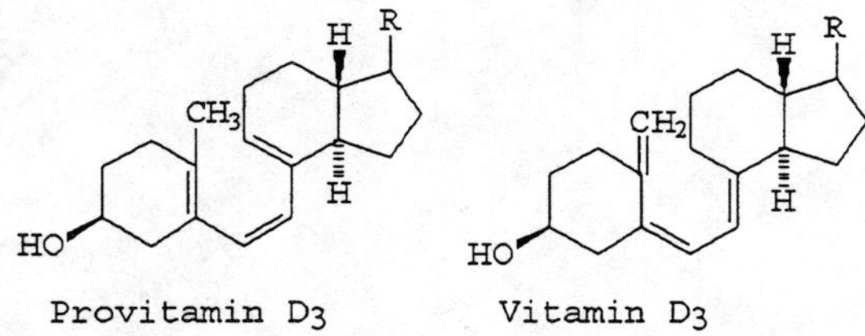

Provitamin D3 Vitamin D3

Answer: c Section: 5

SHORT ANSWER

27. What are the three most common pericyclic reactions? Give examples.

Answer:

a. electrocyclic reactions:

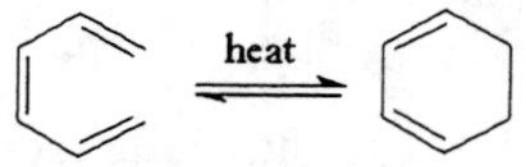

heat

b. cycloaddition reactions:

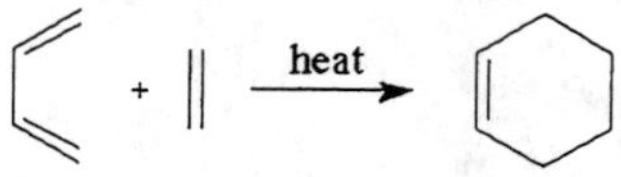

+ heat

c. sigmatropic rearrangement:

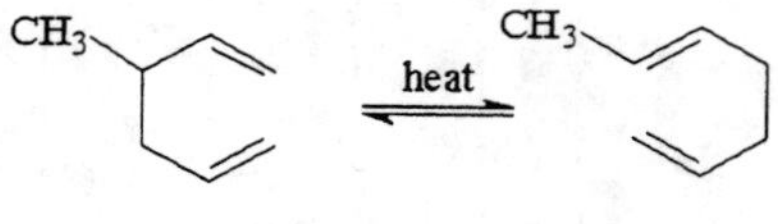

CH_3 heat CH_3

Section: 1

28. Show the mechanism for each of the following pericyclic reactions.

a. heat

b. + heat

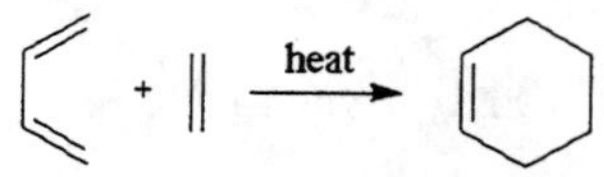

c. CH_3 heat CH_3

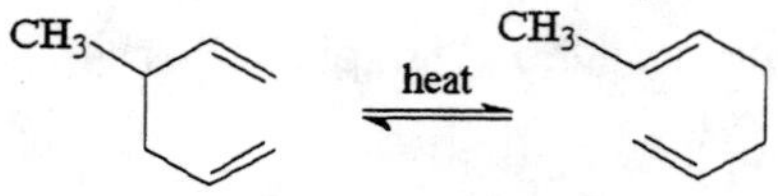

Answer:

a. heat

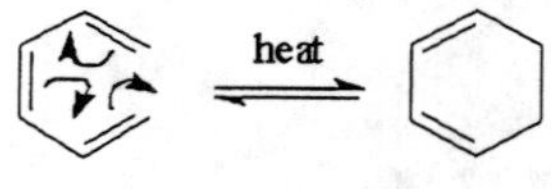

b. + heat

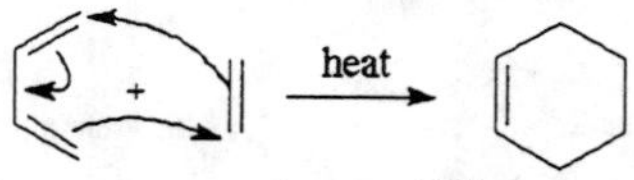

c. CH_3 heat CH_3

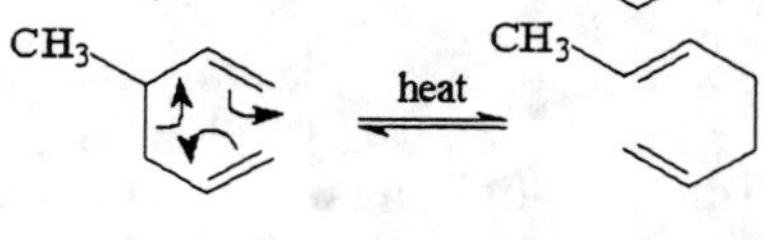

Section: 1

29. What is meant by <u>frontier orbitals</u>?

Answer:
Frontier orbitals are the two molecular orbitals known as the highest occupied molecular orbital (HOMO) and the lowest unoccupied molecular orbital (LUMO).

Section: 1

30. Indicate whether each of the following reactions is an electrocyclic reaction, a cycloaddition reaction, or a sigmatropic rearrangement.

a. heat

b. H, CH_3 heat CH_3

c. OCH_3 + → OCH_3

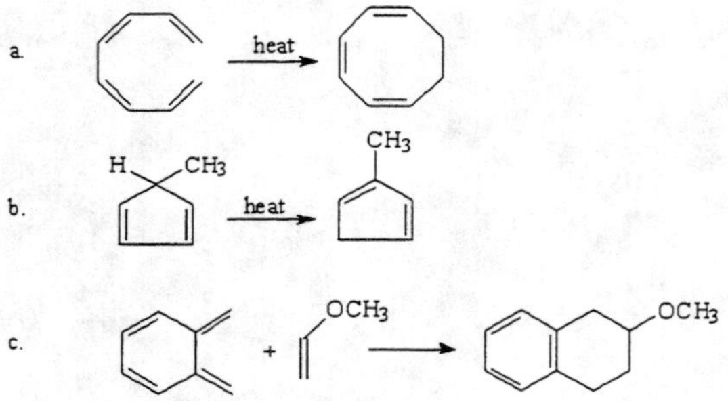

Answer:
a. Is an electrocyclic reaction.
b. Is a sigmatropic rearrangement.
c. Is a cycloaddition reaction.

Section: 1

31. What does LCAO stand for?

Answer: Linear combination of atomic orbitals. Section: 2

32. Which molecular orbital is higher in energy: the in-phase or out-of-phase overlap of atomic orbitals?

Answer:
Out-of-phase overlap of atomic orbitals produces an antibonding molecular orbital which is higher in energy than that of the bonding molecular orbital.

Section: 2

33. Explain what a node is and how it is formed.

Answer:
A node is a plane in which there is no probability of finding an electron. It is formed between overlapping out-of-phase orbitals or through the center of p orbitals.

Section: 2

34. How many nodes would you expect from the following atomic orbital overlap?

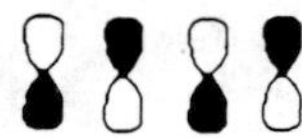

Answer: The molecular orbital is an out-of-phase overlap producing four nodes.

Section: 2

35. (a) Under thermal conditions, will ring closure of (2E, 4Z, 6Z, 8E)-decatetraene be conrotatory or disrotatory? Explain.
(b) Will the product be cis or trans? Explain

Answer:
(a) Since the number of conjugated pi bonds is even and the reaction occurs under thermal conditions, the Woodward-Hoffman rules for electrocyclic reaction predicts that the ring closure is conrotatory.

(b) Since the ring closure is conrotatory and the closure is (E,E) configuration, the product will have a trans configuration.

Section: 3

36. Explain why the following (2+2) cycloaddition reaction would work under photochemical conditions but not under thermal conditions.

+ —heat→ no reaction

+ —hv→

Answer:
Under thermal conditions, suprafacial overlap is not symmetry-allowed while under photochemical conditions, the reaction can take place because it involves symmetry-allowed suprafacial bond formation.

Section: 4

37. Show the mechanism for the following 1,3-hydrogen shift.

hv

Answer:

H

H

Section: 5

MULTIPLE CHOICE

38. In the allyl radical, which π molecular orbital is singly occupied?

a. The bonding π molecular orbital.
b. The nonbonding π molecular orbital.
c. The antibonding π molecular orbital.
d. None of the above.

Answer: b

SHORT ANSWER

39. Give a representation of the bonding π molecular orbital of the allyl cation.

Answer:

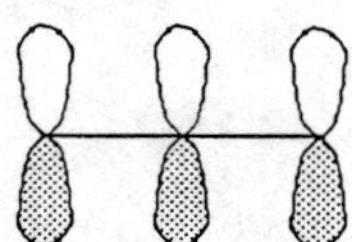

40. Give a representation of the antibonding π molecular orbital of the allyl anion.

Answer:

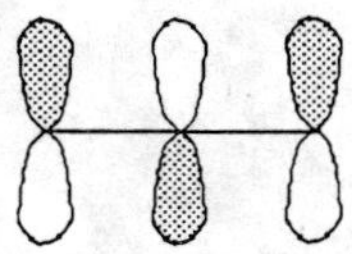

MULTIPLE CHOICE

41. How many electrons are present in the nonbonding π molecular orbital of the allyl anion?

 a. 0
 b. 1
 c. 2
 d. 3
 e. 4

 Answer: c Section: 2

SHORT ANSWER

42. What diene and dienophile would react to give the product below?

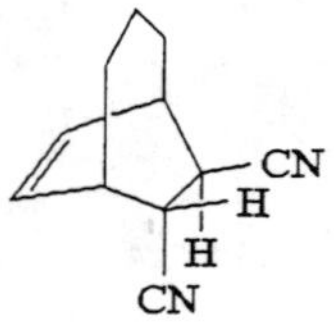

CN H H CN

 Answer:

H CN + NC H

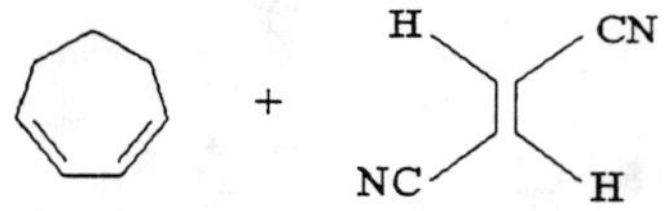

43. Provide the structure of the major organic product in the following reaction.

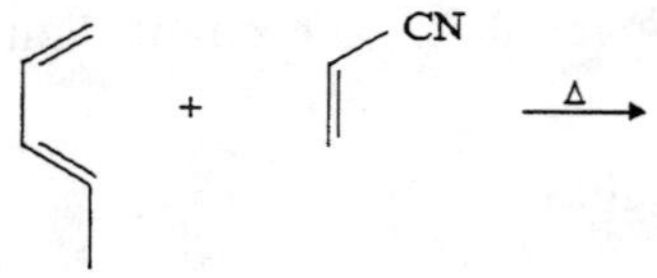

CN + Δ

 Answer:

CN CN_3

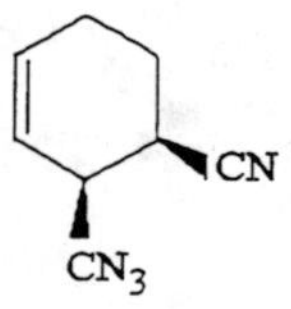

 Section: 4

44. What are the structures of A, B, and C?

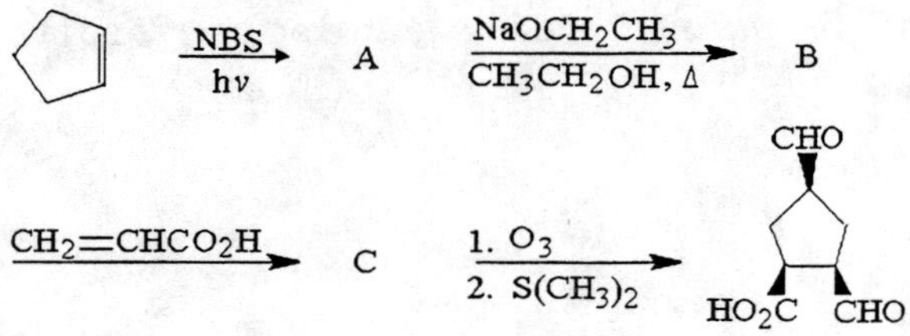

$\xrightarrow[h\nu]{NBS}$ A $\xrightarrow[CH_3CH_2OH,\ \Delta]{NaOCH_2CH_3}$ B

$\xrightarrow{CH_2{=}CHCO_2H}$ C $\xrightarrow[2.\ S(CH_3)_2]{1.\ O_3}$

CHO

HO_2C CHO

Answer:

Br

A B C

H

CO_2H

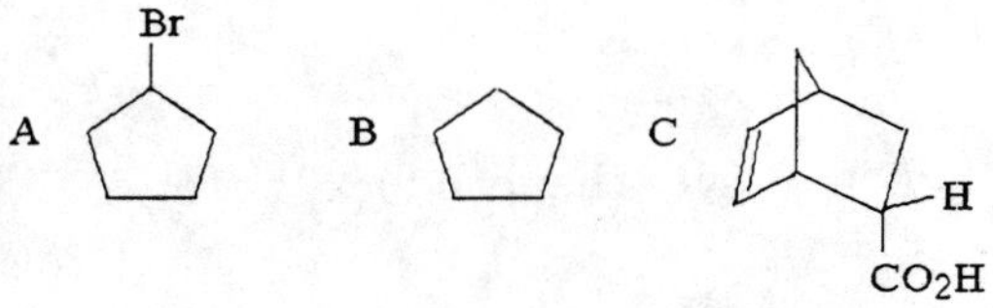

Section: 4

45. Why does the diene shown below fail to undergo a Diels-Alder reaction with even the most reactive dienophiles?

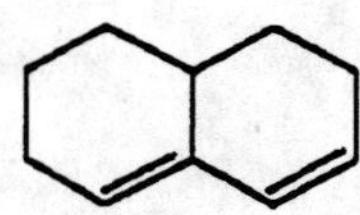

Answer: The diene cannot achieve the necessary s-cis conformation.

46. What characterizes a pericyclic reaction?

Answer:
Pericyclic reactions involve concerted breaking and formation of bonds within a closed ring of interacting orbitals.

47. Consider the possible thermal [4+4] cycloaddition of two molecules of 1,3-butadiene to generate cycloocta-1,5-diene. Show the HOMO/LUMO interaction which would result, and use this interaction to predict whether the proposed cycloaddition could occur.

Answer:

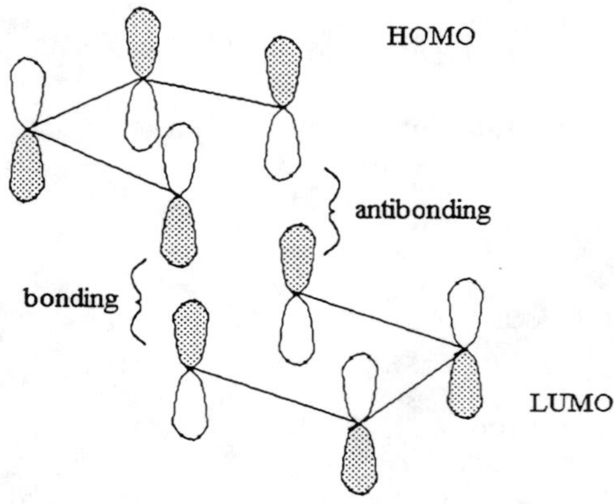

The cycloaddition is thermally forbidden.

48. Is the thermal [4+2] cycloaddition between allylanion and ethylene an allowed one? To answer, draw the HOMO of allyl anion and the LUMO of ethylene, and comment on the symmetry match of the two.

Answer:

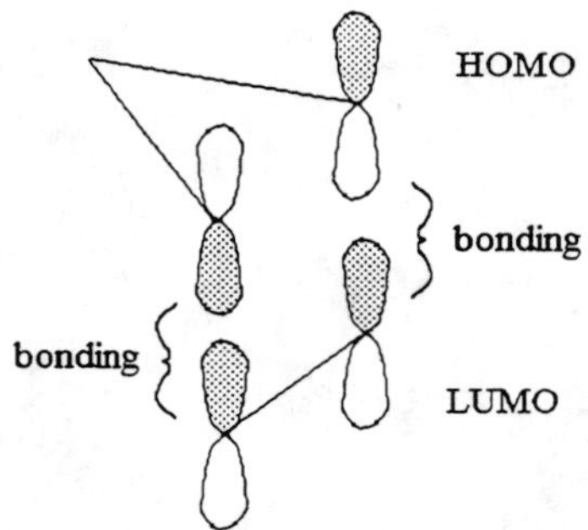

The cycloaddition is thermally allowed.

Chapter 28: More About Multistep Organic Synthesis

MULTIPLE CHOICE

1. Which of the following statements best explains the meaning of <u>natural products</u>?

   a. Organic compounds that are used in organic gardening
   b. Organic compounds that are harmless to nature
   c. Organic compounds synthesized in nature
   d. Organic compounds synthesized in the lab that resemble those found in nature
   e. Organic compounds synthesized in the lab that degrade safely in nature

   Answer: c

2. Why do chemists synthesize organic compounds?

   a. To study their properties
   b. To answer a variety of chemical questions
   c. To study their unusual shape or structure
   d. To provide us with a greater supply of a compound than nature can produce
   e. All of the above

   Answer: e

3. Which of the following reasons would make one synthetic method superior to others?

   a. Produces a higher yield
   b. Utilizes fewer steps
   c. Uses less complicated reactions
   d. Requires cheaper starting materials
   e. All of the above

   Answer: e

4. Which of the following best describes convergent synthesis?

   a. A → B
   b. A → B → C
   c. A → B → C + D → F

                    ↑
                    E

   d. A → B → C → D → E
   e. B → A

   Answer: c

5. Which of the following methods may be used for the synthesis of propyl alcohol,

$$CH_3CH_2CH_2OH\ ?$$

a. $CH_3CH_2\overset{O}{\overset{\|}{C}}H \xrightarrow[2.\ H_2O]{1.\ NaBH_4}$

b. $CH_3CH_2\overset{O}{\overset{\|}{C}}OH \xrightarrow[2.\ H_2O]{1.\ LiAlH_4}$

c. $CH_3CH=CH_2 \xrightarrow[2.\ HO^-,\ H_2O_2,\ H_2O]{1.\ BH_3}$

d. $CH_3CH_2CH_2Br \xrightarrow{^-OH}$

e. All of the above

Answer: e Section: 1

6. Which of the following reduction reactions may be used to synthesize benzyl alcohol,

$$C_6H_5CH_2OH\ ?$$

a. $C_6H_5\overset{O}{\overset{\|}{C}}-Cl \xrightarrow[2\cdot H_2O]{1\cdot LiAlH_4}$

b. $C_6H_5\overset{O}{\overset{\|}{C}}OCH_3 \xrightarrow[2\cdot H_2O]{1\cdot LiAlH_4}$

c. $C_6H_5\overset{O}{\overset{\|}{C}}H \xrightarrow[HCl]{Zn(Hg)}$

d. a and b

e. a and c

Answer: d Section: 1

7. Which of the following methods cannot be used to prepare isobutyl alcohol,

$$CH_3\underset{}{\overset{CH_3}{\overset{|}{C}}}HCH_2OH \text{ ?}$$

a. $CH_3\overset{CH_3}{\overset{|}{C}}{=}CH_2 \xrightarrow[H^+]{H_2O}$

b. $CH_3\overset{CH_3}{\overset{|}{C}}{=}CH_2 \xrightarrow[2.\ HO^-,\ H_2O_2,\ H_2O]{1.\ BH_3}$

c. $CH_3\overset{CH_3}{\overset{|}{C}}HCH_2Br \xrightarrow{HO^-}$

d. $CH_3\underset{CH_3}{\underset{|}{C}}H\overset{O}{\overset{||}{C}}H \xrightarrow[2.\ H_2O]{1.\ NaBH_4}$

e. $CH_3\overset{CH_3}{\overset{|}{C}}HCH_2OCH_3 \xrightarrow[heat]{HI}$

Answer: a Section: 1

8. Which of the following statements best explains the meaning of <u>retrosynthesis</u>?

a. Working forward from the starting material to the desired product
b. Working backward from the desired product to the starting material
c. Working forward using linear synthesis
d. Working forward using convergent synthesis
e. Dissecting the starting material into smaller and smaller pieces until the desired product is reached

Answer: b Section: 2

9. Which of the following statements best explains what is meant by <u>synthons</u>?

a. Fragments of a connection in a convergent synthesis
b. Fragments of a connection in a linear synthesis
c. Fragments of a disconnection in retrosynthesis
d. Bond formation in a retrosynthesis
e. Bond formation in a linear synthesis

Answer: c Section: 2

10. Which of the following statements best explains what is meant by a <u>synthetic</u> <u>equivalent</u>?

    a. Using equal equivalents of reactants
    b. Using equal equivalents of reactants and catalysts
    c. Producing equivalents of the products equal to the equivalents of reactants
    d. Actual reagent used as a source of the synthon
    e. Using a microscale approach to retrosynthesis

    Answer: d Section: 2

11. Which of the following might be considered synthons for the retrosynthetic analysis of the compound shown below?

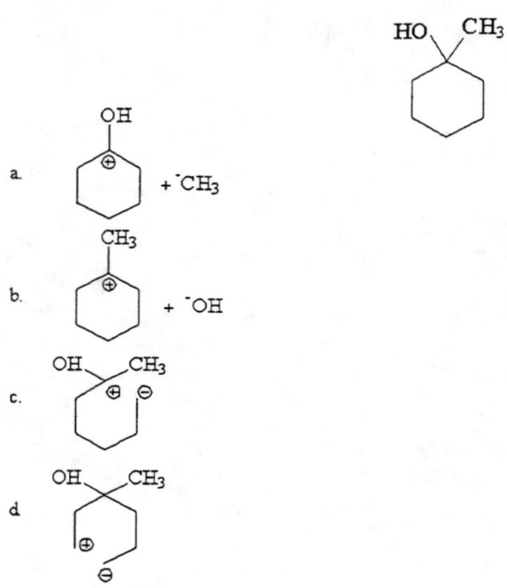

    Answer: e Section: 2

12. Which of the following is the best synthetic equivalent of $:H^-$?

    a. H_2O
    b. $LiAlH_4$
    c. HO^-
    d. $:^-NH_2$
    e. PH_3

    Answer: b Section: 2

13. Which of the following statements best explains the meaning of umpolung?

   a. Retention of configuration in a functional group
   b. Interconversion of a functional group
   c. Retaining the normal polarity of a functional group
   d. Reversing the normal polarity of a functional group
   e. Enhancing the normal polarity of a functional group

   Answer: d Section: 2

14. What is the synthetic equivalent for the following synthon?

   $CH_3\ddot{C}H_2^{\ominus}$

   a. CH_3CH_2Br
   b. $CH_3CH_2\ddot{N}HCH_2CH_3$
   c. CH_3CH_2MgBr
   d. CH_3CH_2Na
   e. $CH_3CH_2\ddot{O}CH_2CH_3$

   Answer: c Section: 2

15. What is the synthetic equivalent for

   $^{\ominus}\!:CH_2CH{=}O$?

   a. $CH_3CH{=}O$
   b. $ClCH_2CH{=}O$
   c. $CH_3CH_2CH{=}O$
   d. 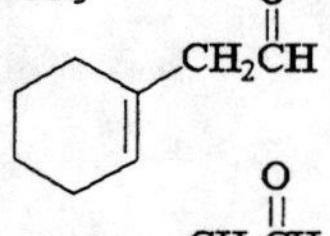 (cyclohexenyl)$CH_2CH{=}O$
   e. 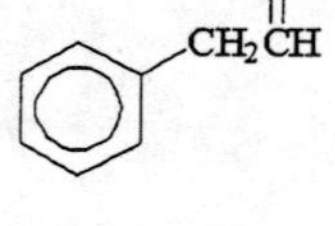 (phenyl)$CH_2CH{=}O$

   Answer: a Section: 2

16. How would you accomplish the following synthesis?

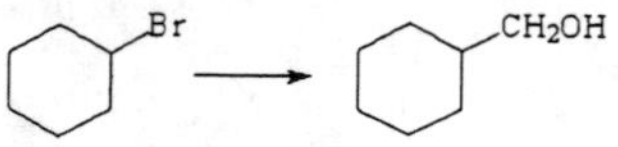

Br → CH_2OH

a. 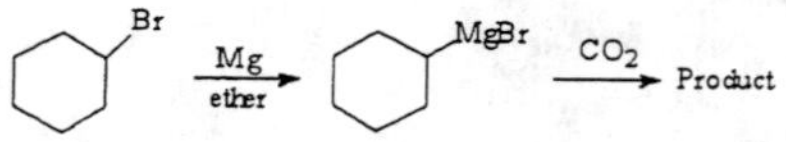 Br, Mg/ether → MgBr, CO_2 → Product

b. 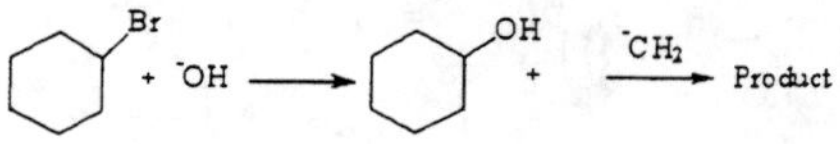 Br + ^-OH → OH + $^-CH_3$ → Product

c. 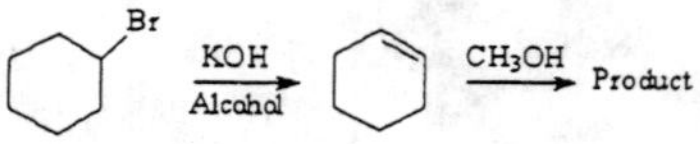 Br, KOH/Alcohol → , CH_3OH → Product

d. 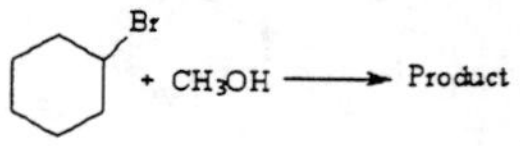 Br + CH_3OH → Product

e. 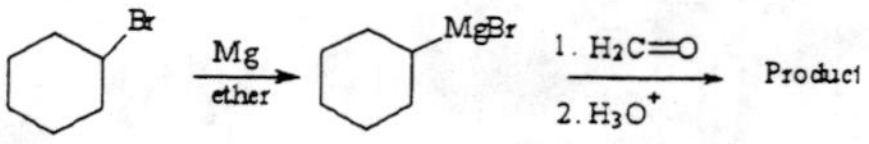 Br, Mg/ether → MgBr, 1. $H_2C=O$ 2. H_3O^+ → Product

Answer: e Section: 2

17. Which of the following will serve as the starting materials (A) and (B) for the following retrosynthesis?

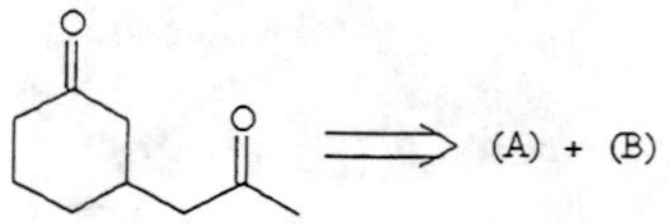

O O ⟹ (A) 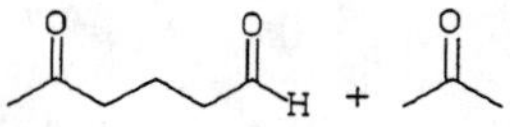 + (B)

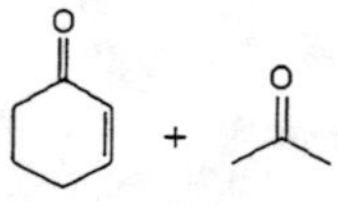

a. O O H + O

b. O + O

c. 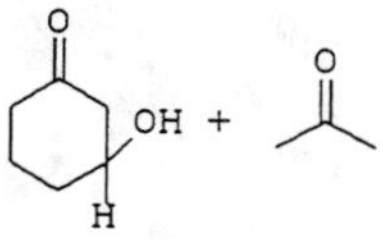 O OH + O H

d. 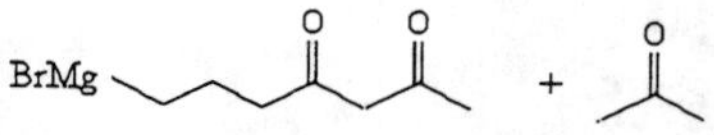 BrMg O O + O

e. All of the above

Answer: e Section: 3

18.

Answer:

19. How can you synthesize the following compound from starting materials containing no more than four carbons?

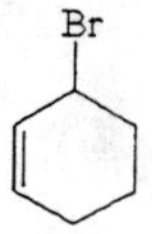

Br

a. 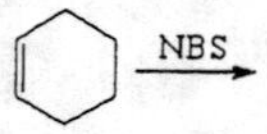 NBS

b. 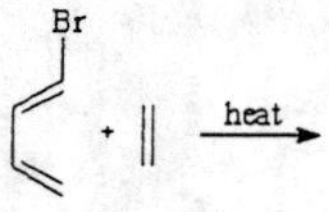 Br heat

c. 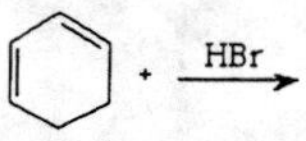 HBr

d. 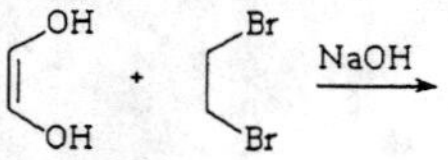 OH OH Br Br NaOH

e. 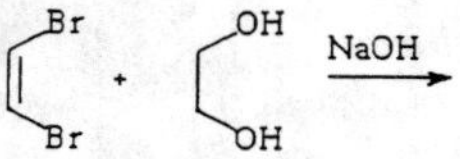 Br Br OH OH NaOH

Answer: b Section: 1

20. Propose a method to synthesize the following compound:

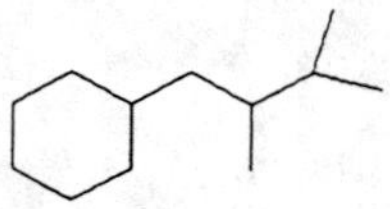

a. 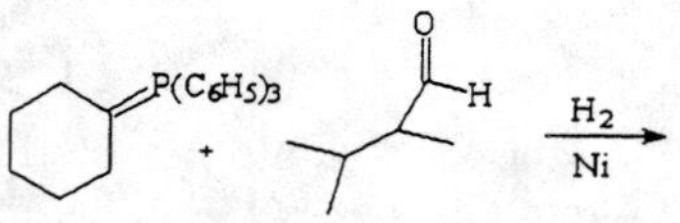 $P(C_6H_5)_3$ + H O H_2 Ni

b. 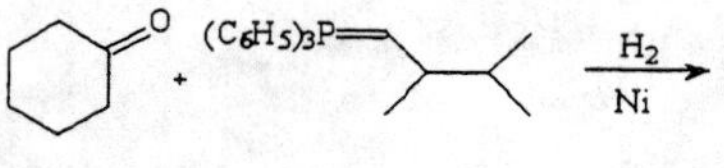 O + $(C_6H_5)_3P$ H_2 Ni

c. O + BrMg H_2O H^+ heat H_2 Ni

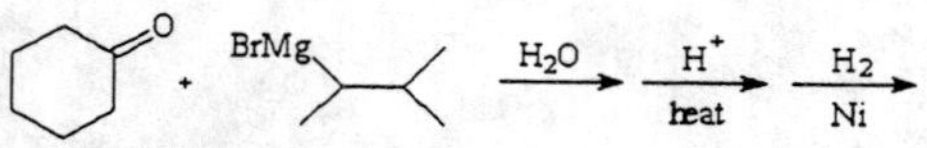

d. a and b

e. a, b, and c

Answer: e Section: 4

21. Which of the following statements explains the meaning of a protecting group?

a. A reagent that interconverts a functional group during a synthetic operation
b. A reagent that increases the reactivity of a functional group during a synthetic operation
c. A reagent that protects a functional group from a synthetic operation
d. A reagent that decreases the reactivity of a functional group during a synthetic conversion
e. A reagent that improves the availability of a functional group for synthetic conversions

Answer: c Section: 4

22. Identify the protecting group in the following synthesis:

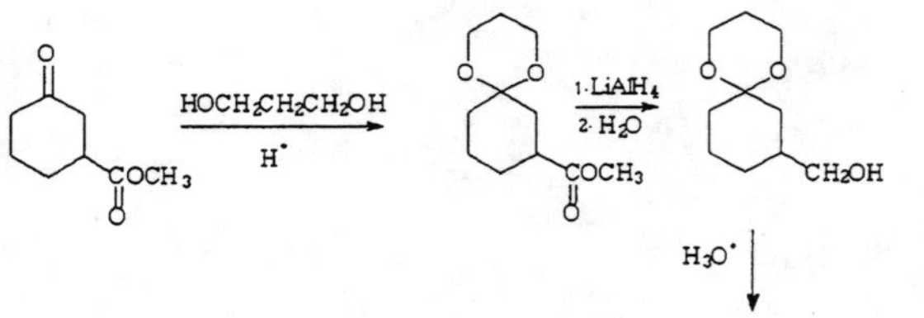

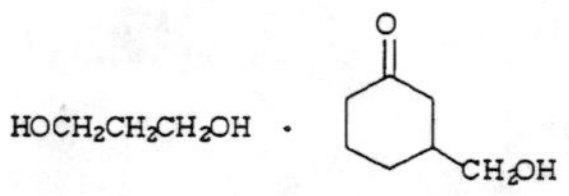

$HOCH_2CH_2CH_2OH$, H^+; 1. $LiAlH_4$, 2. H_2O; H_3O^+

$COCH_3$ CH_2OH

$HOCH_2CH_2CH_2OH$ + CH_2OH

a. 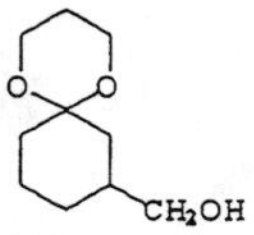 CH_2OH

b. H_3O^+

c. 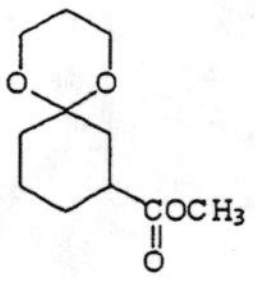 $COCH_3$

d. $HOCH_2CH_2CH_2OH$

e. $LiAlH_4/H_2O$

Answer: d Section: 4

23. Which of the following protecting groups is best used to protect the carboxyl group in the following synthesis?

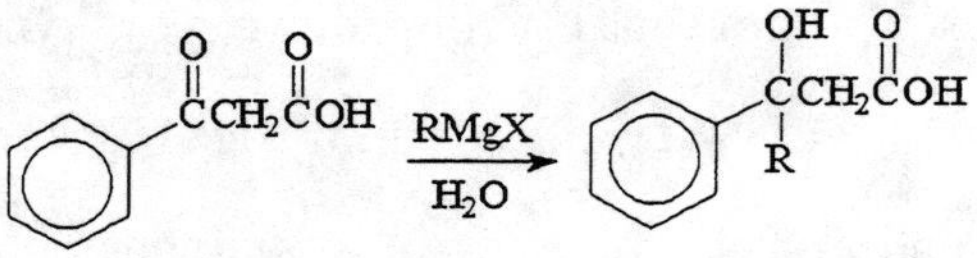

$$C_6H_5\overset{O}{\overset{\|}{C}}CH_2\overset{O}{\overset{\|}{C}}OH \xrightarrow[H_2O]{RMgX} C_6H_5\underset{R}{\overset{OH}{C}}CH_2\overset{O}{\overset{\|}{C}}OH$$

a. $HSCH_2CH_2SH$

b. $HOCH_2CH_2CH_2OH$

c. CH_3CH_2OH

d. $CH_3-C(NH_2)(CH_3)CH_2OH$

e. $CH_3\overset{O}{\overset{\|}{C}}Cl$

Answer: d Section: 4

24. Which of the following protecting groups is best used to protect the amino group in the following reaction?

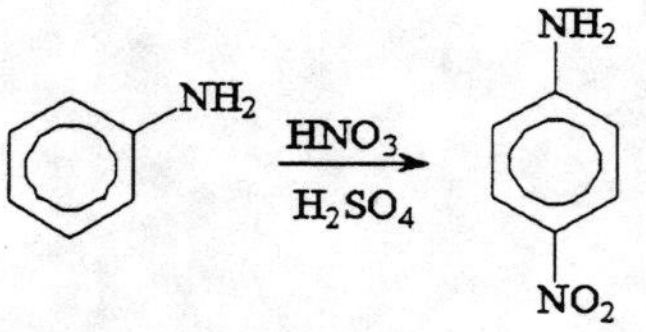

$$C_6H_5NH_2 \xrightarrow[H_2SO_4]{HNO_3} p\text{-}NH_2C_6H_4NO_2$$

a. $HSCH_2CH_2SH$

b. $HOCH_2CH_2CH_2OH$

c. CH_3CH_2OH

d. $CH_3-C(NH_2)(CH_3)CH_2OH$

e. $CH_3\overset{O}{\overset{\|}{C}}Cl$

Answer: e Section: 4

25. Which of the following statements best explains the meaning of an <u>enantioselective reaction</u>?

a. A reaction that forms the S configuration exclusively
b. A reaction that forms a racemic mixture
c. A reaction that forms diastereomers
d. A reaction that forms optically inactive products
e. A reaction that forms an excess of one enantiomer over the other

Answer: e Section: 5

26. How would the following synthesis best be accomplished?

$CH_3CH_2C{\equiv}CH$ (1-Butyne) $\longrightarrow$ $CH_3CH_2(H)C{=}C(H)CH_2CH_3$ (trans-3-hexene)

a. $CH_3CH_2C{\equiv}CH + Na \longrightarrow CH_3CH_2C{\equiv}\ddot{C}^{\ominus}Na^{\oplus} \xrightarrow{CH_3CH_2Br} CH_3CH_2C{\equiv}CCH_2CH_3 \xrightarrow{H_2,\ \text{Lindlar's catalyst}}$ Product

b. $CH_3CH_2C{\equiv}CH + Na \longrightarrow CH_3CH_2C{\equiv}\ddot{C}^{\ominus}Na^{\oplus} \xrightarrow{CH_3CH_2Br} CH_3CH_2C{\equiv}CCH_2CH_3 \xrightarrow{Li,\ NH_3}$ Product

c. $CH_3CH_2C{\equiv}CH + CH_3CH_2MgBr \longrightarrow CH_3CH_2C{\equiv}CCH_2CH_3 \xrightarrow{H_2,\ \text{Lindlar's catalyst}}$ Product

d $CH_3CH_2C{\equiv}CH + CH_3CH_2MgBr \longrightarrow CH_3CH_2C{\equiv}CCH_2CH_3 \xrightarrow{Li,\ NH_3}$ Product

e. $CH_3CH_2C{\equiv}CH + CH_3CH_2 \xrightarrow{Na} CH_3CH_2C{\equiv}CCH_2CH_3 \xrightarrow{Na,\ NH_3}$ Product

Answer: b Section: 1

27. Identify the <u>chiral auxiliary</u> in the following synthesis:

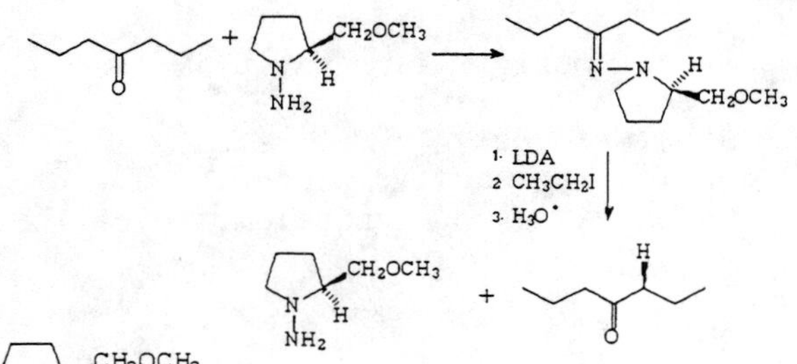

CH_2OCH_3, N, H, NH_2 → N—N, H, CH_2OCH_3

1. LDA
2. CH_3CH_2I
3. H_3O^+

CH_2OCH_3, N, H, NH_2 + H

a. 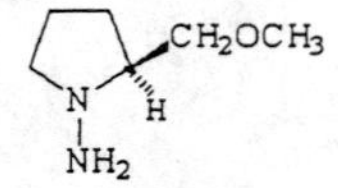 CH_2OCH_3, N, H, NH_2

b. LDA

c. CH_3CH_2I

d. 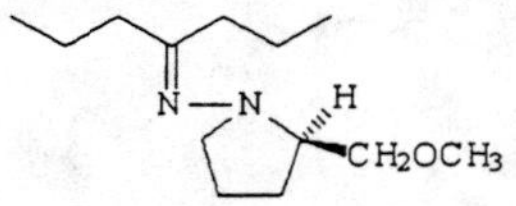 N—N, H, CH_2OCH_3

e. H_3O^+

Answer: a Section: 5

28. Which of the following methods may be used to produce an <u>enantioselective</u> reaction?

a. Using an enzyme-catalyzed reaction
b. Using a chiral auxiliary
c. Using enantiomerically pure starting material
d. Using a reactant that already has the chirality center with the configuration needed in the product
e. All of the above

Answer: e Section: 5

29. What is meant by an <u>iterative</u> synthesis?

a. A reaction in which the product can be easily reversed to the reactant
b. A synthesis in which a catalyst is never used
c. A synthesis in which a reaction sequence is carried out more than once
d. A synthesis in which a reaction sequence is carried out only once
e. A synthesis in which the first step is carried out twice

Answer: c Section: 6

SHORT ANSWER

30. Explain what is meant by a multistep synthesis. Give an example.

Answer:
A synthesis that requires more than one step to accomplish. For instance, adenine can be prepared starting with hydrogen cyanide and ammonia using a multistep synthesis.

31. Explain what is meant by linear synthesis? Give an example.

Answer:
Linear synthesis means building a molecule step-by-step from the starting material.
ex.

X→Y→Z→P

32. Design two different routes for synthesizing 1-aminobutane, $CH_3CH_2CH_2CH_2$-NH_2, using 1-butene, $CH_3CH_2CH{=}CH_2$, as the starting material.

Route I.

$$CH_3CH_2CH{=}CH_2 \xrightarrow[\text{Peroxide}]{HBr} CH_3CH_2CH_2CH_2Br \xrightarrow{^-N_3} CH_3CH_2CH_2CH_2N{=}\overset{+}{N}{=}\overset{-}{N}$$

$$\xrightarrow{H_2/Pt} CH_3CH_2CH_2CH_2NH_2$$

Route II.

$$CH_3CH_2CH{=}CH_2 \xrightarrow[2\text{-}HO^-,\ H_2O_2,\ H_2O]{1\text{-}BH_3} CH_3CH_2CH_2CH_2OH \xrightarrow[H_2SO_4]{K_2Cr_2O_7} CH_3CH_2CH_2\overset{O}{\overset{\|}{C}}OH$$

$$\xrightarrow{SOCl_2} CH_3CH_2CH_2\overset{O}{\overset{\|}{C}}Cl \xrightarrow{NH_3} CH_3CH_2CH_2\overset{O}{\overset{\|}{C}}NH_2 \xrightarrow[2\text{. }H_2O]{1\text{. }LiAlH_4} CH_3CH_2CH_2CH_2NH_2$$

Note: Other routes are also possible.

Answer: route 1 Section: 1

33. What does an open arrow, ⟹, represent in organic synthesis? Give an example.

Ex.
A ⟹ B

Answer:
Open arrows represent a retrosynthesis operation.

Ex. A ⟹ B

This implies that A is the target molecule, and B is the starting material.
Section: 2

34. What is the synthetic equivalent of $^{+}NO_2$?

Answer: HNO_3/H_2SO_4 Section: 2

35. Perform retrosynthetic analysis to accomplish the following:

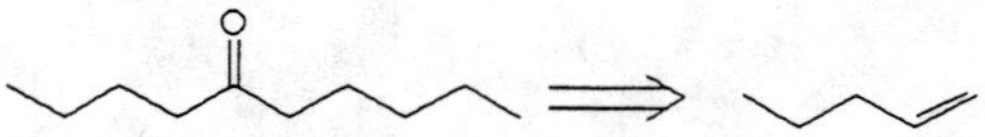

Answer:

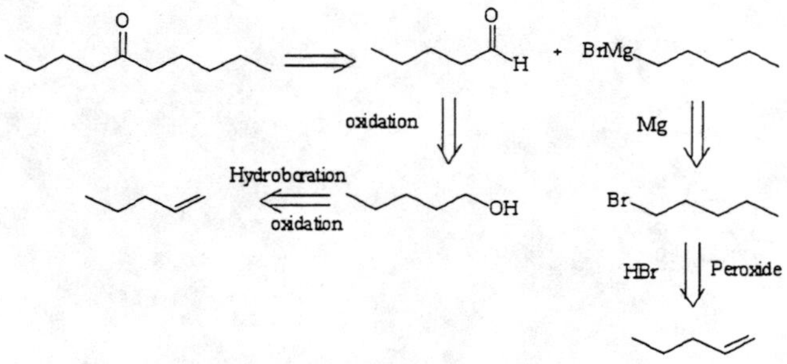

H
+ BrMg
oxidation
Mg
Hydroboration
oxidation
OH
Br
HBr
Peroxide

Section: 2

36. Perform retrosynthetic analysis to accomplish the following:

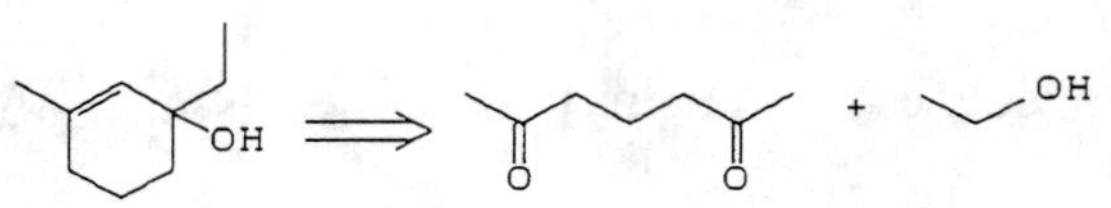

Answer:

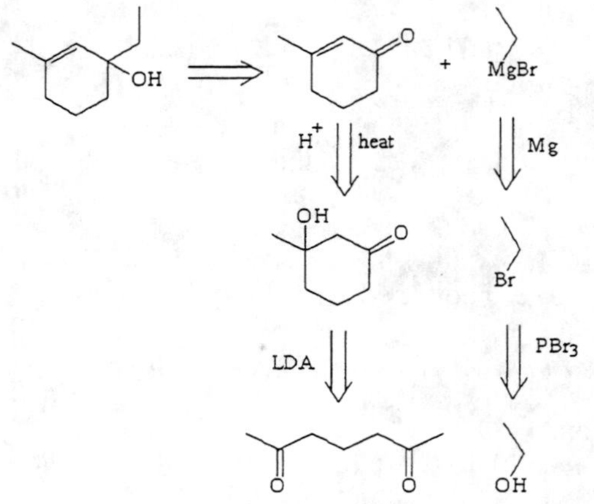

H$^+$ heat

Mg

LDA

PBr_3

Section: 2

37. Suggest a mechanism for the following reaction, the Darzens condensation:

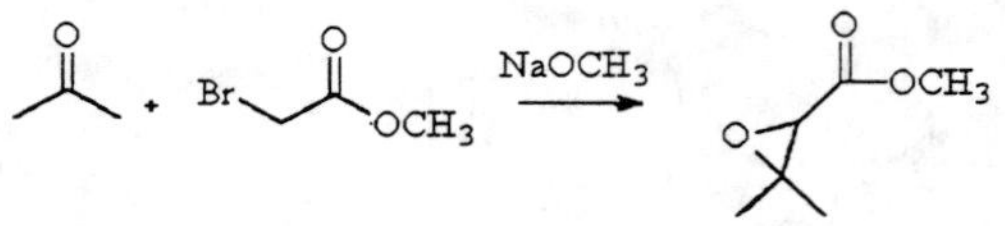

$NaOCH_3$

Answer:

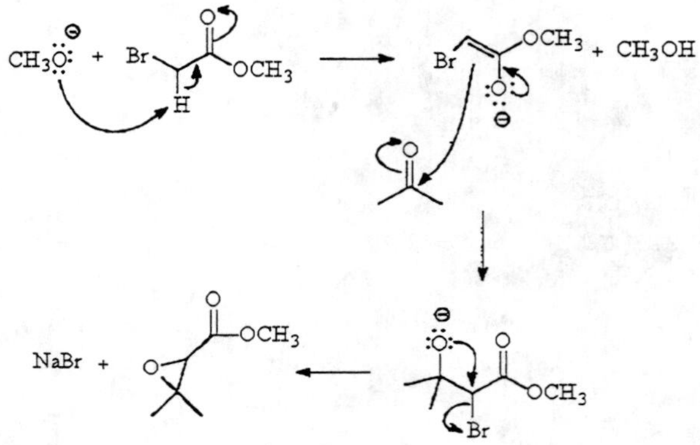

Section: 2

38. What is meant by an <u>extrusion</u> reaction? Give an example.

Answer:
A reaction in which a neutral molecule is eliminated. (e.g., CO_2, CO, N_2, SO_2, etc.)

Ex. → + CO

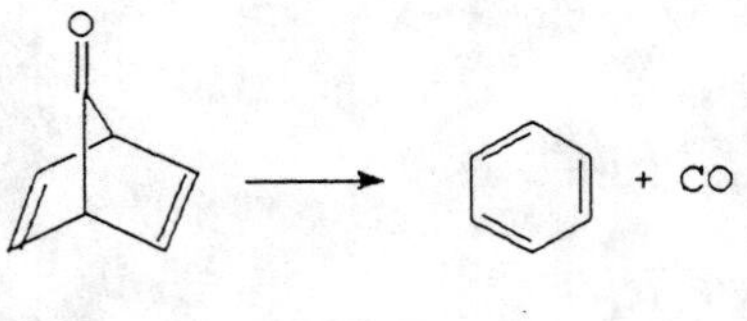

Section: 3

39. Two equivalents of Grignard reagent are required in the following reaction. Give an explanation.

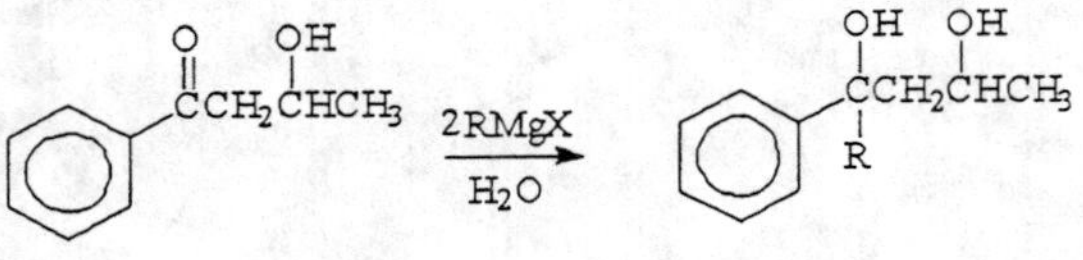

$C_6H_5CCH_2CHCH_3$ (O, OH) $\xrightarrow[H_2O]{2RMgX}$ $C_6H_5C(R)(OH)CH_2CH(OH)CH_3$

Answer:
The first equivalent of RMgX is consumed by reacting with the labile proton of the OH group.

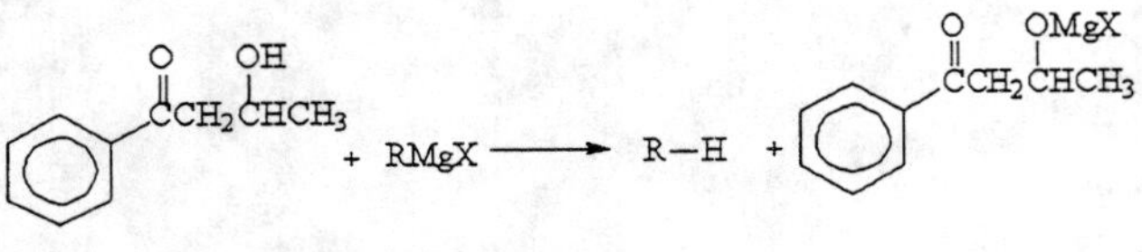

$C_6H_5CCH_2CHCH_3$ (O, OH) + RMgX → R—H + $C_6H_5CCH_2CHCH_3$ (O, OMgX)

The second equivalent of RMgX will attack the carbonyl carbon and later be removed by hydrolysis.

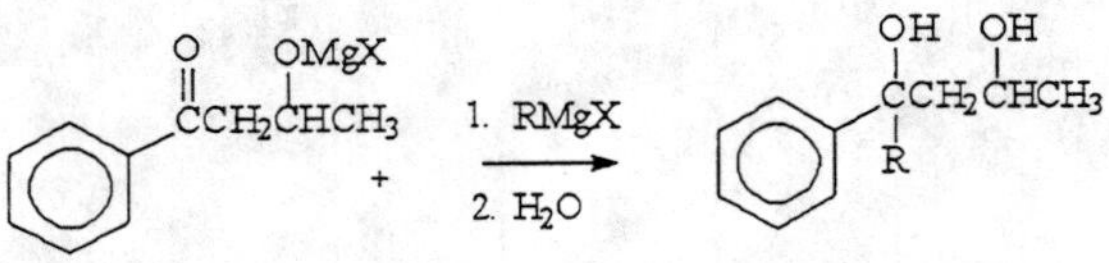

$C_6H_5CCH_2CHCH_3$ (O, OMgX) + $\xrightarrow[2.\ H_2O]{1.\ RMgX}$ $C_6H_5C(R)(OH)CH_2CH(OH)CH_3$

Section: 1

40. Explain what is meant by <u>chiral auxiliary</u>?

Answer:
This is a method of synthesizing a single enantiomer by using an enantiomerically pure compound which, when attached to the reactant, causes the product with the desired stereochemistry to be formed.

Section: 5

41. Propose a mechanism for the following conversion:

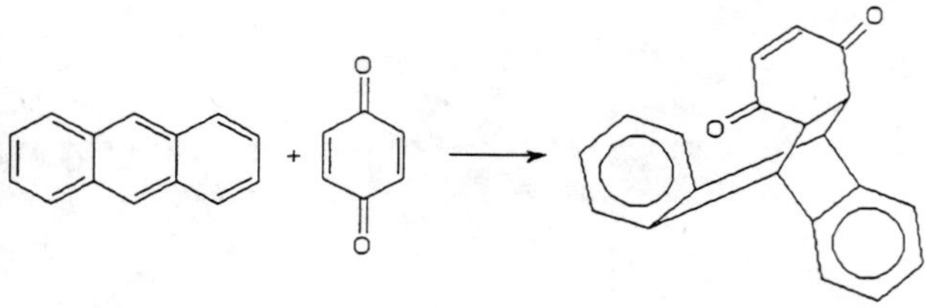

Answer: This is a Diels-Alder (4+2) cycloaddition reaction.

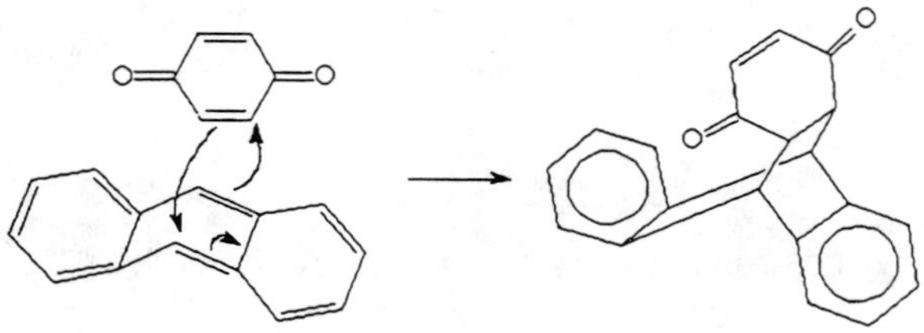

Section: 6

42. Propose a mechanism for the Favorski ring contraction reaction shown below:

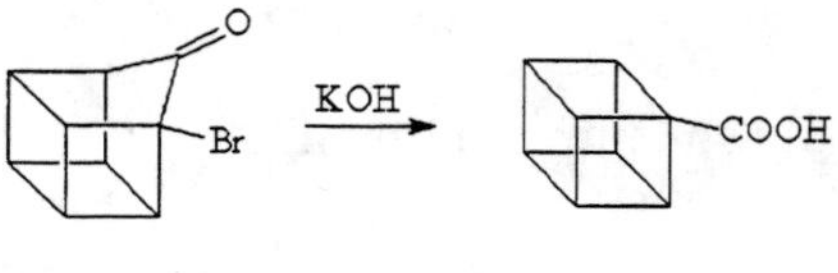

Answer:

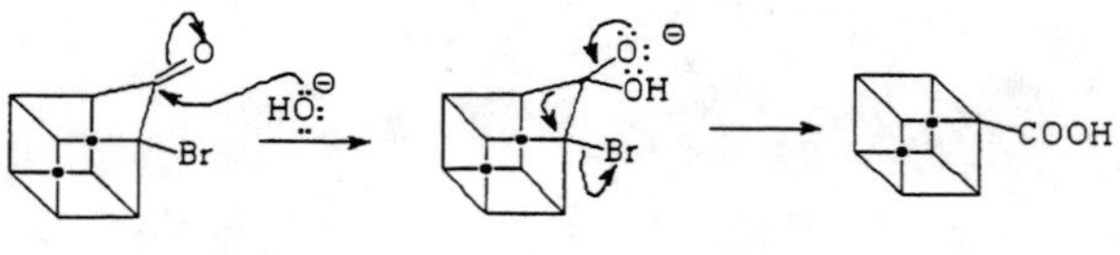

Section: 6

43. Provide the reagents necessary to complete the following transformation.

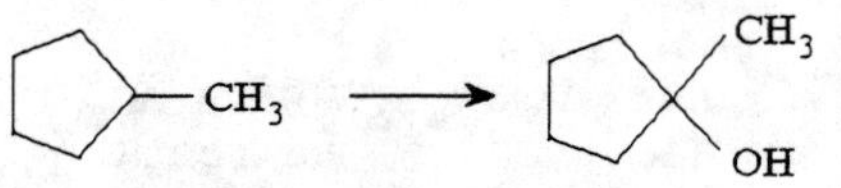

Answer:
1. Br_2, hυ
2. H_2O, Δ
or
1. Br_2, hυ
2. $NaOCH_3$, CH_3OH
3. H_3O^+ or $Hg(OAc)_2$, H_2O; $NaBH_4$

44. Provide the reagents necessary to complete the following transformation.

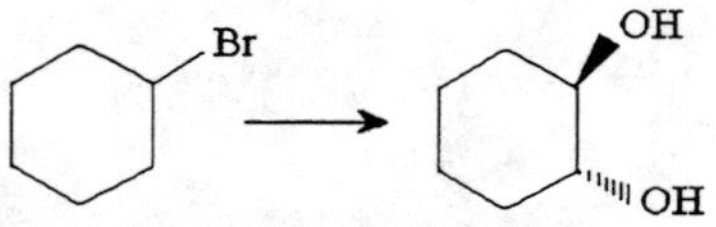

Answer:
1. $NaOCH_3$, CH_3OH
2. MCPBA or CH_3CO_3H
3. H_3O^+ or ^-OH

45. Provide the reagents necessary to convert 3-methyl-2-butanol to 2-methyl-2-butanol.

Answer:
1. conc. H_2SO_4 or PBr_3; $NaOCH_3$, CH_3OH
2. H_3O^+ or $Hg(OAc)_2$; H_2O $NaBH_4$

Section: 1

46. Provide the reagents necessary to accomplish the following transformation.

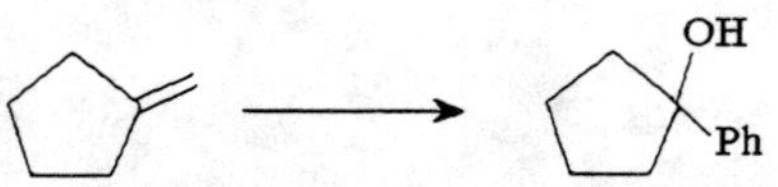

Answer:
1. O_3; CH_3SCH_3 or hot $KMnO_4$, ^-OH
2. PhMgBr
3. H_3O^+

Section: 1

47. Provide the structure of the major organic product in the reaction shown below.

Br

1. Mg, Et_2O
2. H_2CO
3. H_3O^+

Answer:

OH

Section: 1

48. Describe a sequence of reactions by which E-1-bromo-3-pentene can be straightforwardly prepared from propyne.

Answer:
1. $NaNH_2$
2. ethylene oxide
3. Na, NH_3
4. PBr_3

Section: 1

49. Describe a sequence of reactions by which Z-3-hexen-1-ol can be straightforwardly prepared from 1-butyne.

Answer:
1. $NaNH_2$
2. ethylene oxide
3. H_2, Lindlar's catalyst

Section: 1

50. Describe a sequence of reactions by which the compound shown below can be straightforwardly prepared from acetylene.

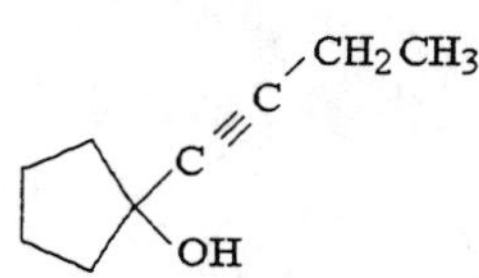

Answer:
1. $NaNH_2$
2. CH_3CH_2Br
3. $NaNH_2$
4. cyclopentanone

Section: 1

51. Provide a series of synthetic steps by which the compound below can be prepared from benzene.

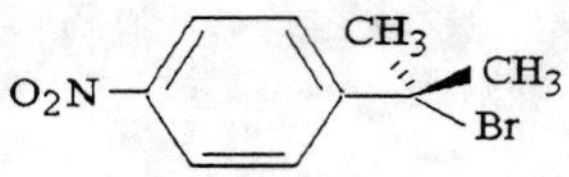

Answer:
1. $(CH_3)_2CHCl$, $AlCl_3$
2. HNO_3, H_2SO_4
3. Br_2, $h\upsilon$

Section: 1

52. Provide a series of synthetic steps by which 4-phenyl-4-heptanol can be prepared from benzene.

Answer:
1. Br_2, $FeBr_3$
2. Mg, ether
3. $CH_3CH_2CH_2COCH_2CH_2CH_3$
4. H_3O^+

53. Propose a Wittig reaction-based synthesis of 3-hexene using propene as the carbon source and any other reagents necessary.

Answer:
Prepare 3-hexene from propene as follows:
1. BH_3 THF
2. H_2O_2, NaOH
3. PCC
4. $CH_3CH_2CH{=}PPh_3$

Prepare the ylide ($CH_3CH_2CH{=}PPh_3$) from propene as follows:

1. HBr, peroxides
2. PPh_3
3. BuLi

54. Provide the sequence of synthetic steps necessary to convert cyclohexanone into the compound shown below.

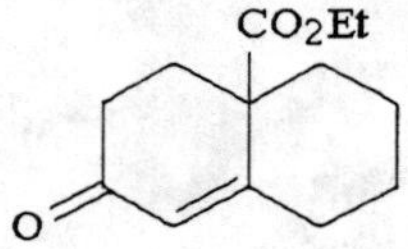

Answer:
1. $NaOCH_2CH_3$, $CO(OCH_2CH_3)_2$
2. $NaOCH_2CH_3$, $CH_2{=}CHCOCH_3$

Section: 1

55. Suggest a sequence of synthetic steps through which phenylacetic acid can be prepared from toluene and in which Grignard chemistry is employed.

Answer:
1. Br_2, hν, or NBS
2. Mg, ether
3. CO_2
4. H^+, H_2O

Section: 1

56. Suggest a sequence of synthetic steps through which p-toluic acid can be prepared from toluene.

Answer:
1. Br_2, $FeBr_3$
2. Mg, ether
3. Co_2
4. H_3O^+

Section: 1

57. Suggest a sequence of synthetic steps through which 2-phenylethanol can be prepared from toluene. One of your intermediates must be carboxylic acid.

Answer:
1. Br_2, hν or NBS
2. NaCN, acetone
3. H^+, H_2O
4. $LiAlH_4$

or

1. Br_2, hν or NBS
2. Mg, ether
3. CO_2
4. H^+, H_2O
5. $LiAlH_4$

Section: 1

58. Propose a reasonable synthetic route to prepare cyclohexylacetic acid from methylenecyclohexane.

Answer:
1. HBr, peroxides
2. Mg, ether
3. Co_2
4. H^+ , H_2O

or

1. HBr peroxides
2. NaCN, acetone
3. H^+, H_2O

59. Propose a synthesis of 2-phenylethylamine from toluene and any other necessary reagents.

Answer:
1. Br_2, hυ <u>or</u> NBS
2. NaCN
3. $LiAlH_4$
4. H_2O

60. Using ethanol as your only source of carbon compounds and using any other necessary inorganic reagents, propose a synthesis of acetonitrile.

Answer:
Beginning with ethanol,
1. $Na_2Cr_2O_7$, H_2SO_4
2. $SOCl_2$
3. NH_3
4. P_2O_5

Section: 1

Chapter 29: The Organic Chemistry of Drugs: Discovery and Design

SHORT ANSWER

1. What is a drug?

   Answer:
   A compound which interacts with a biological molecule and thereby triggers a physiological effect is called a drug.

MULTIPLE CHOICE

2. Which of the following terms is (are) synonymous with the term "proprietary name?"

   a. generic name
   b. trademark
   c. trade name
   d. brand name
   e. both (c) and (d)

   Answer: e

SHORT ANSWER

3. The _____ name of a drug is the name that any pharmaceutical company can use to identify the drug.

   Answer: generic

4. In pharmaceutical research, what is a "lead compound"?

   Answer:
   A lead compound is a pharmaceutically active compound of known structure which is used as the prototype in a search for other, structurally similar, active compounds.

5. In pharmaceutical research, explain what is meant by "molecular modification."

   Answer:
   Molecular modification is the process through which analogs of the lead compound are synthesized by altering its structure (changing functional groups, ring sizes, chain lengths, etc).

6. What compound served as the lead compound in the research which developed today's most commonly utilized local anesthetics?

   a. morphine
   b. cocaine
   c. codeine
   d. sulfanilamide
   e. nitroglycerin

   Answer: b

7. Novocain and Xylocaine are two compounds which are commonly used as local anesthetics. The structures of these two drugs are similar, but where Novocain has an aromatic ester functionality, Xylocaine has an aromatic amide group. Which of these drugs has a longer lifetime in the body and why?

   Answer:
   Xylocaine has the longer lifetime because amides are slower to hydrolyze than esters.

MULTIPLE CHOICE

8. What compound is the most widely used analgesic for severe pain?

   a. Librium
   b. dextromethorphan
   c. Ativan
   d. Benadryl
   e. morphine

   Answer: e

SHORT ANSWER

9. Based on the structures of morphine, codeine, and heroin shown below, explain why heroin is the most rapid acting member of the opiate family.

Structures found together in Section 29.3

   Answer:
   Because heroin is less polar than morphine or codeine, it crosses the blood-brain barrier more rapidly and results in a more rapid physiological response.

10. What is the name of the receptor in the central nervous system which binds morphine and codeine?

    Answer: the opiate receptor

MULTIPLE CHOICE

11. Who is credited with carrying out the first blind screen in his search for a "magic bullet" to cure African sleeping sickness?

    a. Jonas Salk
    b. Louis Pasteur
    c. Edward Jenner
    d. Paul Ehrlich
    e. Gerhard Domagk

    Answer: d

12. An important part of a random screen is the recognition of effectiveness in the compounds screened. What term is used to identify the test by which effectiveness is measured?

    a. assay
    b. dissolution
    c. modeling
    d. combinatorial analysis
    e. synergism

    Answer: a

13. What term is used to describe a screening test which is conducted outside the body, in a test tube for example?

    a. *in toto*
    b. *in tubo*
    c. *in virto*
    d. *in loco*
    e. *in vivo*

    Answer: c

SHORT ANSWER

14. What gave scientists the idea that azo dyes might be candidates as antibacterial agents?

    Answer:
    The fact that these dyes bound to wool fibers which are composed of protein led scientists to speculate that they might also bind to bacterial proteins.

15. Prontosil, a bright red dye, was an active antibacterial agent in vivo but not in vitro. Scientists also noted that mice given Prontosil did not excrete a red compound. What possible explanation of the activity of Prontosil is consistent with these two observations?

    Answer:
    The dye is converted into an active compound via some metabolic process of the mouse.

MULTIPLE CHOICE

16. What term describes a drug that, once taken, inhibits further growth of bacteria?

    a. generic drug
    b. bacteriostatic drug
    c. antiviral agent
    d. vasodiolation agent
    e. bactericidal drug

    Answer: b

17. Which of the following drugs is a member of the benzodiazepine family?

    a. nitroglycerin
    b. acetylcholine
    c. morphine
    d. penicillin
    e. Librium

    Answer: e

18. Which of the following drugs, discovered serendipitiously, is known to relieve angina pectoris by dilating cardiac blood vessels?

    a. nitroglycerin
    b. morphine
    c. acyclovir
    d. sulfanilamide
    e. Librium

    Answer: a

SHORT ANSWER

19. What term is given to the specific cellular binding site with which some drugs interact to trigger their physiological effect?

    Answer: receptor

20. What structural feature do most effective antihistamines have in common with histamine and how is this common feature believed to be related to the effectiveness of antihistamines as drugs?

    Answer:
    Like histamine, antihistamines have a protonated amino group at the end of an alkyl chain which allows these compounds to bind to a negatively charged portion of the histamine receptor. Antihistamines thus bind to the histamine receptor and prevent histamine from doing so.

MULTIPLE CHOICE

21. What name is given to the receptors involved in neurotransmission, memory, and wakefulness?

    a. cholinergic receptors
    b. beta-blockers
    c. histamine receptors
    d. prostaglandin receptors
    e. opiate receptors

    Answer: a

22. What type of enzyme is secreted by penicillin resistant bacteria which hydrolyzes penicillin before it has an opportunity to interfere with cell wall synthesis?

    a. esterase
    b. β-lactamase
    c. protease
    d. oxidase
    e. reductase

    Answer: b

23. What new functional group is introduced to the penicillin structure when it is converted to a derivative by the oxidizing agent MMPP?

    a. N-oxide
    b. oxime
    c. sulfone
    d. amide
    e. carboxylic acid

    Answer: c

24. What term is given to the effect which results when two drugs are used in combination and a response greater than the sum of the drugs' individual effects is observed?

    a. additive
    b. combinatorial
    c. distributive
    d. synergistic
    e. antagonistic

    Answer: d

25. Name the last major class of antibiotics which were discovered. Members of this family work by inhibiting an enzyme required for transcription in bacteria.

    a. sulfa drugs
    b. cephalasporins
    c. tetracyclines
    d. fluoroquinolones
    e. penicillins

    Answer: d

SHORT ANSWER

26. If we have a variety of effective antibiotics, why do pharmaceutical companies continue to search for new ones?

    Answer:
    Bacterial strains develop resistance to an antibiotic within 15 to 20 years. Drug resistant strains will only be fought with new antibiotics or synergistic agents.

27. How is the therapeutic index of a drug defined?

    Answer:
    The therapeutic index of a drug is the ratio of the lethal dose to the therapeutic dose.

28. Given a choice between two drugs of comparable effectiveness, would you choose the drug with the higher or lower therapeutic index? Explain.

    Answer:
    Choose the drug with the higher therapeutic index. The higher the therapeutic index, the greater the margin of safety associated with the drug's use.

29. In what context of drug design and synthesis, what is a suicide substrate?

    Answer:
    A suicide substrate is a drug which binds to the active site of a particular enzyme and subsequently reacts with the enzyme to inactivate it irreversibly.

30. The technique of relating a property of a series of compounds to biological activity is known as a QSAR. For what does this acronym stand?

    Answer: quantitative structure-activity relationship

MULTIPLE CHOICE

31. What name is given to the strategy in pharmaceutical research which employs the concept of mass production, namely that a large group of related compounds can be prepared by covalently linking sets of building blocks?

    a. combinatorial synthesis
    b. serendipity
    c. random screening
    d. blind screening
    e. molecular modeling

    Answer: a

32. Most antiviral drugs are analogs of _____

    a. carbohydrates
    b. steroids
    c. nucleosides
    d. proteins
    e. vitamins

    Answer: c

SHORT ANSWER

33. What name is given to drugs which have limited demand and thus prove costly to develop?

    Answer: orphan drugs